Thyroid Cancer

From Emergent Biotechnologies to Clinical Practice Guidelines

Thyroid Cancer

From Emergent Biotechnologies to Clinical Practice Guidelines

Edited by

Angelo Carpi

Jeffrey I. Mechanick

CRC Press
Taylor & Francis Group
Boca Raton London New York

CRC Press is an imprint of the
Taylor & Francis Group an **informa** business

CRC Press
Taylor & Francis Group
6000 Broken Sound Parkway NW, Suite 300
Boca Raton, FL 33487-2742

Printed in the United States of America on acid-free paper
10 9 8 7 6 5 4 3 2 1

International Standard Book Number: 978-1-4398-6221-6 (Hardback)

Visit the Taylor & Francis Web site at
http://www.taylorandfrancis.com

and the CRC Press Web site at
http://www.crcpress.com

Contents

Foreword

The title of this book, *Thyroid Cancer: From Emergent Biotechnologies to Clinical Practice Guidelines*, is extremely timely in view of the burgeoning frequency of thyroid cancer worldwide, probably due to early detection by neck ultrasound, computed tomography (CT), and magnetic resonance imaging (MRI) studies of the head, neck, and chest for nonthyroid reasons. The contributors are all well-recognized leaders in various aspects of the molecular biology, clinical presentation, diagnostic modalities now available, therapy, and long-term follow-up of the various types of thyroid cancer. The genetic and epigenetic changes in thyroid cancer and how these changes result in the risk of progression of thyroid cancer are carefully discussed. Such molecular and genetic findings will play an increasing role in determining the risk stratification of thyroid cancer, which will then impact the therapy and follow-up of patients with various types of thyroid cancer. Various chapters deal in detail with the diagnostic techniques employed in the detection and follow-up of thyroid cancer, including ultrasound; needle aspiration techniques, such as molecular analyses of biopsy specimens; CT, positron emission tomography (PET), PET-CT, and MRI; ^{131}I scans and serum thyroglobulin concentrations either following withdrawal of thyroid hormone or after recombinant human TSH (rhTSH) to stimulate residual thyroid tissue; and new PET tracers under development. Advances in thyroid cancer therapy are brought up to date, including new surgical approaches such as same-day surgery and surgery performed under local anesthesia, external beam radiation therapy for recurrent, inoperable local disease, chemotherapy for widespread disease, and the explosion in the use of tyrosine kinase inhibitors for metastatic disease. Three chapters are devoted to the molecular pathophysiology of medullary thyroid cancer and molecular targeted therapies of this sporadic and familial malignancy. The authors then propose new clinical practice guidelines, including surgical and nonsurgical management of thyroid cancer.

This book should provide the reader with an up-to-date and comprehensive treatise on thyroid cancer and will provide a useful reference book to both the specialist and referring physician.

Lewis E. Braverman, MD
Professor of Medicine
Section of Endocrinology, Diabetes, and Nutrition
Boston University School of Medicine
Boston Massachusetts

Preface

Thyroid cancer evaluation and treatment has enjoyed very little evolution in comparison with other disease management strategies. The standard treatment algorithms for thyroid cancer have varied only in response to major, though infrequent, biotechnological innovations. These resultant algorithm node insertions produced paradigm changes, but there has been little to advance the care of the patient with advanced thyroid cancer—until very recently. Perhaps the reasons for a relative lack of interest and scientific progress are the overwhelming likelihood of benignity among thyroid nodules, the overwhelmingly favorable clinical outcome even if a patient has thyroid cancer, and the overwhelming long time required for clinical follow-up of these cancer patients. Unfortunately, the problem now is that we may be overmanaging low-risk patients, and we have not achieved the clinical responses we would have liked with high-risk patients.

In order to examine emergent biotechnologies, let us first consider historical examples of major innovations for the thyroid cancer patient that shape our current management algorithms and clinical practice guidelines:

- The advent of therapeutic radioiodine in 1942 with radioisotopes approved for use in 1946
  - Improves postoperative treatment
  - Allows radioactive iodine (RAI) scanning and thyroglobulin (TG) measurements to monitor for recurrence
  - Is associated with decreased mortality risk in many patients
- Recent improvements in thyroid surgery technique, minimally invasive procedures, lymph node dissection, and managements of risks
  - Allow primary surgical management of nearly all cases of thyroid cancer and gross cervical recurrence
- Availability of an accurate serum thyroglobulin assay to detect recurrence
- Availability of an accurate serum thyroid-stimulating hormone (TSH) assay to facilitate "suppression therapy" with levothyroxine
- Development of high-resolution thyroid ultrasonography
- Improvements in the fine-needle aspiration technique and cytological analysis

These diagnostic and therapeutic tools account for the majority of procedures we utilize today.

The current age of clinical medicine is witnessing biotechnological innovation at an unprecedented pace. As a result, the recent popularization of clinical practice guidelines (CPG) as a tool to assist clinical decision making has been unable to keep up with the rate of scientific discovery. Many problems have plagued the development and implementation of CPG; most notably, these include disagreement regarding evidence-based medicine methodologies, lack of transparency in the development process, and difficulty evaluating the utility of CPG in order to improve them with future iterations. The purpose of this book is to surf the wave of medical innovation and attempt to predict what CPG will look like based on emergent biotechnologies in the field of thyroid cancer. This premise, that a forward-looking approach to thyroid cancer will improve patient care today, was spawned mainly by two of our collaborative activities: first, the biennial University of Pisa Multidisciplinary Conference on Advances in Management of Malignancies (presided by A.C.), and second, our longstanding interest in the use of molecular markers, namely, galectin-3, in histological specimens from large-needle aspiration biopsies of the thyroid gland.

Our task will be approached in the following manner. First, basic principles of systems biology, molecular and translational medicine, CPG development, and risk stratification for thyroid cancer will lay the foundation to understand subsequent chapters. Thyroid cancer clinical cases will also be presented to instantiate how current CPG contain knowledge gaps and produce uncertainty and impair decision making. Then, a detailed survey of technologically advanced diagnostic procedures and therapeutic interventions will be presented. Each of these chapters will be structured by presenting the current state of the art for each technology, followed by emergent technologies that are not yet approved, though currently under investigation and considered to be very promising. Finally, by critically analyzing the findings of each of the chapters, a futuristic CPG will be constructed based on the potential impact of these emergent biotechnologies. Thus, a synthesis of ideas by prominent world experts in the field of thyroid cancer research and clinical practice will provide an enlightening glimpse into the near future.

We thank each of the authors involved in the production of this book, as their leadership in the care of patients with thyroid cancer has set the stage for dramatic change in clinical outcomes. We also thank Randy Brehm and Jill Jurgenson of Taylor & Francis for their tremendous assistance in the preparation and publication of this book, which has truly been a global project, with contributors representing multiple nations and continents. We hope that readers of this book will avail themselves of the authors' expert opinions and innovative thinking in the care of their patients.

The Editors

Angelo Carpi, MD, is Clinical Professor of Medicine and Director of the Male Infertility Division of the Department of Reproduction and Aging in the Pisa University Medical School, Pisa, Italy. He earned his MD and postgraduate diplomas in internal medicine and nuclear medicine from the University of Pisa, and his diploma of qualification in peptide hormones from the Collegio Medico Giuridico—Scuola Normale Superiore and the Scuola Superiore Sant'Anna, Pisa, Italy. His clinical practice and research has included thyroid tumors. He has authored more than 200 publications included in PubMed. He currently participates on the editorial boards of the international journals *Biomedicine & Pharmacotherapy* and *Frontiers in Bioscience*; he is editor-in-chief of *Thyroidology, Clinical and Experimental* and *Internal Medicine, Clinical and Laboratory.* Additional information may be obtained at www.carpimed.it.

Jeffrey I. Mechanick, MD, is Clinical Professor of Medicine and Director of Metabolic Support in the Division of Endocrinology, Diabetes, and Bone Disease at Mount Sinai School of Medicine. He received his MD from Mount Sinai School of Medicine, completed internal medicine residency training at Baylor College of Medicine, and then completed endocrinology and metabolism fellowship training at Mount Sinai School of Medicine. Dr. Mechanick has authored over 160 publications in endocrinology, metabolism, and nutrition. He is on the board of directors of the American Association of Clinical Endocrinologists and was elected Secretary in 2010. Dr. Mechanick chairs the AACE Publications Committee and the task force that authored the 2010 updated *Guidelines for Clinical Practice Guidelines.* He is an Associate Editor of *Endocrine Practice* and Section Editor for *Current Opinion in Clinical Nutrition and Metabolic Care* and *Current Opinion in Endocrinology and Diabetes*. Dr. Mechanick has a private practice in endocrinology and metabolic support in New York City and a clinical and research interest in thyroid cancer.

Contributors

Faise Al Bunni
Department of Medical-Diagnostic Sciences and Special Therapies
University of Padua
Padua, Italy

John D. Allendorf
Department of Surgery
Columbia University College of Physicians and Surgeons
New York, New York

Giuseppe Barbesino
Thyroid Unit
Massachusetts General Hospital—Harvard Medical School
Boston, Massachusetts

Agnese Barnabei
Endocrinology Service
IRCCS Regina Elena Cancer Institute
Rome, Italy

Donald A. Bergman
Division of Endocrinology, Diabetes, and Bone Diseases
Mount Sinai School of Medicine
New York, New York

Antonio Bianchini
Department of Endocrinology and Department of Diagnostic Imaging
Ospedale Regina Apostolorum
Albano, Italy

Giancarlo Bizzarri
Department of Endocrinology and Department of Diagnostic Imaging
Ospedale Regina Apostolorum
Albano, Italy

Paolo Bossi
Department of Medical Oncology
Fondazione IRCCS Istituto Tumori
Milano, Italy

Dario Casara
Radiotherapy and Nuclear Medicine Unit
Istituto Oncologico Veneto-IRCSS
Padua, Italy

Walter Choi
Department of Radiation Oncology
Continuum Cancer Centers of New York
Beth Israel Medical Center
St. Luke's and Roosevelt Hospitals
New York, New York

Rhoda H. Cobin
Division of Endocrinology, Diabetes, and Bone Diseases
Mount Sinai School of Medicine
New York, New York

Salvatore M. Corsello
Endocrinology Unit
Catholic University School of Medicine
Rome, Italy

Joseph DiNorcia
Department of Surgery
Columbia University College of Physicians and Surgeons
New York, New York

Emily Jane Gallagher
Department of Medicine
Division of Endocrinology, Diabetes, and Bone Diseases
Mount Sinai Medical Center
New York, New York

Jing Gao
Department of Radiology
New York–Presbyterian Hospital of Weill Cornell Medical College
New York, New York

Rinaldo Guglielmi
Department of Endocrinology and Department of Diagnostic Imaging
Ospedale Regina Apostolorum
Albano, Italy

Louis B. Harrison, FASTRO
Continuum Cancer Centers of New York
Beth Israel Medical Center
St. Luke's and Roosevelt Hospitals
Albert Einstein College of Medicine
New York, New York

William B. Inabnet III
Department of Surgery
Division of Metabolic, Endocrine and Minimally Invasive Surgery
Mount Sinai Medical Center
New York, New York

Ravi Iyengar
Department of Pharmacology and Systems Therapeutics
Mount Sinai School of Medicine
New York, New York

Elias Kazam
Weill Cornell Medical College
New York, New York

Jonathan K. Kazam
Department of Radiology
Weill Cornell Medical College
New York, New York

Paul W. Ladenson
Division of Endocrinology and Metabolism
Johns Hopkins University School of Medicine
Baltimore, Maryland

Laurence Leenhardt
Department of Nuclear Medicine
Pitié Salpêtrière Hospital
Paris, France

Derek LeRoith
Department of Medicine
Division of Endocrinology, Diabetes, and Bone Diseases
Mount Sinai Medical Center
New York, New York

Jean-Christophe Lifante
Service de Chirurgie Générale et Endocrinienne
Centre Hopsitalier Lyon Sud
Université Claude Bernard
Lyon, France

Azi Lipshtat
Department of Pharmacology and Systems Therapeutics
Mount Sinai School of Medicine
New York, New York

Laura D. Locati
Department of Medical Oncology
Fondazione IRCCS Istituto Tumori
Milano, Italy

Josef Machac
Mount Sinai Medical Center
New York, New York

Irene Misischi
Department of Endocrinology and Department of Diagnostic Imaging
Ospedale Regina Apostolorum
Albano, Italy

Andrea Nicolini
Department of Internal Medicine
University of Pisa
Pisa, Italy

Claudio M. Pacella
Department of Endocrinology and Department of Diagnostic Imaging
Ospedale Regina Apostolorum
Albano, Italy

Enrico Papini
Department of Endocrinology and Department of Diagnostic Imaging
Ospedale Regina Apostolorum
Albano, Italy

Laura Papini
Department of Endocrinology and Department of Diagnostic Imaging
Ospedale Regina Apostolorum
Albano, Italy

Rosa Maria Paragliola
Endocrinology Unit
Catholic University School of Medicine
Rome, Italy

Maria Rosa Pelizzo
Department of Medical and Surgical Sciences
Special Surgical Pathology Unit
University of Padua
Padua, Italy

Fabio Pomerri
Department of Medical-Diagnostic Sciences and Special Therapies
University of Padua
and
Oncological Radiology Unit
Istituto Oncologico Veneto-IRCSS
Padua, Italy

Steven I. Sherman
Department of Endocrine Neoplasia and Hormonal Disorders
Division of Internal Medicine
University of Texas M.D. Anderson Cancer Center
Houston, Texas

Eriko Suzuki
Department of Urology
School of Medicine
Keio University
Tokyo, Japan

Francesco Torino
Medical Oncology Division
San Filippo Neri Hospital
Rome, Italy

Jean Tramalloni
Department of Radiology
Necker Hospital
Paris, France

François Tranquart
Bracco Research SA
Plan-Les-Ouates, Switzerland

Kazuo Umezawa
Department of Applied Chemistry
Faculty of Science and Technology
Keio University
Yokohama, Japan

Mark L. Urken
Division of Head and Neck Surgical Oncology
Department of Otolaryngology
Continuum Cancer Centers of New York
and
Institute for Head and Neck and Thyroid Cancer
Beth Israel Hospital
New York, New York

Mingzhao Xing
Division of Endocrinology and Metabolism
Johns Hopkins University School of Medicine
Baltimore, Maryland

Shunichi Yamashita
Department of Molecular Medicine
Nagasaki University
Graduate School of Biomedical Sciences
Nagasaki, Japan

1 Principles of Translational Medicine and Applications to Thyroid Cancer Investigation

Emily Jane Gallagher and Derek LeRoith

CONTENTS

ABSTRACT: Translational medicine is an expanding area of modern medicine. It embodies the collaboration between basic science and clinical medicine, the purpose of which is to deepen our insight into disease pathogenesis and prognosis in order to advance and refine therapeutics. The ultimate aim of translational medicine is to individualize care by tailoring therapy to best fit each person and his or her specific disease. Much of this approach to medicine involves the use of molecular techniques to study genetics, gene expression, and signaling pathways, while incorporating innovations in modern technology from the computer industry. Therefore, this form of medicine is also called molecular medicine. The management of thyroid cancer involves many challenges and uncertainties relating to disease development, diagnosis, and management. In this chapter, we discuss some of the molecular biology techniques involved in the study of thyroid cancer and the important role of translational medicine in revolutionizing thyroid cancer care.

1.1 INTRODUCTION

Translational medicine is the term applied to the integration of basic science and clinical medicine. The collaboration between basic science and clinical medicine, or "bench to bedside" medicine, is a process of clinical inquiry that inspires scientific research, leading to novel discoveries and new insights into disease pathogenesis, diagnosis, and management that can then be incorporated into patient care. These discoveries result in the synergistic advancement of both clinical medicine and basic science, as scientific research leads to new clinical knowledge, which sparks subsequent questions for science to answer and the need for technological advancements to help answer these questions. Therefore, a perpetual flow of discovery circulates from the bench to the bedside and back.

Much of this "molecular medicine" involves the study of genetics, gene expression, and signaling pathways, incorporating micro- and nanotechnologies from the computer industry.[1] It is hoped that translational medicine will allow medical care to be tailored to the individual, based upon one's genetic makeup or the genetics of one's disease. Genetic polymorphisms and mutations can make an individual susceptible to certain illnesses, or more likely to respond to or develop side effects to certain treatments. Similarly, knowledge of the genetic mutations in cancer cells can predict their aggressiveness and response to chemotherapy. With this knowledge, an individual could avail of targeted therapies and avoid potentially harmful treatments, in a new form of medicine called personalized medicine.

While the concept of integrating basic science and clinical medicine is not new, it is seen as an essential part of the evolution of modern medicine, and the term *translational medicine* was adopted to draw attention to and advance this strategy. Historically, cancers were diagnosed by tissue and histological subtype alone, with the choice of therapy determined on the same basis. This approach led to unpredictably varied responses to the same therapeutic agents, in individuals with apparently the same tumor. In more recent years, the genetic and biological basis of many cancers has been determined, which is allowing for great advances in the development of targeted therapies.[2,3] Based on this genetic and biological knowledge, a clinician could predict a patient's prognosis and likelihood of response to treatment, thereby distinguishing those patients who may be cured with surgery alone, so avoiding unnecessary and potentially toxic medication, from those with tumors that are likely to recur or metastasize, and therefore would benefit from adjuvant therapy. The care of patients with thyroid cancer involves many challenges, for example, diagnosing follicular thyroid carcinoma (FTC) with fine-needle aspiration biopsy (FNAB), predicting which patients with papillary thyroid cancer (PTC) are likely to develop recurrent or metastatic disease, how to treat patients who do not respond to radioactive iodine (RAI), and how to alter the poor prognosis associated with anaplastic thyroid cancer (ATC). Genetic breakthroughs since the 1980s and ongoing breakthroughs in molecular biology are confronting these challenges.

1.2 MOLECULAR BIOLOGY METHODS

Observational studies of humans and animals in the early part of the twentieth century described inherited forms of cancer and evoked theories on the presence of genetic mutations inciting cancer development in these families. But scientific evidence was wanting, and these theories remained unproven for some time.[4,5] Advances in basic science followed, with the description of the DNA double helix in 1953, and subsequent depiction of transcription and translation, succeeded by major advances in molecular biology techniques through the latter part of the twentieth century, allowing for the completion of the Human Genome Project in 2003.[6,7] While DNA contains the entire genome, it consists of both coding and noncoding regions. Of the coding regions, certain genes are only expressed at certain times in certain tissues. Studying RNA allows us to identify the genes that are expressed, upregulated, downregulated, or silenced in a particular cell, giving a truer reflection of the biological activity within that cell. RNA undergoes posttranscription modifications that result in splice variants and variations in the amino acid sequence produced during protein translation. The recently discovered microRNAs (miRNAs) add another dimension to the regulation of gene expression, by binding to mRNA transcripts, leading to their degeneration or preventing their translation.[8] Proteins also undergo conformational changes after translation, forming different isoforms and potentially lipoproteins or glycoproteins. Proteomics is an evolving field of translational medicine that studies these proteins and their contributions to cell activity.[9]

Some of the techniques frequently used in molecular biology and their applications are outlined in Table 1.1. Notable breakthroughs in molecular biology include the development of recombinant DNA techniques, as described in 1972.[10] Recombinant DNA involves the integration of a DNA sequence of interest (target DNA) into the self-replicating DNA of a host, which may be a virus (bacteriophage), a bacterial plasmid, a bacterial artificial chromosome (BAC), or a yeast artificial

TABLE 1.1
Molecular Biology Techniques

Technique	Method	Application	Use in Thyroid Cancer
DNA recombination	A DNA segment of interest is inserted into a vector DNA (bacteriophage, plasmid, BAC, or YAC) that can replicate independently, by cutting the target DNA and vector DNA using restriction enzymes and inserting the target DNA into the vector, by compromising the integrity of the cell wall with electrical, chemical, or thermal stress in a process called a transformation	• Cloning of DNA sequences; different vectors allow replication of different lengths of DNA • Transfecting cell lines with mutant DNA to evaluate the effects on cell growth, RNA expression, and protein production • Creation of a transgene for development of transgenic mouse model	• Cell lines transfected with mutated oncogenes used to study the RET/PTC rearrangements on thyroid cell lines[18] • Mouse models of PTC with BRAF and RAS mutations[32,33]
Polymerase chain reaction (PCR)	Double-stranded DNA is denatured into single-stranded DNA; DNA primers anneal to complementary sequences near the target DNA, allowing DNA synthesis to occur in the presence of free deoxynucleotides and DNA polymerase in a repetitive process of denaturation, primer annealing, and strand elongation	• Rapid cloning of DNA sequences, resulting in numerous copies of target DNA RT-PCR to produce cDNA from RNA • Quantification of RNA products by real-time PCR and TaqMan • DNA sequencing through hybridization techniques	• PCR is utilized to identify DNA mutations in fMTC, PTC, and FTC[18,22,28] • RT-PCR is used to study gene expression in BRAF and RAS mutations and involvement in signal transduction pathways[17,37]
Gel electrophoresis	Southern blotting allows DNA fragments to be separated on an agarose gel under the influence of an electric current; DNA fragments migrate away from the negative electrode, with the smallest fragments migrating furthest away from the electrode; RNA is separated in a similar process called Northern blotting and proteins by Western blotting	• Separation of DNA fragments cut by restriction endonucleases (restriction fragments) to identify DNA mutations, insertions and deletions DNA and cDNA sequencing • Protein identification in two-dimensional protein separation by charge and size (SDS-PAGE)	• Identification of BRAF mutations by heteroduplex analysis[29] • Separation of proteins from signal transduction pathways altered in thyroid cancer[47]
Hybridization	DNA oligonucleotides labeled with radioisotopes or fluorophores anneal to complementary DNA target sequences that are fixed; reverse hybridization involves fixing of the oligonucleotides and labeling the target DNA sequences; cDNA probes can be used to hybridize RNA; immunohistochemistry allows for identification of proteins using antibody binding	• DNA can be sequenced by hybridization in an adaptation of PCR, called nucleic acid hybridization; Southern blot hybridization involves hybridization of DNA on electrophoresis gel • Reverse hybridization in DNA microarrays and gene chips • Hybridization of chromosomes includes FISH, CGH, and SKY	• Allowed the identification of BRAF mutations in PTC[29] • Identification of RET/PTC in PTC, PAX8/PPARγ rearrangements in FTC, and multiple translocations with ATC[20,22,27]

chromosome (YAC). The choice of vector is made based on the size of the target DNA sequence. Recombinant DNA technology is a cornerstone of modern molecular biology. It facilitates DNA cloning, the study of transfected cell lines, and the generation of transgenic mouse models of disease.[11–13] Another major advance was the development of polymerase chain reaction (PCR) to clone DNA.[14] PCR allows the rapid generation of many copies of a target DNA sequence. PCR has been adapted to clone RNA in the form of complementary DNA (cDNA) by reverse transcriptase PCR (RT-PCR). It has also been modified to sequence DNA by nucleic acid hybridization and to quantify RNA (real-time PCR, TaqMan®). Most recently, microarray technology has revolutionized DNA hybridization and sequencing through integrating molecular biology techniques with nanotechnology from the computer industry. Microarrays were employed in the sequencing of DNA for the Human Genome Project. They are also being used widely to study gene expression in certain cancers.[9,15]

1.3 MOLECULAR BIOLOGY AND THYROID CANCER

1.3.1 Chromosomal Translocations and Mutations

From the 1900s, chromosomal abnormalities were suspected to be associated with tumor development. It took until the latter part of the twentieth century to prove this with certainty. The first chromosomal translocation was identified in the 1970s. This was the reciprocal translocation between chromosomes 9 and 22 (t(9;22)), named the Philadelphia chromosome, which was found to be a consistent abnormality in chronic myelogenous leukemia (CML). Subsequently, many chromosomal translocations were identified throughout the different hematological malignancies.[16,17]

Chromosomal translocations can alter gene expression in two ways: an activating gene may become situated proximal to a proto-oncogene, leading to activation of the oncogene; alternatively, a break in the gene can create a new fusion gene and result in a novel protein. While these chromosomal abnormalities are salient in hematological malignancies, relatively few have been identified in solid tumors. PTC was the first solid tumor in which a chromosomal translocation was detected. The initial abnormality detected was the fusion of the *RET* gene with the *H4* gene, by a paracentric inversion of the long arm of chromosome 10, known as the *RET/PTC1* rearrangement.[18] The *RET* gene encodes a transmembrane tyrosine kinase that is found in the parafollicular C cells, but is not present in normal thyroid follicular cells. *RET/PTC* rearrangements result in a constitutively activated form of the tyrosine kinase due to the fusion of a promoter region to the *RET* oncogene. Many forms of *RET/PTC* rearrangements have been identified in thyroid cancer cells, resulting from paracentric inversions of the long arm of chromosome 10 (*RET/PTC1, RET/PTC3*) and others from reciprocal translocations, such as the t(10;17) involved in *RET/PTC2*. These rearrangements are particularly common in individuals with thyroid cancer and a history of radiation exposure.[19–21]

FTC cells have been found to have the chromosomal translocation t(2;3) in approximately one- to two-thirds of cases.[22,23] The result of this translocation is the fusion of transcription factor gene *PAX8* with the gene for peroxisome proliferator-activated receptor γ (PPARγ), and leads to upregulation of the PPARγ transcription factors.[22] FTCs with *PAX8/PPARγ* translocation appear to have a more aggressive phenotype, although some adenomas also carry this translocation.[23]

Initial karyotyping identified chromosomes by their distinct banding pattern (G-banding) and could detect gross chromosomal abnormalities, such as the Philadelphia chromosome.[16] However hybridization has allowed for more subtle anomalies to be detected.[24] Fluorescent *in situ* hybridization (FISH) uses large DNA probes to locate specific DNA sequences on chromosomes, and comparative genomic hybridization (CGH) compares the binding pattern on the DNA of interest to normal DNA.[11] FISH revealed the positions of chromosomes during interphase, demonstrating how a single radiation beam could create simultaneous breaks in the *RET/PTC* genes, allowing for the observed rearrangements seen in individuals with thyroid cancer after radiation exposure.[25] Multiplex FISH (M-FISH) and spectral karyotyping (SKY) use the same principles, but use multiple probes for each chromosome. The pattern of probe hybridization undergoes computer analysis so that the origin of each chromosome involved in a rearrangement can be distinguished. ATC cells have complex chromosomal rearrangements that have been demonstrated with SKY.[24,26,27] These techniques also allowed for the identification of the multiple different *RET/PTC* fusion genes in PTC.[20]

Germline point mutations in the *RET* oncogene are present in familial medullary thyroid carcinoma (fMTC), the multiple endocrine neoplasia (MEN) 2A and MEN2B syndromes, and are inherited in an autosomal dominant manner. These mutations were found in MEN2A to be a single base pair change that most commonly altered the amino acid produced by codon 380.[28] Screening different tumor types for point mutations led to the discovery of mutations in the *BRAF* gene in melanoma, colorectal cancer, and ovarian cancer.[29] The RAF proteins are serine/threonine protein kinases that affect cell growth, differentiation, and apoptosis through the mitogen-activated protein kinase (MAPK) pathway signaling. These *RET* and *BRAF* point mutations were detected by DNA sequencing using nucleotide hybridization.[28,29] Heteroduplex analysis was also used to detect the *BRAF* mutations. This process involves the use of PCR to replicate DNA from two sources, normal DNA and DNA with a suspected mutation. The PCR products from the normal and abnormal DNA templates recombine to form heteroduplexes, but the point mutation results in unsuccessful base pairing at that nucleic acid. This leads to a bubble in the DNA double helix and allows it to be separated on gel electrophoresis from the normal double-stranded DNA. The abnormal nucleic acid can then be identified by sequencing.[29] Forty to seventy percent of PTC has been shown to carry the *BRAF* mutation, but it is uncommon in follicular variants of papillary thyroid cancer (FVPTCs) and FTC.[29,30] *BRAF* mutations translate into single amino acid substitutions that alter the activity of BRAF kinase and activate MAPK signaling.[29]

FTC, FVPTCs, ATC, as well as some follicular adenomas have been found to harbor point mutations in the *RAS* gene.[21,23,31] The RAS proteins are signal transducing proteins involved in cell growth and differentiation and are associated with the same signaling pathway as PPARγ. As the *RAS* mutations and *PAX8/PPARγ* translocations are rarely found in the same tumor, it has been suggested that these two mutations may represent two distinct tumor types.[23] Transgenic mouse models of thyroid cancer have been produced to study these genetic mutations *in vivo*. The use of transgenic mouse models allows the introduction of these known mutations of the *RAS* and *BRAF* genes to see if thyroid cancer can be attributed to these mutations, rather than being merely coincidental anomalies. Studies of *BRAF* mutations in transgenic mice confirmed that these mutations induce thyroid tumors with histological findings similar to those of human PTC, and *RAS* mutations have been shown to lead to the development of aggressive FTC.[32,33]

Since the discovery of germline mutations in *RET*, the screening of relatives of affected individuals with MTC and MEN2 involves DNA testing. Previously, stimulated calcitonin (CTN) levels were monitored in all family members. CTN monitoring had suboptimal sensitivity, and there was unnecessary testing and follow-up of unaffected relatives. Since MTC is the major cause of mortality among individuals with MEN2A, a positive DNA test for the mutation should lead to prophylactic thyroidectomy in childhood, as this appears to be curative. Therefore, DNA testing for MEN and fMTC has transformed the care of these patients.[34,35]

The presence of the *BRAF* mutation in PTC is associated with increased disease recurrence and metastases, so testing thyroid FNAB specimens for this mutation has been proposed preoperatively to guide the extent of surgical resection.[36,37] Also, PTC with this mutation has a less avid response to RAI, and therefore may benefit from molecular targeted therapies such as inhibitors of tyrosine kinases.[38]

As mentioned above, the identification of different translocations and mutations within thyroid cancers, such the *PAX8/PPARγ* translocation and *RAS* mutation within FTC, may provide highly useful information regarding tumor behavior. This information may ultimately lead to a reclassification of thyroid cancers, based not on their histological subtype, but on their genetics. However, even though these mutations may be present in the majority of thyroid cancers, noncancerous tissues have also been found to harbor some of these genetic anomalies. Therefore, detecting these mutations is not "diagnostic"; they are still unable to completely distinguish benign from malignant disease.[39]

Hence, further clinical insights can be gained by studying additional molecular biological mechanisms, such as gene expression and patterns of protein synthesis and interactions.

1.3.1.1 Gene Expression

Gene expression is studied by analyzing the RNA produced by gene transcription. Although mutations and translocations affecting DNA can lead to thyroid tumors, the disease phenotype can vary due to differences in gene expression. For example, *BRAF* mutations lead to more aggressive forms of PTC than *RET/PTC* translocations, even though they both act through the same signaling pathway. Microarray and real-time PCR are the two methods most commonly used to analyze RNA. Microarrays have been used to analyze gene expression in many cancers, but they have been best studied in breast cancer, where the expression profiles of genes regulating cell cycle, metastases, and angiogenesis can be used to predict tumor behavior and guide therapy. Gene expression has also been used in colorectal cancer to establish whether tumors will respond to certain chemotherapeutic agents.[9] Microarray allows for the detection of known sequences of RNA or its cDNA produced by RT-PCR. The microarray has predefined sequences fixed to the glass chip. Labeled cDNA hybridizes to the complementary sequences on the chip in a process called reverse hybridization. The gene expression is demonstrated by the signal intensity, which is related to the strength of hybridization when compared to the normal tissue.[40] In real-time PCR, the quantity of RNA is detected by the quantification of fluorescent molecules that are bound to the PCR product. The TaqMan® probe contains a fluorescent molecule and a quencher that masks the fluorescence. It is used as a third primer that binds downstream to one of the conventional DNA primers during PCR. As the DNA is elongating, DNA polymerase encounters the quencher and releases it, allowing the DNA strand to fluoresce. The intensity of the fluorescence increases with greater quantity of the PCR product.[41]

In PTC, gene expression analysis has shown different expression patterns in *BRAF*- and *RET/PTC*-associated tumors. The *BRAF* mutation has been shown to promote methylation and silencing of a number of tumor suppressor genes and upregulation of the angiogenic vascular endothelial growth factor (VEGF), matrix metalloproteinases (MMPs) that promote destruction of the interstitial matrix and promote angiogenesis, and c-Met, a cell surface tyrosine kinase that stimulates mitogenesis and is associated with a number of tumors. Iodide metabolizing genes are downregulated in *BRAF* mutations, perhaps accounting for the poor response to iodide treatment in patients with metastatic PTC.[37] Comparing the gene expression profiles of *RET/PTC*, *BRAF*- and *RAS*-associated PTC have shown a defining gene expression profile associated with each mutation, involving genes associated with signal transduction and the immune response.[19] Microarrays could potentially be used in thyroid cancer to determine therapy based on these different gene expression profiles.

1.3.1.2 MicroRNA

In 1993, miRNAs first appeared in the genomics cast. They were initially discovered in the nematode *Caenorhabditis elegans*, but were later recognized in many eukaryotes, including humans.[8,42] In oncology, they were first noted to be involved in chronic lymphocytic leukemia (CLL), where a deletion on chromosome 13 was found to cause downregulation of microRNA 15 (miR15) and miR16, prompting further research into these small molecules and their role in cancer.[43] miRNAs are transcribed from nonprotein coding regions of DNA, to form the precursor pri-miRNA. Pri-miRNA is then enzymatically cut to form pre-miRNA in the nucleus and is transported to the cytoplasm as a double strand. One strand forms the mature miRNA that enters a protein complex, mi-ribonucleoprotein (mi-RNP). The second strand was thought to be degraded, but may also form functional mature miRNA.[44] The miRNA binds to the 3' untranslated region (3'UTR) of mRNA, that is, the region of mRNA between the 3' end of the translated mRNA and the polyadenylated tail. How precisely the miRNA binds to the 3'UTR region determines whether it leads to mRNA degradation or inhibition.[8,45]

Different miRNAs may be upregulated or downregulated in different tumor types. In thyroid cancer, there appears to be distinct patterns of miRNA upregulation, linked to the different gene mutations and translocations.[46] In addition, single nucleotide polymorphisms (SNPs) in the genome have been associated with altered regulation of miRNAs and an increased risk of PTC.[47] It appears that the potential for tumorigenesis is related to the mRNA affected by the miRNA. In PTC, miRNA also affects expression of the protein KIT, a tyrosine kinase receptor, whereas in B cell leukemia, miRNAs activate the oncogene *Myc*.[8] It has been suggested that these miRNAs also mediate drug resistance. Therefore, miRNAs may be useful for diagnostics and guiding therapeutics. It is hoped that expression profiles of miRNAs may assist in FNAB diagnoses. Additionally, pharmacological research continues to study the forms of anti-miRNAs that can bind to and inhibit the action of miRNAs.[45]

1.3.1.3 Proteomics

The product of gene transcription and translation and the ultimate effector of any alterations in DNA or RNA is protein. Proteomics studies these molecular products of genetic mutations. Proteins can be studied by 2D electrophoresis and mass spectrometry. Mass spectrometry has a greater sensitivity for protein detection than electrophoresis. Two common types of mass spectrometry are matrix-assisted laser desorption ionization (MALDI) and electrospray ionization (ESI). Analysis of whole tissue sections is possible using MALDI, and reveals which proteins are present and where in the tissue they are functioning. In thyroid cancer, mutations frequently lead to continuous activation of kinases, increased signaling through the (MAPK) pathway, and angiogenesis resulting from the production of VEGF.[23,48] Therefore, inhibitors of tyrosine kinase, VEGF and PPARγ agonists are potential therapeutic agents.[49,50]

As well as identifying the alterations in signaling pathways that result from genetic mutations or altered gene expression, proteomics can identify biomarkers for cancer that might predict response to therapies and monitor response to treatment.[9] Protein profiling of thyroid FNAB specimens is an area under investigation, with the hope that markers will be identified that can define the malignant potential of the thyroid tissue and its likely response to treatment.[51]

1.3.2 Clinical Implications

Ultimately, the aim of translational medicine is to apply scientific knowledge to clinical medicine and thereby improve patient care. As mentioned, DNA testing in patients with MEN syndromes and fMTC has allowed for curative surgery in those who carry this germline mutation, and no need for ongoing screening of unaffected relatives. In thyroid cancers of follicular cell origin, insights into the associated mutations have allowed a new understanding of cancer development and its progression, and may redefine how we think of this disease. The identification of potential treatment targets through gene expression has led to pharmacological trials of novel treatments, such as tyrosine kinase inhibitors in advanced cancers that were previously untreatable.

Eventually, gene expression analysis of mRNA, miRNA, or biomarker detection could be performed on FNAB samples, and the results could determine the malignant potential of the lesion and the surgical intervention required. In addition, a patient's likelihood of disease progression could be determined, as well as the optimal treatment strategy for that individual. New therapeutic targets are being investigated, with new molecular targets, such as miRNAs. Perhaps with the development of proteomics, sensitive and specific biomarkers will be available that can detect tumors before they are clinically apparent and allow for intervention before the disease progresses.

1.4 CONCLUSION

Through integrating basic science and clinical medicine, translational medicine has enhanced our understanding of thyroid cancer in recent times. Innovations in technology have been embraced by science and have accelerated the rate of scientific discoveries. The momentum of scientific advancement has facilitated breakthroughs in diagnostics and therapeutics. Concurrently, these new discoveries have motivated clinicians to collaborate with basic scientists, to decipher further clinical enigmas. In this way, harnessing the strengths of science and medicine continues to transform our understanding of diseases and revolutionize their treatments. Therefore, looking ahead, it seems that under the united umbrella of translational medicine, this union of science and medicine has tremendous potential for an illustrious future.

REFERENCES

1. Fontanarosa PB, DeAngelis CD. 2003. Translational medical research. *JAMA* 289(16):2133.
2. Campeau P, Foulkes W, Tischkowitz M. 2008. Hereditary breast cancer: New genetic developments, new therapeutic avenues. *Hum Genet* 124(1):31–42.
3. Hauser SL, Waubant E, Arnold DL, et al. 2008. B-cell depletion with rituximab in relapsing-remitting multiple sclerosis. *N Engl J Med* 358(7):676–88.
4. Strong LC. 1926. On the occurrence of mutations within transplantable neoplasms. *Genetics* 11(3):294–303.
5. Crabtree JA. 1941. Observations on the familial incidence of cancer. *Am J Public Health Nations Health* 31(1):49–56.
6. Watson JD, Crick FH. 1953. Molecular structure of nucleic acids; a structure for deoxyribose nucleic acid. *Nature* 171(4356):737–38.
7. Human Genome Sequencing Consortium. 2004. Finishing the euchromatic sequence of the human genome. *Nature* 431(7011):931–45.
8. Sotiropoulou G, Pampalakis G, Lianidou E, Mourelatos Z. 2009. Emerging roles of microRNAs as molecular switches in the integrated circuit of the cancer call. *RNA* 15:1443–61.
9. Lin J, Li M. 2008. Molecular profiling in the age of cancer genomics. *Expert Rev Mol Diagnostics* 8(3):263–76.
10. Jackson DA, Symons RH, Berg P. 1972. Biochemical method for inserting new genetic information into DNA of Simian virus 40: Circular SV40 DNA molecules containing lambda phage genes and the galactose operon of *Escherichia coli*. *Proc Natl Acad Sci USA* 69(10):2904–9.
11. Strachan T, Read A. 2003. *Human molecular genetics 3*. 3rd ed. Garland Science/Taylor & Francis Group, London.
12. Takahashi M, Ritz J, Cooper GM. 1985. Activation of a novel human transforming gene, ret, by DNA rearrangement. *Cell* 42(2):581–88.
13. Palmiter RD, Brinster RL. 1986. Germ-line transformation of mice. *Annu Rev Genet* 20(1):465–99.
14. Mullis KB, Faloona FA, Ray W. 1987. Specific synthesis of DNA *in vitro* via a polymerase-catalyzed chain reaction. *Methods Enzymol* 155:335–50.
15. Sears C, Armstrong SA, Garret M, Hampton KSGFVW, George K. 2006. Microarrays to identify new therapeutic strategies for cancer. *Adv Cancer Res* 96:51–74.
16. Rowley JD. 1973. Letter: A new consistent chromosomal abnormality in chronic myelogenous leukaemia identified by quinacrine fluorescence and Giemsa staining. *Nature* 243(5405):290–93.
17. Mitelman F, Johansson B, Mertens F. 2007. The impact of translocations and gene fusions on cancer causation. *Nat Rev Cancer* 7(4):233–45.
18. Pierotti MA, Santoro M, Jenkins RB, et al. 1992. Characterization of an inversion on the long arm of chromosome 10 juxtaposing D10S170 and RET and creating the oncogenic sequence RET/PTC. *Proc Natl Acad Sci USA* 89(5):1616–20.
19. Giordano TJ, Kuick R, Thomas DG, et al. 2005. Molecular classification of papillary thyroid carcinoma: Distinct BRAF, RAS, and RET/PTC mutation-specific gene expression profiles discovered by DNA microarray analysis. *Oncogene* 24(44):6646–56.
20. Ciampi R, Nikiforov YE. 2007. RET/PTC rearrangements and BRAF mutations in thyroid tumorigenesis. *Endocrinology* 148(3):936–41.
21. DeLellis RA. 2006. Pathology and genetics of thyroid carcinoma. *J Surg Oncol* 94(8):662–69.
22. Kroll TG, Sarraf P, Pecciarini L, et al. 2000. PAX8-PPARgamma 1 fusion in oncogene human thyroid carcinoma. *Science* 289(5483):1357–60.
23. Nikiforova MN, Lynch RA, Biddinger PW, et al. 2003. RAS point mutations and PAX8-PPAR{gamma} rearrangement in thyroid tumors: Evidence for distinct molecular pathways in thyroid follicular carcinoma. *J Clin Endocrinol Metab* 88(5):2318–26.
24. Padilla-Nash HM, Barenboim-Stapleton L, Difilippantonio MJ, Ried T. 2007. Spectral karyotyping analysis of human and mouse chromosomes. *Nat Protocols* 1(6):3129–42.

25. Gandhi M, Medvedovic M, Stringer JR, Nikiforov YE. 2005. Interphase chromosome folding determines spatial proximity of genes participating in carcinogenic RET/PTC rearrangements. *Oncogene* 25(16):2360–66.
26. Patel AS, Hawkins AL, Griffin CA. 2000. Cytogenetics and cancer. *Curr Opin Oncol* 12(1):62–67.
27. Lee J-J, Foukakis T, Hashemi J, et al. 2007. Molecular cytogenetic profiles of novel and established human anaplastic thyroid carcinoma models. *Thyroid* 17(4):289–301.
28. Mulligan LM, Kwok JB, Healey CS, et al. 1993. Germ-line mutations of the RET proto-oncogene in multiple endocrine neoplasia type 2A. *Nature* 363(6428):458–60.
29. Davies H, Bignell GR, Cox C, et al. 2002. Mutations of the BRAF gene in human cancer. *Nature* 417(6892):949–54.
30. Kimura ET, Nikiforova MN, Zhu Z, Knauf JA, Nikiforov YE, Fagin JA. 2003. High prevalence of BRAF mutations in thyroid cancer: Genetic evidence for constitutive activation of the RET/PTC-RAS-BRAF signaling pathway in papillary thyroid carcinoma. *Cancer Res* 63(7):1454–57.
31. Nikiforov YE, Steward DL, Robinson-Smith TM, et al. 2009. Molecular testing for mutations in improving the fine needle aspiration diagnosis of thyroid nodules. *J Clin Endocrinol Metab* 94(6):2092–98.
32. Knauf JA, Ma X, Smith EP, et al. 2005. Targeted expression of BRAFV600E in thyroid cells of transgenic mice results in papillary thyroid cancers that undergo dedifferentiation. *Cancer Res* 65(10):4238–45.
33. Vitagliano D, Portella G, Troncone G, et al. 2006. Thyroid targeting of the N-ras(Gln61Lys) oncogene in transgenic mice results in follicular tumors that progress to poorly differentiated carcinomas. *Oncogene* 25(39):5467–74.
34. Kouvaraki MA, Shapiro SE, Perrier ND, et al. 2005. RET proto-oncogene: A review and update of genotype-phenotype correlations in hereditary medullary thyroid cancer and associated endocrine tumors. *Thyroid* 15(6):531–44.
35. Skinner MA, Moley JA, Dilley WG, Owzar K, DeBenedetti MK, Wells SA, Jr. 2005. Prophylactic thyroidectomy in multiple endocrine neoplasia type 2A. *N Engl J Med* 353(11):1105–13.
36. Xing M, Westra WH, Tufano RP, et al. 2005. BRAF mutation predicts a poorer clinical prognosis for papillary thyroid cancer. *J Clin Endocrinol Metab* 90(12):6373–79.
37. Xing M. 2007. BRAF mutation in papillary thyroid cancer: Pathogenic role, molecular bases, and clinical implications. *Endocr Rev* 28(7):742–62.
38. Kloos RT, Ringel MD, Knopp MV, et al. 2009. Phase II trial of sorafenib in metastatic thyroid cancer. *J Clin Oncol* 27(10):1675–84.
39. Shibru D, Chung KW, Kebebew E. 2008. Recent developments in the clinical application of thyroid cancer biomarkers. *Curr Opin Oncol* 20(1):13–18.
40. Trevino V, Falciani F, Barrera-Saldana HA. 2007. DNA microarrays: A powerful genomic tool for biomedical and clinical research. *Mol Med* 13(9–10):527–41.
41. Kubista M, Andrade JM, Bengtsson M, et al. 2006. The real-time polymerase chain reaction. *Mol Aspects Med* 27(2–3):95–125.
42. Lee RC, Feinbaum RL, Ambros V. 1993. The *C. elegans* heterochronic gene lin-4 encodes small RNAs with antisense complementarity to lin-14. *Cell* 75(5):843–54.
43. Calin GA, Dumitru CD, Shimizu M, et al. 2002. Frequent deletions and down-regulation of micro-RNA genes miR15 and miR16 at 13q14 in chronic lymphocytic leukemia. *Proc Natl Acad Sci USA* 99(24):15524–29.
44. Jazdzewski K, Liyanarachchi S, Swierniak M, et al. 2009. Polymorphic mature microRNAs from passenger strand of pre-miR-146a contribute to thyroid cancer. *Proc Natl Acad Sci USA* 106(5):1502–5.
45. Visone R, Croce CM. 2009. MiRNAs and cancer. *Am J Pathol* 174(4):1131–38.
46. Nikiforova MN, Tseng GC, Steward D, Diorio D, Nikiforov YE. 2008. MicroRNA expression profiling of thyroid tumors: Biological significance and diagnostic utility. *J Clin Endocrinol Metab* 93(5):1600–8.
47. Jazdzewski K, Murray EL, Franssila K, Jarzab B, Schoenberg DR, de la Chapelle A. 2008. Common SNP in pre-miR-146a decreases mature miR expression and predisposes to papillary thyroid carcinoma. *Proc Natl Acad Sci USA* 105(20):7269–74.
48. Gorla L, Mondellini P, Cuccuru G, et al. 2009. Proteomics study of medullary thyroid carcinomas expressing RET germ-line mutations: Identification of new signaling elements. *Mol Carcinogenesis* 48(3):220–31.
49. Copland JA, Marlow LA, Kurakata S, et al. 2005. Novel high-affinity PPAR[gamma] agonist alone and in combination with paclitaxel inhibits human anaplastic thyroid carcinoma tumor growth via p21WAF1//CIP1. *Oncogene* 25(16):2304–17.

50. Sherman SI. 2009. Advances in chemotherapy of differentiated epithelial and medullary thyroid cancers. *J Clin Endocrinol Metab* 94(5):1493–99.
51. Giusti L, Iacconi P, Ciregia F, et al. 2008. Fine-needle aspiration of thyroid nodules: Proteomic analysis to identify cancer biomarkers. *J Proteome Res* 7(9):4079–88.

2 Systems Biology—An Overview

Ravi Iyengar and Azi Lipshtat

CONTENTS

ABSTRACT: Many diseases, including cancers, arise from a combination of molecular, cellular, and environmental factors. Products of genes with mutations or other changes often interact to form large functional networks. Systems biology helps us to understand how these networks are organized and function. New experimental technologies can provide large data sets for the study of various systems. Systems biology is an interdisciplinary field that helps researchers understand the fundamental biological principles underlying multiple changes seen in these large data sets. Instead of trying to understand each function separately, systems biology attempts to explain how all the different functions are interconnected to form a unified system. In this chapter, we review the experimental and computational techniques currently used in systems biology research. With respect to thyroid cancer, systems biology research of two interacting networks of signaling pathways, B-Raf/MAPK and PTEN/AKT/PI3K, has already yielded strategic sites for molecular targeted therapy. This novel research paradigm holds great promise for the continued understanding of thyroid cancer biology and therapeutics.

2.1 INTRODUCTION

There has been a substantial increase in information about biological systems from the most ancient bacteria to humans since the mid-twentieth century. Obtaining such information has been the focus of many disciplines within biological and biomedical sciences. The fields of biochemistry and molecular biology have focused on cataloging a parts list of what living organisms are made of. This has included empirically identifying molecules and interactions within cells,[1,2] sequencing genomes from yeast to flies to humans,[3] as well as characterizing interactions on a genome-wide basis.[4–6] These types of molecular interactions have been complemented by the growth of descriptive knowledge of the varied processes that occur at the cell, tissue, and organismal levels. The study of these dynamic processes, broadly termed cell biology and physiology, had proceeded for a while largely independently of the molecular parts list characterization.

Over the past 30 years, these fields of study have started to merge, and there has been increasing recognition that important physiological processes can be directly correlated to the function of individual molecules. The ability to produce genetically modified animals[7–9] highlighted such relationships. As the study of these relationships proceeded, it has become increasingly clear that individual molecules do not in most cases drive processes in living organisms. Rather, it is groups of molecules acting in a coordinated fashion that give rise to multilevel physiological processes. This realization has catalyzed the development of an integrated field in biological and biomedical sciences that is termed systems biology. Since most processes in biological systems, from the level of atomic interactions within molecules to complex physiology in whole organisms, involve many interacting entities, it is not possible to study these processes without the aid of computer-based calculations and simulations. Thus, computations have become an integral part of biological and biomedical sciences.

Systems biology, in contrast to computational biology, refers to a conceptual framework rather than a technical approach. Whereas computational biology offers new tools and methods for answering complex biological questions, systems biology asks deep questions about how function arises from organization, and uses a broad range of methods to look for the answers. Although the overall scope of systems biology is still developing, it can be said that systems biology is an integrative approach, which is focused on the performance of a system as a whole, from the characteristics of the individual components.[10–13] The system can vary in size, from cellular systems up to the level of populations and ecology. Accordingly, the objects under investigation are the whole genome (all the genes in a given system), proteome (all the protein in the cell), or interactome (ensemble of all the interactions), rather than a particular gene, protein, or interaction. Various mathematical and statistical approaches are employed in order to explore, characterize, and analyze these systems. Due to its global view, much of the analysis refers to the structure of the system, in terms of the mutual relationships between its components. Thus, systems biology analysis can often be done in a top-down approach, where individual components are considered only with respect to their effect on the entire system.

The impressive advance in experimental technologies in the recent decades, accompanied by modern computers, provides large data sets, which were unavailable in the past. This new situation evokes new challenges for researchers. On the one hand, new systematic methods for mining, organization, and analysis of data are required in order to make this data useful. On the other hand, despite the huge databases we have, there are very few systems that are fully characterized, even in those rare cases where the available data are qualitative in nature. In fact, it is a common situation, where a reasonably complete detailed description of a system is not available. We cannot claim that we know each and every single chemical reaction that takes place in a particular cell, nor can we draw a complete evolutionary tree. In most cases, not only don't we have the entire knowledge, but also we cannot tell what fraction of the data is still missing. We are still far away from the desired goal of systematic quantitative description of even a unicellular organism. Under conditions of missing data, it is very hard to provide reliable large-scale comprehensive analyses. This is the place where systems biology can help. Assuming that the partial data we have is a representative sampling of the entire system, then there can be useful insights to the global structure of the system. These insights may highlight the important parts of the system and direct us when looking for more data and designing more experiments. Thus, systems biology deals with analyses of biological data, in terms of how progressively higher-order function arises from organization and modularity.

2.2 GENOMIC AND PROTEOMIC APPROACHES

A system, by definition, is composed of many components. Analysis of systems requires experimental examination of these components. Experimental work is needed for defining the system, identifying its structure, and then further analyzing its characteristics and functions. In order to get the large amount of data that is needed, new technologies have been developed with significant international effort. The large data sets that are available now are the accumulated result of many experiments done by many labs. These data sets encompass our knowledge about the various levels of organization—from molecular structure, through intracellular reaction, and up to higher organism functions. The various experimental approaches used in systems biology are summarized schematically in Figure 2.1.

2.2.1 SEQUENCING WHOLE GENOMES

Genetic information is stored in the nucleotide sequence of DNA. Knowing the exact order of nucleotide bases of a given gene enables one to analyze its structure and function on the one hand, and

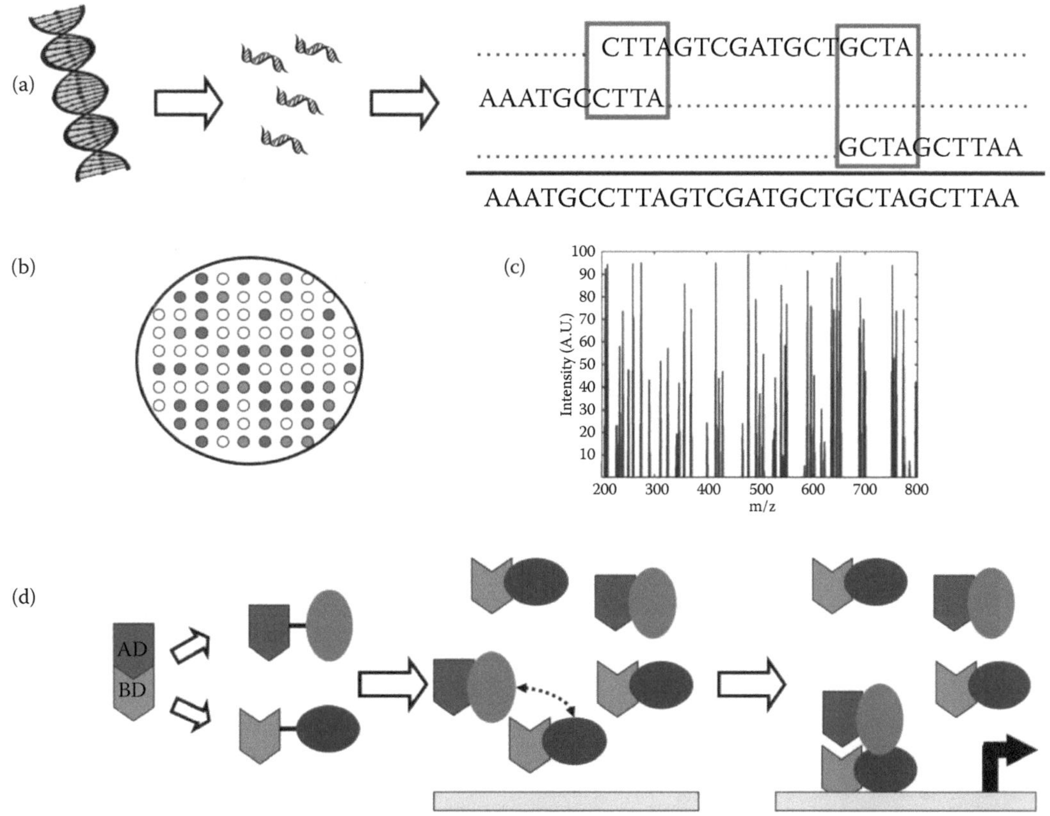

FIGURE 2.1 (See color insert.) Schematic summary of various experimental approaches. (a) DNA sequencing: Partially overlapping fragments of DNA are sequenced separately and then merged into a unified sequence. (b) Measuring gene expression profiles: The different colors in microarray indicate the differences in gene expression. (c) Mass spectrometry to obtain the relative abundance of proteins. A typical m/z spectrum obtained from mass spectrometry experiments. High intensity indicates high abundance of the respective component. From the mass, the sequence of the fragment and the protein can be deduced. (d) Measuring protein-protein interactions: yeast 2 hybrid experiments: the activation and binding domains (AD and BD) of a transcription factor are attached to the oval proteins. If there is an interaction between those proteins, then the transcription factor can bind to the promoter and activate it. Increase in gene expression is an indication for the interaction.

design drugs that can bind the gene product and manipulate its activity on the other hand. Thus, sequencing of single genes, namely, finding the order of their nucleotides, is important for understanding the fundamental processes of genetics. Sequencing is typically done by cutting the DNA into small pieces, with partial overlap between them.[14–16] Each fragment is cloned and sequenced independently, and the sequences are then combined computationally based on their overlapping. This process is used for the sequence of the human genome and is called shotgun sequencing. Another option is a consecutive reading of a long DNA in a piecewise manner, where the end of one sequence is used as a primer for the reading of the next sequence (primer walking). This method is simpler, but is also less efficient. Newer, high-throughput methods are being continuously developed.

Sequencing of the whole genome of an organism is a big challenge. Due to the large number of genes, automation and parallelization are required, as well as efficient algorithms for alignment of the sequence pieces. Today, we have the whole genome of many organisms, including the severe acute respiratory syndrome (SARS) virus,[17] *Drosophila*,[18] human,[19,20] and others. In 2008, the genome of a human individual was completely sequenced, using massively parallel DNA sequencing.[21]

The whole genome database is used in systems biology in various ways. It enables one to compare genes in terms of structure and function, identify the chemical basis for cellular processes, and identify possible targets for drugs. Variations as well as damage in the DNA are often associated with diseases, and thus knowledge about the differences between normal and damaged sequences is the first step toward drug development.

2.2.2 mRNA Profiling by Microarrays

Experiments in systems biology require simultaneous examination of many components of a system. Microarrays are useful devices that enable one to analyze the expression of many different genes under similar conditions.[22] The device is a small glass slide or a silicon chip with an array of hundreds or thousands of microscopic wells containing fragments of gene sequences. Microarrays measure the change in the expression of a large number of genes, and thus provide information about correlations between distinct components and about the differences in activities of normal and mutated genes.

2.2.3 Protein–Protein Interactions by Yeast 2 Hybrid

Cellular systems are described as a list of their components and the interactions among these components. Yeast 2 hybrid (Y2H) is a technique used to reveal interactions between two protein components.[23,24] The method is based on a conditional transcription of a reporter gene. This gene is expressed only when its transcription factor is bound to the promoter and activates it. The binding domain of the transcription factor can be separated and fused into a protein of interest. The activation domain is fused into another protein. Since both binding and the activation domain are required for activation of the gene expression, transcription of the reporter gene is an indicator of an interaction between the two proteins. Yeast 2 hybrid experiments have been used to obtain genome-wide interactions in several organisms.

2.2.4 Mass-Spectrometry-Based Proteomics

Mass spectrometry (MS) is a general method for determining the composition of a sample. In this technique, a sample is charged and then transferred through an electromagnetic field. The mass-to-charge (m/z) ratio affects the dynamics of the compound, which is then measured. In biology, MS is used to identify proteins.[25] The standard procedure includes fragmentation of a sample into peptides (or even smaller pieces), and measurement of the m/z ratio of the fragments. The results are then compared with databases that include known values of m/z for many sequences. This way, the proteome of a given system can be revealed. Advanced techniques enable quantification of the relative abundance ratio of the different proteins in a system, and determine posttranslational modification on a genome-wide basis.[25–27]

2.2.5 Graph Theory and Network Analyses

Biological systems are typically composed of a large number of components.[28] These can be metabolites or proteins in a cell,[29–32] cells in a tissue,[33] drugs and their targets in a body,[34,35] individuals within a population,[36,37] or different populations in an ecological system.[38] In all the examples mentioned above, the large system can be described by a list of the components and the relationships between those components. A convenient mathematical framework for that type of data is graph theory.[39,40] A mathematical graph (also known as a network) is defined by a set of nodes (or vertices) that represent the individual components and a set of edges (arches or links) between the nodes. The existence of an edge between two nodes indicates some kind of relationship between those nodes, whereas the absence of such an edge indicates that these relationships do not hold between the respective nodes. For example, one of the famous networks in systems biology is the protein-protein interaction (PPI) network.[29,41] This network represents the interactome.[42] Each node represents one intracellular component, like a protein or an enzyme. An edge between two nodes means that the two respective proteins interact with each other. The exact nature of the interaction (binding, activation, etc.), as well as the kinetic parameters or other details, is not included in the basic definition of the network. However, this network provides a global overview on the cell interactome as a system. Are there distinct modules? Are there components that are involved in more interactions than the others? What is the effect of knocking out or duplicating a certain protein? These kinds of questions can be answered by simple analysis of the network.

An important advantage of using graphs for data representation is their convenient visualization. Drawing a symbol for each node and connecting them by lines produces a network, which gives a good idea of its structure in a glance. By more sophisticated visualization, one can embed in the network much more information than is found in the formal definition of the graph. For example, families of nodes can be distinguished by color or shape, quantitative information can be expressed by size of the node or width of the edge, etc.

There are several quantitative measurements that characterize the global structure of a network. The degree of a node is the number of edges connected to that node. In case the edges are directed (i.e., an edge from node A to node B is not considered the same as from B to A), there are in-degree and out-degree for the number of incoming and outgoing edges, respectively. The degree distribution of a graph is an important characteristic.[39,43] If a graph has a narrow distribution, then all of its nodes have approximately the same number of connections. This means that there are no nodes that are more significant than the others, in terms of their contribution to the entire system performance.[44,45] On the other hand, a broad distribution, where there are few nodes with high degree and many nodes with fewer connections, implies a different organization of the system. In these cases, there are a few important nodes, which control to some extent the performance of the entire system. Any change in those components has a crucial effect that can relate to most of the other nodes.

Analysis of networks, which are based on real biological data, shows that in many cases the connectivity exhibits a power law distribution. This is a broad distribution where the number of nodes with degree n is proportional to $n^{-\alpha}$, where α is some positive number. This distribution is also called a scale-free distribution, because in this distribution there is no typical number of connections per node. The evolutionary advantages of such an organization have been analyzed. Scale-free networks can tolerate a higher frequency of random error.[44] The reason is that most of the nodes are poorly connected, and damaging these nodes has a minor effect on the overall performance of the network. The drawback is, of course, the existence of several highly connected and extremely important nodes. Error in performance of these few nodes can be fatal to the global network. Since random mutations are by far more prevalent than targeted changes, there is an evolutionary advantage in a scale-free network design.

Another measure of networks is the "small world" feature.[46,47] A small world network is organized in a way that there is a relatively short path of edges between any two nodes. The famous six degrees of separation between people is an example of the small world network of personal relationships. The existence of high-degree nodes (hubs) facilitates this feature because all the neighbors of the hub have a short path (consists of two links and goes via the hub) between them. This feature emphasizes the connectivity of the graph and the effect of one region on others. It is tightly related to the clustering coefficient, which represents the local density of the graph. The clustering coefficient of a node is defined as the fraction of existing edges out of all possible edges within the neighborhood of the node (i.e., the set of its neighbors). It measures how close this neighborhood is to being a clique (a set of nodes where all possible edges do exist). The clustering coefficient of a graph is the average over all nodes of their local clustering coefficients. Dense clusters have short paths between their nodes, and thus networks with large clustering coefficients are very often small world networks.

There are many ways to produce networks with a given number of nodes and edges.[48] Comparison between real biologic networks and randomly generated networks may be insightful.[49] It highlights the unique features of those networks that have been developed along evolution, and may teach us about evolutionary mechanisms, and functional patterns that appear in the evolutionary-designed network. For example, it was found in many networks that there are certain circuits, of three to five nodes, that appear in these networks in much higher abundance than expected from random design.[50] This finding implies that the overrepresented circuits, also called network motifs, have a role in the system, and are designed to perform a certain function.[1,51] For example, it was shown that feed-forward loops and bifan motifs are in large excess in the gene regulation network of *E. coli* and *S. cerevisiae*. These findings motivated analyses of those motifs, which revealed their role in filtering out noise and fluctuations.[52] Key features of network analyses are summarized in Figure 2.2.

2.2.6 Dynamics in Networks

All the analyses mentioned above refer to the topology (organization) of the network, but ignore its dynamical performance. Dynamics is important both in the node and in the network scales. The abstract nodes represent actual biological entities, and as such can change their properties in time.[53] For example, proteins can change their concentrations, receptors may be active or not, populations can be replaced, etc. At the network level, we are interested in the different types of dynamics that may emerge from the given network. Does the system always arrive at a steady state? Is there more than one steady state?[54,55] Can the system oscillate?[56] Are there chaotic dynamics? These system dynamics are crucial for understanding the functionality of a system. Different topologies in the network representing the system can lead to different dynamics, and in some cases, the exact details and kinetic parameters are also important.[57–59]

There are several ways to incorporate temporal changes into the graph theory framework. The simplest one is Boolean dynamics.[60] In this approach, each node is assigned a Boolean variable, which can have two possible values, like 0 or 1. These two values correspond to the two possible states of the cellular component represented by the node (high vs. low concentration, active vs. non-active, free vs. bound, etc.). These values are then repeatedly updated as functions of the activity of the neighboring nodes change. The exact dependence of one node on its neighbors can be different for each node and reflect the biological relationships among them. Each updating cycle represents a pseudo-time unit, and the overall process produces pseudodynamics of the system. This is a simple way of simulating the dynamics of a network and getting a qualitative understanding of the possible effect of one component (or motif) of the system on the rest of the network.

More quantitative modeling of system dynamics can be done by constructing and integrating the respective ordinary differential equations (ODEs).[53,61] Here, each node is associated with a number

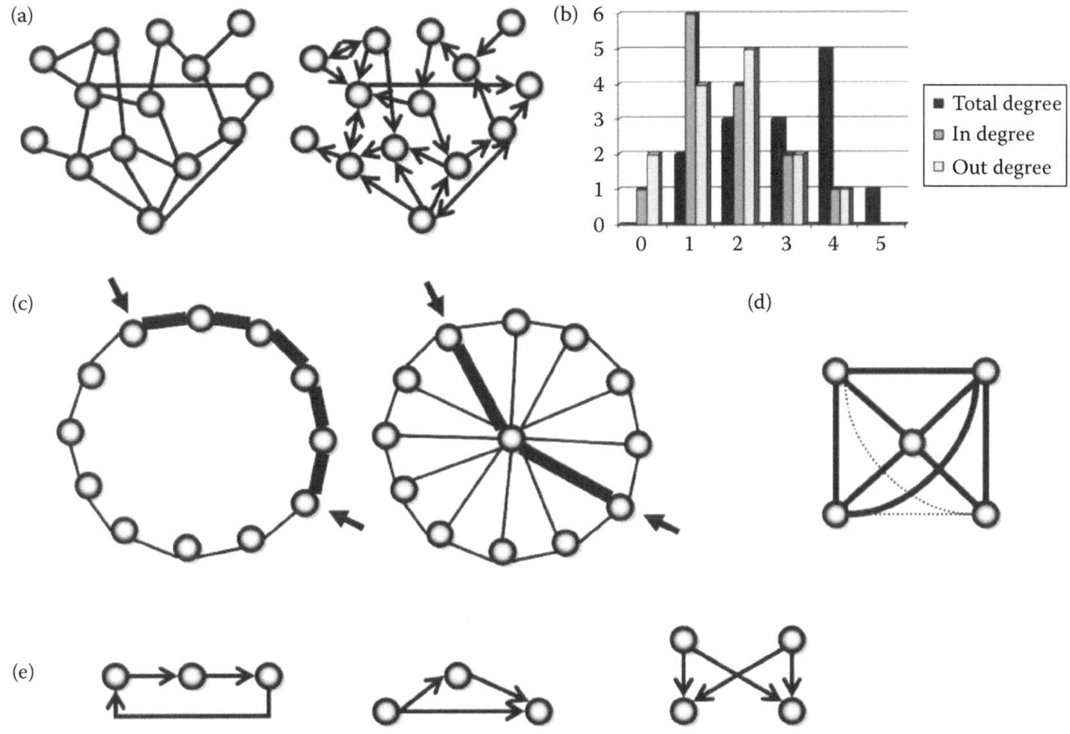

FIGURE 2.2 Key features of network analyses. (a) Examples of undirected and directed graphs. Each graph consists of 14 nodes and 21 links. Note that some edges can be directed in both directions. The nodes may represent proteins and the links, interaction. (b) Degree distribution for the two graphs. In this example the networks display broad distribution and there are relatively many nodes with high degree, i.e., many connections. This is not a scale-free network. (c) The right graph is a small world network, since the length of the shortest path between any two nodes is at most 2. In the left graph the distance between the marked nodes is 5. (d) The clustering coefficient of the node in the center is 0.67. Out of six possible links between its four neighbors, four links are present (solid lines) and two are missing (dotted lines). (e) Important network motifs (left to right): feedback loop, feed-forward loop, and bifan.

that may represent the concentration of the respective component, population size, or activity. These numbers are the variables of the ODEs, and their derivatives with respect to time are functions of their relationships with their neighbors. The solution of the equations provides real numbers and potentially a reliable description of the dynamics. However, in this approach, the graph representation is no more than a convenient way of visualization. There is no real use in any of the graph theory tools. Such quantitative analyses, although very useful for understanding temporal dynamics, often obscure the regulatory topology of networks and do not provide insight into the relationship between different network motifs and the regulatory capabilities that may arise from interactions among network motifs. Another problem with this approach is the computational resources that are needed. For large systems, it is still not feasible to simulate the dynamics of the entire system for long times. It is not a rare situation when the network consists of thousands of nodes, with tens of thousands of interactions. The typical timescales for interaction (kinetic rates) can span over several orders of magnitude, making the computational challenge even harder. Stronger and faster computers can, of course, deal with larger systems. However, the computation is nonlinear, which means that a computer that works 10 times faster is not expected to be capable of simulating a system that is 10 times bigger, but only a much smaller one (like two or three times larger than the system that was simulated by the slower computer).

2.2.7 Statistical Tools

Systems biology deals with large sets of data. In cases where the data refer to relationships among entities, such as in interactome analyses, graph theory can be employed. However, there are cases where the data are such that graph representation does not apply. For example, microarray experiments yield plenty of data that are not structured as a graph. For these cases, different methods of data organization are required. A useful approach for these cases is hierarchical clustering.[62,63] In this approach, the data are divided into clusters, where each cluster contains data that has something in common. Each cluster is then divided into subclusters, and the process continues recursively. There is no single correct way to divide a large data set, but the division can vary with the specific needs of that particular research. For example, in one case, few large clusters are desirable, but for other purposes, many smaller clusters (which are expected to be more homogeneous) would be better. There are many algorithms for data clustering, and one should match the appropriate algorithm for each specific case.[62–65] A crucial step for any algorithm is a definition of a metric, or a distance function. This is a quantization of the distance or similarity between data elements. There may be many different distance functions for the same data set. For example, consider the case that the data contain time courses of many proteins under various conditions. One possible metric is the temporal correlation, which measures the similarity between two concentrations (for example, to what extent A increases together with B, or when B decreases). Another metric may be the difference in the maximal concentration (regardless of the time in which this maxima was measured). A third option is the difference between average concentrations. In any case, the metric assigns a number to data element pairs, and smaller numbers indicate closer data in some sense. The aim of the clustering process is to define clusters with minimal distance between the elements within each cluster, but large inter-cluster distance.

Clustering algorithms can be divided into two groups: bottom-up and top-down approaches. In the bottom-up approaches, all data elements are grouped into small clusters, with minimal distance within each cluster. Then, similar clusters are merged into one large cluster and so on, until the desired number of clusters or some other criterion is achieved. In the top-down approach, we start with one large cluster that is the entire data set. This cluster is then divided into smaller clusters, and the division process continues within each cluster recursively. A summary of the types of co-relations and their uses is given in Table 2.1.

2.3 FROM ORGANIZATION TO FUNCTION

The mathematical methods described above deal with the architecture and organization of the system. However, the aim of all these analyses is not limited to mathematical characterization of the organization, but rather to understand the relationship between organization and function. Generally,

TABLE 2.1
Summary of Clustering Algorithm Co-Relations

Clustering Criteria	Explanation
Temporal correlation	System components that change their activity at the same time are assumed to be involved in similar processes.
Spatial correlation	System components have to be located in close proximity in order to be able to interact with each other. Similar localization indicates that the components are related.
Topological proximity	Protein-protein interaction network summarizes the cell interactome. A group of highly connected nodes forms a module that has a specific role in the global system.
Structure similarity	Components with similar structure are assumed to perform similar functions.

large systems have been designed through evolution to perform some functions. The functions can be either local and specific, like turning on and off a certain pathway under predefined conditions, or higher-level functions, like control of the cell cycle at the cellular level or filtration in the kidney or heartbeat at the organ level. In all cases, the biological function is tightly related to the organization of the system. Changes in the network, such as addition or deletion of interactions, are expected to affect the dynamics of the system and thus its function. This concept applies to large networks, such as the interactome of an organism, as well as to small subnetworks. For example, there are several subnetworks that are known to be prevalent in biological networks (network motifs).[50] Their high abundance is believed to indicate that they may perform a regulatory function in signaling network, such as filtering noise, causing oscillations, etc.[52] Much effort is invested in understanding the relations between organization and function in larger scales.[49] Understanding topological mechanisms that control biological functions prompts the next level of identifying key components and manipulating the system's performance.

Despite the enormous importance of topology and organization to the function of the system, there are other factors that must also be considered. Function is almost always associated with dynamics—changes in activity with respect to time or space. Thus, kinetic rates and other quantitative parameters can have a dramatic effect on the dynamics and function of any system. Systems can operate in one mode within a certain range of parameters, but have completely different dynamics when the parameters change. For example, a two-state system can be used as a switch under a very low enzyme/substrate ratio, but not in another range of parameters.[66] Thus, every study of organization-function relationships has to be accompanied by analysis of the range in which these relationships apply. If this range is narrow, then the system is sensitive to any change in the parameters. Robustness, i.e., a lack of sensitivity to exact values of parameters, is considered an advantageous design principle, as it enables the system to operate properly even if the system is perturbed.

2.4 STOCHASTICITY

Living systems, at either the subcellular level or population scale, are stochastic by nature.[67–70] Similar systems under similar conditions exhibit in many cases different behaviors. Variables as concentrations of intracellular components or population sizes are subject to fluctuations. Environmental conditions also fluctuate, and as a result, there are no two systems that can be considered completely identical.

Different aspects of stochasticity and its consequences are studied in systems biology.[71] Understanding and characterizing the various origins of noise is an important systems-level issue. There is a distinction between two types of noise. Extrinsic noise is the variability that results from the fact that there are never identical conditions for similar systems. Intrinsic noise originates from the stochastic nature of the biochemical reactions that govern biological processes.[72,73] The different mechanisms by which biological systems are designed in order to reduce noise and stochasticity are also an important area of research in systems biology.[52,74,75] Regulation of complex processes is much easier when the system behaves in a deterministic manner, and thus noise suppression is a design principle of many biological networks. Some of the most abundant network motifs were shown to be efficient filters of noise, and this may be a possible explanation of their overrepresentation.[76–78] Paradoxically, stochasticity and noise can facilitate and initiate processes. At the cellular level there is evidence for phenomena like bistability or oscillations, which are noise driven, which would not emerge from deterministic dynamics.[10,54,70,79,80] Thus, under different conditions, stochasticity can be either a problem or the solution. Modeling and simulation of stochastic systems is a mathematical challenge. There are both analytical and numerical methods for this problem. The mathematical basis for both approaches is the master equation, which calculates the probability of finding the system in a given state (as a function of time). This equation can be analytically or numerically integrated. Monte Carlo simulations are another method to find the solution of the master equation. The basic algorithm for Monte Carlo simulations in a biological context was given by Gillespie more than three decades ago, and various improvements have been suggested since then.[81–84]

2.5 DEVELOPING SYSTEMS BIOLOGY

Systems biology provides a useful approach as well as valuable analysis tools to enhance our understanding about large and complex systems. However, one has to be aware of its current limitations. The most common limitation is the lack of comprehensive data sets. Other than genome sequence data, all other data sets are partial at best. Although partial data are often sufficient to get statistical characteristics of the system, these characteristics are not enough when we want to conduct a top-down investigation and derive a mechanistic understanding about the function of particular components in systems-level functions. A useful system biology study consists of two phases: First, we deduce the organization of the system from all pairwise interaction data (bottom-up approach). Then, we analyze the integrated system and derive conclusions as to how particular components contribute to the observed systems behavior. Whereas the first stage can be done even with partial data, the second phase requires many details at differing levels of organization. A lack of data limits the ability to provide high-resolution, multiscale analysis. Furthermore, even in the few cases where the topology and architecture of the system are fully known, these data alone may not be sufficient. Most applications require dynamics analysis and examination of temporal behavior. This requires a host of kinetic parameters that are generally not available. Much more experimental work has to be done before there is a full dynamic description of a system, and before that happens, we cannot take full advantage of systems biology tools and concepts.

Even in the case where there is enough information, systems biology research often produces useful hypotheses but does not allow us to draw definite conclusions about particular components. Consider the example of a drug-target network, where drugs are connected to their biochemical targets. Since many drugs affect more than one target, and each target can be affected by several drugs, these connections form a complex network. Analysis of this network may rank the targets and highlight new candidates with high probability to be good targets for a new drug. The underlying assumption is that a component that is in close proximity (in the drug-target network) to many known targets is a good candidate as a potential new target. Thus, components can be ranked based on their proximity and similarity to a known target. The higher the rank is, the higher is the probability that this component is a good target. However, systems biology is largely useful in identifying and ranking targets in a conceptual sense. Examination of each candidate has to be done using other approaches, which do not necessarily consider the whole system, but rather take a deeper look at one or a few components at a time. Thus, systems biology can play a significant role in enhancing our understanding of complex systems, but it is unlikely to fully replace more focused studies.

2.6 SYSTEMS BIOLOGY AND CANCER

Systems biology approaches have been particularly useful in cancer diagnosis and show great potential in cancer therapy. Microarray analyses of tumors have been useful in developing clinical prognosticators. Several diagnostic kits for breast cancer progression utilize microarray patterns. In thyroid cancer the use of single gene or protein markers to classify benign and malignant thyroid tumors has not been successful. The use of microarray technologies has been useful in classifying thyroid tumors as malignant or benign.[85] Using sophisticated statistical analyses, 6- and 10-gene models have been able to classify thyroid tumors.[86] How the changes in these genes reflect the underlying biological processes is a question for future research in thyroid cancer. Mutations in Raf-MAP-kinase 1,2 signaling pathways appear to be important determinants of malignancy in thyroid cancer.[87] Mutations and amplifications in the components of the phosphoinositide 3-kinase (PI3K) pathway, such as phosphatase and tensin homolog (PTEN), serine/threonine-specific protein kinases (AKT), and PI3K catalytic subunit, have also been implicated in thyroid tumors.[88] Together, these two pathways are well networked and required for both proliferation and survival in different thyroid cancer cell types. MAP-kinase-PI3-kinase network contain many potential targets for

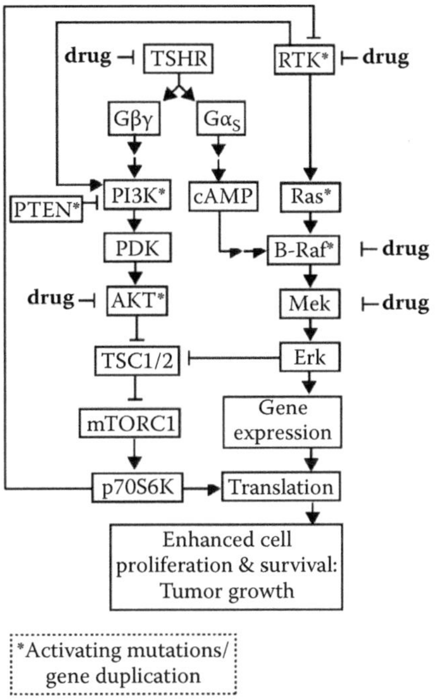

FIGURE 2.3 Interactions between the B-Raf (B-RAF) and phosphoinositide 3-kinase (PI3K) pathways in thyroid cells. This is a simplified network of signaling pathways involved in certain thyroid cancers. Both the G protein pathways, such as the TSH receptor pathway, and receptor tyrosine kinase pathways regulate expression and translation to promote cell growth and survival. Signaling components with mutations and gene duplications involved in cancer are marked with asterisks. Components against which drugs are available are also indicated. (We thank Dr. Aislyn Wist for this figure. With permission.)

treatment (Figure 2.3), and a systems-level understanding of how this network functions is likely to be very useful in developing therapeutic strategies for the treatment of thyroid cancer.

2.7 PERSPECTIVE AND CONCLUSIONS

The availability of large data sets, accompanied by powerful computational resources, has catalyzed the development of new approaches for unraveling the complexity in biological systems. Systems biology research utilizes these approaches. The information that can be obtained from systems biology analyses relates to the system as a whole. It can explain the various mechanisms by which the system functions, and quantifies the effects of organization of the different components within the system. This information is crucial for understanding therapeutics for complex diseases, especially novel and emergent modalities. Comprehensive understanding of the relations between the interacting components is the first step toward finding ways of manipulating the system. In disease states, such understanding enables successful therapeutic strategies. However, there will always be more steps and questions that cannot be answered by current systems biology approaches. With complex data, systems biology can draw a map of many physiological functions and how variations of these functions occur in individuals. The identification of many variants in the genome, including single nucleotide polymorphisms and copy number variations, provides a basis for understanding how the genomic variations give rise to varied physiology and pathophysiology. In these types of studies, systems biology approaches are likely to make critical contributions. The broad systems biology approaches are also likely to be combined with in-depth analyses since the exact details of many processes are highly affected by local parameters, such as kinetic rates, concentrations of components, and their location within the cell. Knowing these parameters is critical for understanding drug action. Thus, the importance of systems biology is likely to be in mapping the forest of information and identifying the important trees. Understanding how the trees contribute to the overall ecosystem will require other approaches as well.

ACKNOWLEDGMENTS

We thank Dr. Aislyn Wist for Figure 2.3. We are supported by the Systems Biology Center Grant GM 071558.

REFERENCES

1. Bhalla US and Iyengar R. 1999. Emergent properties of networks of biological signaling pathways. *Science* 283:381–87.
2. Weng G, Bhalla US, and Iyengar R. 1999. Complexity in biological signaling systems. *Science* 284:92–96.
3. Myers EW, Sutton GG, Delcher AL, Dew IM, Fasulo DP, Flanigan MJ, Kravitz SA, Mobarry CM, Reinert KHJ, Remington KA, et al. 2000. A whole-genome assembly of *Drosophila. Science* 287:2196–204.
4. Yu H, Braun P, Yildirim MA, Lemmens I, Venkatesan K, Sahalie J, Hirozane-Kishikawa T, Gebreab F, Li N, Simonis N, et al. 2008. High-quality binary protein interaction map of the yeast interactome network. *Science* 322:104–10.
5. Cornell M, Paton NW, and Oliver SG. 2004. A critical and integrated view of the yeast interactome. *Comp Funct Genomics* 5:382–402.
6. Giot L, Bader JS, Brouwer C, Chaudhuri A, Kuang B, Li Y, Hao YL, Ooi CE, Godwin B, Vitols E, et al. 2003. A protein interaction map of *Drosophila melanogaster. Science* 302:1727–36.
7. Cohen SN, Chang ACY, Boyer HW, and Helling RB. 1973. Construction of biologically functional bacterial plasmids *in vitro. Proc Natl Acad Sci USA* 70:3240–44.
8. Costantini F and Lacy E. 1981. Introduction of a rabbit [beta]-globin gene into the mouse germ line. *Nature* 294:92–94.
9. Rubin GM and Spradling AC. 1982. Genetic transformation of *Drosophila* with transposable element vectors. *Science* 218:348–53.
10. Kitano H. 2002. Systems biology: A brief overview. *Science* 295:1662–64.
11. Kitano H. 2002. Computational systems biology. *Nature* 420:206–10.
12. Nicholson JK and Wilson ID. 2003. Understanding "global" systems biology: Metabonomics and the continuum of metabolism. *Nat Rev Drug Discov* 2:668–76.
13. Alon U. 2006. *An introduction to systems biology: Design principles of biological circuits.* Chapman & Hall Mathematical and Computational Biology Series. London: Chapman & Hall.
14. Maxam AM and Gilbert W. 1977. A new method for sequencing DNA. *Proc Natl Acad Sci USA* 74:560–64.
15. Sanger F and Coulson AR. 1975. A rapid method for determining sequences in DNA by primed synthesis with DNA polymerase. *J Mol Biol* 94:441–46.
16. Anderson S. 1981. Shotgun DNA sequencing using cloned DNase I-generated fragments. *Nucl Acids Res* 9:3015–27.
17. Marra MA, Jones SJM, Astell CR, Holt RA, Brooks-Wilson A, Butterfield YSN, Khattra J, Asano JK, Barber SA, Chan SY, et al. 2003. The genome sequence of the SARS-associated coronavirus. *Science* 300:1399–404.
18. Adams MD, Celniker SE, Holt RA, Evans CA, Gocayne JD, Amanatides PG, Scherer SE, Li PW, Hoskins RA, Galle RF, et al. 2000. The genome sequence of *Drosophila melanogaster. Science* 287:2185–95.
19. Lander ES, Linton LM, Birren B, Nusbaum C, Zody MC, Baldwin J Devon K, Dewar K, et al. 2001. Initial sequencing and analysis of the human genome. *Nature* 409:860–921.
20. Venter JC, Adams MD, Myers EW, Li PW, Mural RJ, Sutton GG, Smith HO, Yandell M, Evans CA, Holt RA, et al. 2001. The sequence of the human genome. *Science* 291:1304–51.
21. Wheeler DA, Srinivasan M, Egholm M, Shen Y, Chen L, McGuire A, He W, Chen Y-J, Makhijani V, Roth GT, et al. 2008. The complete genome of an individual by massively parallel DNA sequencing. *Nature* 452:872–76.
22. Schena M, Shalon D, Davis RW, and Brown PO. 1995. Quantitative monitoring of gene expression patterns with a complementary DNA microarray. *Science* 270:467–70.

23. Fields S, Song O-K. 1989. A novel genetic system to detect protein–protein interactions. *Nature* 340:245–46.
24. Young KH. 1998. Yeast two-hybrid: So many interactions, (in) so little time. *Biol Reprod* 58:302–11.
25. Aebersold R and Mann M. 2003. Mass spectrometry-based proteomics. *Nature* 422:198–207.
26. Demirev PA, Lin JS, Pineda FJ, and Fenselau C. 2001. Bioinformatics and mass spectrometry for microorganism identification: Proteome-wide post-translational modifications and database search algorithms for characterization of intact *H. pylori*. *Anal Chem* 73:4566–73.
27. Blackstock WP and Weir MP. 1999. Proteomics: Quantitative and physical mapping of cellular proteins. *Trends Biotechnol* 17:121–27.
28. Strogatz SH. 2001. Exploring complex networks. *Nature* 410:268–76.
29. Ma'ayan A, Jenkins SL, Neves S, Hasseldine A, Grace E, Dubin-Thaler B, Eungdamrong NJ, Weng G, Ram PT, Rice JJ, et al. 2005. Formation of regulatory patterns during signal propagation in a mammalian cellular network. *Science* 309:1078–83.
30. Jeong H, Tombor B, Albert R, Oltvai ZN, and Barabasi AL. 2000. The large-scale organization of metabolic networks. *Nature* 407:651–54.
31. Ravasz E, Somera AL, Mongru DA, Oltvai ZN, and Barabasi AL. 2002. Hierarchical organization of modularity in metabolic networks. *Science* 297:1551–55.
32. Maslov S and Sneppen K. 2002. Specificity and stability in topology of protein networks. *Science* 296:910–13.
33. Sahiner B, Heang-Ping C, Petrick N, Datong W, Helvie MA, Adler DD, and Goodsitt MM. 1996. Classification of mass and normal breast tissue: A convolution neural network classifier with spatial domain and texture images. *IEEE Trans Med Imaging* 15:598–610.
34. Yildirim MA, Goh K-I, Cusick ME, Barabasi A-L, and Vidal M. 2007. Drug—target network. *Nat Biotechnol* 25:1119–26.
35. Hood L, Heath JR, Phelps ME, and Lin B. 2004. Systems biology and new technologies enable predictive and preventative medicine. *Science* 306:640–43.
36. Girvan M and Newman MEJ. 2002. Community structure in social and biological networks. *Proc Natl Acad Sci USA* 99:7821–26.
37. Newman MEJ, Watts DJ, and Strogatz SH. 2002. Random graph models of social networks. *Proc Natl Acad Sci USA* 99:2566–72.
38. Ulanowicz RE. 2004. Quantitative methods for ecological network analysis. *Comput Biol Chem* 28:321–39.
39. Albert-Laszlo B. 2003. Linked: The new science of networks. *Am J Physics* 71:409–10.
40. Bollobás B. 1998. *Modern graph theory*. Springer, New York.
41. Stelzl U, Worm U, Lalowski M, Haenig C, Brembeck FH, Goehler H, Stroedicke M, Zenkner M, Schoenherr A, Koeppen S, et al. 2005. A human protein-protein interaction network: A resource for annotating the proteome. *Cell* 122:957–68.
42. Cusick ME, Klitgord N, Vidal M, and Hill DE. 2005. Interactome: Gateway into systems biology. *Hum Mol Genet* ddi335.
43. Barabasi A-L and Albert R. 1999. Emergence of scaling in random networks. *Science* 286:509–12.
44. Albert R, Jeong H, and Barabasi A-L. 2000. Error and attack tolerance of complex networks. *Nature* 406:378–82.
45. Goh K-I, Oh E, Jeong H, Kahng B, and Kim D. 2002. Classification of scale-free networks. *Proc Natl Acad Sci USA* 99:12583–88.
46. Watts DJ and Strogatz SH. 1998. Collective dynamics of "small-world" networks. *Nature* 393:440–42.
47. Cohen R and Havlin S. 2003. Scale-free networks are ultrasmall. *Phys Rev Lett* 90:058701.
48. Callaway DS, Hopcroft JE, Kleinberg JM, Newman MEJ, and Strogatz SH. 2001. Are randomly grown graphs really random? *Phys Rev E* 64:041902.
49. Ma'ayan A, Lipshtat A, and Iyengar R. 2006. Topology of resultant networks shaped by evolutionary pressure. *Phys Rev E* 73:061912.
50. Milo R, Shen-Orr S, Itzkovitz S, Kashtan N, Chklovskii D, and Alon U. 2002. Network motifs: Simple building blocks of complex networks. *Science* 298:824–27.
51. Shen-Orr SS, Milo R, Mangan S, and Alon U. 2002. Network motifs in the transcriptional regulation network of *Escherichia coli*. *Nat Genet* 31:64–68.
52. Lipshtat A, Purushothaman SP, Iyengar R, and Ma'ayan A. 2008. Functions of bifans in context of multiple regulatory motifs in signaling networks. *Biophys J* 94:2566–79.
53. Tyson JJ, Chen K, and Novak B. 2001. Network dynamics and cell physiology. *Nat Rev Mol Cell Biol* 2:908–16.

54. Lipshtat A, Loinger A, Balaban NQ, and Biham O. 2006. Genetic toggle switch without cooperative binding. *Phys Rev Lett* 96:188101.
55. Ozbudak EM, Thattai M, Lim HN, Shraiman BI, and van Oudenaarden A. 2004. Multistability in the lactose utilization network of *Escherichia coli*. *Nature* 427:737–40.
56. Elowitz MB and Leibler S. 2000. A synthetic oscillatory network of transcriptional regulators. *Nature* 403:335–38.
57. Tyson JJ, Chen KC, and Novak B. 2003. Sniffers, buzzers, toggles, and blinkers: Dynamics of regulatory and signaling pathways in the cell. *Curr Opin Cell Biol* 15:221–31.
58. Barkai N and Leibler S. 1997. Robustness in simple biochemical networks. *Nature* 387:913–17.
59. Kholodenko BN. 2006. Cell-signalling dynamics in time and space. *Nat Rev Mol Cell Biol* 7:165–76.
60. Aldana M. 2003. Boolean dynamics of networks with scale-free topology. *Physica D Nonlinear Phenomena* 185:45–66.
61. Machne R, Finney A, Muller S, Lu J, Widder S, and Flamm C. 2006. The SBML ODE Solver Library: A native API for symbolic and fast numerical analysis of reaction networks. *Bioinformatics* 22:1406–7.
62. Sturn A, Quackenbush J, and Trajanoski Z. 2002. Genesis: Cluster analysis of microarray data. *Bioinformatics* 18:207–8.
63. Getz G, Levine E, and Domany E. 2000. Coupled two-way clustering analysis of gene microarray data. *Proc Natl Acad Sci USA* 97:12079–84.
64. Datta S and Datta S. 2003. Comparisons and validation of statistical clustering techniques for microarray gene expression data. *Bioinformatics* 19:459–66.
65. Quackenbush J. 2001. Computational analysis of microarray data. *Nat Rev Genet* 2:418–27.
66. Goldbeter A and Koshland DE. 1981. An amplified sensitivity arising from covalent modification in biological systems. *Proc Natl Acad Sci USA* 78:6840–44.
67. Blake WJ, Kaern M, Cantor CR, and Collins JJ. 2003. Noise in eukaryotic gene expression. *Nature* 422:633–37.
68. Elowitz MB, Levine AJ, Siggia ED, and Swain PS. 2002. Stochastic gene expression in a single cell. *Science* 297:1183–86.
69. Losick R and Desplan C. 2008. Stochasticity and cell fate. *Science* 320:65–68.
70. Rao CV, Wolf DM, and Arkin AP. 2002. Control, exploitation and tolerance of intracellular noise. *Nature* 420:231–37.
71. Raser JM and O'Shea EK. 2005. Noise in gene expression: Origins, consequences, and control. *Science* 309:2010–13.
72. Rosenfeld N, Young JW, Alon U, Swain PS, and Elowitz MB. 2005. Gene regulation at the single-cell level. *Science* 307:1962–65.
73. Swain PS, Elowitz MB, and Siggia ED. 2002. Intrinsic and extrinsic contributions to stochasticity in gene expression. *Proc Natl Acad Sci USA* 99:12795–800.
74. Ozbudak EM, Thattai M, Kurtser I, Grossman AD, and van Oudenaarden A. 2002. Regulation of noise in the expression of a single gene. *Nat Genet* 31:69–73.
75. Vilar JMG, Kueh HY, Barkai N, and Leibler S. 2002. Mechanisms of noise-resistance in genetic oscillators. *Proc Natl Acad Sci USA* 99:5988–92.
76. Hayot F, Jayaprakash C. 2005. A feedforward loop motif in transcriptional regulation: Induction and repression. *J Theor Biol* 234:133–43.
77. Bhaswar G, et al. 2005. Noise characteristics of feed forward loops. *Phys Biol* 2:36.
78. Brandman O, Ferrell JE, Jr., Li R, and Meyer T. 2005. Interlinked fast and slow positive feedback loops drive reliable cell decisions. *Science* 310:496–98.
79. Hanggi P. 2002. Stochastic resonance in biology: How noise can enhance detection of weak signals and help improve biological information processing. *ChemPhysChem* 3:285–90.
80. Astumian RD and Frank M. 1998. Overview: The constructive role of noise in fluctuation driven transport and stochastic resonance. *Chaos Interdisciplinary J Nonlinear Sci* 8:533–38.
81. Gillespie DT. 1977. Exact stochastic simulation of coupled chemical reactions. *J Phys Chem* 81:2340–61.
82. Lipshtat A and Biham O. 2004. Efficient simulations of gas-grain chemistry in interstellar clouds. *Phys Rev Lett* 93:170601.
83. Lipshtat A. 2007. "All possible steps" approach to the accelerated use of Gillespie's algorithm. *J Chem Physics* 126:184103.
84. Gillespie DT and Petzold LR. 2003. Improved leap-size selection for accelerated stochastic simulation. *J Chem Physics* 119:8229–34.
85. Kato MA and Fahey TJ. 2009. Molecular markers in thyroid cancer diagnostics. *Surg Clin N Am* 89:1139–55.

86. Manzzanti C, Zeiger MA, Costouros NG, et al. 2004. Using gene expression profiling to differentiate benign vs. malignant thyroid tumors. *Cancer Res* 64:2898–903.
87. Ringel MD. 2009. Molecular markers of aggressiveness of thyroid cancer. *Curr Opin Endocrinol Diabetes Obes* 16:361–66.
88. Paes JE, Ringel MD. 2008. Dysregulation of the phosphatidyl inositol-3 kinase pathway in thyroid neoplasia. *Endocrinol Metab Clin N Am* 37:375–87.

3 Genetic and Epigenetic Alterations in the MAP Kinase and PI3K/Akt Pathways in Thyroid Cancer

Mingzhao Xing and Paul W. Ladenson

CONTENTS

ABBREVIATIONS:

ATC	Anaplastic thyroid cancer
EGFR	Epidermal growth factor receptor
ERK	Extracellular signal-regulated kinase
FTC	Follicular thyroid cancer
FVPTC	Follicular variant PTC
MAP	Mitogen-activated protein
MAPK	MAP kinase
MEK	MAP-ERK kinase
MTC	Medullary thyroid cancer
NIS	Sodium-iodide symporter
PDGFR	Platelet-derived growth factor receptor
PDK	Phosphoinositide-dependent kinase
PDTC	Poorly differentiated thyroid cancer
PI3K	Phosphatidylinositol-3 kinase
PIP3	Phosphatidylinositol-3,4,5-trisphosphate
PTC	Papillary thyroid cancer
RAI	Radioiodine
TA	Thyroid adenoma
TG	Thyroglobulin

TPO	Thyroperoxidase
TSH	Thyroid-stimulating hormone
TSHR	TSH receptor
VEGF	Vascular epithelial growth factor

ABSTRACT: Remarkable progress has occurred in recent years in understanding the molecular derangements in major signaling pathways in thyroid cancer, particularly genetic and epigenetic alterations in the MAP kinase and PI3K/Akt pathways. The previously established *Ras* and *RET/PTC* mutations and the recently discovered *BRAF* mutation (V600E) play a fundamental role in thyroid tumorigenesis through MAP kinase pathway activation. The prevalent and oncogenically potent *BRAF* mutation plays a unique role, particularly in promoting the aggressiveness of papillary thyroid cancer. Genetic alterations, such as mutations in the genes for Ras, PIK3CA, PTEN, and receptor tyrosine kinases, and genomic amplifications of some of these genes are also common, which play their roles in thyroid tumorigenesis through aberrantly activating the PI3K/Akt pathway. Consequent or associated epigenetic alterations, such as aberrant methylation of tumor suppressor and thyroid-specific genes, are also common in the molecular pathogenesis of thyroid cancer, including aberrant silencing of thyroid genes and resultant loss of radioiodine avidity of thyroid cancer. These genetic and epigenetic alterations in the two major signaling pathways provide a strong basis for the current development of novel molecular-based diagnostic, prognostic, and therapeutic strategies for thyroid cancer.

3.1 INTRODUCTION

There has been a rapid rise globally in the incidence of thyroid cancer in recent decades.[1–4] In the United States, it is estimated that there will be 37,200 new cases and 1,630 deaths in 2009, with a prevalence of more than 400,000 cases.[3,4] The vast majority of thyroid cancers are derived from follicular thyroid cells and include, histologically, mainly the two well-differentiated thyroid cancers—papillary thyroid cancer (PTC) and follicular thyroid cancer (FTC)—and undifferentiated anaplastic thyroid cancer (ATC).[1–5] Thyroid adenoma (TA), a benign and common endocrine tumor, is also derived from follicular thyroid cells. Medullary thyroid cancer (MTC), a relatively rare malignancy, is derived from parafollicular cells and will not be discussed here. ATC is a rapidly progressive cancer that is among the most deadly human cancers,[6,7] whereas PTC and FTC are generally indolent and highly curable with current treatments. There are also poorly differentiated thyroid cancers (PDTCs) that have a high incidence of incurability and mortality, albeit with a better prognosis than ATC. It is generally believed that PDTCs can progress into ATC, and both can derive from PTC and FTC or occur *de novo.*

This large and still growing number of thyroid cancer patients poses a significant challenge. There is currently no cure for ATC and, in most cases, for PDTC. Even differentiated thyroid cancers (PTC and FTC), which are generally curable with current treatments, are often difficult to manage because of their wide spectrum of clinical behavior, which remains incompletely predictable. Similar to ATC and PDTC, some cases of differentiated thyroid cancer can lose radioiodine (RAI) avidity and be resistant to RAI treatment. RAI-resistant cases of thyroid cancer are incurable once they become surgically inoperable. Understanding the molecular derangements and mechanisms in thyroid cancer may provide novel and effective management strategies. There have been extensive recent studies and considerable advances in exploring the genetic and epigenetic derangements in several major signaling pathways in thyroid cancer, particularly the MAP kinase and PI3K/Akt pathways.

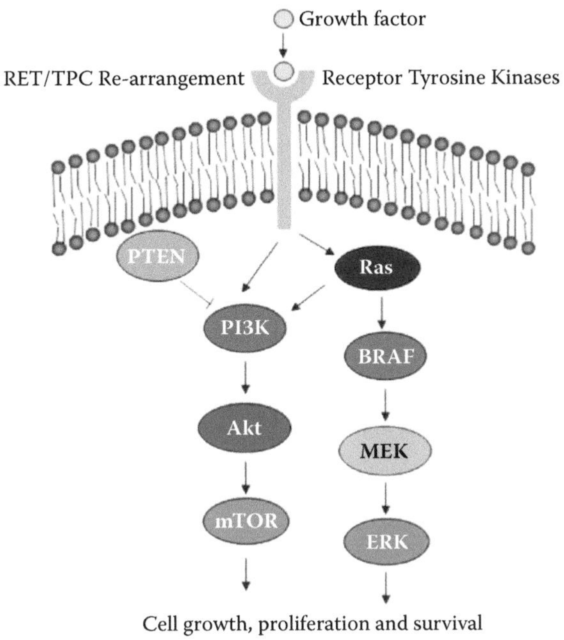

FIGURE 3.1 Schematic illustration of the MAPK and PI3K/Akt pathways. (Adapted from Xing, M., Genetic alterations in the phosphatidylinositol-3 kinase/Akt pathway in thyroid cancer. *Thyroid*, 20(7) 697–706, 2010. With permission of Mary Ann Liebert, Inc., Publishers.)

3.2 *BRAF* MUTATION IN THE MAP KINASE SIGNALING PATHWAY

As illustrated in Figure 3.1, the Ras → Raf → MEK → MAP kinase/ERK pathway (MAPK pathway) is a conserved intracellular signaling pathway that plays a fundamental role in cell proliferation, differentiation, survival, and when aberrantly activated, tumorigenesis.[8,9] The classical activators of this pathway in thyroid cancer include *RET/PTC* rearrangements[10,11] and activating *Ras* mutations.[12] In recent years, the exciting discovery of the *BRAF* mutation as a major genetic event activating the MAP kinase pathway in thyroid cancer has revolutionized our understanding of the role of this signaling cascade.[13–15] There are three Raf protein kinases, including A-Raf, B-Raf (BRAF), and C-Raf, and among these, BRAF is the most potent activator of the MAP kinase pathway.[8] *BRAF* mutation plays a major role in aberrantly activating the MAP kinase pathway in several other human cancers, particularly melanoma and colon cancer.[16] The T1799A point *BRAF* mutation accounts for more than 90% of all *BRAF* mutations in human cancers,[17] and there are now approximately 50 *BRAF* mutations that have been identified. The T1799A *BRAF* mutation causes a V600E amino acid change in the BRAF protein, resulting in constitutive and oncogenic activation of the BRAF kinase.[16,17]

The T1779A *BRAF* mutation is the most commonly known genetic alteration in thyroid cancer. It is a somatic rather than germline mutation for this cancer.[18,19] Among various types of thyroid cancer, the T1799A *BRAF* mutation uniquely occurs in 45% of PTC and 25% of ATC and does not occur in FTC or other types of thyroid tumors.[13–15] Several other types of *BRAF* mutations are also rarely seen in thyroid cancer. Examples include the BRAF V599ins,[20] BRAF V600E+K601del,[21,22] BRAF K601E,[23] AKAP9-BRAF,[24] and V600D+FGLAT601–605ins, which results from an insertion of 18 nucleotides at nucleotide T1799.[22] However, the T1799A *BRAF* mutation (hereafter referred to as the *BRAF* mutation) is far and away the predominant *BRAF* mutation in PTC and apparently PTC-derived ATC.

The oncogenic role of the *BRAF* mutation in thyroid cancer has been well established in many studies, including transgenic mouse,[25] cell line, and xenograft tumor models.[26–29] These investigations have demonstrated the tumorigenic ability of the *BRAF* mutation as well as its essential role in maintaining cancer cell growth and proliferation.

The important pathogenic role of the *BRAF* mutation is also supported by its strong association with aggressive clinicopathological outcomes, including tumor invasion, metastasis, and recurrence of PTC.[13] This initial report has now been confirmed by many more recent studies that also demonstrate a strong association of the *BRAF* mutation with poorer clinicopathological outcomes.[14,30–36] The *BRAF* mutation is more commonly seen in recurrent PTC, with a prevalence of 80%–85%,[30,34] versus primary PTC, with a prevalence of 45%,[13] consistent with its role in the aggressiveness, progression, and pathogenesis of PTC. The aggressiveness conferred by the *BRAF* mutation in thyroid tumorigenesis occurs even in low-stage and papillary thyroid microcarcinomas, as this mutation was found to be highly associated with high-risk clinicopathological characteristics of these cancers, such as extrathyroidal extension and lymph node metastasis, in several studies, as recently summarized.[37] It has therefore been proposed that the *BRAF* mutation is a powerful prognostic molecular marker for poorer prognosis of thyroid cancer and an effective therapeutic target in this cancer.[14,15,37,38] The high prognostic value of the *BRAF* mutation has recently been demonstrated in preoperative FNAB specimens to predict lymph node metastasis, extrathyroidal extension, and tumor recurrence of PT.[39]

The *BRAF* mutation is also associated with loss of RAI avidity in PTC and its failure to be cured.[40–43] This corresponds well to the close association of the *BRAF* mutation with dedifferentiation of PTC as reflected by decreased expression of thyroid-specific genes in PTC, including the genes for thyroid-stimulating hormone (TSH) receptor (TSHR), thyroperoxidase (TPO),[41,42,44–46] pendrin,[42] the sodium-iodide symporter (NIS),[41,44,47] and thyroglobulin (TG).[41] The role of the *BRAF* mutation in aberrantly silencing thyroid iodide-handling genes in thyroid tumorigenesis is functionally confirmed by the *in vitro* demonstration that induced expression of the *BRAF* mutation promoted silencing of thyroid genes in thyroid cells, and that terminating the expression of this mutation restored the expression of thyroid genes.[26]

3.3 GENETIC ALTERATIONS IN THE PI3K/AKT SIGNALING PATHWAY

As illustrated in Figure 3.1, the phosphatidylinositol-3 kinase (PI3K)/Akt pathway is another conserved intracellular signaling pathway that plays a fundamental role in the regulation of cell growth, proliferation, and survival in human tumorigenesis.[48,49] Of the several classes of PI3Ks, class I is composed of heterodimers of a regulatory subunit, p85, and one of the several p110 catalytic subunits. The most important p110 subunits are the a-type (PIK3CA) and b-type (PIK3CB) that are widely expressed in tissues. PIK3CA and PIK3CB are members of class IA, which is activated by receptor tyrosine kinases. The p110 catalytic subunits contain a Ras-binding site through which Ras is involved in the PI3K/Akt signaling. PI3K can be activated by various membrane growth factor receptors, such as platelet-derived growth factor receptor (PDGFR), epidermal growth factor receptor (EGFR), vascular epithelial growth factor (VEGF) receptor, c-KIT, and c-MET. Upon activation, PI3K phosphorylates phosphatidylinositol-4,5-bisphosphate, producing phosphatidylinositol-3,4,5-trisphosphate (PIP_3). The latter localizes the Ser/Thr kinase Akt to the cell membrane, where it becomes phosphorylated and activated by the phosphoinositide-dependent kinases (PDKs), particularly PDK-1. Activated Akt phosphorylates downstream protein effectors and amplifies the signaling cascade. The three types of Akts—Akt-1, Akt-2, and Akt-3—may play different roles in different tissues. In this PI3K/Akt pathway signaling, the *PTEN* gene product, PTEN, which is a phosphatase that dephosphorylates PIP_3, plays a critical role in terminating this pathway's signaling.[50] Therefore, PTEN is an important tumor suppressor.

Many studies show common activation of the PI3K/Akt signaling in thyroid cancers.[51] Among the three Akt isoforms, Akt-1 and Akt-2 are the most abundant and important in thyroid cancer.[52] Genomic copy gain and amplification of the *PIK3CA* are common in thyroid tumors, particularly FTC and ATC.[53–55] *PIK3CA* mutations also occur in thyroid cancer, particularly in ATC.[53–56]

Ras mutation is commonly found in thyroid tumors, particularly FTC, follicular variant PTC (FVPTC), and some thyroid adenomas.[12] In two previous studies analyzing a number of genetic alterations in the PI3K/Akt pathway, including *PIKCA* mutation and amplification, *Ras* mutation, and *PTEN* mutation in various thyroid tumors, a high prevalence was found, particularly in FTC and ATC.[54,55]

Furthermore, coexistence of these genetic alterations with each other and with *BRAF* mutations was more frequently seen in aggressive thyroid cancers, particularly ATC.[54,57] The study by Liu et al.[57] analyzed a large panel of genetic alterations, including mutations and genetic amplifications of *EGFR*, *PDGFR*, *c-MET*, *KIT*, *PDK-1*, and *Akt-2* genes. In this study, at least one genetic alteration was found in 96% (46/48) of ATC, and coexistence of two or more genetic alterations in 77% (37/48) of ATC. These results established a strong genetic basis for the fundamental role of the PI3K/Akt pathway in thyroid tumorigenesis. Interestingly, in this study, genetic alterations that could potentially activate both the MAP kinase and PI3K pathways, with correspondingly elevated phosphorylation of ERK and Akt, were found in the vast majority of ATC. This result strongly supported the concept that both the MAP kinase and PI3K/Akt pathways play an extensive role in the pathogenesis of ATC.[14] This is consistent with the report that simultaneously targeting these two pathways had synergistic antithyroid cancer cell effects.[29] A recent study demonstrated that the genetic alterations in the PI3K/Akt pathway conferred sensitivity of thyroid cancer cells to Akt and mTOR inhibitors.[58] These results not only confirmed the importance of the genetic alterations in the PI3K/Akt pathway in thyroid cancer cell survival, but also provided an important preclinical basis for clinical trials to test further genetic-targeted therapies for thyroid cancer using inhibitors of the PI3K/Akt pathway.

3.4 UNIQUE DUAL INVOLVEMENT OF *RET/PTC* AND *RAS* MUTATIONS IN THE MAP KINASE AND PI3K/AKT PATHWAYS

Compared with other genetic alterations, the *RET/PTC* and *Ras* mutations are unique in that they can each dually activate the MAP kinase and PI3K/Akt pathways, albeit with a preference for one or the other. *RET/PTC* represents a somatic genomic DNA rearrangement as a result of the fusion of the 3' portion of the *RET* gene to the 5' portion of an unrelated gene.[59] Due to various types of partner genes involved in the rearrangement, there are different types of *RET/PTC*, with at least 11 types having been identified.[11] Among them, *RET/PTC1* and *RET/PTC3*, which represent combinations of *RET* with the *H4* gene and the *NCOA4* gene, respectively, are the most common, accounting for the vast majority of *RET/PTC* in PTC. It has been shown that chromosomal folding with coils of approximately 8 Mb in length positions the *RET* gene close to both the H4 and NCOA4 loci, suggesting a structural basis for preferential recombinogenic events at these loci.[60] The rearrangement of *RET/PTC* results in a recombinant protein that contains constitutively activated, hence oncogenic, RET tyrosine kinase. *RET/PTC* is classically a PTC-associated genetic alteration, but recently it was shown that 20% of primary peritoneal cancer harbored *RET/PTC1*.[61]

RET/PTC occurs in about 20% of adult PTC patients and 60–70% of pediatric PTC patients. Thus, contrary to the *BRAF* mutation, *RET/PTC* is mainly a pediatric-associated genetic alteration. Occurrence of RET/PTC is associated with radiation exposure. Therefore, *RET/PTC* is commonly seen in victims of nuclear events, such as the victims of the Chernobyl nuclear accident. *RET/PTC* is also commonly seen in sporadic pediatric PTC. The overall prevalence of *RET/PTC* seems to be similar in sporadic and radiation-associated pediatric PTC,[13] suggesting that young age is an important risk factor for the development of *RET/PTC*.

The ability of both *RET/PTC1* and *RET/PTC2* to cause PTC was well demonstrated in a transgenic mouse model in which their targeted overexpression in the thyroid gland induced development of PTC.[62–64] Signaling through the MAP kinase pathway is an established mechanism for the role of RET/PTC in PTC tumorigenesis.[65,66] In differentiated PTC, *RET/PTC* and

BRAF mutations are mutually exclusive,[13] consistent with a role of *RET/PTC* being independently capable of promoting the development of PTC through activation of the MAP kinase pathway. The molecular details in how RET/PTC specifically fits into the MAP kinase pathway are largely unclear, but as a membrane-associated tyrosine kinase, RET/PTC is generally known to join this pathway at an early point, upstream of Ras/Raf/MEK and through Y1062 signaling.[65] Like many receptor tyrosine kinases, RET/PTC is also coupled to the PI3K/Akt pathway, as shown by the activation of PI3K/Akt signaling upon expression of RET/PTC in cells.[67,68] This coupling has recently been shown to involve the adaptor protein XB130.[69] Thus, unlike BRAF, a classical and exclusive activator of the MAP kinase pathway, RET/PTC is a dual activator of the MAP kinase and PI3K/Akt pathways. Interestingly, RET/PTC preferentially activated the PI3K/Akt pathway over the MAP kinase pathway when overexpressed in the FRTL-5 rat thyroid cells,[67] suggesting that this oncoprotein may play its role in thyroid tumorigenesis predominantly through the PI3K/Akt pathway. This concept is consistent with the results from microarray studies in human thyroid cancer showing that *RET/PTC* rearrangement was associated with expression of genes downstream of the PI3K/Akt pathway.[45] Although earlier studies showed that *RET/PTC* and *BRAF* mutations did not coexist in differentiated PTC, recent studies reported their coexistence in advanced or recurrent PTC and ATC.[34,57,70] This genetic pattern may provide a basis for coactivation of the MAP kinase and PI3K/Akt signaling pathways, and thus contributes to the progression of thyroid cancer.

Ras mutation is not uncommon in thyroid cancer, particularly in FTC and FVPTC, although it is virtually absent in tall-cell PTC and conventional PTC.[12,71] It is also often seen in follicular adenoma. Therefore, *Ras* mutation is mainly a genetic alteration in follicular thyroid tumors. There are three types of Ras: H-Ras, K-Ras, and N-Ras. In thyroid tumor, *N-Ras* mutation is most common and other types of *Ras* mutations are relatively uncommon.[12] Signaling through the MAP kinase pathway is also a well-established and classical molecular mechanism for the role of *Ras* mutations in thyroid tumorigenesis. This is consistent with the finding that *Ras* and *BRAF* mutations were mutually exclusive in differentiated PTC,[13] suggesting that, like *BRAF* mutation, *Ras* mutation is also able to independently cause PTC through the MAP kinase pathway.

Ras mutants also activate the PI3K/Akt pathway in human cancers. Therefore, like RET/PTC, Ras mutant dually activates the MAP kinase and PI3K/Akt pathways. Coexistence among *Ras* mutation, *RET/PTC*, and *BRAF* mutation, although uncommon in differentiated thyroid cancer, can occur in ATC,[57] representing an important mechanism for simultaneous activation of multiple signaling pathways in the aggressiveness and progression of this cancer. Different isoforms of Ras mutants may play a different role in activating the PI3K/Akt and BRAF pathways in human cancer. For example, the N-Ras mutant is a preferential activator of the PI3K/Akt pathway, whereas K-Ras is a preferential activator of the MAP kinase pathway.[72] This also seems to be the case in follicular thyroid tumors, as *N-Ras* mutation is the most common among various types of *Ras* mutations, and aberrant PI3K/Akt signaling is more important than aberrant MAP kinase pathway signaling in these tumors.

3.5 ABERRANT METHYLATION OF GENES AROUND THE MAP KINASE AND PI3K/AKT PATHWAYS

Gene methylation is an epigenetic modification of DNA in which a methyl group is covalently added to the fifth carbon of the cytosine residue in a CpG dinucleotide. This usually occurs in CpG islands, which are characterized with high contents of CpG dinucleotides and usually located in the promoter and the first exon areas of the 5′ flanking region of a gene. In general, gene methylation is associated with gene silencing,[73,74] which involves the recruitment of DNA methyl-binding transcription repressors that block the binding of the gene with its transcriptional machinery. An alternative mechanism is the direct blockade of the binding of the gene with transcription factors by

DNA methylation. This is exemplified by the rat *TSHR* gene, in which the transcription factor GA was able to normally bind the nonmethylated promoter, but could not bind the methylated promoter of the gene.[75] Aberrant gene methylation commonly occurs in human cancers, including thyroid cancer, resulting in inappropriate silencing of functionally important genes, such as tumor suppressor genes.[73,74,76]

A large number of functionally important tumor suppressor and thyroid-specific genes have been found to be frequently hypermethylated in thyroid cancer.[76] These include the tissue inhibitor of metalloproteinase-3 (TIMP3), SLC5A8, death-associated protein kinase (DAPK), and retinoic acid receptor-beta-2 (RARβ2). Methylation of these genes was associated with aggressive clinicopathological outcomes and *BRAF* mutation in PTC.[77] The *BRAF* mutation is now well known to be associated with aggressiveness and poorer prognosis of PTC.[14] Therefore, methylation, and hence silencing of these tumor suppressor genes, is an important mechanism in the aggressive pathogenesis of PTC promoted by the *BRAF* mutation-driven MAP kinase pathway signaling. In further support of this concept, a recent study demonstrated an association of methylation of DNA repair genes, particularly the *hMLH1* gene, with *BRAF* mutation and aggressiveness of PTC.[78]

Methylation of the *RASSF1A* gene was inversely associated with the *BRAF* mutation in PTC,[79] suggesting that epigenetic disruption of this tumor suppressor gene may play a role in thyroid tumorigenesis through signaling pathways that may be equally important to the MAP kinase pathway. Methylation of the *RASSF1A* gene is seen in FTC with the highest prevalence and extent, but can also occur in follicular adenomas, suggesting an early role of *RASSF1A* methylation in the tumorigenesis of follicular thyroid tumors. The fact that *RASSF1A* methylation is most common and extensive in FTC, in which the PI3K/Akt pathway plays a fundamental role, suggests a potential role of this epigenetic inactivation of the *RASSF1A* gene through the PI3K/Akt pathway.

Aberrant methylation of the *PTEN* gene was also recently identified.[80] In this interesting study, *PTEN* methylation was found to be associated with activating genetic alterations in the PI3K/Akt pathway in thyroid tumors, suggesting a self-enhancing mechanism for the PI3K/Akt signaling through the epigenetic silencing of the *PTEN* gene as a consequence of activation of this pathway by genetic alterations.

Aberrant silencing of several thyroid iodide-handling genes, including *NIS*,[81] *TSHR*,[82–84] *SLC26A4*,[85] and *Tg*,[86] provides an explanation for the commonly seen silencing of these genes in thyroid cancer. Aberrant methylation of some of these genes, such as *TSHR*, was interestingly associated with *BRAF* mutation-driven MAP kinase pathway signaling. The role of the BRAF/MAP kinase pathway in the regulation of the *TSHR* gene was directly tested using a Lucifer reporting system.[26] In this study, the *TSHR* gene could be partially silenced through *BRAF* mutation-promoted signaling of the MAP kinase pathway involving methylation of the *TSHR* promoter. Therefore, functional disruption of an important gene through epigenetic methylation can be directly linked to the aberrant activation of a major signaling pathway in thyroid cancer.

3.6 IMPAIRED IODIDE METABOLISM COUPLED TO ABERRANT MAP KINASE AND PI3K/AKT SIGNALING

Utilizing iodide as a substrate for thyroid hormone synthesis is a unique and fundamental function of follicular thyroid cells. This process involves several thyroid iodide-handling protein molecules.[87] A key and first step is the transportation of iodide from the bloodstream into the thyroid cell through NIS in the basal membrane, followed by its transportation into the follicular lumen through pendrin (also called SLC26A4), a putative iodide transporter in the apical membrane of the cell. Iodide is oxidized, organified, and incorporated into TG through iodination of tyrosine residues in TG to form the iodothyronines, a process catalyzed by TPO. Several thyroid-specific transcription factors, including TTF-1, TTF-2, and Pax-8, play important roles in the regulation of these thyroid genes.

This thyroid iodide-metabolizing process is well coordinated and upregulated by TSH, which acts by binding to TSHR on the cell membrane.

Loss of expression of the genes for TSHR, NIS, TPO, TG, and pendrin is often seen in thyroid cancers.[52,81,88–91] Consequently, the ability of thyroid cancer cells to take up and concentrate RAI may be impaired or lost. Such cases of thyroid cancer are then unresponsive to RAI therapy, with disease persistence, progression, or recurrence. Currently, this represents a major therapeutic challenge in thyroid cancer management, particularly in PDTC, ATC, and certain cases of differentiated thyroid cancer as well.[6,7]

Although the specific molecular mechanisms underlying this dedifferentiation of thyroid cells is largely unclear, available data suggest that aberrant signaling of the MAP kinase and PI3K/Akt pathways plays an important role, as discussed above. The *BRAF* mutation was closely associated with loss of RAI avidity in PTC.[40] *In vitro* studies demonstrated that thyroid-specific genes could be silenced by induced expression of the BRAF mutant and the consequent overactivation of the MAP kinase pathway in thyroid cell lines.[26,47] Expression of thyroid genes could be restored by removing the BRAF mutant.[26] As discussed above, there are also data showing that gene methylation played a role in the silencing of the *TSHR* gene promoted by the *BRAF* mutant.[26] Interestingly, in melanoma cells, dual suppression of the MAP kinase and PI3K/Akt pathways could induce the expression of thyroid iodide-metabolizing genes, such as *NIS*, *TSHR*, *TPO*, and *TG*.[92] More recently, it was demonstrated that, similar to the MAP kinase pathway, the PI3K/Akt pathway also played an important role in the regulation of thyroid genes in thyroid cancer cells (Hou and Xing, unpublished data). Several previous studies showed that inhibition of histone deacetylase with concurrent increase in histone acetylation could induce thyroid gene expression,[93–95] although many of the cells used in these studies are now known to be nonthyroid in origin.[96] In fact, a recent study demonstrated that dual suppression of the MAP kinase and PI3K/Akt pathways could synergize the expression of thyroid genes induced by histone deacetylase inhibitors in thyroid cancer cells (Hou and Xing, unpublished data). Therefore, like aberrant gene methylation, aberrant histone modifications in chromatin remodeling are another important epigenetic mechanism coupled to the MAP kinase and PI3K/Akt pathways that may play an important role in aberrant thyroid gene silencing in thyroid cancer. Aiming to reverse these molecular changes coupled to the MAP kinase and PI3K/Akt pathways and restore thyroid gene expression and RAI avidity may prove to be an effective therapeutic strategy for thyroid cancer.

3.7 SUMMARY

There has been considerable progress in understanding the molecular basis of thyroid cancer in recent years. Particularly, characterization and understanding of the genetic and epigenetic alterations in two major thyrocyte signaling pathways, the MAP kinase and PI3K/Akt pathways, have advanced our comprehension of thyroid tumorigenesis. As in many other human cancers, aberrant activation of the two pathways is central to thyroid tumorigenesis. The discovery of several major mutations in recent years, such as *BRAF* mutation in the MAP kinase pathway, and *PIK3CA* mutation and amplification, as well as other common genetic alterations in the PI3K/Akt pathway, have established a strong genetic basis for these two pathways in thyroid tumorigenesis. Epigenetic alterations coupled to the MAP kinase and PI3K/Akt pathways have added a new dimension to the molecular pathogenesis of thyroid cancer. Furthermore, the coupling of thyroid gene silencing to epigenetic alterations as a mechanism responsible for RAI treatment failure in thyroid cancer patients is an especially important development. These exciting advances in understanding the molecular biology of thyroid cancer provide great opportunities for molecular-based strategies to combat this cancer, including the development of novel diagnostic, prognostic, and therapeutic strategies for this increasingly common endocrine malignancy.

REFERENCES

1. Leenhardt L, Grosclaude P, Cherie-Challine L. 2004. Increased incidence of thyroid carcinoma in France: A true epidemic or thyroid nodule management effects? Report from the French Thyroid Cancer Committee. *Thyroid* 14:1056–60.
2. Davies L, Welch HG. 2006. Increasing incidence of thyroid cancer in the United States, 1973–2002. *JAMA* 295:2164–67.
3. Horner MJ, Ries LAG, Krapcho M, et al., eds. *SEER cancer statistics review, 1975–2006*. Bethesda, MD: National Cancer Institute. Based on November 2008 SEER data submission, posted to the SEER website: http://seer.cancer.gov/csr/1975_2006/ (accessed September 17, 2009).
4. Jemal A, Siegel R, Ward E, Hao Y, Xu J, Thun MJ. 2009. Cancer statistics, 2009. *CA Cancer J Clin* 59:225–49.
5. Hundahl SA, Fleming ID, Fremgen AM, Menck HR. 1998. A National Cancer Data Base report on 53,856 cases of thyroid carcinoma treated in the U.S., 1985–1995. *Cancer* 83:2638–48.
6. Ain KB. 2000. Management of undifferentiated thyroid cancer. *Baillieres Best Pract Res Clin Endocrinol Metab* 14:615–29.
7. Cornett WR, Sharma AK, Day TA, et al. 2007. Anaplastic thyroid carcinoma: An overview. *Curr Oncol Rep* 9:152–58.
8. Mercer KE, Pritchard CA. 2003. Raf proteins and cancer: B-Raf is identified as a mutational target. *Biochim Biophys Acta* 1653:25–40.
9. Kohno M, Pouyssegur J. 2006. Targeting the ERK signaling pathway in cancer therapy. *Ann Med* 38:200–11.
10. Santoro M, Melillo RM, Fusco A. 2006. RET/PTC activation in papillary thyroid carcinoma: *European Journal of Endocrinology* Prize Lecture. *Eur J Endocrinol* 155:645–53.
11. Ciampi R, Nikiforov YE. 2007. RET/PTC rearrangements and BRAF mutations in thyroid tumorigenesis. *Endocrinology* 148:936–41.
12. Vasko V, Ferrand M, Di Cristofaro J, Carayon P, Henry JF, de Micco C. 2003. Specific pattern of RAS oncogene mutations in follicular thyroid tumors. *J Clin Endocrinol Metab* 88:2745–52.
13. Xing M. 2005. BRAF mutation in thyroid cancer. *Endocr Relat Cancer* 12:245–62.
14. Xing M. 2007. BRAF mutation in papillary thyroid cancer: Pathogenic role, molecular bases, and clinical implications. *Endocr Rev* 28:742–62.
15. Xing M. 2008. Recent advances in molecular biology of thyroid cancer and their clinical implications. *Otolaryngol Clin North Am* 41:1135–46.
16. Davies H, Bignell GR, Cox C, et al. 2002. Mutations of the BRAF gene in human cancer. *Nature* 417:949–54.
17. Garnett MJ, Marais R. 2004. Guilty as charged: B-RAF is a human oncogene. *Cancer Cell* 6:313–19.
18. Xing M. 2005. The T1799A BRAF mutation is not a germline mutation in familial nonmedullary thyroid cancer. *Clin Endocrinol* 63:263–66.
19. Hou P, Xing M. 2006. Absence of germline mutations in genes within the MAP kinase pathway in familial non-medullary thyroid cancer. *Cell Cycle* 5:2036–39.
20. Moretti S, Macchiarulo A, De Falco V, et al. 2006. Biochemical and molecular characterization of the novel BRAF(V599Ins) mutation detected in a classic papillary thyroid carcinoma. *Oncogene* 25:4235–40.
21. Oler G, Ebina KN, Michaluart P Jr, Kimura ET, Cerutti J. 2005. Investigation of BRAF mutation in a series of papillary thyroid carcinoma and matched-lymph node metastasis reveals a new mutation in metastasis. *Clin Endocrinol* 62:509–11.
22. Hou P, Liu D, Xing M. 2007. Functional characterization of the T1799–1801del and A1799–1816ins BRAF mutations in papillary thyroid cancer. *Cell Cycle* 6:377–79.
23. Trovisco V, Vieira de Castro I, Soares P, et al. 2004. BRAF mutations are associated with some histological types of papillary thyroid carcinoma. *J Pathol* 202:247–51.
24. Ciampi R, Knauf JA, Kerler R, et al. 2005. Oncogenic AKAP9-BRAF fusion is a novel mechanism of MAPK pathway activation in thyroid cancer. *J Clin Invest* 115:94–101.
25. Knauf JA, Ma X, Smith EP, et al. 2005. Targeted expression of BRAFV600E in thyroid cells of transgenic mice results in papillary thyroid cancers that undergo dedifferentiation. *Cancer Res* 65:4238–45.
26. Liu D, Hu S, Hou P, Jiang D, Condouris S, Xing M. 2007. Suppression of BRAF/MEK/MAP kinase pathway restores expression of iodide-metabolizing genes in thyroid cells expressing the V600E BRAF mutant. *Clin Cancer Res* 13:1341–49.
27. Liu D, Liu Z, Condouris S, Xing M. 2007. BRAF V600E maintains proliferation, transformation and tumorigenicity of BRAF mutant papillary thyroid cancer cells. *J Clin Endocrinol Metab* 92:2264–71.

28. Liu D, Liu Z, Jiang D, Dackiw AP, Xing M. 2007. Inhibitory effects of the mitogen-activated protein kinase inhibitor CI-1040 on the proliferation and tumor growth of thyroid cancer cells with BRAF or RAS mutations. *J Clin Endocrinol Metab* 92:4686–95.
29. Liu D, Xing M. 2008. Potent inhibition of thyroid cancer cells by the MEK inhibitor PD0325901 and its potentiation by suppression of the PI3K and NF-kappaB pathways. *Thyroid* 18:853–64.
30. Nakayama H, Yoshida A, Nakamura Y, et al. 2007. Clinical significance of BRAF (V600E) mutation and Ki-67 labeling index in papillary thyroid carcinomas. *Anticancer Res* 27:3645–49.
31. Abubaker J, Jehan Z, Bavi P, et al. 2008. Clinicopathological analysis of papillary thyroid cancer with PIK3CA alterations in a Middle Eastern population. *J Clin Endocrinol Metab* 93:611–18.
32. Frasca F, Nucera C, Pellegriti G, et al. 2008. BRAF(V600E) mutation and the biology of papillary thyroid cancer. *Endocr Relat Cancer* 15:191–205.
33. Elisei R, Ugolini C, Viola D, et al. 2008. BRAFV600E mutation and outcome of patients with papillary thyroid carcinoma: A 15-year median follow-up study. *J Clin Endocrinol Metab* 93:3943–49.
34. Henderson YC, Shellenberger TD, Williams MD, et al. 2009. High rate of BRAF and RET/PTC dual mutations associated with recurrent papillary thyroid carcinoma. *Clin Cancer Res* 15:485–91.
35. Kim SK, Song KH, Lim SD, et al. 2009. Clinical and pathological features and the BRAF(V600E) mutation in patients with papillary thyroid carcinoma with and without concurrent hashimoto thyroiditis. *Thyroid* 19:137–41.
36. Oler G, Cerutti JM. 2009. High prevalence of BRAF mutation in a Brazilian cohort of patients with sporadic papillary thyroid carcinomas: Correlation with more aggressive phenotype and decreased expression of iodide-metabolizing genes. *Cancer* 115:972–80.
37. Xing M. 2009. BRAF mutation in papillary thyroid microcarcinoma: The promise of better risk management. *Ann Surg Oncol* 16:801–3.
38. Xing M. 2009. Genetic-targeted therapy of thyroid cancer: A real promise. *Thyroid* 19:805–9.
39. Xing M, Clark D, Guan H, et al. 2009. BRAF mutation testing of thyroid fine-needle aspiration biopsy specimens for preoperative risk stratification in papillary thyroid cancer. *J Clin Oncol* 27:2977–82.
40. Xing M, Westra WH, Tufano RP, et al. 2005. BRAF mutation predicts a poorer clinical prognosis for papillary thyroid cancer. *J Clin Endocrinol Metab* 90:6373–79.
41. Durante C, Puxeddu E, Ferretti E, et al. 2007. BRAF mutations in papillary thyroid carcinomas inhibit genes involved in iodine metabolism. *J Clin Endocrinol Metab* 92:2840–43.
42. Mian C, Barollo S, Pennelli G, et al. 2008. Molecular characteristics in papillary thyroid cancers (PTCs) with no (131) I uptake. *Clin Endocrinol* (Oxf) 681:108–16.
43. Ricarte-Filho JC, Ryder M, Chitale DA, et al. 2009. Mutational profile of advanced primary and metastatic radioactive iodine-refractory thyroid cancers reveals distinct pathogenetic roles for BRAF, PIK3CA, and AKT1. *Cancer Res* 69:4885–93.
44. Romei C, Ciampi R, Faviana P, et al. 2008. BRAFV600E mutation, but not RET/PTC rearrangements, is correlated with a lower expression of both thyroperoxidase and sodium iodide symporter genes in papillary thyroid cancer. *Endocr Relat Cancer* 15:511–20.
45. Giordano TJ, Kuick R, Thomas DG, et al. 2005. Molecular classification of papillary thyroid carcinoma: Distinct BRAF, RAS, and RET/PTC mutation-specific gene expression profiles discovered by DNA microarray analysis. *Oncogene* 24:6646–56.
46. Di Cristofaro J, Silvy M, Lanteaume A, Marcy M, Carayon P, De Micco C. 2006. Expression of TPO mRNA in thyroid tumors: Quantitative PCR analysis and correlation with alterations of RET, Braf, Ras and PAX8 genes. *Endocr Relat Cancer* 13:485–95.
47. Riesco-Eizaguirre G, Gutiérrez-Martínez P, García-Cabezas MA, Nistal M, Santisteban P. 2006. The oncogene BRAF V600E is associated with a high risk of recurrence and less differentiated papillary thyroid carcinoma due to the impairment of Na+/I– targeting to the membrane. *Endocr Relat Cancer* 13:257–69.
48. Vivanco I, Sawyers CL. 2002. The phosphatidylinositol 3-kinase AKT pathway in human cancer. *Nat Rev Cancer* 2:489–501.
49. Fresno Vara JA, Casado E, de Castro J, Cejas P, Belda-Iniesta C, González-Barón M. 2004. PI3K/Akt signalling pathway and cancer. *Cancer Treat Rev* 30:193–204.
50. Sansal I, Sellers WR. 2004. The biology and clinical relevance of the PTEN tumor suppressor pathway. *J Clin Oncol* 22:2954–63.
51. Paes JE, Ringel MD. 2008. Dysregulation of the phosphatidylinositol 3-kinase pathway in thyroid neoplasia. *Endocrinol Metab Clin North Am* 37:375–87.
52. Ringel MD, Hayre N, Saito J, et al. 2001. Overexpression and overactivation of Akt in thyroid carcinoma. *Cancer Res* 61:6105–11.

53. Wu G, Mambo E, Guo Z, et al. 2005. Uncommon mutation, but common amplifications, of the PIK3CA gene in thyroid tumors. *J Clin Endocrinol Metab* 90:4688–93.
54. Hou P, Liu D, Shan Y, et al. 2007. Genetic alterations and their relationship in the phosphatidylinositol 3-kinase/Akt pathway in thyroid cancer. *Clin Cancer Res* 13:1161–70.
55. Wang Y, Hou P, Yu H, et al. 2007. High prevalence and mutual exclusivity of genetic alterations in the PI3K/Akt pathway in thyroid tumors. *J Clin Endocrinol Metab* 92:2387–90.
56. García-Rostán G, Costa AM, Pereira-Castro I, et al. 2005. Mutation of the PIK3CA gene in anaplastic thyroid cancer. *Cancer Res* 65:10199–207.
57. Liu Z, Hou P, Ji M, et al. 2008. Highly prevalent genetic alterations in receptor tyrosine kinases and phosphatidylinositol 3-kinase/akt and mitogen-activated protein kinase pathways in anaplastic and follicular thyroid cancers. *J Clin Endocrinol Metab* 93:3106–16.
58. Liu D, Hou P, Liu Z, Wu G, Xing M. 2009. Genetic alterations in the phosphoinositide 3-kinase/Akt signaling pathway confer sensitivity of thyroid cancer cells to therapeutic targeting of Akt and mammalian target of rapamycin. *Cancer Res* 69:7311–19.
59. Fusco A, Grieco M, Santoro M, et al. 1987. A new oncogene in human thyroid papillary carcinomas and their lymph-nodal metastases. *Nature* 328:170–72.
60. Gandhi M, Medvedovic M, Stringer JR, Nikiforov YE. 2006. Interphase chromosome folding determines spatial proximity of genes participating in carcinogenic RET/PTC rearrangements. *Oncogene* 25:2360–66.
61. Flavin R, Jackl G, Finn S, et al. 2009. RET/PTC rearrangement occurring in primary peritoneal carcinoma. *Int J Surg Pathol* 17:187–97.
62. Jhiang SM, Sagartz JE, Tong Q, et al. 1996. Targeted expression of the ret/PTC1 oncogene induces papillary thyroid carcinomas. *Endocrinology* 137:375–78.
63. Santoro M, Chiappetta G, Cerrato A, et al. 1996. Development of thyroid papillary carcinomas secondary to tissue-specific expression of the RET/PTC1 oncogene in transgenic mice. *Oncogene* 12:1821–26.
64. Powell DJ Jr, Russell J, Nibu K, et al. 1998. The RET/PTC3 oncogene: Metastatic solid-type papillary carcinomas in murine thyroids. *Cancer Res* 58:5523–8.
65. Knauf JA, Kuroda H, Basu S, Fagin JA. 2003. RET/PTC-induced dedifferentiation of thyroid cells is mediated through Y1062 signaling through SHC-RAS-MAP kinase. *Oncogene* 22:4406–12.
66. Melillo RM, Castellone MD, Guarino V, et al. 2005. The RET/PTC-RAS-BRAF linear signaling cascade mediates the motile and mitogenic phenotype of thyroid cancer cells. *J Clin Invest* 115:1068–81.
67. Miyagi E, Braga-Basaria M, Hardy E, et al. 2004. Chronic expression of RET/PTC 3 enhances basal and insulin-stimulated PI3 kinase/AKT signaling and increases IRS-2 expression in FRTL-5 thyroid cells. *Mol Carcinog* 41:98–107.
68. Jung HS, Kim DW, Jo YS, et al. 2005. Regulation of protein kinase B tyrosine phosphorylation by thyroid-specific oncogenic RET/PTC kinases. *Mol Endocrinol* 19:2748–59.
69. Lodyga M, De Falco V, Bai XH, et al. 2009. XB130, a tissue-specific adaptor protein that couples the RET/PTC oncogenic kinase to PI3-kinase pathway. *Oncogene* 28:937–49.
70. Wang YL, Wang JC, Wu Y, et al. 2008. Incidentally simultaneous occurrence of RET/PTC, H4-PTEN and BRAF mutation in papillary thyroid carcinoma. *Cancer Lett* 263:44–52.
71. Zhu Z, Gandhi M, Nikiforova MN, Fischer AH, Nikiforov YE. 2003. Molecular profile and clinical-pathologic features of the follicular variant of papillary thyroid carcinoma. An unusually high prevalence of ras mutations. *Am J Clin Pathol* 120:71–7.
72. Haigis KM, Kendall KR, Wang Y, et al. 2008. Differential effects of oncogenic K-Ras and N-Ras on proliferation, differentiation and tumor progression in the colon. *Nat Genet* 40:600–8.
73. Bird A. 2002. DNA methylation patterns and epigenetic memory. *Genes Dev* 16:6–21.
74. Yoo CB, Jones PA. 2006. Epigenetic therapy of cancer: Past, present, and future. *Nat Rev Drug Discov* 5:37–50.
75. Yokomori N, Tawata M, Saito T, Shimura H, Onaya T. 1998. Regulation of the rat thyrotropin receptor gene by the methylation-sensitive transcription factor GA-binding protein. *Mol Endocrinol* 12:1241–49.
76. Xing M. 2007. Gene methylation in thyroid tumorigenesis. *Endocrinology* 148:948–53.
77. Hu S, Liu D, Tufano RP, et al. 2006. Association of aberrant methylation of tumor suppressor genes with tumor aggressiveness and BRAF mutation in papillary thyroid cancer. *Int J Cancer* 119:2322–29.
78. Guan H, Ji M, Hou P, et al. 2008. Hypermethylation of the DNA mismatch repair gene hMLH1 and its association with lymph node metastasis and T1799A BRAF mutation in papillary thyroid cancer. *Cancer* 113:247–55.
79. Xing M, Cohen Y, Mambo E, et al. 2004. Early occurrence of RASSF1A hypermethylation and its mutual exclusion with BRAF mutation in thyroid tumorigenesis. *Cancer Res* 64:1664–68.

80. Hou P, Ji M, Xing M. 2008. Association of the PTEN gene methylation with genetic alterations in the phosphatidylinositol 3-kinase/AKT signaling pathway in thyroid tumors. *Cancer* 113:2440–47.
81. Venkataraman GM, Yatin M, Marcinek R, Ain KB. 1999. Restoration of iodide uptake in dedifferentiated thyroid carcinoma: Relationship to human Na+/I– symporter gene methylation status. *J Clin Endocrinol Metab* 84:2449–57.
82. Xing M, Usadel H, Cohen Y, et al. 2003. Methylation of the thyroid-stimulating hormone receptor gene in epithelial thyroid tumors: A marker of malignancy and a cause of gene silencing. *Cancer Res* 63:2316–21.
83. Hoque MO, Rosenbaum E, Westra WH, et al. 2005. Quantitative assessment of promoter methylation profiles in thyroid neoplasms. *J Clin Endocrinol Metab* 90:11–18.
84. Schagdarsurengin U, Gimm O, Dralle H, Hoang-Vu C, Dammann R. 2006. CpG island methylation of tumor-related promoters occurs preferentially in undifferentiated carcinoma. *Thyroid* 16:633–42.
85. Xing M, Tokumaru Y, Wu G, Westra WB, Ladenson PW, Sidransky D. 2003. Hypermethylation of the Pendred syndrome gene SLC26A4 is an early event in thyroid tumorigenesis. *Cancer Res* 63:2312–15.
86. Berlingieri MT, Musti AM, Avvedimento VE, Di Lauro R, Di Fiore PP, Fusco A. 1989. The block of thyroglobulin synthesis, which occurs upon transformation of rat thyroid epithelial cells, is at the transcriptional level and it is associated with methylation of the 5′ flanking region of the gene. *Exp Cell Res* 183:277–83.
87. Nilsson M. 2001. Iodide handling by the thyroid epithelial cell. *Exp Clin Endocrinol Diabetes* 109:13–17.
88. Sheils OM, Sweeney EC. 1999. TSH receptor status of thyroid neoplasms—TaqMan RT-PCR analysis of archival material. *J Pathol* 188:87–92.
89. Lazar V, Bidart JM, Caillou B, et al. 1999. Expression of the Na+/I– symporter gene in human thyroid tumors: A comparison study with other thyroid-specific genes. *J Clin Endocrinol Metab* 84:3228–34.
90. Arturi F, Russo D, Bidart JM, Scarpelli D, Schlumberger M, Filetti S. 2001. Expression pattern of the pendrin and sodium/iodide symporter genes in human thyroid carcinoma cell lines and human thyroid tumors. *Eur J Endocrinol* 145:129–35.
91. Mirebeau-Prunier D, Guyétant S, Rodien P, et al. 2004. Decreased expression of thyrotropin receptor gene suggests a high-risk subgroup for oncocytic adenoma. *Eur J Endocrinol* 150:269–76.
92. Hou P, Liu D, Ji M, et al. 2009. Induction of thyroid gene expression and radioiodine uptake in melanoma cells: Novel therapeutic implications. *PLoS One* 4:e6200.
93. Kitazono M, Robey R, Zhan Z, et al. 2001. Low concentrations of the histone deacetylase inhibitor, depsipeptide (FR901228), increase expression of the Na(+)/I(–) symporter and iodine accumulation in poorly differentiated thyroid carcinoma cells. *J Clin Endocrinol Metab* 86:3430–35.
94. Furuya F, Shimura H, Suzuki H, et al. 2004. Histone deacetylase inhibitors restore radioiodide uptake and retention in poorly differentiated and anaplastic thyroid cancer cells by expression of the sodium/iodide symporter thyroperoxidase and thyroglobulin. *Endocrinology* 145:2865–75.
95. Puppin C, D'Aurizio F, D'Elia AV, et al. 2005. Effects of histone acetylation on sodium iodide symporter promoter and expression of thyroid-specific transcription factors. *Endocrinology* 146:3967–74.
96. Schweppe RE, Klopper JP, Korch C, et al. 2008. Deoxyribonucleic acid profiling analysis of 40 human thyroid cancer cell lines reveals cross-contamination resulting in cell line redundancy and misidentification. *J Clin Endocrinol Metab* 93:4331–41.

4 Genetic and Molecular Pathophysiology of Medullary Thyroid Carcinoma

Rosa Maria Paragliola, Salvatore M. Corsello, Francesco Torino, and Agnese Barnabei

CONTENTS

ABSTRACT: Medullary thyroid carcinoma (MTC) is a neuroendocrine tumor that accounts for 5 to 10% of all thyroid cancers. A unique feature of MTC is the production of multiple tumor markers, such as calcitonin (CTN), carcinoembryonic antigen, and chromogranin A. The typical diagnosis is represented by a solitary nodule, often associated with nodal metastases, and elevated basal or stimulated CTN levels. However, even this infrequent malignancy has a variable phenotype. The majority of MTCs are sporadic, but about 20% result from a germline mutation in the proto-oncogene *RET*. Dominant-activating or gain-of-function mutations in *RET* lead to the constitutive activation of receptor tyrosine kinases and downstream pathways involved in cell survival and proliferation. These mutations are related to an autosomal dominant cancer syndrome that can manifest either as a single disorder (familial MTC (fMTC)) or as multiple endocrine neoplasia (MEN) syndrome—type 2A or 2B. Despite the rarity of these syndromes, early diagnosis is very important since MTC is a potentially lethal disease if not promptly treated. In this context, genetic evaluation can swiftly diagnose fMTC and MEN and, furthermore, can identify at-risk subjects who require prophylactic surgery. Thanks to the introduction of *RET* genetic screening in the workup of all patients with MTC (both hereditary and apparently sporadic types), the number and type of recognized *RET* mutations have grown over the last 10 years. Even though MTC is an infrequent human cancer, it has long been considered a model for the study of molecular abnormalities that underlie the initiation and progression of neoplasia. These advances in understanding the molecular pathophysiology of MTC have incited research and development of molecular targeted therapies.

4.1 INTRODUCTION

Medullary thyroid carcinoma (MTC) is a neuroendocrine tumor of the parafollicular calcitonin (CTN)-secreting C-cells of the thyroid gland. Currently, MTC accounts for 5 to 10% of all thyroid cancers,[1] with a 1 to 2% incidence in nodular thyroid disease.[2] This tumor was initially described at the turn of the twentieth century by Jaquet as a "malignant goiter with amyloid," while the definitive description was provided by Hazard et al.[3] in 1959, who referred to the tumor as a "solid nonfollicular carcinoma with co-existing amyloid." In about 20% of cases, MTC is inherited as an autosomal dominant trait[4] related to a germline mutation of *RET*, with a variable degree of expressivity and an age-related penetrance. There are three different genetic syndromes that include MTC: familial MTC (fMTC), multiple endocrine neoplasia (MEN) type 2A (MEN-2A), and MEN-2B, in which other endocrine glands may be involved. MTC was first described in association with pheochromocytoma (MEN-2B) by Sippel in 1961.[5]

A unique feature of MTC is the production of tumor markers, such as calcitonin (CTN), carcinoembryonic antigen (CEA), and chromogranin A (CGA). A diagnosis of MTC is based on the presence of a solitary thyroid nodule, which can be associated with nodal metastases, and elevated CTN levels. Genetic testing is performed in suspected fMTC and to identify at-risk subjects who require prophylactic surgery.

Treatment of MTC is based on complete surgical removal: total thyroidectomy with level VI neck dissection, and when image or biopsy proven, a lateral compartment neck dissection. On microscopic examination, MTC can show variable features, but immunostaining is typically positive for CTN, CEA, and CGA, and negative for thyroglobulin (TG). Amyloid is also present in the majority of cases.

Systemic chemotherapies for advanced or metastatic MTC have only limited effectiveness. In fact, the activation of *RET* in MTC might inhibit drug-induced apoptosis and induce the resistance of these tumors to systemic chemotherapy. MTC might be rendered more responsive to chemotherapeutic agents by the coadministration of small molecule targeted therapies, such as RET kinase inhibitors.[6] The somatostatin analogs, octreotide and lanreotide, may produce objective tumor shrinkage in patients with advanced MTC, albeit infrequently, and they may simply be more effective in reducing disease-related symptoms. Despite the fact that MTC is a relatively uncommon human cancer, it can be a productive model for the study of molecular abnormalities that underlie the initiation and progression of neoplasia. In fact, during disease progression, MTC cells could develop a defect in their ability to maintain their endocrine phenotype, altering the clinical outcome.[7] In this chapter, the molecular pathophysiology of MTC will be reviewed in order to provide the context for a later chapter on emergent molecular targeted therapies.

4.2 PARAFOLLICULAR CELLS AND CTN

MTC arises from C-cells that secrete CTN (Figure 4.1), as first described by Williams, who defined the histology of this tumor in 1966.[8] C-cells originate from the embryonic neural crest. They are part of the amine precursor uptake and decarboxylation (APUD) system, so MTC can frequently manifest clinical and histological features of other neuroendocrine tumors. For example, severe diarrhea in some patients with advanced disease has been attributed to prostaglandin secretion by the tumor.[9] Thus, paraneoplastic syndromes frequently occur with MTC, owing to the cosecretion of various humoral factors.

Cushing syndrome due to MTC is rare and can be due to ectopic secretion of adrenocorticotropic hormone (ACTH) or corticotrophin-releasing factor (CRF).[10] Barbosa et al.[11] retrospectively

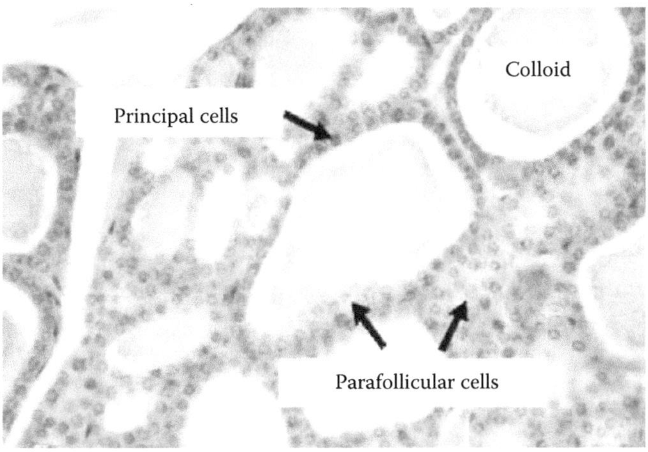

FIGURE 4.1 (See color insert.) Thyroid histology. Principal thyroid follicular cells secrete thyroid hormones. Parafollicular cells secrete calcitonin and may be found as single cells in the epithelial lining of the follicle or in groups in the connective tissue between follicles. Parafollicular cells usually appear as large, clear cells since they do not stain well with hematoxylin and eosin. (From http://instruction.cvhs.okstate.edu/.)

reported ten cases of MTC with ectopic ACTH-dependent Cushing syndrome, treated with medical or surgical therapy (adrenolytic therapy alone or in association with somatostatin analog, somatostatin analog alone, or bilateral adrenalectomy). Lal et al.[12] discussed the case of a 59-year-old man who presented with a hypercoagulable state, nonbacterial endocarditis, and recurrent strokes as paraneoplastic features of MTC. However, the first description of hypercoagulable state associated with MTC was in 1994, when Tiede et al.[13] reported a 69-year-old woman with a superior sagittal sinus thrombosis, thrombocytosis, right lower extremity deep venous thrombosis, and subsequent pulmonary embolus, in which markedly elevated values of CTN and CEA suggested the diagnosis of MTC. Surgical debulking of the tumor resulted in substantial reduction of the CTN and CEA levels in a matter of days, with a progressive resolution of the hypercoagulable state and paraneoplastic cholestatic pattern.[13] A tumor-related circulating factor causing eosinophilia by inducing eosinophilic differentiation and inhibiting myeloid differentiation was postulated by Balducci et al.[14] in 1989 in the first description of the association between eosinophilia and metastatic MTC. In this patient the eosinophil count normalized during remission after chemotherapy and increased at the time of relapse.[14]

Even if MTC is classified as a neuroectodermal tumor, some patients have MTC with elements of endodermal follicular epithelium.[15] Most likely, a small number of MTCs originate from a stem cell of the branchial body and can progress into a mixed medullary-follicular or medullary-papillary carcinoma.[15] The neuroectodermal C-cells are also able to produce other molecules, such as katacalcin (a CTN equimolar fragment), neuron-specific enolase (NSE), tissue polypeptide antigen (TPA), and parathyroid hormone (PTH). MTC cells can also have multidirectional differentiation and produce nonhormonal products, such as mucin[16] and melanin.[17] It is important to note, though, that metastatic melanoma must be excluded before diagnosing a melanin-producing MTC because melanoma is the most common metastatic tumor in the thyroid.

CTN is a 32-amino-acid peptide hormone with an intramolecular disulfide bridge between cysteine residues at positions 1 and 7. This structure is obtained by cleavage and posttranslational processing of procalcitonin, a precursor peptide derived from preprocalcitonin (Figure 4.2). CTN is thought to play a minor role in calcium and phosphorous metabolism, essentially opposing the effects of PTH.[18] CTN has other actions: inhibition of osteoclastic bone resorption,[19] inhibition of prolactin secretion,[20] effects on prostate cancer cell growth,[21] effects on breast cancer carcinomas,[22] and effects on maternal-fetal calcium homeostasis.[23]

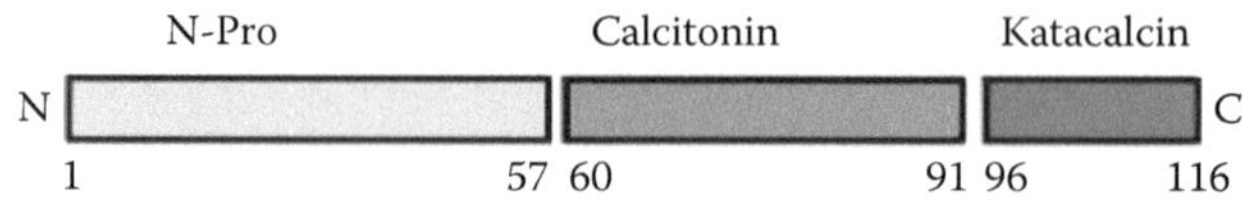

FIGURE 4.2 Structure of calcitonin.

The CTN receptor is a G-protein-coupled receptor that is widely expressed (Figure 4.3). Even if CTN can lower serum calcium levels, patients with MTC (elevated CTN[24]) and status postthyroidectomy (low to absent CTN[25]) are eucalcemic. On the other hand, the role of CTN in regulating 1,25-dihydroxyvitamin D3 (1,25-D) levels in normocalcemic conditions has been recently confirmed.[26] Specifically, even though secondary hyperparathyroidism resulting from hypocalcemia stimulates renal 1α-hydroxylase gene expression, under normocalcemic conditions, 1α-hydroxylase gene expression is primarily induced by CTN and not PTH.[26]

In C-cell disease, which includes both C-cell hyperplasia (CCH) and MTC, serum CTN levels rise proportionately with tumor size and are predictive of metastatic disease.[27] However, elevation of serum CTN levels has also been associated with other conditions, such as chronic kidney disease, sepsis, neuroendocrine tumors of the lung or gastrointestinal tract, hypergastrinemia, mastocytosis, autoimmune thyroid disease, and type 1A pseudohypoparathyroidism[28] (Table 4.1). False positive increases of CTN levels can be caused by the presence of heterophilic antibodies,[29] which can be confirmed or excluded with dilution techniques.

With the CTN immunoradiometric assay (IRMA), most normal subjects show values of <10 pg/ml.[30] The sensitivity of the assay is approximately 2 pg/ml. With the formerly used CTN radioimmunoassay (RIA), the maximum normal baseline value was about 100 pg/ml. Therefore, the mean value and range of CTN assessed with IRMA or similar assays, such as immunochemiluminometric assay (ICMA) or immunoluminometric assay (ILMA), are approximately one-tenth of RIA values. The lack of marked elevation of CTN or other tumor markers poses a difficulty with respect to disease surveillance. There are only a few cases of patients with CTN-negative MTC,[31] and in these cases, CGA and CEA should be used as complementary markers of disease.[32]

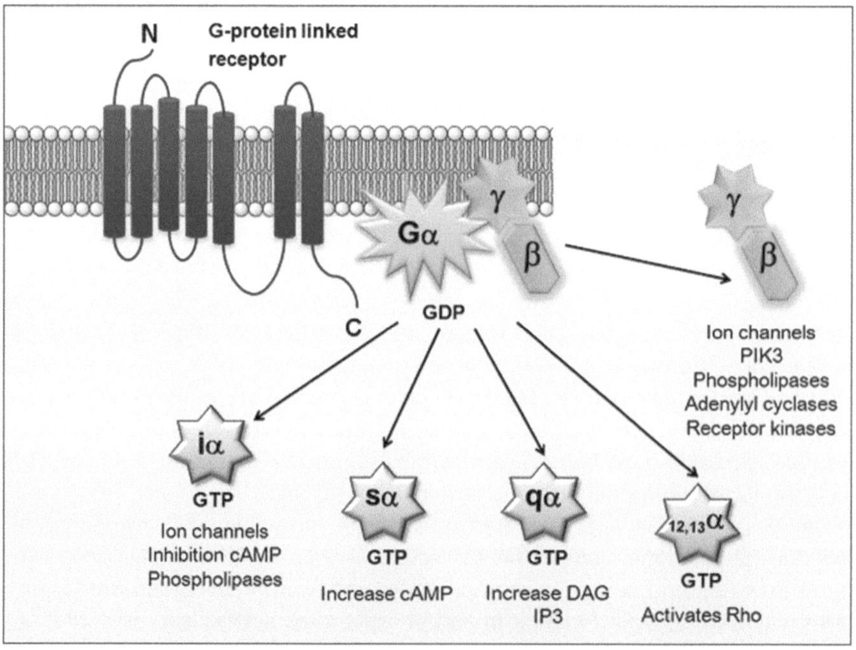

FIGURE 4.3 (See color insert.) Calcitonin receptor signal transduction by G-proteins. (From http://medschool.creighton.edu.)

TABLE 4.1
Causes of Increased Calcitonin Levels

C-Cell Disease	Other Conditions
Medullary thyroid cancer	Chronic renal failure
C-cell hyperplasia	Sepsis
Neoplastic	Neuroendocrine tumors
Nonneoplastic	Hypergastrinemia
	Autoimmune thyroid disease
	Mastocytosis
	Pseudohypoparathyroidism
	False positive increases due to heterophilic antibodies

Stimulated CTN is a useful tool in the diagnosis of MTC. The most widely used method to stimulate CTN secretion is based on the slow intravenous administration of pentagastrin (0.5 μg/kg). Serum CTN levels are measured before infusion and then at 3 and 5 min after initiation of the pentagastrin infusion. Peak-stimulated CTN levels of <10 and < 30 pg/ml are found in 80% and 95% of healthy individuals, respectively. Stimulated CTN levels of >100 pg/ml suggest C-cell disease, while moderate elevations between 30 and 100 pg/ml are described in other adult thyroid diseases. In patients with MTC who have elevated basal CTN levels, peak CTN levels after pentagastrin are usually 5–10 times higher than their basal levels.[33] CTN secretion can also be provoked by a short intravenous calcium gluconate infusion (2.5 mg/kg). In both healthy individuals and patients with C-cell disease, serum CTN levels obtained after a 30-second infusion of calcium are similar to those induced by pentagastrin.[33]

Predictive values of CTN in MTC are considered 100% when basal levels are >100 pg/ml or when stimulated levels after pentagastrin are >1000 pg/ml. Scheuba et al.[34] evaluated the clinical and surgical consequences of calcitonin screening in a series of patients with mildly elevated basal and pentagastrin-stimulated CTN levels. Their conclusion was that these tests can predict MTC. In fact, all patients with basal CTN levels of >64 pg/ml and stimulated CTN levels of >560 pg/ml have MTC, while only 20% of patients with basal levels between 10 and 64 pg/ml and stimulated levels between 100 and 560 pg/ml have MTC.[34] In general, basal CTN levels of >100 pg/ml in the absence of thyroid autoimmunity are an indication for surgery.

The American Thyroid Association clinical practice guidelines state that the positive predictive values of basal CTN to diagnose MTC with values between 20 and 50 pg/ml, 50 and 100 pg/ml, and >100 pg/ml were 8%, 25%, and 100%, respectively.[35] In a recent review, Costante et al.[18] proposed cutoff values of pentagastrin-stimulated CTN that are suspicious for C-cell disease. Practically, stimulated CTN levels of <10 pg/ml exclude C-cell disease, while values between 10 and 100 pg/ml are indeterminate, notwithstanding false positivity. Stimulated CTN levels between 100 and 500 pg/ml and between 500 and 1000 pg/ml are suspicious for CCH and MTC, respectively. A result of >1000 pg/ml is essentially diagnostic for MTC.[18]

The CTN assay imposes several limitations on the diagnostic evaluation of MTC. This is due to its concentration-dependent biphasic half-life and the presence of different CTN isoforms. For these reasons, measurement of procalcitonin, the prohormone of calcitonin currently used as a sepsis marker, may serve as a complementary MTC tumor marker due to its greater analytical stability.[36] D'Herbomez et al.[37] suggest that the interpretation of CTN assay results must be evaluated on the basis of the method used, the patient's gender, age, and weight, and the potential influence of cigarette smoking, which can increase the concentration of CTN. The limited data available on children suggest that CTN levels are relatively high during the first 6 months of life, while they progressively decline, subsequently reaching adult levels during the third year of life.[38]

Fine-needle aspiration biopsy (FNAB) is also an essential tool in the diagnostic evaluation of thyroid nodules suspicious for MTC. This technique is usually satisfactory (80–90%) in the diagnosis of MTC, especially when performed with CTN immunocytochemistry.[39] The most important cytologic criteria of MTC with FNAB are dispersed cell pattern of polygonal or triangular cells, azurophilic cytoplasmatic granules, and extremely eccentrically placed nuclei with coarse granular chromatin and amyloid.[39] In situations where the FNAB cytological analysis is negative for malignancy in a thyroid nodule, especially in patients >40 years with nodules of <10 mm, routine preoperative CTN may be useful,[40] though this practice remains controversial.

4.3 GENETICS OF C-CELL DISEASE

Hereditary MTC is usually bilateral and multifocal. Despite the rarity of these syndromes, early diagnosis is important since MTC is a lethal disease if not promptly treated. In particular, early prophylactic thyroidectomy performed before the age of one year in MEN-2A patients with abnormal serum CTN levels and a total thyroidectomy with central neck dissection within the first weeks of life in MEN-2B patients offer the best chance for a normal life expectancy.[41] More than 97% of patients affected by hereditary MTC carry a germline missense mutation in *RET*. The presence of a somatic *RET* mutation correlates, independent of other clinical and pathological features, with a worse outcome for MTC patients and with the presence of lymph node metastases at diagnosis.[42] In addition, somatic *RET* mutations are involved in tumorigenesis in a relatively large percentage of sporadic MTC cases.[42]

4.3.1 Multiple Endocrine Neoplasia

MEN-2 is an autosomal dominant inherited disorder with an estimated prevalence of 2.5 per 100,000 in the general population and characterized by the development of multiple endocrine tumors, including MTC.

4.3.1.1 MEN-2A

MEN-2A syndrome is the most common form of MEN-2 and is characterized by MTC (90%) in combination with pheochromocytoma (40–50%) or parathyroid hyperplasia or single parathyroid adenoma (10–20%) in a single patient, or the presence of two or more tumor types in multiple members of a single family.[43] MTC is generally the first manifestation of MEN-2A and develops between the ages of 5 to 25 years.[44] There is a variant of MEN-2A associated with lichen amyloidosis with an earlier onset.[45] Mutations in *RET* codon 634, and to a lesser extent in *RET* codons 609, 611, 618, and 620, are related to the classic MEN-2 phenotype.[46] However, mutations in *RET* codons 630, 790, 791, 804, and 891 are also capable of producing the MEN-2A phenotype.[46]

4.3.1.2 MEN-2B

MEN-2B syndrome is the more aggressive form of MEN-2, but fortunately the less common, and includes 5–10% of all cases. MEN-2B is characterized by MTC and pheochromocytoma, usually without hyperparathyroidism. Other physical stigmata are mucosal neuromas, intestinal ganglioneuromas, and marfanoid habitus with skeletal deformations and joint laxity, but without the vascular and ophthalmologic abnormalities that may be seen with other phakomatoses, such as von Hippel-Lindau syndrome.

The skeletal abnormalities of MEN-2B may be due to overexpression of chondromodulin, a regulator of cartilage and bone growth, which is reduced in this syndrome.[46] Germline mutations in *RET* codon 918 were found to be related to the classic MEN-2B phenotype in 95% of cases.[46] The onset of disease usually occurs during the first year of life. MEN-2B patients have a more aggressive form of MTC with a higher morbidity and mortality rate than patients with MEN-2A. They often lack a family history of MTC and harbor a *de novo* mutation.[44]

4.3.1.3 Familial MTC

fMTC is considered the mildest variant of MEN-2, with a clinical course more benign than MEN-2A and MEN-2B. In fact, these patients have a strong predisposition to develop MTC but a very low incidence of other manifestations of MEN-2A. Typically, fMTC patients have a late onset. fMTC can be diagnosed based on at least four family members affected. However, a family history is often inadequate in establishing a diagnosis of familial disease, and biochemical and genetic screening often reveal fMTC in patients originally thought to have the sporadic form of the disease. The prognosis for fMTC is relatively good. However, aggressive MTC tumors have been reported in some cases harboring codon 804 mutations.[47] The International *RET* Mutation Consortium Analysis reported that only 10% of fMTC families showed a germline mutation in the intracellular domain of the *RET*,[48] while the French Calcitonin Tumors Study Group found that a mutation located within the intracellular domain of *RET* (at codons 768, 790, 791 (exon 13), or 804 (exon 14)) is present in about half of fMTC families.[49]

4.3.1.4 The *RET* Proto-Oncogene

RET is localized in chromosome subband 10q11.2 (Figure 4.4) and comprises 21 exons. *RET* encodes the protein RET (rearranged during transfection), a receptor tyrosine kinase (RTK) expressed in tissues and tumors derived from neural crest origin. *RET* is normally expressed in neural-crest-derived tissue in the adult, including basal ganglia, enteric ganglia, adrenal medullary chromaffin cells, autonomic neurons, and thyroid C-cells.[50] Dominant-activating or gain-of-function mutations in *RET* lead to the constitutive activation of RTKs and downstream pathways involved in cell survival and proliferation (Figures 4.5 and 4.6). Because *RET* is a proto-oncogene, only a single point mutation is required for malignant transformation. The first germline mutation of *RET* was identified in patients in 1993.[51]

RTK modulates the extracellular signals for processes as diverse as cell growth, differentiation, survival, and programmed cell death. In response to binding of extracellular ligands, RTKs generally form homodimers or heterodimers. On dimerization, autophosphorylation occurs, followed by intracellular signal transduction through effectors that recognize and interact with the phosphorylated form of the RTK.[52] The ligand-receptor interaction itself is very specific, but

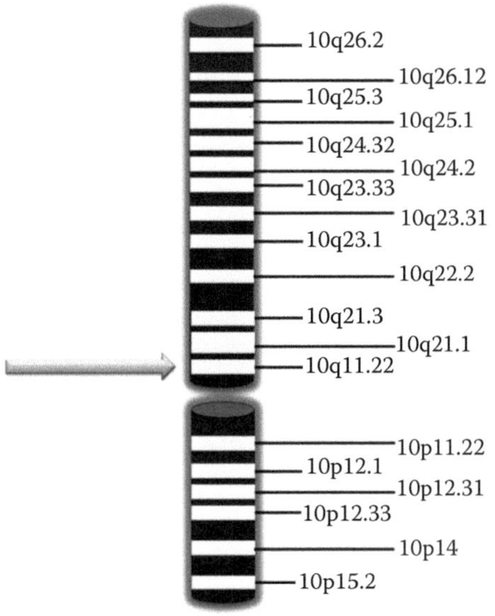

FIGURE 4.4 The *RET* gene is located on the long (q) arm of chromosome 10. (From http://ghr.nlm.nih.gov/gene=ret.)

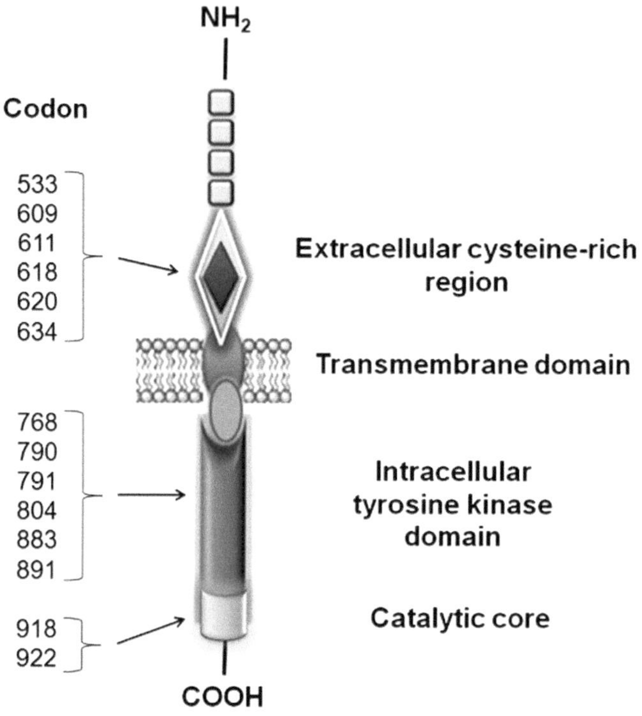

FIGURE 4.5 Schematic representation of the *RET* tyrosine kinase receptor.

in some cases this action can be modulated by the presence of other, low-affinity, nonsignaling accessory molecules at the cell surface. The RET RTK is activated by dimerization facilitated by its binding to the heterodimeric complex formed by glial-cell-derived neurotrophic factor (GDNF)–family ligand (GFL) and GNDF–family receptor (GFR). Both GDNFs and their coreceptors are a small protein family.[53]

The number and type of recognized *RET* mutations have grown over the last ten years, especially after the introduction of *RET* genetic screening in the workup of all patients with MTC, both hereditary and apparently sporadic types. In 98% of MEN-2A patients and in 80–90% of fMTC patients, the germinal *RET* mutation occurs in one of five cysteine codons of the extracellular domain of the RET RTK protein: 609, 611, 618, 629 (exon 10), and 634 (exon 11).[49] Codon 634 (exon 11) is the most commonly affected (mutated in 80% of MEN-2A cases).[54]

In the remaining cases, cysteine codon mutations may occur in codons 610, 620, or 630.[55] In addition to these genetic changes, other, rarer noncysteine mutations located within the intracellular catalytic domain of *RET* have been described. These mutations can give rise to fMTC (codons 768, 790, 791, 804, 806, and 891) and MEN-2B (codons 833 and 918).[56] More than 95% of patients with MEN-2B have a mutation at codon 918, causing receptor autophosphorylation and activation. The M918T mutation is also found as a somatic mutation in about 30–40% of sporadic MTC cases.

Point mutations involving the extracellular *RET* codons 609, 618, and 620 may exert a dual effect, causing both loss of function, resulting from a decrease in RET RTK levels at the cell surface (as in Hirschsprung's disease), and gain of function, resulting from impaired disulfide bonding of two adjacent RET RTK molecules owing to steric hindrance (fMTC, MEN-2A, or MEN-2B). The mechanisms of the gain-of-function mutations depend on the position of the *RET* germline mutation. In fact, mutations involving extracellular domain codons 609, 611, 618, 620, 630, and

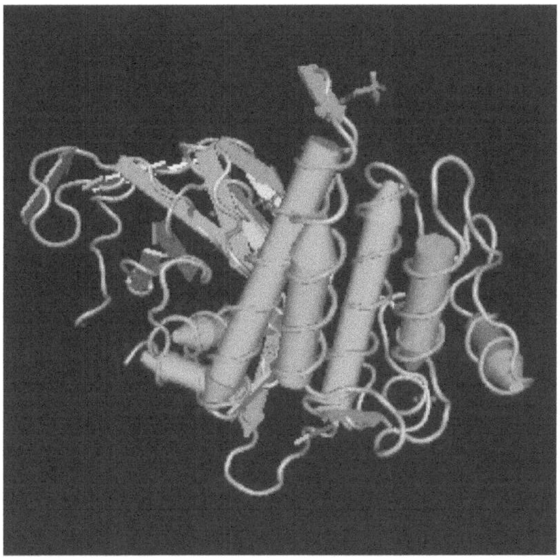

FIGURE 4.6 Crystal structure of phosphorylated *RET* tyrosine kinase domain. (From http://www.ncbi.nlm.nih.gov.)

634 activate the RTK by ligand-independent dimerization and cross-phosphorylation. Intracellular domain mutations affect only codons 768, 790, 791, 804, and 891 and may interfere with intracellular ATP binding of the RTK.[54]

RET mutations can be associated with different degrees of aggressiveness and different outcomes of the disease. Colombo-Benkmann et al.[57] describe 15 individuals from five kindreds who were identified as carriers of a *RET* mutation in exon 11 codon 649 TCG/TTG. Their analysis indicates that the S649L mutation in the transmembrane domain of the RET RTK is characterized by low penetrance of MTC and by a relatively low aggressive potential of the disease.[57]

Some observations have suggested that the V804L *RET* mutation in codon 804 causes low-penetrance disease, with late onset and a relatively indolent course,[58] while other studies reported that the V804L mutation or other mutations in the same codon (substitution of valine with methionine) have an aggressive potential.[47] Patients with the 804 *RET* mutation rarely develop parathyroid disease, the highest frequency being with the codon 634 *RET* mutation.[48]

Lombardo et al.[58] studied 61 heterozygotes from 5 independent families, with fMTC harboring the V804L *RET* mutation. Their data confirmed that a codon 804 *RET* mutation has a weaker activating capability and a lesser transforming capacity than other *RET* mutations. Moreover, the rare and late occurrence of pheochromocytoma in patients carrying this mutation suggests that this has a low penetrance and aggressive form. Therefore, the aggressive presentation of MTC described in some cases with the V804L mutation could be explained through investigation of other genetic or environmental factors. These factors could cause a different outcome from an identical *RET* mutation. In addition, an extensive gene analysis performed on an aggressive MTC tumor related to the V804L mutation showed a second somatic RET mutation.[58] These two mutations were probably responsible for the observed aggressive clinical course.[58] Several studies have confirmed the presence in MEN-2A/fMTC families of other somatic RET mutations, supporting the hypothesis that these patients require a second germline or somatic mutation of RET in addition to the preexisting germline mutation for the disease to be manifested. Even more important is the observation that mutations at *RET* codon 804 can induce resistance to certain molecular targeted (tyrosine kinase inhibitor) therapies.[59]

TABLE 4.2
RET Mutations and Risk Stratification[46]

Risk Levels	Risk	RET Genotype	Recommended Age for Surgery: Thyroidectomy before Age (Years)
3	Highest	883, 918, 922	0.5
2	High	609, 611, 618, 620, 630, 634	5
1	Least high	768, 790, 791, 804, 891	5–10

4.3.1.5 *RET* and Risk Stratification

On the basis of which *RET* codon is mutated, three levels of risk stratification to predict MTC development and aggressiveness are known (Table 4.2). Level 3 mutations (codons 883, 918, and 922) have the most aggressive course, with metastatic disease presenting in the first years of life. Because of the high risk for malignancy at an early age, thyroidectomy is recommended within the first 6 months of life, and preferably within the first month of life. Level 2 mutations (codons 609, 611, 618, 620, 630, and 634) are considered at high risk for MTC, and the current recommendation is that these patients undergo thyroidectomy before the age of 5 years. Level 1 mutations (codons 768, 790, 791 804, and 891) are still considered at high risk for MTC, but are the lowest risk of the *RET* mutations. MTC in these patients tends to develop later in life and takes on a more indolent course. Because clinically apparent disease is rarely reported prior to 10 years of age, many recommend waiting until then to perform a thyroidectomy. However, some variability and unpredictability is reported in some families; hence many surgeons recommend treating all patients with MEN-2A in the same way, performing their prophylactic operation by the age of 5 years.

Fazioli et al.[60] described a new germline point mutation in *RET* exon 8 (Cys515Ser) in a family with MTC. In a member of this family, clinical evaluation revealed bilateral multifocal microscopic MTC and papillary thyroid carcinoma (PTC). *In vitro* and *in vivo* analysis indicated ligand-independent activation of the *RET*-Cys515Ser mutant due to aberrant disulfide homodimerization, increased mitogenic activity, and ability to induce anchorage-independent growth in NIH-3T3 cells in comparison to wild-type *RET*, suggesting a possible role of Cys515Ser in tumor development.[60]

4.3.2 C-Cell Hyperplasia

CCH corresponds to a MTC preneoplastic condition and is defined as an increased number of normal C-cells, more commonly with a diffuse pattern, in a number greater than that found in most normal subjects. Because CCH is a fairly common finding in adult thyroids (about 30% of thyroid glands on autopsy), several criteria have been proposed to discriminate between physiological and neoplastic CCH. Physiological/reactive CCH has been reported in neonates, the elderly, hyperparathyroidism, Hashimoto's thyroiditis, and follicular thyroid adenomas[61] and can be defined as focal, diffuse, or nodular. Reactive CCH usually requires immunohistochemistry for its identification. Neoplastic CCH is characterized by the presence of several cytological alterations (presence of large, mildly to moderate atypical, round, polygonal, or spindle-shaped cells with nuclear pleomorphism[62]), which are differentiated from follicular cells on hematoxylin and eosin sections. However, the pathological definition and clinical correlation of reactive and neoplastic CCH are still blurred.[63]

In hereditary MTC, the finding of *RET* mutations frequently prompts a prophylactic thyroidectomy in the CCH stage because it has a major impact on the prognosis. In fact, CCH is considered, albeit not universally, to be a possible precancerous condition even in the absence of *RET* mutations, even if no conclusive data exist. Diaz-Cano et al.[64] described a neoplastic CCH subtype related to germline *RET* mutations affecting codon Cys634, which led to MEN-2A-related CCH, while sporadic CCH (reactive or associated with nonfamilial MTC) seems to not have point mutations, and therefore should not be regarded as a predisposing condition for nonfamilial MTC.[65]

Very recently Verga et al.[66] demonstrated, in a small group of patients with multinodular goiter (MNG), that those with pentagastrin-stimulated CT of >50 pg/ml had MTC or CCH displaying a neoplastic phenotype on histological examination. The number of C-cells was higher than that found in MNG with normal basal CTN levels and was comparable with that observed in fMTC patients studied as controls. No germline/somatic mutations were found. The authors suggest the possible preneoplastic potential of CCH in this setting, even in the absence of *RET* mutations.[66]

FNAB may play a role in the differential diagnosis between CCH and MTC. Aulicino et al.[67] in 1998 suggested that in patients with elevated serum CTN and the absence of a discrete thyroid nodule, the finding of clusters of CTN-positive cells intermixed with normal follicular cells by FNAB may provide a means of making a presurgical diagnosis of CCH. In a recent study, Rossi et al.[68] evaluated the relevance of immunocytochemistry on thin-layer cytology in thyroid lesions suspicious for MTC. Their conclusions were that combined findings of morphology and a small immunopanel including TG, CTN, and CEA yield a 100% diagnostic efficacy for MTC.[68]

Despite the fact that MTC is a relatively uncommon human cancer, it has properties that make it an important model for understanding the molecular steps in the genesis of neoplasia and the events that can occur during tumor progression. In fact, during the course of the disease, MTC cells can develop a defect in their ability to maintain their mature endocrine phenotype, and these abnormalities correlate with clinical outcome.

4.4 MOLECULAR PATHOGENESIS OF MTC

It has been previously suggested that the mechanism of tumorigenesis involving chromosome 10 in MTC may be different from that proposed for other heritable cancers and particularly for retinoblastoma. In hereditary human cancers, the initial role in tumor development is the loss of the recessive tumor suppressor function, as described by Knudson in his "two hit" hypothesis of carcinogenesis for retinoblastoma.[69] In this model, the germline chromosomic defect is assumed to be phenotypically silent as long as the normal gene allele on the opposite chromosome is functionally intact. The second hit frequently occurs through structural changes of the chromosome (as extensive chromosomal deletions or loss of the entire chromosome) and can cause clonal tumor growth from the cells that have suffered these two hits. For MTC, several studies suggested that the first hit might not be silent, but may determine a growth stimulus on the target thyroid C-cells or adrenal chromaffin cells (MEN-2). In fact, tumor development in these syndromes is probably preceded by a C-cell or adrenal chromaffin cell hyperplasia. The MTCs and pheochromocytomas originate as multifocal clonal tumors, which rise from a different cell clone. The stimulus for this polyclonal hyperplasia could result from the original chromosome 10 abnormality and other subsequent genetic events. These events might cause the progression of clonal carcinomas and involve chromosomal loci other than the germline defect on chromosome 10.[7]

Maintenance of telomere length has been reported to be an absolute requirement for unlimited growth of human tumor cells, and in about 85% of cases, this is achieved by reactivation of telomerase, the enzyme that elongates telomeres. In the case of human MTC, it seems that telomerase activity (TA) is low, supporting the idea that this enzyme is not important for the tumorigenesis of MTC. This is in contrast to the data obtained in other carcinomas, where high TA has been correlated with malignant transformation. Indeed, abrogation of tumor suppressor protein function such as p53 and pRb has been described in MTC and could explain the low apoptotic rate of this tumor also in the absence of activating a telomere-stabilizing mechanism. On the other hand, it is possible that the low TA could also exert a protective role against cell death or stabilize the karyotype.[70]

There are probably other genetic events in association with *RET* mutations that could be responsible for the different features of the disease. However, these additional mutations in both hereditary and sporadic MTC development are largely unknown. Van Veelen et al.[71] described the presence of somatic mutations in the cell cycle regulator P18 in human *RET*-associated MTCs, which cause an amino acid substitution in the cyclin-dependent kinase-interacting region of P18 (INK4C). This inhibits P18 function and reduces its stability.[70] In fact, in human MTC, both hereditary and sporadic, it is frequent to detect the loss of a specific part or the entire short arm of chromosome 1, with the most common break point on 1p32, where the tumor suppressor gene CDKN2C8p18 is located.

Metalloproteinase 2 (MMP-2) and tissue inhibitor of metalloproteinase 2 (TIMP-2) participate in the degeneration of the extracellular matrix and are associated with carcinogenesis in MTC. MMP-2 is one of the principal metalloproteinases active in neoplasia, and it is a marker of the malignant phenotype and novel prognosticator for this cancer.[72]

Several signal transduction pathways have been identified as mediating the oncogenic properties of MTC, including both the phosphatidylinositol 3-kinase (PI3K)/Akt3 and Raf-1/MEK/extracellular signal-regulated kinase (ERK) pathways (also referred to as Ras/mitogen-activated protein kinase (MAPK)).[73] The *RET* mutation upregulates PI3K/Akt signaling.[74] This pathway may function by suppressing apoptosis even if recent evidence suggests that regulation of cell cycle progression is a parallel mechanism resulting in prolonged cellular survival.[73] The PI3K/Akt signal transduction pathway has also been implicated in the control of CTN and CGA production.[73] Other signaling molecules have also been implicated in the development of MTC, such as Notch 1/hairy enhancer of Split 1/achaete-scute complex like-16 and glycogen synthase kinase-3b.[75,76]

Interestingly, Greenblatt et al.[77] examined the effects of valproic acid, a histone deacetylase (HDAC) inhibitor, on cell proliferation in human MTC and particularly on Notch 1 expression. Notch 1 is a multifunctional protein that regulates cellular differentiation, development, proliferation, and survival in a variety of contexts. Its signaling pathway controls cell fate in multiple developmental programs, and its dysregulation has been implicated in the oncogenesis of several types of cancer. It has been previously demonstrated that overexpression of Notch 1 in MTC cells inhibits cell growth and hormone secretion.[75] Notch 1 is absent in MTC cells at baseline, but valproic acid treatment leads to an activation of Notch 1 protein that, in turn, induces cell apoptosis inhibiting tumor growth.[77] However, the exact mechanism by which valproic acid activates Notch 1 signaling is yet to be determined. The induction of the Notch 1 signaling cascade is also related to the antiproliferative and pro-apoptotic effect of the suberoyl bis-hydroxamic acid (SBHA), a relatively new HDAC inhibitor.[78] Nude mice injected with human MTC and then treated with SBHA demonstrated a mean 55% inhibition of tumor growth.[78] In particular, SBHA caused an increase in p21(CIP1/WAF1)—a p53-inducible protein, p27(KIP1)—a cyclin-dependent kinase inhibitor, cleaved caspase-9, cleaved caspase-3, and cleaved polyADP-ribose polymerase, with a concomitant decrease in cyclin D1 and cyclin B1.[78] This indicates that the demonstrable growth inhibition was due to both cell cycle arrest and apoptosis. Moreover, SBHA downregulated the cell survival proteins Bcl-2 and Bcl-X(L), but upregulated the apoptotic proteins Bax, Bad, and Bmf.[79] These findings suggest that Notch 1 activation with HDAC inhibitors may present a promising new form of molecular targeted therapy for the treatment of patients with MTC.

Another important consideration involves the correlation between type 2 deiodinase and MTC cells. The type 1 (D1) and type 2 (D2) deiodinases are responsible for catalyzing deiodination of T4 to T3 and, consequently, playing a critical role in regulating intracellular T3 concentrations. D2 is expressed in normal and stimulated human thyroid tissue and has been evaluated as a possible marker of thyroid follicular cell differentiation. D2 is underexpressed in PTC[80] but increased in FTC.[81] The role of iodine metabolism in MTC was demonstrated by Souza Meyer et al.,[82] who showed that D2 is expressed in MTC tissues at levels comparable with normal human follicular thyroid cells. The biochemical and molecular properties of D2 in the human MTC cell line TT seem to be preserved despite C-cell dedifferentiation. The presence of this D2 in TT suggests a possible role of thyroid hormones in human C-cell metabolism. Furthermore, the expression of the thyroid hormone receptor was demonstrated in all MTC samples and in human MTC cell lines, suggesting a potential role of locally produced T3 by D2 in this neoplastic tissue.

4.5 METASTATIC MTC

MTC is associated with higher mortality rates when there are distant metastases. MTC commonly metastasizes to lymph nodes in the central and lateral cervical compartments, but can also spread to distant organs, such as lung, liver, and bone, especially in cervical vertebrae.[83] The frequency of metastases is often related to the primary tumor size.[84]

Santarpia et al. describe an unusual case and peculiar presentation of pituitary metastasis from MTC with mild diabetes insipidus and optic chiasmatic compression,[85] as well as four patients with cutaneous metastases from sporadic MTC. Cutaneous metastases from thyroid cancer, especially MTC, are rare and are generally localized to the upper part of the body. This pattern is related to the rich capillary network of the scalp, chest, and neck, and the presence of local vascular and growth factors essential for metastases formation.[86] Orbital metastasis from MTC was also described in a patient with MEN-2A.[87] Rare cases of breast metastasis of MTC have also been reported.[88]

Recently it has been reported that activation of the Raf-1/MEK/ERK signaling pathway in MTC cells results in morphologic changes. Ning et al.[89] hypothesized that the metastatic potential of MTC cells could be due to alteration in cell–cell contact molecules caused by Raf-1-induced morphologic changes. In fact, activation of the Raf-1/MEK/ERK pathway reduces the expression of essential cell–cell contact molecules, including E-cadherin, beta-catenin, and occluding, and dramatically inhibits adhesion and migration in MTC cells. Furthermore, treatment of Raf-1-activated cells with U0126, a specific inhibitor of MEK, abrogated these Raf-1-induced effects, indicating that the suppression of the metastatic phenotype in MTC cells is a MEK-dependent pathway.[89]

Diagnosis and management of recurrent or metastatic MTC are very difficult because the course of the disease can often be variable and unpredictable. Generally, patients with postoperative normal levels of CTN and CEA are considered surgically cured.[90] Unfortunately, despite aggressive surgery, about 10% of patients with undetectable postoperative serum CTN develop a recurrence, while about 40% of patients have persistence of the disease.[91,92] The diagnostic imaging workup includes ultrasound, computed tomography, magnetic resonance imaging, and bone scan aimed at investigating the more frequent sites of recurrence or metastases.[93] Unfortunately, many lesions cannot be detected because of their small size,[94] and in this case, scintigraphy plays an important role in the diagnostic evaluation and therapeutic plan. Multimodality testing can identify recurrent tumors in about 40% of patients.[95]

^{111}In octreotide (OCT) is a labeled somatostatin analog with high affinity for somatostatin receptor subtypes 2 and 5, which are overexpressed in several tumors originating from the neural crest. However, in MTC the amount of somatostatin receptors is lower than in other neuroendocrine tumors, and for this reason, ^{111}In OCT scintigraphy is not as useful. The sensitivity of ^{111}In OCT scintigraphy varies from 37[96] to 75%[97] and is higher in the presence of neck and mediastinal lymph nodes and lower for distant metastases.[98] Moreover, ^{111}In OCT scintigraphy can be used to select patients eligible for somatostatin analog therapy, currently employed in the treatment of neuroendocrine tumors and mainly to obtain symptom relief.[98] Recently, ^{111}In-labeled somatostatin analogs have been employed to select patients with MTC suitable for treatment with ^{90}Y [DOTA] Tyr^3 octreotide (DOTATOC) and ^{177}Lu [DOTA] Tyr^3 octreotate (DOTATATE).[93,99]

^{18}Fluorodeoxyglucose (^{18}FDG) positron emission tomography (PET) is largely used in the MTC diagnostic evaluation because of its higher image quality and spatial resolution. However, its sensitivity is correlated to the biological characteristics of the tumor, being higher in more aggressive ones.[100] For this, positive findings in ^{18}FDG PET correlate with high progression rate and unfavorable outcome. ^{18}FDG PET sensitiviy is high for neck and mediastinum lymph node metastases, intermediate for lung metastases, and low for bone metastases.[93] Other PET radiopharmaceuticals, like ^{18}fluoro-L-dihydroxyphenylalanine (^{18}F-DOPA), seem to have higher sensitivity and are currently under evaluation.[93]

Radioiodinated metaiodobenzylguanidine (MIBG) is a guanethidine derivative structurally similar to norepinephrine and accumulated in tissue arising from the neural crest. ^{131}I or ^{123}I MIBG has been employed since 1980 in the diagnosis and follow-up of neuroendocrine tumors. However, the diagnostic accuracy of MIBG in MTC is poor because of its very low sensitivity (25–30%) in spite of very high specificity (95%).[93] Moreover, published reports on a small series of patients treated with therapeutic doses of ^{131}I MIBG showed a response rate of approximately 40%[10] and symptom relief in 60% of patients.[102]

Normal and malignant C-cells do not concentrate radioactive iodine, so MTC is not responsive to radioiodine therapy in contrast to PTC and FTC. Some reports have noted a decrease of postoperative CTN levels after RAI in MTC,[103] based on normal thyroid remnant follicular cell ^{131}I uptake and destruction of neighboring non-iodine-trapping cells; however, other studies failed to demonstrate any decrease in recurrence rate or improvement in survival.[104]

4.6 CONCLUSIONS

MTC is a heterogeneous disease in terms of biological behavior with a clinical course that is extremely variable and unpredictable. Overall, the MTC is generally more aggressive than differentiated thyroid cancer arising from the thyroid follicular cell. Understanding the molecular biology behind specific mutations of the *RET* and their prognostic implications has led to the development of early treatment modalities for certain patients, especially in the context of MEN. In fact, the correlation of genotype and phenotype already permits identification of patients at risk, in which prophylactic thyroidectomy should be performed. In MTC, surgery is the treatment of choice, but although early diagnosis and extensive surgical treatment may favor a good clinical outcome, they do not always guarantee definitive cure. Thus, the knowledge gaps in the genetic and molecular pathophysiology of MTC, notwithstanding tremendous advances in recent years, translate into uncertainties in our diagnostic and therapeutic clinical algorithms. Fortunately, there may now be sufficient information and investigative experience to devise practical molecular targeted therapies for MTC that can truly be a platform for real cures.

REFERENCES

1. Ball DW, Baylin SB, De Butros AC. 2000. Medullary thyroid carcinoma. In *Werner and Ingbar's the thyroid*, ed. LE Braverman, RD Utiger, 930–43. 8th ed. Philadelphia: Lippincott Williams and Wilkins.
2. Niccoli-Sire P, Conte-Devolx B. 2007. Medullary thyroid carcinoma. *Ann Endocrinol* (Paris) 68:325–31.
3. Hazard JB, Hawk WA, Crile G Jr. 1959. Medullary (solid) carcinoma of the thyroid; a clinicopathologic entity. *J Clin Endocrinol Metab* 19:152–61.
4. Keiser HR, Beven MA, Doppman J, Wells SA Jr, Buja LM. 1973. Sipple's syndrome: Medullary thyroid carcinoma, pheochromocytoma, and parathyroid disease. *Ann Intern Med* 88:561–79.
5. Sippel J. 1961. The association of pheocromocytoma with carcinoma of the thyroid gland. *Am J Med* 31:163–66.
6. Skinner MA, Lackey KE, Freemerman AJ. 2008. RET activation inhibits doxorubicin-induced apoptosis in SK-N-MC cells. *Anticancer Res* 28:2019–25.
7. Nelkin BD, de Bustros AC, Mabry M, Baylin SB. 1989. The molecular biology of medullary thyroid carcinoma. A model for cancer development and progression. *JAMA* 261:3130–35.
8. Williams ED. 1966. Histogenesis of medullary carcinoma of the thyroid. *J Clin Pathol* 19:114–18.
9. Roberts LJ 2nd, Hubbard WC, Bloomgarden ZT, Bertagna XY, McKenna TJ, Rabinowitz D, Oates JA. 1979. Prostaglandins: Role in the humoral manifestations of medullary carcinoma of the thyroid and inhibition by somatostatin. *Trans Assoc Am Phys* 92:286–91.
10. Chrisoulidou A, Pazaitou-Panayiotou K, Georgiou E, Boudina M, Kontogeorgos G, Iakovou I, Efstratiou I, Patakiouta F, Vainas I. 2008. Ectopic Cushing's syndrome due to CRH secreting liver metastasis in a patient with medullary thyroid carcinoma. *Hormones* 7:259–62.
11. Barbosa SL, Rodien P, Leboulleux S, Niccoli-Sire P, Kraimps JL, Caron P, Archambeaud-Mouveroux F, Conte-Devolx B, Rohmer V. 2005. Ectopic adrenocorticotropic hormone-syndrome in medullary carcinoma of the thyroid: A retrospective analysis and review of the literature. *Thyroid* 15:618–23.

12. Lal G, Brennan TV, Hambleton J, Clark OH. 2003. Coagulopathy, marantic endocarditis, and cerebrovascular accidents as paraneoplastic features in medullary thyroid cancer—Case report and review of the literature. *Thyroid* 13:601–5.
13. Tiede DJ, Tefferi A, Kochhar R, Thompson GB, Hay ID. 1994. Paraneoplastic cholestasis and hypercoagulability associated with medullary thyroid carcinoma. Resolution with tumor debulking. *Cancer* 73:702–5.
14. Balducci L, Chapman SW, Little DD, Hardy CL. 1989. Paraneoplastic eosinophilia. Report of a case with *in vitro* studies of hemopoiesis. *Cancer* 64:2250–53.
15. Kovacs CS, Masé RM, Kovacs K, Nguyen GK, Chik CL. 1994. Thyroid medullary carcinoma with thyroglobulin immunoreactivity in sporadic multiple endocrine neoplasia type 2-B. *Cancer* 74:928–32.
16. Fernandez BJ, Bedard YC, Rossen I. 1982. Mucus-producing medullary carcinoma of the thyroid gland. *Am J Clin Pathol* 7:536–40.
17. Marcus JN, Dise CA, LiVolsi VA. 1982. Melanin production in a medullary thyroid carcinoma. *Cancer* 49:2518–26.
18. Costante G, Durante C, Francis Z, Schlumberger M, Filetti S. 2009. Determination of calcitonin levels in C-cell disease: Clinical interest and potential pitfalls. *Nat Clin Pract Endocrinol Metab* 5:35–44.
19. Zaidi M, Inzerillo AM, Moonga BS, Bevis PJ, Huang CL. 2002. Forty years of calcitonin—Where are we now? A tribute to the work of Iain Macintyre, FRS. *Bone* 30:655–63.
20. Wang YQ, Yuan R, Sun YP, Lee TJ, Shah GV. 2003. Antiproliferative action of calcitonin on lactotrophs of the rat anterior pituitary gland: Evidence for the involvement of transforming growth factor beta 1 in calcitonin action. *Endocrinology* 144:2164–71.
21. Thomas S, Chigurupati S, Anbalagan M, Shah G. 2006. Calcitonin increases tumorigenicity of prostate cancer cells: Evidence for the role of protein kinase A and urokinase-type plasminogen receptor. *Mol Endocrinol* 20:1894–911.
22. Nakamura M, Han B, Nishishita T, Bai Y, Kakudo K. 2007. Calcitonin targets extracellular signal-regulated kinase signaling pathway in human cancers. *J Mol Endocrinol* 39:375–84.
23. Kovacs CS, Chafe LL, Woodland ML, McDonald KR, Fudge NJ, Wookey PJ. 2002. Calcitropic gene expression suggests a role for the intraplacental yolk sac in maternal-fetal calcium exchange. *Am J Physiol Endocrinol Metab* 282:E721–32.
24. Emmertsen K, Melsen F, Mosekilde L, Lund B, Lund B, Sørensen OH, Nielsen HE, Sølling H, Hansen HH. 1982. Altered vitamin D metabolism and bone remodelling in patients with medullary thyroid carcinoma and hypercalcitoninemia. *Metab Bone Dis Relat Res* 4:17–23.
25. Hoff AO, Catala-Lehnen P, Thomas PM, Priemel M, Rueger JM, Nasonkin I, Bradley A, Hughes MR, Ordonez N, Cote GJ, Amling M, Gagel RF. 2002. Increased bone mass is an unexpected phenotype associated with deletion of the calcitonin gene. *J Clin Invest* 110:1849–57.
26. Zhong Y, Armbrecht HJ, Christakos S. 2009. Calcitonin: A regulator of the 25-hydroxyvitamin D3 1alpha hydroxylase gene. *J Biol Chem*, March 4 (epub ahead of print).
27. Cohen R, Campos JM, Salaün C, Heshmati HM, Kraimps JL, Proye C, Sarfati E, Henry JF, Niccoli-Sire P, Modigliani E. 2000. Preoperative calcitonin levels are predictive of tumor size and postoperative calcitonin normalization in medullary thyroid carcinoma. Groupe d'Etudes des Tumeurs a Calcitonine (GETC). *J Clin Endocrinol Metab* 85:919–22.
28. Becker KL, Nylén ES, White JC, Müller B, Snider RH Jr. 2004. Procalcitonin and the calcitonin gene family of peptides in inflammation, infection, and sepsis: A journey from calcitonin back to its precursors. *J Clin Endocrinol Metab* 89:1512–25.
29. Tommasi M, Brocchi A, Cappellini A, Raspanti S, Mannelli M. 2001. False serum calcitonin high levels using a non-competitive two-site IRMA. *J Endocrinol Invest* 24:356–60.
30. Barbott N, Calmettes C, Schuffenecker I, Saint-André JP, Franc B, Rohmer V, Jallet P, Bigorgne JC. 1994. Pentagastrin stimulation test and early diagnosis of medullary thyroid carcinoma using an immunoradiometric assay of calcitonin: Comparison with genetic screening in hereditary medullary thyroid carcinoma. *J Clin Endocrinol Metab* 78:114–20.
31. Wang TS, Ocal IT, Sosa JA, Cox H, Roman S. 2008. Medullary thyroid carcinoma without marked elevation of calcitonin: A diagnostic and surveillance dilemma. *Thyroid* 18:889–94.
32. Giovanella L, Crippa S, Cariani L. 2008. Serum calcitonin-negative medullary thyroid carcinoma: Role of CgA and CEA as complementary markers. *Int J Biol Markers* 23:129–31.
33. Baloch Z, Carayon P, Conte-Devolx B, Demers LM, Feldt-Rasmussen U, Henry JF, LiVosli VA, Niccoli-Sire P, John R, Ruf J, Smyth PP, Spencer CA, Stockigt JR, Guidelines Committee, National Academy of Clinical Biochemistry. 2003. Laboratory medicine practice guidelines. Laboratory support for the diagnosis and monitoring of thyroid disease. *Thyroid* 13:3–126.

34. Scheuba C, Kaserer K, Moritz A, Drosten R, Vierhapper H, Bieglmayer C, Haas OA, Niederle B. 2009. Sporadic hypercalcitoninemia: Clinical and therapeutic consequences. *Endocr Relat Cancer* 16:243–53.
35. American Thyroid Association Guidelines Task Force, Kloos RT, Eng C, Evans DB, Francis GL, Gagel RF, Gharib H, Moley JF, Pacini F, Ringel MD, Schlumberger M, Wells SA Jr. 2009. Medullary thyroid cancer: Management guidelines of the American Thyroid Association. *Thyroid* 19:565–612.
36. Algeciras-Schimnich A, Preissner CM, Theobald JP, Finseth MS, Grebe SK. 2009. Procalcitonin: A marker for the diagnosis and follow-up of patients with medullary thyroid carcinoma. *J Clin Endocrinol Metab* 94:861–68.
37. d'Herbomez M, Caron P, Bauters C, Cao CD, Schlienger JL, Sapin R, Baldet L, Carnaille B, Wémeau JL, French Group GTE (Groupe des Tumeurs Endocrines). 2007. Reference range of serum calcitonin levels in humans: Influence of calcitonin assays, sex, age, and cigarette smoking. *Eur J Endocrinol* 157:749–55.
38. Basuyau JP, Mallet E, Leroy M, Brunelle P. 2004. Reference intervals for serum calcitonin in men, women, and children. *Clin Chem* 50:1828–30.
39. Papaparaskeva K, Nagel H, Droese M. 2000. Cytologic diagnosis of medullary carcinoma of the thyroid gland. *Diagnostic Cytopathol* 22:351–58.
40. Papi G, Corsello SM, Cioni K, Pizzini AM, Corrado S, Carapezzi C, Fadda G, Baldini A, Carani C, Pontecorvi A, Roti E. 2006. Value of routine measurement of serum calcitonin concentrations in patients with nodular thyroid disease: A multicenter study. *J Endocrinol Invest* 29:427–37.
41. Zenaty D, Aigrain Y, Peuchmaur M, Philippe Chomette P, Baumann C, Cornelis F, Hugot JP, Chevenne D, Barbu V, Guillausseau JP, Schlumberger M, Carel JC, Travagli JP, Leger J. 2009. Medullary thyroid carcinoma identified within first year of life in children with multiple endocrine neoplasia type 2A (codon 634) and 2B. *Eur J Endocrinol*, February 24 (epub ahead of print).
42. Elisei R, Cosci B, Romei C, Bottici V, Renzini G, Molinaro E, Agate L,Vivaldi A, Faviana P, Basolo F, Miccoli P, Berti P, Pacini F, Pinchera A. 2008. Prognostic significance of somatic RET oncogene mutations in sporadic medullary thyroid cancer: A 10-year follow-up study. *J Clin Endocrinol Metab* 93:682–87.
43. Kouvaraki MA, Shapiro SE, Perrier ND, Cote GJ, Gagel RF, Hoff AO, Sherman SI, Lee JE, Evans DB. 2005. RET proto-oncogene: A review and update of genotype-phenotype correlations in hereditary medullary thyroid cancer and associated endocrine tumors. *Thyroid* 15:531–44.
44. Raue F, Frank-Raue K. 2009. Genotype-phenotype relationship in multiple endocrine neoplasia type 2. Implications for clinical management. *Hormones* 8:23–28.
45. Corsello SM, Lovicu RM, Migneco MG, Rufini V, Summaria V. 2000. Diagnostic approach, genetic screening and prognostic factors of medullary thyroid carcinoma. *Rays* 25:257–66.
46. Machens A, Ukkat J, Brauckhoff M, Gimm O, Dralle H. 2005. Advances in the management of hereditary medullary thyroid cancer. *J Intern Med* 257:50–59.
47. Fattorusso O, Quadro L, Libroia A, Verga U, Lupoli G, Cascone E, Colantuoni V. 1998. A GTG to ATG novel point mutation at codon 804 in exon 14 of the RET proto-oncogene in two families affected by familial medullary thyroid carcinoma. *Hum Mut* Suppl 1:S167–71.
48. Eng C, Clayton D, Schuffenecker I, Lenoir G, Cote G, Gagel RF, van Amstel HK, Lips CJ, Nishisho I, Takai SI, Marsh DJ, Robinson BG, Frank-Raue K, Raue F, Xue F, Noll WW, Romei C, Pacini F, Fink M, Niederle B, Zedenius J, Nordenskjold M, Komminoth P, Hendy GN, Gharib H, Thibodeau SN, Lacroix A, Frilling A, Ponder BAJ, Mulligan LM. 1996. The relationship between specific RET proto-oncogene mutation and disease phenotype in multiple endocrine neoplasia type 2: International RET Mutation Consortium Analysis. *JAMA* 276:1575–79.
49. Niccoli-Sire P, Murat A, Rohmer V, Franc S, Chabrier G, Baldet L, Maes B, Savagner F, Giraud S, Bezieau S, Kottler ML, Morange S, Conte-Devolx B. 2001. Familial medullary thyroid carcinoma with noncysteine ret mutations: Phenotype-genotype relationship in a large series of patients. *J Clin Endocrinol Metab* 86:3746–53.
50. Ball DW. 2007. Medullary thyroid cancer: Monitoring and therapy. *Endocrinol Metab Clin North Am* 36:823–37.
51. Mulligan LM, Kwok JB, Healey CS, Elsdon MJ, Eng C, Gardner E, Love DR, Mole SE, Moore JK, Papi L, et al. 1993. Germ-line mutations of the RET proto-oncogene in multiple endocrine neoplasia type 2A. *Nature* 363:458–60.
52. Eng C. 1999. RET proto-oncogene in the development of human cancer. *J Clin Oncol* 17:380–393.
53. Jurvansuu JM, Goldman A. 2008. Recent inventions on receptor tyrosine kinase RET modulation. *Recent Pat Biotechnol* 2:47–54.

54. Machens A, Niccoli-Sire P, Hoegel J, Frank-Raue K, van Vroonhoven TJ, Roeher HD, Wahl RA, Lamesch P, Raue F, Conte-Devolx B, Dralle H, European Multiple Endocrine Neoplasia (EUROMEN) Study Group. 2003. Early malignant progression of hereditary medullary thyroid cancer. *N Engl J Med* 16:1517–25.
55. Bugalho MJ, Domingues R, Santos JR, Catarino AL, Sobrinho L. 2007. Mutation analysis of the RET proto-oncogene and early thyroidectomy: Results of a Portuguese cancer centre. *Surgery* 141:90–95.
56. You YN, Lakhani VT, Wells SA Jr. 2007. The role of prophylactic surgery in cancer prevention. *World J Surg* 31:450–64.
57. Colombo-Benkmann M, Li Z, Riemann B, Hengst K, Herbst H, Keuser R, Gross U, Rondot S, Raue F, Senninger N, Pützer BM, Frank-Raue K. 2008. Characterization of the RET protooncogene transmembrane domain mutation S649L associated with nonaggressive medullary thyroid carcinoma. *Eur J Endocrinol* 158:811–16.
58. Lombardo F, Baudin E, Chiefari E, Arturi F, Bardet S, Caillou B, Conte C, Dallapiccola B, Giuffrida D, Bidart JM, Schlumberger M, Filetti S. 2002. Familial medullary thyroid carcinoma: Clinical variability and low aggressiveness associated with RET mutation at codon 804. *J Clin Endocrinol Metab* 87:1674–80.
59. Carlomagno F, Guida T, Anaganti S, Vecchio G, Fusco A, Ryan AJ, Billaud M, Santoro M. 2004. Disease associated mutations at valine 804 in the RET receptor tyrosine kinase confer resistance to selective kinase inhibitors. *Oncogene* 12:6056–63.
60. Fazioli F, Piccinini G, Appolloni G, Bacchiocchi R, Palmonella G, Recchioni R, Pierpaoli E, Silvetti F, Scarpelli M, Bruglia M, Melillo RM, Santoro M, Boscaro M, Taccaliti A. 2008. A new germline point mutation in Ret exon 8 (cys515ser) in a family with medullary thyroid carcinoma. *Thyroid* 18:775–82.
61. Guyetant S, Wion-Barbot N, Rousselet MC, Franc B, Bigorgne JC, Saint-Andre JP. 1994. C-cell hyperplasia associated with chronic lymphocytic thyroiditis: A retrospective quantitative study of 112 cases. *Hum Pathol* 25:514–21.
62. Perry A, Molberg K, Albores-Saavedra J. 1996. Physiologic versus neoplastic C-cell hyperplasia of the thyroid: Separation of distinct histologic and biologic entities. *Cancer* 77:750–56.
63. Hinze R, Gimm O, Brauckhoff M, Schneyer U, Dralle H, Holzhausen HJ. 2001. "Physiological" and "neoplastic" C-cell hyperplasia of the thyroid. Morphologically and biologically distinct entities? *Pathologe* 22:259–65.
64. Diaz-Cano SJ, de Miguel M, Blanes A, Tashjian R, Wolfe HJ. 2001. Germline RET634 mutation positive MEN 2A-related C-cell hyperplasias have genetic features consistent with intraepithelial neoplasia. *J Clin Endocrinol Metab* 86:3948–57.
65. Saggiorato E, Rapa I, Garino F, Bussolati G, Orlandi F, Papotti MN, Volante M. 2007. Absence of RET gene point mutations in sporadic thyroid C-cell hyperplasia. *J Mol Diagn* 9:214–19.
66. Verga U, Ferrero S, Vicentini L, Brambilla T, Cirello V, Muzza M, Beck-Peccoz P, Fugazzola L. 2007. Histopathological and molecular studies in patients with goiter and hypercalcitoninemia: Reactive or neoplastic C-cell hyperplasia? *Endocr Relat Cancer* 14:393–403.
67. Aulicino MR, Szporn AH, Dembitzer R, Mechanick J, Batheja N, Bleiweiss IJ, Burstein DE. 1998. Cytologic findings in the differential diagnosis of C-cell hyperplasia and medullary carcinoma by fine needle aspiration. A case report. *Acta Cytol* 42:963–67.
68. Rossi ED, Raffaelli M, Mulè A, Zannoni GF, Pontecorvi A, Santeusanio G, Minimo C, Fadda G. 2008. Relevance of immunocytochemistry on thin-layer cytology in thyroid lesions suspicious for medullary carcinoma: A case-control study. *Appl Immunohistochem Mol Morphol* 16:548–53.
69. Knudson AG. 1996. Hereditary cancer: Two hits revisited. *J Cancer Res Clin Oncol.* 122:135–40.
70. Stadler G, Wieser M, Streubel B, Stift A, Friedl J, Gnant M, Niederle B, Beham A, Katinger H, Pfragner R, Grillari J, Voglauer R. 2008. Low telomerase activity: Possible role in the progression of human medullary thyroid carcinoma. *Eur J Cancer* 44:866–75.
71. van Veelen W, Klompmaker R, Gloerich M, van Gasteren CJ, Kalkhoven E, Berger R, Lips CJ, Medema RH, Höppener JW, Acton DS. 2009. P18 is a tumor suppressor gene involved in human medullary thyroid carcinoma and pheochromocytoma development. *Int J Cancer* 124:339–45.
72. Cavalheiro BG, Junqueira CR, Brandão LG. 2008. Expression of matrix metalloproteinase 2 (MMP-2) and tissue inhibitor of metalloproteinase 2 (TIMP-2) in medullary thyroid carcinoma: Prognostic implications. *Thyroid* 18:865–71.
73. Pitt SC, Chen H. 2008. The phosphatidylinositol 3-kinase/akt signaling pathway in medullary thyroid cancer. *Surgery* 144:721–24.
74. Segouffin-Cariou C, Billaud M. 2000. Transforming ability of MEN2A-RET requires activation of the phosphatidylinositol 3-kinase/Akt signaling pathway. *J Biol Chem* 275:3568–76.

75. Kunnimalaiyaan M, Vaccaro AM, Ndiaye MA, Chen H. 2006. Overexpression of the NOTCH1 intracellular domain inhibits cell proliferation and alters the neuroendocrine phenotype of medullary thyroid cancer cells. *J Biol Chem* 281:39819–30.
76. Kunnimalaiyaan M, Vaccaro AM, Ndiaye MA, Chen H. 2007. Inactivation of glycogen synthase-3b, a downstream target of he raf-1 pathway, is associated with growth suppression in medullary thyroid cancer cells. *Mol Cancer Ther* 6:1151–58.
77. Greenblatt DY, Cayo MA, Adler JT, Ning L, Haymart MR, Kunnimalaiyaan M, Herbert Chen H. 2008. Valproic acid activates Notch1 signaling and induces apoptosis in medullary thyroid cancer cells. *Ann Surg* 247:1036–40.
78. Ning L, Greenblatt DY, Kunnimalaiyaan M, Chen H. 2008. Suberoyl bis-hydroxamic acid activates Notch-1 signaling and induces apoptosis in medullary thyroid carcinoma cells. *Oncologist* 13:98–104.
79. Ning L, Jaskula-Sztul R, Kunnimalaiyaan M, Chen H. 2008. Suberoyl bishydroxamic acid activates notch1 signaling and suppresses tumor progression in an animal model of medullary thyroid carcinoma. *Ann Surg Oncol* 15:2600–5.
80. Arnaldi LA, Borra RC, Maciel RM, Cerutti JM. 2005. Gene expression profiles reveal that DCN, DIO1, and DIO2 are underexpressed in benign and malignant thyroid tumors. *Thyroid* 15:210–21.
81. Souza Meyer EL, Dora JM, Wagner MS, Maia AL. 2005. Decreased type 1 iodothyronine deiodinase expression might be an early and discrete event in thyroid cell dedifferentation towards papillary carcinoma. *Clin Endocrinol* (Oxf) 62:672–78.
82. Souza Meyer EL, Goemann IM, Dora JM, Wagner MS, Maia AL. 2008. Type 2 iodothyronine deiodinase is highly expressed in medullary thyroid carcinoma. *Mol Cell Endocrinol* 16:16–22.
83. Oudoux A, Salaun PY, Bournaud C, Campion L, Ansquer C, Rousseau C, Bardet S, Borson-Chazot F, Vuillez JP, Murat A, Mirallie E, Barbet J, Goldenberg DM, Chatal JF, Kraeber-Bodere F. 2007. Sensitivity and prognostic value of positron emission tomography with f-18-fluorodeoxyglucose and sensitivity of immunoscintigraphy in patients with medullary thyroid carcinoma treated with anticarcinoembryonic antigen-targeted radioimmunotherapy. *J Clin Endocrinol Metab* 92:4590–97.
84. Scollo C, Baudin E, Travagli JP, Caillou B, Bellon N, Leboulleux S, Schlumberger M. 2003. Rationale for central and bilateral lymph node dissection in sporadic and hereditary medullary thyroid cancer. *J Clin Endocrinol Metab* 88:2010–75.
85. Santarpia L, Gagel RF, Sherman SI, Sarlis NJ, Evans DB, Hoff AO. 2008. Diabetes insipidus and panhypopituitarism due to intrasellar metastasis from medullary thyroid cancer. *Head Neck* 31:419–23.
86. Santarpia L, El-Naggar AK, Sherman SI, Hymes SR, Gagel RF, Shaw S, Sarlis NJ. 2008. Four patients with cutaneous metastases from medullary thyroid cancer. *Thyroid* 18:901–5.
87. Seiff BD, Seiff SR. 2008. Orbital metastasis from medullary thyroid carcinoma. *Ophthal Plast Reconstr Surg* 24(6):484–85.
88. Marcy PY, Thariat J, Peyrottes I, Dassonville O. 2009. Bilateral breast involvement in medullary thyroid carcinoma. *Thyroid* 19:197–99.
89. Ning L, Kunnimalaiyaan M, Chen H. 2008. Regulation of cell-cell contact molecules and the metastatic phenotype of medullary thyroid carcinoma by the Raf-1/MEK/ERK pathway. *Surgery* 144:920–24.
90. Orlandi F, Caraci P, Mussa A, Saggiorato E, Pancani G, Angeli A. 2001. Treatment of medullary thyroid carcinoma: An update. *Endocr Relat Cancer* 8:135–47.
91. Kebebew, Ituarte PH, Siperstein AE, Duh QY, Clark OH. 2000. Medullary thyroid carcinoma: Clinical characteristics, treatment, prognostic factors, and a comparison of staging systems. *Cancer* 88:1139–48.
92. Pellegriti G, Leboulleux S, Baudin E, Bellon N, Scollo C, Travagli JP, Schlumberger M. 2003. Long-term outcome of medullary thyroid carcinoma in patients with normal postoperative medical imaging. *Br J Cancer* 88:1537–42.
93. Castellani MR, Seregni E, Maccauro M, Chiesa C, Aliberti G, Orunesu E, Bombardieri E. 2008. MIBG for diagnosis and therapy of medullary thyroid carcinoma: Is there still a role? *Q J Nucl Med Mol Imaging* 52(4):430–40.
94. Wang Q, Takashima S, Fukuda H, Takayama F, Kobayashi S, Sone S. 1999. Detection of medullary thyroid carcinoma and regional lymph node metastases by magnetic resonance imaging. *Arch Otolaryngol Head Neck Surg* 125:842–48.
95. Rufini V, Castaldi P, Treglia G, Perotti G, Gross MD, Al-Nahhas A, Rubello D. 2008. Nuclear medicine procedures in the diagnosis and therapy of medullary thyroid carcinoma. *Biomed Pharmacother* 62:139–46.
96. Baudin E, Lumbroso J, Schlumberger M, Leclere J, Giammarile F, Gardet P, Roche A, Travagli JP, Parmentier C. 1996. Comparison of octreotide scintigraphy and conventional imaging in medullary thyroid carcinoma. *J Nucl Med* 37:912–16.

97. Arslan N, Ilgan S, Yuksel D, Serdengecti M, Bulakbasi N, Ugur O, Ozguven MA. 2001. Comparison of In-111 octreotide and Tc-99m (V) DMSA scintigraphy in the detection of medullary thyroid tumor foci in patients with elevated levels of tumor markers after surgery. *Clin Nucl Med* 26:683–88.
98. Behr TM, Béhé M, Becker W. 1999. Diagnostic applications of radiolabeled peptides in nuclear endocrinology. *Q J Nucl Med* 43(3):268–80.
99. Bodei L, Handkiewicz-Junak D, Grana C, Mazzetta C, Rocca P, Bartolomei M, Lopera Sierra M, Cremonesi M, Chinol M, Mäcke HR, Paganelli G. 2004. Receptor radionuclide therapy with 90Y-DOTATOC in patients with medullary thyroid carcinomas. *Cancer Biother Radiopharm* 19:65–71.
100. Rufini V, Treglia G, Perotti G, Leccisotti L, Calcagni ML, Rubello D. 2008. Role of PET in medullary thyroid carcinoma. *Minerva Endocrinol* 33:67–73.
101. Orlandi F, Caraci P, Mussa A, Saggiorato E, Pancani G, Angeli A. 2001. Treatment of medullary thyroid carcinoma: An update. *Endocr Relat Cancer* 8:135–47.
102. Hoefnagel CA. 1995. MIBG and radiolabeled octreotide in neuroendocrine tumors. *Q J Nucl Med* 39:137–39.
103. Hellman DE, Kartchner M, Van Antwerp JD, Salmon SE, Patton DD, O'Mara R. 1979. Radioiodine in the treatment of medullary carcinoma of the thyroid. *J Clin Endocrinol Metab* 48:451–55.
104. Saad MF, Guido JJ, Samaan NA. 1983. Radioactive iodine in the treatment of medullary carcinoma of the thyroid. *J Clin Endocrinol Metab* 57:124–28.

5 Risk Stratification in Differentiated Thyroid Cancer

Giuseppe Barbesino

CONTENTS

ABSTRACT: The incidence of differentiated thyroid cancer is rapidly increasing in the Western world. While the majority of patients with this disease will experience normal life span and quality, a few will die from more severe forms of the disease. Traditional demographic and pathological variables provide reasonable risk stratification after primary treatment. However, due to the imperfection of these tools, many patients still receive unnecessary aggressive adjuvant therapy, while the early identification of patients at risk for relapses and death remains unsatisfactory. Recent developments in the understanding of mutations underlying tumoral behavior offer new promise in the development of risk stratification tools. In addition, the availability of sensitive tumor markers for disease persistence or relapse and of newer imaging techniques provides new methods for risk stratification during patient follow-up.

5.1 INTRODUCTION

Thyroid cancer of follicular epithelium derivation is the most common endocrine malignancy. Some 35,000 new cases are diagnosed every year in the United States, and this incidence is growing rapidly as a possible result of either increased detection or yet to be unveiled environmental factors.[1]

TABLE 5.1
Summary of Mortality Risk Assessment in Differentiated Thyroid Cancer

Risk Group	Baseline Assessment	Longitudinal Assessment
Low risk	Patients < 45 • Tumor < 4 cm Patients > 45 • Tumor < 1 cm	Classical PTC or FTC with undetectable sensitive TG after thyroidectomy and I-131 treatment Classical PTC or FTC with stable low sensitive TG after thyroidectomy alone
Intermediate risk	Patients < 45 • Tumor > 4 cm • Tumor 1–4 cm with either • Worrisome histology • Minimal extrathyroidal extension • Vascular invasion Patients > 45 • Tumor 1–4 cm • Tumor < 1 cm with lymph node metastases	Worrisome histology with undetectable sensitive TG after thyroidectomy and I-131 treatment Low and stable sensitive TG after thyroidectomy and I-131 treatment Patients > 45 with lymph node metastases after primary treatment
High risk	Patients < 45 • Tumor > 4 cm with either • Worrisome histology • Extrathyroidal extension • Vascular invasion • Incomplete resection • Distant metastases Patients > 45 • Tumor > 4 cm • Incomplete resection • Distant metastases	Distant metastases Rising sensitive TG levels
Very high risk		Non-iodine-avid, FDG-avid distant metastases Symptomatic unresectable local recurrence

Note: While postsurgery mortality risk is fairly well assessed with traditional variables as listed, longitudinal risk factors are less well defined, and therefore the proposed algorithm should be considered a guideline awaiting confirmation.

The combination of a low aggressiveness of most thyroid cancers with highly effective treatment modalities yields very high overall survival figures, well above 90% at 20 years.[2] In spite of a generally favorable outlook, a few patients experience a much more aggressive behavior of thyroid cancer, with mortality rates up to 50% at five years in some subgroups.[2] The number of thyroid cancer survivors is large, close to 400,000 in the United States. These patients are subjected to some type of lifelong surveillance, as very late relapses are one of the peculiar features of this indolent cancer. It is also notable that while relapses after initial treatment are fairly common, the mortality from relapses remains quite low. Therefore the mortality risk must be distinguished from relapse risk (Table 5.1).

In this scenario, endocrinologists need to tailor highly expensive diagnostic and therapeutic procedures according to specific risk profiles of patients, sparing ever-scarcer resources and securing at the same time the best possible outcome for all patients. Like with almost all human cancers, this task starts with staging patients based on demographic and pathological variables. Available staging systems based on these variables provide relatively accurate risk prediction and offer guidance in the selection of initial treatment (Table 5.2). However, given the long survival of patients with thyroid cancer, caregivers face the challenge of integrating the initial staging process with longitudinal data based on sophisticated tools, such as sensitive thyroglobulin (TG) assays, novel methods

TABLE 5.2
Comparison of the Four Most Commonly Used Staging Systems in the United States

System	Variables	Advantages	Disadvantages
MACIS	Age, tumor size, completeness of resection, distant metastases	Most accurate in predicting survival; simple, objective variables	Does not include lymph node status; may not be useful in predicting loco-regional relapses
AJCC/UICC	Age, tumor size, lymph node metastases, extrathyroidal extension, distant metastases	Incorporates the TNM system; analytical use of variables; adopted by the ATA	Age is used dichotomously; stage III is very heterogeneous
NTCTCS	Age, tumor size, multifocality, degree of differentiation, lymph node status, distant metastases	Distinguishes FTC and PTC	Age is used dichotomously; stage III is very heterogeneous
GAMES	Age, tumor size, distant metastases	Simple	Lymph node status not used

from radioiodine (RAI) scanning, positron emission tomography (PET) scanning, ultrasound (US), and computerized tomography (CT) imaging. This provides an ongoing risk assessment to optimize outcomes, save resources, and reduce unnecessary treatments. More specifically, risk stratification, both preoperatively and perhaps during follow-up, can be engineered from evidence pertaining to individual risk factors and their respective predictive power.

In this chapter, tools available to accomplish this task will be reviewed. It is worth noting how most of the data available in this field derive from retrospective data and are therefore possibly affected by several types of bias. However some of these data sets are quite large and long in terms of follow-up and therefore relatively reliable.

5.2 SPECIFIC BASELINE CLINICOPATHOLOGICAL RISK FACTORS FOR THYROID CANCER

5.2.1 Demographic Factors

The Surveillance Epidemiology and End Result (SEER) registry collects ongoing epidemiological data on a large and statistically representative group of Americans. Published analyses from the registry have provided important epidemiological information on several cancer types, including thyroid.[3]

5.2.2 Gender

In the most recently published SEER data, the incidence of thyroid cancer is 2.6 times higher in women than in men: 9.2 versus 3.6/100,000. However, absolute mortality rates were similar: 0.5/100,000. One can deduce that either there is underdetection of nonlethal thyroid cancer in men, or that men experience relatively higher mortality when diagnosed with thyroid cancer. However, no multivariate analysis has been published from the SEER registry, so it is not clear whether the male gender is an independent risk factor. Among the several predictive staging systems published, some but not all include gender as a risk modifier. In particular the Metastasis, Age at presentation, Completeness of surgical resection, Invasion, Size (MACIS) system, considered to be the most accurate predictive model in several publications, does not use gender as a factor.[4] In contrast, the European Organization for Research on Treatment of Cancer (EORTC) system attributes a relatively high score to male gender: 12 points, over a total maximum score of 77 + age.[5] A recent retrospective analysis was not able to demonstrate an independent effect of gender on mortality and recurrence rate.[6]

5.2.3 Age

In the SEER data, 5-year survival is 99.3% in patients aged 45 or less, followed by a progressive decline to 77.8% in patients aged 75 or older. Adjusted mortality rates were 4.8/100,000 in patients 85 or older versus 0.1/100,000 in patients younger than 45. It therefore appears that mortality is influenced by age.[1] Indeed, all most commonly used staging systems include age as a modifier. The effect of age is complex, in that it seems to be evident mostly in large or otherwise moderately aggressive tumors. Several recent studies looking at the outcome of small (<2 cm) intrathyroidal papillary thyroid cancer (PTC) did not show a significant independent effect of age on outcome.[7] Therefore, selecting a more aggressive treatment of small thyroid cancers in the older patient does not seem to be warranted. The effect of age on the risk of relapse is more complex. Indeed, children and younger adults have a higher risk of relapse than middle-aged patients, without this being associated with higher mortality, while in older adults, a higher relapse rate is directly associated with a higher mortality.

5.2.4 Ethnicity

Most available series have failed to suggest that ethnicity plays an important role in the incidence and outcome of thyroid cancer.

5.2.5 Iodine Deficiency

A large fraction of the world is exposed to mild to moderate iodine deficiency. This is clearly associated with a large increase in the incidence of benign thyroid nodular disease. However, benign thyroid nodular disease does not seem to either predispose to or protect from thyroid cancer. The incidence of thyroid cancer has been reported to be either unchanged or increased in areas of iodine deficiency, but negative studies predominate. Some studies have reported a higher incidence of thyroid cancer in areas of high iodine intake, but this effect could be linked to other geological factors, such as the radiation background. A number of studies have reported that the ratio of PTC to follicular thyroid cancer (FTC) increased after the introduction of iodine supplementation.[8] Since FTC is generally more aggressive than PTC, a worse overall outcome should be expected in populations with low iodine intake. However, it should be observed that the criteria for the histological classifications of differentiated thyroid cancer have changed over time, and this could result in a perceived lower incidence of FTC in favor of the follicular variant of PTC (FVPTC). In fact, there is no sufficient evidence to suggest an influence of iodine deficiency on thyroid cancer outcomes. One point of interest that has not been studied is the theoretical possibility that a chronically high dietary content of iodine is associated with a reduced efficacy of RAI scanning and treatment for metastatic disease.

5.2.6 Tumor Size

Tumor size is probably the most reproducible outcome predictor in thyroid cancer in most studies. This is probably due to the simplicity of the variable. Several studies clearly show 20-year survival close to 100% with tumors smaller than 1 cm in size, compared to 70–80% with tumors of >4 cm. Tumor size seems to be independent of other tumor features, with the exception of age: elderly patients are more likely to have larger tumors at diagnosis, and therefore the effect of age on prognosis cannot be easily dissected from the effect of size.[4,5,9]

5.2.7 Lymph Node Metastasis

The controversy on whether the presence or absence of lymph node metastasis at the time of thyroidectomy is capable of affecting survival remains unresolved. Most retrospective data suggest no effect on survival, but a few studies have suggested an increased risk of death in older patients. It seems clear that the presence of lymph node involvement signals an increased risk of nodal recurrences during follow-up.[10,11]

5.2.8 Distant Metastases

The presence of distant metastases is clearly associated with significant mortality, roughly 60% at 10 years, in the Mayo Clinic data.[2] This effect is modulated by age, with a much more favorable outcome in younger patients, even with lung metastases.[4] The use of fluorodeoxyglucose (FDG)-PET scanning and RAI scanning can further identify a subgroup of patients with metastatic disease and shorter survival. Lung and brain metastases are associated with the highest mortality, while bone metastases (typical of FTC) have a more indolent course.

5.2.9 Histological Subtype

Differentiated thyroid cancer is broadly divided into two major subtypes: FTC and PTC. FTC appears to be more aggressive and associated with higher mortality rates. However, the distinction between the two subtypes is sometimes equivocal, as many PTCs may have extensive follicular features on histology. More stringent pathologic criteria have been proposed recently.[12] As a consequence, older studies that may have classified certain thyroid cancers as FTC would nowadays classify some of these tumors as PTC.

5.2.9.1 Prognostic Histological Factors in PTC

Several histological subtypes of PTC are recognized, but FVPTC is the most common "nonclassical" variant. While its subsumption under the PTC groups is controversial, most authorities agree that this indolent form of cancer is related to PTC.[13] The outcome in patients with FVPTC does not appear to be different than that in patients with the classical form. However, in some studies, the behavior of FVPTC has been found to be closer to minimally invasive FTC, with a very low incidence of lymphatic spread and with occasional hematogenous spread. Some authors have therefore proposed a subclassification of FVPTC into nonencapsulated (infiltrative/diffuse) forms that are similar to PTC and encapsulated forms that are similar to FTC.[12] While this controversy should be potentially addressed by future molecular studies, the bulk of available data suggest that FVPTC carries a prognosis similar to that of PTC.

Other, less common variants of PTC include the diffuse sclerosing, tall-cell, and columnar variants. Because of the relative rarity of these variants, available data derive from small series or single case reports. The diffuse sclerosing variant occurs most often in children or young adults. This tumor is almost always found in the setting of Hashimoto's thyroiditis, and extensive lymph node metastases are the rule, while lung metastases are found in almost 50% of cases at the time of the diagnosis. In spite of this worrisome presentation, the outcome is usually favorable, with good response to RAI treatment, probably because of the young age of most patients.[14] In contrast, tall-cell and columnar variants are more often encountered in older patients. The tall-cell variant of PTC is probably the more aggressive of the two. It has a high rate of local recurrence, nodal metastases, and distant metastases. Part of this effect is related to the older age, the larger mass, and the more frequent extrathyroidal extension at presentation. However, the tall-cell histology seems to be an independent biological adverse prognostic factor, at least in increasing the risk of lymph node metastases.[15] The columnar cell variants seem to have intermediate aggressiveness. These relatively uncommon variants of PTC share a tendency to secrete little or no TG and are typically not iodine avid. Because of these characteristics, the recognition of these histological features should prompt more aggressive monitoring of these patients, with PET and CT scanning in addition to or in substitution of RAI scanning. Even in these moderately aggressive variants, size remains the most important prognostic factor, and it remains unclear whether patients with intrathyroidal forms of these variants have shorter survival than patients with classical PTC.[15]

5.2.9.2 Prognostic Histological Factors in FTC

Within the subgroup of FTC, various pathological features, together with tumor size, appear to play a strong role in determining the risk of relapse and metastasis. The most important prognostic feature is the presence of extensive vascular invasion by the tumor, which carries a significant risk of death and hematogenous spread.[16] A high number of mitoses and a solid pattern are also associated with a poor outcome, but these features are also strongly associated with widespread vascular invasion and are therefore not truly independent factors. Capsular invasion was much less, or not at all, important in determining prognosis. Because of the importance of vascular invasion, FTC is now divided into minimally invasive and widely invasive based on the extent of this feature. While not all authorities agree on the criteria distinguishing the two groups, it appears that small, minimally invasive (as defined by the presence of transcapsular invasion and minimal, if any, vascular invasion) has an excellent prognosis, with survival rates of 95% at 20 years.[4] In one study, the presence of four or more foci of vascular invasion dramatically decreased the five-year relapse-free survival from 100% to 20%.[17] In spite of the generally reported good outcome of patients with only minimal transcapsular invasion, in at least one study there were five deaths (11%) in a group of patients with this feature only.[16] Since these prognostic factors are specific to FTC, epidemiologists have sought to understand whether commonly used scoring systems provide prognostic information for this disease, in spite of the fact that they have been derived from large populations including only a minority of patients with FTC.[18] In spite of this limitation, many of the most widely used scoring systems provide acceptable risk stratification when applied to FTC patient groups, independently of whether histological variables are considered.

5.3 STAGING SYSTEMS BASED ON CLINICOPATHOLOGICAL RISK FACTORS

Over the years, several staging systems have been proposed for thyroid cancers. All of these systems rely on retrospective analyses of some combination of demographic and pathological data analyzed above. The main limitation of such an approach is that patients included in these analyses have then received a wide variety of treatment and monitoring procedures, capable of influencing outcome independently. In spite of these limitations, several of these systems have been validated in other series and proved extremely useful in rapidly assigning newly diagnosed patients to a risk tier. Summarizing all of the available systems here would be lengthy and unnecessary. Some systems are really institution dependent in that they rely on features not easily reproducible.

It is worth mentioning the two most widely used systems in North America. The MACIS system, derived from the Mayo Clinic series, is credited with being the most accurate in predicting disease-specific mortality. The MACIS system attributes a score to each patient based on age, the presence of extrathyroidal invasion, the completeness of resection, the presence of distant metastases, and the size of the tumor. The MACIS score is simple and rapid, employing readily available and objective demographic and pathologic variables. Remarkably, the presence or absence of metastatic lymph nodes is not included in the score. This may explain why the MACIS score has been found to be accurate in predicting mortality but not relapses.

The American Thyroid Association (ATA) has relied on the American Joint Commission on Cancer/Union Internationale Contre le Cancer–Tumor Node Metastasis (AJCC/UICC–TNM) staging system when issuing societal guidelines for thyroid cancer treatments.[19] This score integrates the TNM classification for head and neck cancer with age in assigning patients to one of four stages, represented in Table 5.1. While this score is close to that of MACIS in terms of predictive accuracy, it is more complicated and treats age as a dichotomous variable (less than 45 years vs. 45 years and older). As a consequence, a patient who is 44 and has 1.5 cm papillary thyroid cancer with a paratracheal lymph node metastasis will be classified as stage I, while she would fall in stage III if she were 46. It is also unclear whether this staging system is adequate for FTC staging, while the MACIS system has been validated for FTC.[18] Both systems find their strength in using objective and easy-to-obtain variables, with increased reproducibility.

While the use of any staging system in clinical practice is highly encouraged, it should be recognized that these systems are imperfect. In a recent validation study, the MACIS explained less than 20% of the total outcome variability of the PTC population being analyzed, even though it ranked higher than all of the other staging systems.[20] In other words, no staging system at this time can identify a population with zero risk of relapse or death, or a population in which aggressive treatment will be guaranteed to be beneficial. Moreover, one could conjecture that there must be important prognostic information related to thyroid cancer that is simply not being incorporated into these staging systems.

5.4 MOLECULAR RISK FACTORS

Over the past few years, a large body of evidence has accumulated, partially clarifying the genetic changes underlying the tumorigenesis of differentiated thyroid cancer. The recognition that different molecular changes may be present in thyroid cancers (e.g., classical PTC) with similar pathological phenotypes has stimulated a large body of studies dedicated to answering the question whether these changes may be responsible for a quota of the variability in clinical behavior, beyond the available clinicopathological variable. While a detailed review of these changes is beyond the purpose of this chapter, it is useful to summarize the most important findings.

In approximately 70% of PTC cases, mutations of factors affecting the mitogen-activated protein kinase (MAPK) pathway are found. These include the *RET/PTC* gene rearrangements and mutations of the *BRAF* gene. Rearrangements of the *RET/PTC* genes are found in 13–45% of PTC and seem to predominate in radiation-induced PTC. The second most important mutations involve the *BRAF* oncogene and are found in roughly 30–60% of PTC. Some specific *BRAF* mutations seem to confer the FVPTC phenotype, further justifying the classification of this tumor under the PTC group.[21] There is accumulating but still incomplete evidence that the presence of one or the other genotype may have important consequences on the tumor phenotype in PTC. In transgenic mice, the *RET/PTC* genotype alone is not capable of causing metastasis. *RET/PTC* translocations are found commonly in papillary thyroid microcarcinoma and rarely in aggressive poorly differentiated PTC. In contrast, *BRAF* mutations are observed in PTC with unfavorable histology with high frequency, and several studies have shown a correlation between the presence of the *BRAF* mutation and higher tumor stages, extrathyroidal invasion, lymph node metastases, and even mortality.[22] In summary, genotyping of PTC is emerging as a promising candidate molecular risk stratification tool. However, the data at the time are still largely incomplete, owing to the lack of large longitudinal studies.

While several tertiary referral centers are starting to offer routine *BRAF* and *RET/PTC* genotyping on surgical specimens, a number of questions remain unanswered. The benignity of PTC carrying *RET/PTC* translocations needs to be confirmed, as firm evidence confirming the presence of this translocation would identify a very low-risk population with little need for aggressive treatment and monitoring. While the adverse effect of *BRAF* mutations seems to be a qualitatively consistent phenomenon across several studies, its magnitude also needs to be fully understood, before it can be adopted in clinical decision making. Many initial studies were enriched with highly aggressive tumors, therefore carrying significant selection bias. Considering that *BRAF* mutations are quite prevalent in PTC, a tumor with a low overall mortality, one must conclude that the majority of patients carrying such mutations still have an excellent outcome. This suggests that other epigenetic phenomena or secondary mutations may be necessary for the full expression of the malignant potential of *BRAF* positive tumors, and that these events occur relatively rarely. Mutations in the *P53* and *CTNNB1* genes are potential candidates for this role under active investigation.[21] Finally, from the clinician's perspective, it would be very important to test the integration of genotypic information with predictive tools that are already available. In other words, the question whether tumor genotyping adds predictive value beyond and independently of the currently used staging systems needs to be answered. The suggestion that *BRAF* positive tumors may have a more aggressive propensity, even when they are small, awaits confirmation.[23]

In FTC, a more complex picture has emerged so far. Parallel to the difficulty in distinguishing benign from malignant follicular neoplasms, the genetic changes seen in FTC are not univocal. Mutations in the *RAS* oncogenes are seen in 40–53% of FTC, but also in a variable number of lesions histologically classified as benign follicular adenomas. The presence of aneuploidy and of chromosomal loss or gain seems to correlate with RAS mutations and a poor outcome, but is similarly found in some adenomas. The second most common change is a rearrangement event between the thyroid-specific transcription factor *PAX-8* and the *PPAR-γ* genes, found in 25–60% FTC. Contrary to earlier reports, this rearrangement is also found in follicular adenomas. There is little information on the prognostic value of *PPAR-γ-PAX-8* rearrangements, although some authors have suggested a more aggressive phenotype and younger age of patients. However while RAS mutations have been observed in poorly differentiated tumors, the *PPAR-γ-PAX-8* rearrangement has only been detected rarely in such tumors.[21]

5.5 LONGITUDINAL RISK ASSESSMENT

While baseline epidemiological, histological, and molecular risk assessments provide satisfactory guidance in selecting the most appropriate initial treatment for patients with thyroid cancer, tools for ongoing risk assessment during follow-up are more difficult to analyze. In recent years the field has been significantly changed by the advent of ultra-sensitive TG assays, high-resolution neck US, and to a lesser extent, FDG-PET scanning. Older published longitudinal series relied on RAI scanning and TG levels measurable down to 2 or 3 ng/ml. Greater sensitivity of the TG assay was achieved using thyroid-stimulating hormone (TSH) stimulation, either by withdrawing L-thyroxine therapy or, more recently, through the administration of recombinant human TSH (rhTSH).[24] Most of the longitudinal available data are based on first-generation TG assays and relatively insensitive US techniques. Current TG assays allow detection limits of 0.1–0.2 ng/ml,[25] while lymph nodes as small as a few millimeters can now be detected with high-frequency neck US and classified as benign, indeterminate, or clearly pathologic.[26] In this setting, early patient restaging after primary treatment (thyroidectomy with or without RAI treatment) readily assigns patients to either one of three groups.

The first group includes patients who achieve undetectable TG levels as assayed by novel sensitive assays and with no evidence of residual disease on high-resolution neck US. All patients in this group will have undergone total thyroidectomy, and most have received remnant ablation with RAI treatment. The significance of this result of primary treatment is dependent on the preexisting risk status of the patient. It is very likely that in a low-risk PTC patient, such findings will indicate very low or nil risk of disease recurrence and death. These patients will undergo some sort of follow-up monitoring owing to the anecdotic reports of very late (up to 40 years) relapses and deaths in patients many years after surgery who were thought to be "cured." This is the largest group of thyroid cancer survivors, the group therefore using the largest share of healthcare resources dedicated to this disease. Clinicians and epidemiologists need an estimate of the risk for recurrence and death in patients with TG levels that are undetectable using assays more sensitive than those used in large studies of the 1990s. Preliminary data on this group of patients have been recently presented in the literature.[25] In the population presented in this study, only 2 of 80 patients with TG levels of <0.1 ng/ml had TG levels of >2 ng/ml following rhTSH stimulation. Moreover, neither of these 2 patients had otherwise detectable disease. This is in striking contrast with figures from the late 1990s, in which up to 30% of patients with residual or metastatic disease had unstimulated TG levels of <1 ng/ml, rising to >2ng/ml in response to TSH stimulation in almost all cases.[27] This suggests that periodic expensive and time-consuming TSH stimulation may no longer be necessary in the era of sensitive TG assays, when monitoring low-risk patients. If these results will be reproduced in larger studies,[28] we will witness a paradigm shift in the ongoing risk assessment of most patients with thyroid cancer. The long-term risk of death and mortality in patients with undetectable sensitive TG is unknown, but presumably very low. We should also recognize that TG measurements do not detect all cases of persistent or relapsing disease.[29] A recent report employing a high-sensitivity assay described 52 cases with evidence of disease and stimulated TG of <2 ng/ml. Unfortunately, while the assay employed had the potential for detecting TG levels of 0.2 ng/ml and higher, the number of patients with truly undetectable TG in this group was not reported.[30] The phenomenon of TG negative disease is more prevalent in tumors with aggressive histology, once again underlying the importance of tumor grading in risk stratification. In any case, these reports clearly suggest that even in these low-risk cases, some form of periodic imaging may be necessary. Therefore, this group will require yearly US and TG measurement initially. Whether surveillance can be interrupted altogether after some years of persistently negative findings remains to be understood. In recent analysis of the SEER data, thyroid cancer and breast cancer were the only two neoplasias for which a minimum follow-up could not be derived to declare a "cure."[31] In many of the late relapsing cases in the literature, there is very little information on the longitudinal TG status of these patients; therefore it is difficult to use these cases for an estimate of long-term risk in fully treated, persistently TG negative patients.

The second group includes patients in whom primary treatment does not achieve undetectable TG levels. This group is heterogeneous, as it will include not only low-risk patients with small tumors who did not receive total thyroidectomy or RAI remnant ablation, but also patients who, after receiving total thyroidectomy and RAI, had persistently detectable TG levels or nonspecific neck US findings. While this group is only second to the first group in terms of size, it can be expected to increase with increased TG assay sensitivity. In these patients, longitudinal risk assessment is crucial in detecting the few with true persistent or progressive disease. In general, the presence of low and stable TG levels during the first two years after primary treatment defines patients at low risk of progression who will need constant monitoring but no aggressive workup.[32] In many of these patients, TG levels will drop over time, possibly as a consequence of TSH suppression, further supporting meticulous vigilance alone. It also seems reasonable to include in this group formerly high-risk patients (large tumors or adverse histology on presentation) who then become TG negative after primary treatment. Whether at some point in time this group of patients can reenter in the very low-risk group described above is unknown. In a recent comprehensive review on this topic, risk reassignment time points have been empirically proposed at 2, 5, and 10 years.[33] In contrast, a rising TG level clearly signals progression of disease in these patients and will prompt more extensive investigations, including CT, PET-FDG, and repeat RAI scanning.[34]

The third group includes patients with persistent or rapidly relapsing macroscopic disease immediately after primary treatment. These are typically patients presenting with high-risk stages at the time of diagnosis, whose risk cannot be rapidly downstaged after surgery and RAI treatment, due to persistence of detectable disease or high TG levels. Defining precise prognostic factors and homogeneous protocols in this group is quite difficult because of the multiple variables involved. These cases are also relatively rare, so that controlled trials or large retrospective series are scanty. The concurrent use of RAI and PET-FDG scanning has emerged as a powerful tool in separating patients into two groups. The first is composed by aggressive tumors with a moderate degree of dedifferentiation, likely to be incurable but expected to follow a relatively indolent course. The hallmark of this group of tumors is persistent iodine-trapping activity on RAI scanning. While RAI may not be curative in these generally bulky metastatic tumors, it may limit progression to some extent. The second group includes rapidly progressing, poorly differentiated tumors with no RAI uptake. The hallmark of this group is prominent FDG uptake on PET scanning, as a sign of intense proliferative and metabolic activity. In this setting, the presence of FDG uptake is a powerful independent predictive factor. In the Memorial Sloan-Kettering series of patients with metastatic disease, the group with positive FDG scans experienced a median survival of 53 months, while almost no deaths were seen in the FDG negative group.[35] FDG uptake is also predictive of RAI treatment failure[36] and *BRAF* positivity.[37] In summary, FDG-PET scanning identifies patients with higher mortality. Patients identified with this technique may, in the future, be considered for early systemic therapy with the novel targeted therapies being studied in current clinical trials.

5.6 CONCLUSION

While thyroid cancer mortality has not changed much over the past 30 years, our understanding of the biology and natural history of this tumor has greatly improved. Novel selective treatments are being developed and hold great promise in the fight against the most severe forms of the disease. While established demographic and pathological data allow risk stratification to an acceptable extent, too many patients with very low chance of recurrent or persistent disease are still overmanaged: they undergo potentially harmful and expensive treatments as well as too frequent monitoring procedures. The recent recognition of genetic indicators of aggressiveness, the development of very sensitive TG assays, and the use of FDG-PET scanning for monitoring may allow, in the near future, an integration of available staging procedures. This development promises the fulfillment of two goals in the thyroid cancer clinic: limiting the use of RAI treatment and expensive follow-up procedures to patients likely to benefit from them, and identifying early patients with aggressive, potentially lethal disease, likely to benefit from novel targeted therapies.

REFERENCES

1. Enewold L, Zhu K, Ron E, et al. 2009. Rising thyroid cancer incidence in the United States by demographic and tumor characteristics, 1980–2005. *Cancer Epidemiol Biomarkers Prev* 18:784–91.
2. Hay ID, Thompson GB, Grant CS, et al. 2002. Papillary thyroid carcinoma managed at the Mayo Clinic during six decades (1940–1999): Temporal trends in initial therapy and long-term outcome in 2444 consecutively treated patients. *World J Surg* 26:879–85.
3. Davies L, Welch HG. 2006. Increasing incidence of thyroid cancer in the United States, 1973–2002. *JAMA* 295:2164–67.
4. Brennan MD, Bergstralh EJ, van Heerden JA, McConahey WM. 1991. Follicular thyroid cancer treated at the Mayo Clinic, 1946 through 1970: Initial manifestations, pathologic findings, therapy, and outcome. *Mayo Clin Proc* 66:11–22.
5. Byar DP, Green SB, Dor P, et al. 1979. A prognostic index for thyroid carcinoma. A study of the E.O.R.T.C. Thyroid Cancer Cooperative Group. *Eur J Cancer* 15:1033–41.
6. Toniato A, Boschin I, Casara D, Mazzarotto R, Rubello D, Pelizzo M. 2008. Papillary thyroid carcinoma: Factors influencing recurrence and survival. *Ann Surg Oncol* 15:1518–22.
7. Perrino M, Vannucchi G, Vicentini L, et al. 2009. Outcome predictors and impact of central node dissection and radiometabolic treatments in papillary thyroid cancers < or =2 cm. *Endocr Relat Cancer* 16:201–10.
8. Feldt-Rasmussen U. 2001. Iodine and cancer. *Thyroid* 11:483–86.
9. Tsuchiya A, Suzuki S, Kanno M, Kikuchi Y, Ando Y, Abe R. 1995. Prognostic factors associated with differentiated thyroid cancer. *Surg Today* 25:778–82.
10. Jonklaas J, Sarlis NJ, Litofsky D, et al. 2006. Outcomes of patients with differentiated thyroid carcinoma following initial therapy. *Thyroid* 16:1229–42.
11. Grebe SK, Hay ID. 1996. Thyroid cancer nodal metastases: Biologic significance and therapeutic considerations. *Surg Oncol Clin North Am* 5:43–63.
12. Ghossein R. 2009. Problems and controversies in the histopathology of thyroid carcinomas of follicular cell origin. *Arch Pathol Lab Med* 133:683–91.
13. Albores-Saavedra J, Wu J. 2006. The many faces and mimics of papillary thyroid carcinoma. *Endocr Pathol* 17:1–18.
14. Fukushima M, Ito Y, Hirokawa M, Akasu H, Shimizu K, Miyauchi A. 2009. Clinicopathologic characteristics and prognosis of diffuse sclerosing variant of papillary thyroid carcinoma in Japan: An 18-year experience at a single institution. *World J Surg* 33:958–62.
15. Ghossein RA, Leboeuf R, Patel KN, et al. 2007. Tall cell variant of papillary thyroid carcinoma without extrathyroid extension: Biologic behavior and clinical implications. *Thyroid* 17:655–61.

16. D'Avanzo A, Treseler P, Ituarte PH, et al. 2004. Follicular thyroid carcinoma: Histology and prognosis. *Cancer* 100:1123–29.
17. Ghossein RA, Hiltzik DH, Carlson DL, et al. 2006. Prognostic factors of recurrence in encapsulated Hürthle cell carcinoma of the thyroid gland: A clinicopathologic study of 50 cases. *Cancer* 106:1669–76.
18. Lang BH, Lo CY, Chan WF, Lam KY, Wan KY. 2007. Staging systems for follicular thyroid carcinoma: Application to 171 consecutive patients treated in a tertiary referral centre. *Endocr Relat Cancer* 14:29–42.
19. Cooper DS, Doherty GM, Haugen BR, et al. 2006. Management guidelines for patients with thyroid nodules and differentiated thyroid cancer. *Thyroid* 16:109–42.
20. Lang BH, Lo CY, Chan WF, Lam KY, Wan KY. 2007. Staging systems for papillary thyroid carcinoma: A review and comparison. *Ann Surg* 245:366–78.
21. Kondo T, Ezzat S, Asa SL. 2006. Pathogenetic mechanisms in thyroid follicular-cell neoplasia. *Nat Rev Cancer* 6:292–306.
22. Xing M. 2007. BRAF mutation in papillary thyroid cancer: Pathogenic role, molecular bases, and clinical implications. *Endocr Rev* 28:742–62.
23. Xing M. 2009. BRAF mutation in papillary thyroid microcarcinoma: The promise of better risk management. *Ann Surg Oncol* 16:801–3.
24. Pacini F, Molinaro E, Castagna MG, et al. 2003. Recombinant human thyrotropin-stimulated serum thyroglobulin combined with neck ultrasonography has the highest sensitivity in monitoring differentiated thyroid carcinoma. *J Clin Endocrinol Metab* 88:3668–73.
25. Smallridge RC, Meek SE, Morgan MA, et al. 2007. Monitoring thyroglobulin in a sensitive immunoassay has comparable sensitivity to recombinant human tsh-stimulated thyroglobulin in follow-up of thyroid cancer patients. *J Clin Endocrinol Metab* 92:82–87.
26. Fish SA, Langer JE, Mandel SJ. 2008. Sonographic imaging of thyroid nodules and cervical lymph nodes. *Endocrinol Metab Clin North Am* 37:401–17, ix.
27. Mazzaferri EL, Robbins RJ, Spencer CA, et al. 2003. A consensus report of the role of serum thyroglobulin as a monitoring method for low-risk patients with papillary thyroid carcinoma. *J Clin Endocrinol Metab* 88:1433–41.
28. Giovanella L, Maffioli M, Ceriani L, De Palma D, Spriano G. 2009. Unstimulated high sensitive thyroglobulin measurement predicts outcome of differentiated thyroid carcinoma. *Clin Chem Lab Med* 47:1001–4.
29. Giovanella L. 2009. False-negative thyroglobulin measurement in recurrent/metastatic thyroid carcinomas. *Eur J Nucl Med Mol Imaging* 36:326–27; author reply, 328.
30. Park EK, Chung JK, Lim IH, et al. 2009. Recurrent/metastatic thyroid carcinomas false negative for serum thyroglobulin but positive by posttherapy I-131 whole body scans. *Eur J Nucl Med Mol Imaging* 36:172–79.
31. Tai P, Yu E, Cserni G, et al. 2005. Minimum follow-up time required for the estimation of statistical cure of cancer patients: Verification using data from 42 cancer sites in the SEER database. *BMC Cancer* 5:48.
32. Huang SH, Wang PW, Huang YE, et al. 2006. Sequential follow-up of serum thyroglobulin and whole body scan in thyroid cancer patients without initial metastasis. *Thyroid* 16:1273–78.
33. Tuttle RM, Leboeuf R, Martorella AJ. 2007. Papillary thyroid cancer: Monitoring and therapy. *Endocrinol Metab Clin North Am* 36:753–78, vii.
34. Pacini F, Agate L, Elisei R, et al. 2001. Outcome of differentiated thyroid cancer with detectable serum Tg and negative diagnostic (131)I whole body scan: Comparison of patients treated with high (131)I activities versus untreated patients. *J Clin Endocrinol Metab* 86:4092–97.
35. Robbins RJ, Wan Q, Grewal RK, et al. 2006. Real-time prognosis for metastatic thyroid carcinoma based on 2-[18F]fluoro-2-deoxy-D-glucose-positron emission tomography scanning. *J Clin Endocrinol Metab* 91:498–505.
36. Wang W, Larson SM, Tuttle RM, et al. 2001. Resistance of [18f]-fluorodeoxyglucose-avid metastatic thyroid cancer lesions to treatment with high-dose radioactive iodine. *Thyroid* 11:1169–75.
37. Ricarte-Filho JC, Ryder M, Chitale DA, et al. 2009. Mutational profile of advanced primary and metastatic radioactive iodine-refractory thyroid cancers reveals distinct pathogenetic roles for BRAF, PIK3CA, and AKT1. *Cancer Res* 69:4885–93.

6 Clinical Practice Guidelines for Thyroid Cancer
Where Are We Now?

Jeffrey I. Mechanick

CONTENTS

ABSTRACT: Clinical practice guidelines (CPG) are systematically developed statements that assist physicians in clinical decision making. The development of CPG has evolved over the last several decades and incorporates evidence-based medicine (EBM). Since 2003, at least seven CPG have been developed that focus on aspects of thyroid cancer evaluation and management. There are many commonalities among these CPG: clinical risk stratification, nodule risk stratification, treatment algorithms for intermediate-risk scenarios, and follow-up algorithms commensurate with risk stratification. However, there are also divergences and "fuzzy" areas in the CPG. Although these might be accounted for by differences in, or absence of, EBM methodology, they most likely reflect knowledge gaps. These fuzzy areas are principally in the low-risk patient, where "watchful waiting" is an option, and the high-risk patient recalcitrant to conventional therapy, where novel diagnostic tools and therapies are needed. Emergent biotechnologies may clarify these fuzzy areas and generate a new brand of thyroid cancer CPG. Among these emergent biotechnologies, molecular medicine holds the greatest promise.

6.1 INTRODUCTION

Current standard of care practices for the evaluation and management of thyroid cancer are derived from many sources. These include word of mouth or informal communications among physicians, formalized teaching in training programs, and self-study from print, computerized, and conference materials. There are important differences between informal and formal sources of information and the way in which they are presented to the physician. Clinical practice guidelines (CPG) are becoming a valuable formal resource for physicians, and several have been published on thyroid cancer. However, there are problems with the CPG paradigm. First, when multiple societies or writing groups produce competing CPG, the reader may be confused by discrepancies, thus diluting their impact. Second, specific CPG on the same topic may be produced using different methodologies, which can confound the recommendations. An extension of this problem is that consistent implementation of a rigid set of recommendations may not be possible across the global spectrum of medical practice. In other words, all recommendations regarding the use of medical technology may not be possible in areas of the world where, for example, ultrasonography, expert cytology, positron emission tomography (PET) imaging, and biological therapies are unavailable. Third, CPG are not perfect and there are many philosophical problems with CPG development and implementation. Besides, these instruments for education and clinical practice are not sufficiently durable to keep pace with the rapid evolution of medical knowledge. With respect to thyroid cancer, the current CPG fail to resolve many of the abounding controversies. In this chapter, the current state of thyroid cancer CPG will be reviewed to set the stage for modification in later chapters. This will occur as new information is assimilated and new CPG algorithms emerge. These forward-looking CPG will ultimately be presented and discussed at the end of this book.

6.2 THE PHILOSOPHY AND SYNTHESIS OF CPG

Clinical practice guidelines have been directly influenced by evidence-based medicine (EBM). Indeed, clinical medicine has undergone an accelerated succession of paradigm shifts, with the emergence of EBM occurring in the early 1990s.[1] At that time, it was recognized that clinical decision making, based upon an inconsistently applied amalgam of experience, theory, intuition, and common sense, was patently insufficient to manage the sophisticated medical problems facing the average physician. Consequently, an emphasis on scientific evidence was promulgated and, over the last 20 years, has permeated almost all aspects of medical education and practice.

Clinical practice guidelines actually predated EBM by nearly a decade. The incorporation of EBM principles was relatively gradual. Simply put, CPG are "systematically developed statements to assist practitioner and patient decisions about appropriate health care for specific clinical circumstances."[2] Philosophically, physicians *must* act based on an incomplete set of information, and as a result, subjectivity, creativity, and even error are introduced.[3] Since the roots of clinical medicine are firmly planted in science, evidence-based CPG serve to bridge the gap between purely objective reasoning and impressionism. This is accomplished by creating a CPG development methodology that incorporates both rigid evidence-rating protocols and various physician subjective factors that result in a final recommendation grade.[4] Other components of a successful CPG are relevance, transparency, and a rigorous vetting process by credentialed practitioners.[4] On the other hand, there are shortcomings of CPG (Table 6.1), not the least of which is that CPG alone simply cannot consistently and sufficiently address complex human conditions. CPG are designed to guide and not

TABLE 6.1
Shortcomings of Clinical Practice Guidelines

EBM Methodology	Logistics	Other
Complicated, confusing	How to incorporate subjective factors	Difficult to implement
Inconsistent	Credentialing of writers and reviewers	Difficult to evaluate
Some are not intuitive	Avoiding industry influence	Geographic variations
Biased literature searching	Vetting writers' biases	
Controversial superiority of RCT		
Laborious and costly due diligence		
Semantic imprecision		

Note: RCT = randomized controlled trial.

dictate. They are not rigid rules, mandates, or laws, but a resource to assist physicians. With these concepts in mind, and notwithstanding the need for ongoing improvement in the way in which CPG are developed and implemented, this chapter will focus on the content and implications of relevant and contemporary thyroid cancer CPG.

6.3 CURRENT THYROID CANCER CPG

In this section, various thyroid cancer, as well as relevant thyroid nodule, CPG will be reviewed in chronological order of their publication date. They will be discussed in terms of their stated intent, methodology, principle findings (and recommendation grades or ratings), and limitations. The principle findings are selected based on their impact on a general clinical algorithm for the management of thyroid cancer in adults. Minor recommendations regarding details of treatment will not be listed here. In addition, redundancy is minimized, as descriptions of more recent CPG in this list will not include restatements of common recommendations, but just include updated recommendations and key divergences. A summary of these CPG is provided in Table 6.2.

6.3.1 American Institute of Ultrasound in Medicine (AIUM) Practice Guideline for the Performance of a Thyroid and Parathyroid Ultrasound (US) Examination[5]

In this CPG, the AIUM collaborated with the American College of Radiology to provide assistance to practitioners performing US of the thyroid and parathyroid glands. The methodology is not given, and although references are provided at the end of the text, the individual recommendations are not explicitly evidence based. Thus, this CPG takes the form of a consensus report and is based on expert opinion and interpretation of the medical literature. The principle findings in this CPG regarding thyroid US are:

- Indications for thyroid US: The evaluation and follow-up of (all) thyroid nodules, thyroid lesions detected by other imaging modalities, and suspected regional nodal metastases in patients with thyroid cancer
- Technical aspects of the thyroid US examination
- Equipment specifications

These CPG provide little in terms of algorithmic guidance for the evaluation of patients with thyroid nodules or thyroid cancer beyond the mandate to utilize US imaging early in the workup.

TABLE 6.2
Summary of Clinical Practice Guidelines Related to Thyroid Cancer

Year	Society	Topic	EBM Protocol
2003	AIUM	Thyroid and parathyroid US	No stated protocol
2006	AACE/AME	Thyroid nodules	Yes: AACE protocol[a]
2006	ATA	Thyroid nodules and DTC	Yes: USPSTF protocol[b]
2008	EANM	RAI therapy of DTC	No stated protocol
2008	NCI	FNA	No stated protocol
2009	NCCN	Thyroid carcinoma	Yes: NCCN protocol
2009	ATA	MTC	Yes: USPSTF protocol[c]
2009	ATA	Thyroid nodules and DTC—revision	Yes: USPSTF protocol[c]

Note: Abbreviations: EBM, evidence-based medicine; AIUM, American Institute of Ultrasound in Medicine; US, ultrasound; AACE/AME, American Association of Clinical Endocrinologists/ Associazione Medici Endocrinologi; ATA, American Thyroid Association; DTC, differentiated thyroid cancer; EANM, European Association of Nuclear Medicine; RAI, radioiodine; NCI, National Cancer Institute; FNA, fine-needle aspiration; NCCN, National Comprehensive Cancer Network; MTC, medullary thyroid carcinoma.

[a] See J. I. Mechanick et al., *Endocrine Pract.* 10, 353–61, 2004.

[b] See *U.S. Preventive Services Task Force ratings: Strength of recommendations and quality of evidence. Guide to clinical preventive services. Periodic updates, 2000–2003*, 3rd ed., U.S. Preventive Services, Rockville, MD.

[c] See D. S. Cooper et al. and the American Thyroid Association Guidelines Task Force, *Thyroid*, 19, 1167–214, 2009.

6.3.2 American Association of Clinical Endocrinologists (AACE) and Associazione Medici Endocrinologi (AME) Medical Guidelines for Clinical Practice for the Diagnosis and Management of Thyroid Nodules[6]

In this CPG, AACE, and AME collaborate to provide a clinically relevant comprehensive and question-oriented list of recommendations and clinical algorithm for the diagnosis and management of thyroid nodules. The treatment of thyroid cancer is not covered in this CPG. The methodology adheres to the 2004 AACE Protocol for Standardized Production of Clinical Practice Guidelines.[4] In this "guideline for guidelines" (G4G),[4] evidence is rated according to four intuitive levels of scientific substantiation (none, 0; weak, 1; intermediate, 2; and strong, 3) with types of experimental methodology explicitly assigned to their respective evidence levels (ELs). In addition, the best ELs (BELs) are then mapped to recommendations, which are graded according to four intuitive levels (expert opinion or not evidence based, D; weak evidence based, C; intermediate evidence based, B; and strong evidence based, A). Recommendation grades may deviate from the BEL mapping if subjective factors have significant impact. If this happens, then the subjective factors and rationale for grade deviation are explicitly described (transparency). Examples of subjective factors include cost-effectiveness, risk-benefit, physician preferences, relevance, and resource availability. Also, clinical actions may be linked with recommendation grades: A, based on strong evidence, should do, or first-line therapy; B, based on intermediate evidence, may do, or second-line therapy; C, based on weak evidence or no objection to doing; D, not evidence based or action not recommended.[7] The principle findings in this CPG are:

- Review risk factors for thyroid cancer, especially since an absence of symptoms does not rule out a malignant lesion (grade C).

- All high-risk patients or those with a palpable thyroid nodule should have a thyroid US (grade C); however, thyroid US should not be performed as a screening test (grade C).
- US-guided fine-needle aspiration (UG-FNA) should be performed for all thyroid nodules of ≥10 mm and suggested for nodules of <10 mm if the clinical information or US is suspicious (grade C).
- Thyroid UG-FNA cytology should be interpreted by an experienced cytologist (grade D); benign findings should be followed and malignant findings treated surgically (grade D).
- Thyroid-stimulating hormone (TSH) should be measured, followed by a free thyroxine if the TSH level is low or a thyroid peroxidase antibody if the TSH is high (grade C).
- Radioactive iodine (RAI) scintigraphy is useful if the TSH is low or suppressed (grade C).
- Percutaneous ethanol injection of the nodule is useful for cystic lesions (grade B).
- Routine serum calcitonin (CTN) measurements are not recommended (grade D).

These findings are summarized and incorporated into a clinical algorithm, which is based on the initial TSH level, US appearance, and UG-FNA cytological interpretation. In the case of thyroid nodules with indeterminate UG-FNA findings, surgery is recommended. This particular recommendation is based primarily on consensus among the primary writers and represents not only a limitation of this CPG, but of a critical knowledge gap for thyroid cancer. The 10 mm cutoff for mandatory UG-FNA versus optional UG-FNA based on risk stratification is another topic with relatively little supporting evidence. Conceivably, there may be high-risk thyroid nodules that elude accurate risk assessment with TSH, US, and UG-FNA and are subsequently managed nonsurgically.

6.3.3 American Thyroid Association (ATA) Management Guidelines for Patients with Thyroid Nodules and Differentiated Thyroid Cancer[8]

In this CPG, the ATA provides a clinically relevant, comprehensive, and question-oriented list of recommendations and clinical algorithm for the management of thyroid nodules and differentiated thyroid cancer. The methodology adheres with the U.S. Preventive Services Task Force protocol for recommendation ratings in CPG.[9] In this G4G protocol, recommendation ratings are given by:

- A—Strongly recommends, based on good evidence that the action can improve outcome.
- B—Recommends, based on fair, sufficient evidence that the action can improve outcome.
- C—Recommends, based on expert opinion.
- D—Recommends against, based on expert opinion.
- E—Recommends against, based on fair, sufficient evidence that the action does not improve outcome or that harm outweighs benefit.
- F—Strongly recommends against, based on good evidence that the action does not improve outcome or that harm outweighs benefit.
- I—Recommends neither for nor against, based on insufficient, poor, or conflicting evidence.

The principle findings of this CPG are:

- TSH should be performed in all patients with a thyroid nodule (rating C).
- Thyroid US should be performed in all patients with a thyroid nodule (rating B).
- There is no recommendation for or against obtaining a serum CTN in patients with a thyroid nodule (rating I).
- FNA is the procedure of choice in the evaluation of thyroid nodules (rating A).
- If the cytology is diagnostic of malignancy, then surgery is recommended (rating A).

- Specific molecular markers to improve the diagnostic accuracy of indeterminate nodules are not recommended (rating I).
- In patients with follicular lesions and indeterminate cytology findings, a RAI scintigraphy should be performed; if a concordant autonomously functioning nodule is not seen, then surgery should be considered (rating I).
- In patients with papillary or Hurthle cell lesions and indeterminate cytology findings, a RAI scintigraphy is not necessary and surgery is recommended (rating A).
- If the cytology is benign, further immediate diagnostic studies are not routinely needed (rating A).
- In patients with >1 nodule larger than 1 to 1.5 cm, those with suspicious US appearance should be aspirated (rating B); if none are suspicious, then the largest nodule(s) should be aspirated (rating C).
- For patients undergoing surgery for malignant cytological findings, preoperative US of the contralateral lobe and central/lateral lymph node (LN) compartments is recommended (rating B); the routine preoperative use of other imaging modalities is not recommended (rating E).
- Thyroid lobectomy alone may be sufficient for small, low-risk, isolated, intrathyroidal papillary thyroid carcinoma (PTC) in the absence of cervical nodal involvement (rating A).
- Total thyroidectomy is indicated for those patients with large tumors (>4 cm) with marked atypia or "suspicious for papillary carcinoma" on FNA cytology, family history of thyroid cancer, a history of radiation exposure to the head or neck, and with bilateral nodular disease wanting to avoid possibility of a second surgery (completion thyroidectomy) (rating A).
- In patients with PTC and suspected Hurthle cell thyroid carcinoma (HTC), (near) total thyroidectomy with central LN dissection, or with RAI therapy if no central LN dissection is recommended; in patients with follicular thyroid carcinoma (FTC), (near) total thyroidectomy may be appropriate without central LN dissection (rating B).
- Lateral LN dissection should be performed in patients with cervical LN involvement by biopsy or imaging, especially if unlikely to respond to RAI treatment based on other factors (rating B).
- American Joint Committee on Cancer/Union Internationale Contre le Cancer (AJCC/UICC) staging (Table 6.3[10,11]) is recommended for all patients with differentiated thyroid cancer (DTC) (rating B).
- RAI treatment is recommended for patients with stages III and IV DTC, stage II DTC if younger than 45 years, most patients with stage II DTC if 45 years and older, and selected patients with stage I DTC, especially those with multifocality, nodal metastases, extrathyroidal or vascular invasion, or more aggressive histologies (rating B).
- Remnant ablation can be performed with thyroxine withdrawal or recombinant human thyroid-stimulating hormone (rhTSH) stimulation (rating B).
- External beam radiotherapy (EBRT) should be considered in patients older than 45 years with gross extrathyroidal extension at time of surgery and high risk for microscopic residual disease, or in patients with gross residual disease in whom further surgery or RAI would be ineffective (rating B).
- There is no role for the routine adjunctive use of chemotherapy in patients with DTC (rating F).

In addition, the ATA CPG recommends consideration of EBRT, chemotherapy, radiofrequency ablation, chemoembolization, or other novel therapies and clinical trials for patients with DTC who are thyroglobulin (TG) positive–RAI uptake negative or with advanced disease recalcitrant or unresponsive to standard therapy (above) (rating C). One limitation of this CPG is related to the staging

TABLE 6.3
AJCC/UICC TNM Staging System for Thyroid Cancer[a]

Stage	10-Year DFS (%)[b]	10-Year DSS (%)[b]	Age (years)	Description
I	90.9	99.4	< 45	No distant metastases (tumor may be any size, with or without nodal metastases)
			≥ 45	Tumor ≤ 2 cm without nodal or distant metastases
II	77.3	100	< 45	Distant metastases (tumor may be any size, with or without nodal metastases)
			≥ 45	Tumor > 2 to 4 cm without nodal or distant metastases
III	66.7	100	≥ 45	Tumor > 4 cm, limited to the thyroid or with minimal extrathyroidal extension, without nodal or distant metastases; or tumor ≤ 4 cm, with level IV (central compartment) involvement but no distant metastases
IVa	44.4	74.5	≥ 45	Presence of capsular extension and invasion to neighboring tissue, except prevertebral fascia or encasement of carotid artery/mediastinal vessels, with no nodal metastases or just level IV involvement; or, if minimal or no extension but nodal metastases beyond level IV involvement
IVb	10	10	≥ 45	Invasion of prevertebral fascia or encasement of carotid artery/mediastinal vessel, with or without nodal involvement, but no distant metastases
IVc	0	27.3	≥ 45	Distant metastases (tumor may be any size, with or without nodal metastases)

Note: Abbreviations: AJCC, American Joint Committee on Cancer; DFS, disease-free survival; DSS, disease-specific survival; UICC, Union Internationale Contre le Cancer.

[a] See L. H. Sobin and C. H. Wittekind,*UICC: TNM Classification of Malignant Tumors*, 6th ed., Wiley-Liss, New York, 2002.

[b] See N. Wada et al., *J. Clin. Endocrinol. Metab.*, 92, 215–18, 2007.

system used. There are several systems available, and each has advantages and disadvantages. In addition, the EBM methodology directly maps levels of scientific substantiation to the recommendation rating and does not transparently incorporate subjective factors. Another limitation is the lack of clear direction on how to manage patients who are at very low or very high mortality risk due to DTC. This is particularly relevant to the wide range of presentations of patients with papillary thyroid microcarcinoma (PTMC) and nodal involvement.

6.3.4 European Association of Nuclear Medicine (EANM) Guidelines for Radioiodine Therapy of Differentiated Thyroid Cancer[12]

In this CPG, the EANM provides advice to nuclear medicine clinicians and other physicians treating DTC on how to ablate a thyroid remnant, treat metastatic disease, and treat inoperable advanced DTC. There is no description of a specific EBM protocol for rating the evidence and transparently identifying the role of expert opinion. Specific recommendation grades, or ratings, are not provided. The principle findings of this CPG are that RAI treatment indications include:

- Unresectable iodine avid LN, pulmonary, bone, or soft tissue metastases
- Known or suspected metastatic disease where iodine avidity is not known, especially if the TG is detectable or increasing

- Anaplastic thyroid carcinoma (ATC) or poorly differentiated thyroid carcinoma with relevant well-differentiated areas or demonstrable TG expression, even if symptomatic or progressive

The limitations of this CPG are related to the absence of any defined EBM methodology. It is not clear how much scientific substantiation figured into each of the recommendations. Additional recommendations are provided in a companion paper on blood and bone marrow dosimetry.[13]

6.3.5 National Cancer Institute (NCI) Thyroid Fine-Needle Aspiration State-of-the-Science Conference[14]

In this CPG, the NCI provides details of recommendations on the use of thyroid FNA based on a two-day conference, held in 2007. There were nine working committees, a dedicated permanent website for vetting and education, and ultimately a final summary document. There is no specific EBM methodology provided, and there is no transparency on how scientific substantiation and expert opinion contribute to final recommendations. The principle findings of this CPG are:

- Every patient with a thyroid nodule is a candidate for an FNA.
- Thyroid nodules 1 cm or greater can be detected by palpation and are clinically significant.
- Prior to FNA, patients should have a complete history, physical examination of the thyroid and cervical LN, TSH level, and thyroid US; specifically, if the TSH is normal or elevated, then a thyroid US is obtained to determine if FNA should be performed.
- If the TSH is suppressed, then RAI scintigraphy should be performed and results correlated with US findings; asymptomatic functioning nodules do not require FNA.
- Nodules that are iso- or hypofunctioning by RAI scintigraphy should be considered for FNA based on US findings.
- Incidentally discovered nodules with focal fluorodeoxyglucose (FDG)–positron emission tomography (PET) uptake, or focal sestamibi uptake, should have an FNA; diffuse FDG-PET uptake does not require FNA unless the US demonstrates a discrete nodule.
- Incidentally discovered nodules by US, computerized tomography (CT), or magnetic resonance imaging (MRI) should have a dedicated thyroid US examination; those with maximal diameter greater than 1 to 1.5 cm with solid components should have FNA.
- Any nodule with suspicious US findings should have FNA; this includes potential microcarcinoma.
- Palpation-guided FNA may be considered for nodules of >1 cm in maximal diameter, confirmed by US.
- UG-FNA is preferred for nonpalpable nodules, those that are >25% cystic, or those that have been biopsied previously with a nondiagnostic result.
- A six-category diagnostic scheme (with percent risk of malignancy) is provided: benign (<1%), follicular lesion of undetermined significance (5 to 10%), follicular neoplasm (20 to 30%), suspicious for malignancy (50 to 75%), malignant (100%), and nondiagnostic.
- Ancillary tests include special stains (CTN, TG, carcinoembryonic antigen (CEA), chromogranin) and molecular markers (no specific recommendations).

Additional recommendations are provided for the follow-up of thyroid nodules based on previous FNA results. The limitations of this CPG include absence of specific EBM methodology and knowledge gaps for the ancillary testing and clinical decision making for indeterminate cytological interpretations.

6.3.6 National Comprehensive Cancer Network (NCCN) Clinical Practice Guidelines in Oncology: Thyroid Carcinoma[15]

In this CPG, the updated 2009 NCCN version provides a comprehensive discussion of thyroid carcinoma management predicated on the belief that best management for any cancer patient is in a clinical trial; thus participation in clinical trials is especially encouraged. The EBM protocol used is explicitly provided as categories of evidence and consensus:

- 1—Recommendation based on high-level evidence with uniform NCCN consensus.
- 2A—Recommendation based on lower-level evidence with uniform NCCN consensus.
- 2B—Recommendation based on lower-level evidence with nonuniform NCCN consensus (but no major disagreement).
- 3—Recommendation based on any level of evidence but reflects major disagreement.

The CPG explicitly states that all recommendations are category 2A unless otherwise noted; the overwhelming majority of the decision tree connections are category 2A. The principle findings are organized by clinical scenario algorithms and are consistent with the above CPG except for the following variations:

- Nodule evaluation: Risk stratification, then TSH/US, including central and lateral neck/FNA nodule and suspicious LN; nodules 1 cm or less and low risk may be followed with (including lateral compartment) or without US.
- FNA cytology: Follicular or Hurthle cell neoplasm and nonsuppressed TSH referred for surgery; indeterminate follicular lesions and nonsuppressed TSH should either be considered for surgery or have repeat FNA.
- The level of TSH suppression therapy should be commensurate with the risk for recurrence.
- PTC: Preoperatively, consider chest XR, perform US if not yet done, consider lateral neck US, CT/MR for fixed, bulky, or substernal lesions (without iodinated contrast, unless essential), and evaluate vocal cord mobility. Total thyroidectomy (cf. lobectomy) recommended for patients at higher risk for recurrence. RAI dosing gradations are based on TG level and uptake. Consider bisphosphonate therapy for bone metastases. For clinically progressive or symptomatic disease, enroll patient into clinical trials for non-RAI-avid tumors, or if trial not available, consider small molecule kinase inhibitors or systemic chemotherapy.
- FTC: Lobectomy/isthmusectomy for minimally invasive cancer; total thyroidectomy and dissection levels II to IV and VI for invasive, metastatic, or LN positive cancers; similar risk determined followed as with PTC.
- MTC: Initial evaluation includes CTN, CEA, pheochromocytoma screening, calcium, genetic counseling, *RET* proto-oncogene mutations, US, vocal cord mobility, and contrast-enhanced chest/mediastinum CT. If < 1 cm, then total thyroidectomy and bilateral level VI dissection; if bilateral or ≥ 1 cm, then total thyroidectomy and bilateral level VI dissection. Consider ipsilateral level II to V dissection and contralateral level II, III, IV or V dissection, and consider adjuvant EBRT for T4 disease involving major neck structures. Additional recommendations based on specific *RET* codon mutations. Disseminated symptomatic disease to enroll in clinical trial, or if not available, palliative EBRT, small molecule kinase inhibitors, dacarbazine-based chemotherapy, or bisphosphonates.
- ATC: Initial evaluation includes complete blood count (CBC), calcium, head/neck/chest/abdomen/pelvis CT, and TSH; consider FDG-PET and bone scans. If locally respectable, perform (near) total thyroidectomy with selective resection of involved loco-regional structures and LN; if not locally respectable, enroll in clinical trials or, if not available, palliative EBRT (consider hyperfractionated) plus chemotherapy.

The limitations of these evidence-based clinical algorithms are the absence of transparency in how scientific substantiation and expert opinion contribute to a final recommendation. In addition, high-risk patients, recalcitrant to standard therapies with surgery, RAI, TSH suppression, and EBRT, are referred for clinical trials or biological therapies, but there are no specific recommendations or protocols based on their clinicopathological presentation and clinical course.

6.3.7 Medullary Thyroid Cancer: Management Guidelines of the American Thyroid Association[16]

In this CPG, the ATA provides EBM recommendations for the management of MTC. The methodology adheres with the U.S. Preventive Services Task Force protocol for recommendation ratings in clinical practice guidelines, described in Section 6.3.3.[9] The principle findings in this CPG are:

- All patients with primary C-cell hyperplasia, MTC, familial MTC (fMTC) with autosomal dominant inheritance, or multiple endocrine neoplasia type 2 (MEN-2) should be offered germline *RET* testing (rating A).
- Specific evaluations and therapeutics are recommended based on specific codon mutations (ratings A and B).
- Devascularized normal parathyroid glands from patients with MEN-2B or fMTC at high risk for primary hyperparathyroidism (PHP) should be autografted into the sternocleidomastoid muscle; patients at low-risk for PHP and fMTC may have autografts into the forearm (rating C).
- A basal or stimulated CTN level of ≥100 pg/ml is suspicious for MTC and merits further evaluation and treatment (rating A).
- Treatment with radiolabeled molecules may be considered in selected patients, ideally in the setting of a well-designed clinical trial (rating C).

The limitations of this CPG are essentially confined to the specifics of which clinical trials should be sought based on the specifics of the individual patient's clinicopathological presentation and course.

6.3.8 Revised American Thyroid Association (ATA) Management Guidelines for Patients with Thyroid Nodules and Differentiated Thyroid Cancer[17]

This update was planned at the time of the writing of the 2006 ATA CPG,[8] discussed in Section 6.3.3, and uses the same EBM methodology. A committee of surgeons was also utilized for authoritative commentary and recommendations regarding central neck dissection.

The principle revisions of this CPG, compared with the 2006 version, are:

- When to perform FNA
  - Patient with high-risk *and*
    - Nodule with suspicious US findings and >5 mm (rating A)
    - Nodule without suspicious US findings and >5 mm (rating I)
    - All nodules associated with abnormal cervical LN (rating A)
    - Nodules with microcalcifications and ≥10 mm (rating B)
  - Solid nodule *and*
    - Hypoechoic and >10 mm (rating B)
    - Iso- or hyperechoic and ≥10 to 15 mm (rating C)
  - Mixed cystic-solid nodule *and*
    - Any suspicious US features and ≥15 to 20 mm (rating B)

    - Without suspicious US features and ≥20 mm (rating C)
  - Spongiform nodule and ≥20 mm (rating C)
  - Interpretation: US technology has a sufficiently advanced sensitivity so that FNA can be deferred for low-risk nodules.
- UG-FNA, as opposed to using palpation to guide the FNA, is recommended for nodules that are nonpalpable, predominantly (>50%) cystic, located posteriorly in the thyroid lobe (rating B), or when the initial cytology result is nondiagnostic (rating A).
  - Interpretation: UG-FNA technology has sufficiently advanced to improve performance for these nodule scenarios.
- Molecular markers (*BRAF*, *RAS*, *RET/PTC*, Pax8-PPARγ, or galectin-3) may be considered in patients with indeterminate FNA cytology to help guide therapy (rating C).
  - Interpretation: Molecular markers are now available and of sufficient sensitivity/specificity to add important decision-making information.
- Deferring prophylactic central neck dissection may be appropriate for T1 or T2 (small) noninvasive, clinically node-negative PTC and most FTC (rating C).
  - Interpretation: This recommendation reflects the higher surgical complication rate with central neck dissection, which should now be reserved for larger primary tumors that have a higher risk for recurrence; the explicit omission of RAITx as a tool to specifically treat level VI LN in FTC (obviating the need for central neck dissection) is consistent with a trend toward using less RAI in low-risk scenarios.
- Explicit statement that the AJCC/IUCC staging system predicts mortality risk should be used in all patients (rating B), but does not predict recurrence risk; for a recurrence risk, the following stratification is recommended:
  - Low risk
    - No local or distant metastases
    - All macroscopic disease is resected
    - No loco-regional or vascular invasion
    - No aggressive histology
    - If RAI given, there is no uptake beyond the thyroid bed
  - Intermediate risk
    - Microscopic invasion of perithyroidal soft tissue at initial surgery
    - Cervical LN involvement
    - RAI uptake beyond thyroid bed
    - Aggressive histology
    - Vascular invasion
  - High risk
    - Macroscopic tumor invasion
    - Incomplete tumor resection
    - Distant metastases
    - Elevated TG beyond expectation from posttreatment RAI scan
  - Interpretation: This is an expert opinion by the ATA to address recurrence risk.
- RAI ablation
  - Recommended for all patients with known distant metastases (rating A), gross extrathyroidal extension (rating B), or primary tumor of >4 cm (rating B)
  - Recommended for selected patients with primary tumor of 1 to 4 cm with intermediate to high risk for recurrence (ratings C to I)
  - Not recommended for patients with unifocal or multifocal primary tumor(s) of <1 cm without higher risk for recurrence features (rating E)
  - Interpretation: The decreased use of RAITx is based on a more critical analysis of the evidence demonstrating lack of benefit in low-risk patients as well as increased recognition of adverse effects, such as sialoadenitis and secondary malignancy risk.

- TSH suppression
    - Persistent disease: <0.1 mU/L indefinitely (rating B)
    - High risk but not free of disease: 0.1 to 0.5 mU/L for five to ten years (rating C)
    - Low risk and free of disease: 0.3 to 2 mU/L (rating B)
    - No remnant ablation, free of disease, undetectable TG, and normal US: 0.3 to 2 mU/L (rating C)
    - Interpretation: Relaxation of the level of TSH suppression in low-risk patients based on evidence that this biochemical target may not be associated with an outcome advantage in low-risk patients.
- ^{18}FDG-PET or ^{18}FDG-PET/CT fusion scanning initially when nonlocalizing with 100 to 200 mCi RAI, especially with unstimulated TG of >10 to 20 ng/ml (rating B)
    - Interpretation: Earlier use of PET technology can identify patients unlikely to respond to RAITx, prompting earlier referral to medical centers offering EBRT, chemotherapy, molecular targeted therapies, or other novel interventions.
- Future directions
    - Earlier use of novel therapies in high-risk patients not expected to respond to conventional therapies (including chemotherapy)
    - Less intervention in low-risk patients
    - Clarification of the impact of small cervical LN metastases
    - Improved risk stratification
- Interpretation: The conspicuous inclusion of referring patients for investigational clinical trials of novel therapies is based on promising evidence regarding their potential benefit as well as a current paucity of effective conventional therapies for high-risk patients.

6.4 WHERE THYROID CANCER CPG REQUIRE OPTIMIZATION

A review of key findings of the above CPG discloses several commonalities. There is relatively uniform agreement that current clinical algorithms for the evaluation of a thyroid nodule—palpable, symptomatic, or incidental—begin with risk stratification based on history and physical examination. Then a serum TSH, thyroid US, thyroid scintigraphy if indicated, and UG-FNA are performed. Based on these findings, a composite risk profile can be formulated and a therapeutic plan designed.

A review of the limitations of the above CPG also discloses several commonalities, but in two broad categories. First, some, but not all, of the CPG are transparently evidence based and may confer increased adherence compared with those that take the form of expert opinion. Nevertheless, there are sufficient CPG that are consistent with one another, and more importantly, those CPG reflecting expert opinion are still authoritative and merit a reasonable element of adherence. Second, each CPG failed to explicitly outline which novel or investigational diagnostic tools and therapies should be employed for recalcitrant or advanced thyroid cancer. Specifically, with a host of clinical trials available for thyroid cancer (see http://www.cancer.gov/search/ResultsClinicalTrials.aspx?protocolsearchid=6800211 for a listing of National Cancer Institute clinical trials for thyroid cancer), the question arises as to which trials should be recommended for which clinical scenario of thyroid cancer. Another fuzzy area addressed by some of the CPG is exactly how aggressive physicians should be when caring for low-risk patients with thyroid cancer, such as PTMC with varying degrees of nodal involvement. Various clinical criteria for risk stratification have been published as well as thresholds for US characterization of thyroid nodule risk. Current EBM CPG are still unable to provide specific guidance regarding watchful waiting or minimal intervention for these low-risk patients.

Ultimately, the areas of EBM CPG that are unclear, or fuzzy, correspond to knowledge gaps. Optimistically, these gaps progressively close with time, and CPG reflect this with periodic updates, revisions, and at times, overhauling. The best example of this CPG evolution is provided by the updated ATA CPG.[17] Nevertheless, the question remains as to how CPG evolution can be

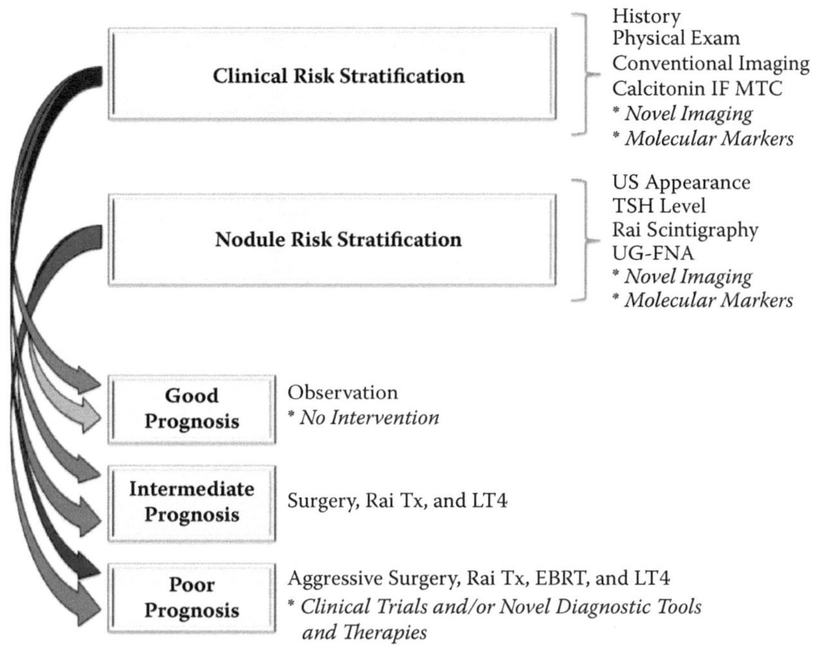

FIGURE 6.1 A generic template of CPG for thyroid cancer. The asterisks identify areas where emergent biotechnologies may potentially influence future CPG. For the purpose of this discussion, "risk" and "prognosis" pertain to a composite of "recurrence" and "mortality." Abbreviations: MTC, medullary thyroid carcinoma; US, ultrasound; TSH, thyroid-stimulating hormone; RAI, radioactive iodine; UG-FNA, ultrasound-guided fine-needle aspiration; RAI TX, radioactive iodine treatment; LT4, levothyroxine; EBRT, external beam radiotherapy.

optimized to keep pace with medical science and the net accumulation of clinical knowledge and experience.

If a current (composite) clinical algorithm is generated for all types of thyroid cancer, then specific nodes can be identified for optimization with emergent biotechnologies (Figure 6.1). These novel diagnostic and therapeutic procedures will be heavily influenced by breakthroughs in molecular medicine and biomedical engineering. They will affect clinical algorithms and CPG at each step depicted in Figure 6.1. There will also be CPG cascades, in which alternative algorithms can be employed in medical settings that do not yet have access to contemporary biotechnologies. In the subsequent chapters of this book, contemporary and emergent biotechnologies will be discussed in detail, and then a potential, newly envisioned CPG strategy will be presented.

REFERENCES

1. Evidence-Based Medicine Working Group. 1992. Evidence-based medicine: A new approach to teaching the practice of medicine. *JAMA* 268:2420–25.
2. Field MJ, Lohr KN, eds. (Committee on Clinical Practice Guidelines, Institute of Medicine). 1990. *Clinical practice guidelines: Directions for a new program*, Washington, DC: National Academies Press.
3. Mechanick JI. 1987. Methods of creative cognition in medical diagnosis. *Mt Sinai J Med* 54:348–54.
4. Mechanick JI, Bergman DA, Braithwaite SS, et al. 2004. American Association of Clinical Endocrinologists protocol for standardized production of clinical practice guidelines. *Endocrine Pract* 10:353–61.
5. American Institute of Ultrasound in Medicine. 2003. AIUM practice guideline for the performance of a thyroid and parathyroid ultrasound examination. *J Ultrasound Med* 22:1126–30.

6. Gharib H, Papini E, Valcavi R, and the AACE/AME Task Force on Thyroid Nodules. 2006. American Association of Clinical Endocrinologists and Associazione Medici Endocrinologi Medical Guidelines for clinical practice for the diagnosis and management of thyroid nodules. *Endocrine Pract* 12:63–102.
7. Mechanick JI, Brett EM, Chausmer AB, et al. 2003. American Association of Clinical Endocrinologists medical guidelines for the clinical use of dietary supplements and nutraceuticals. *Endocrine Pract* 9:417–70.
8. Cooper DS, Doherty GM, Haugen BR, and the American Thyroid Association Guidelines Taskforce. 2006. Management guidelines for patients with thyroid nodules and differentiated thyroid cancer. *Thyroid* 16:1–33.
9. *U.S. Preventive Services Task Force ratings: Strength of recommendations and quality of evidence. Guide to clinical preventive services. Periodic updates, 2000–2003.* 3rd ed. U.S. Preventive Services, Rockville, MD.
10. Sobin LH, Wittekind CH. 2002 *UICC: TNM classification of malignant tumors.* 6th ed. New York: Wiley-Liss.
11. Wada N, Nakayama H, Suganuma N, et al. 2007. Prognostic value of the sixth edition AJCC/UICC TNM classification for differentiated thyroid carcinoma with extrathyroid extension. *J Clin Endocrinol Metab* 92:215–18.
12. Luster M, Clarke SE, Dietlein M, et al. 2008. Guidelines for radioiodine therapy of differentiated thyroid cancer. *Eur J Nucl Med Mol Imaging* 35:1941–59.
13. Lassmann M, Hanscheid H, Chiesa C, et al. 2008. EANM Dosimetry Committee series on standard operational procedures for pre-therapeutic dosimetry I: Blood and bone marrow dosimetry in differentiated thyroid cancer therapy. *Eur J Nucl Med Mol Imaging* 35:1405–12.
14. Baloch ZW, Cibas ES, Clark DP, et al. 2008. The National Cancer Institute thyroid fine-needle aspiration state-of-the-science conference: A summation. *CytoJournal* 5:6. doi:10.1186/1742-6413-5-6.
15. Sherman SI, Tuttle RM, Ball DW, and the NCCN Thyroid Carcinoma Panel members. 2009. NCCN clinical practice guidelines in oncology: Thyroid carcinoma. V.1.2009, updated version. http://www.nccn.org/professionals/physician_gls/PDF/thyroid.pdf (accessed on September 28, 2009).
16. Kloos RT, Eng C, Evans DB, and the American Thyroid Association Guidelines Task Force. 2009. Medullary thyroid cancer: Management guidelines of the American Thyroid Association. *Thyroid* 19:565–612.
17. Cooper DS, Doherty GM, Haugen BR, et al., and the American Thyroid Association Guidelines Task Force. 2009. Management guidelines for patients with thyroid nodules and differentiated thyroid cancer. *Thyroid* 19:1167–214.

7 Illustrative Case Studies of Thyroid Carcinoma and the Emerging Role of Molecular Medicine

Donald A. Bergman and Rhoda H. Cobin

CONTENTS

ABSTRACT: Four clinical cases—familial papillary carcinoma of the thyroid, micropapillary carcinoma, Hürthle cell carcinoma, and medullary carcinoma—are presented that illustrate the role for emerging biotechnologies, namely, molecular medicine, to guide decision making. In each instance, the current understanding of a thyroid cancer's biologic behavior is constrained by histological and clinical investigation of the tumor and its host. This information did not optimally determine the appropriate extent of therapy. In familial papillary thyroid carcinoma, the identification of germline mutations could lead to active screening of families to prevent disease. In papillary thyroid microcarcinoma, molecular features could identify the small number of cases that behave aggressively and therefore merit more aggressive therapy. In Hürthle cell thyroid carcinoma, molecular features could differentiate virulence among tumors with similar histology. Medullary thyroid carcinoma may present in both familial and sporadic forms, and both germline and somatic mutations have already been identified with considerable genotype-phenotype correlations. These mutations could optimize familial screening and the prediction of tumor behavior and associated clinical features, such as pheochromocytomas. These four cases will be revisited in Chapter 22, where the impact of emergent biotechnologies will be demonstrated.

7.1 INTRODUCTION

Current clinical practice guidelines (CPG) for the management of thyroid nodules differentiate risk for thyroid cancer on the basis of clinical history, physical examination, ultrasonographic findings, thyroid-stimulating hormone (TSH) level, and ultrasound-guided fine-needle aspiration biopsy (UG-FNAB). Management decisions regarding the timing and extent of surgery as well as the use and dose of radioiodine (RAI) are logically based on the premise that more aggressive tumors require more proactive, early, and aggressive therapy. However, missing from these current paradigms are differences in the molecular signatures of these tumors, which may profoundly impact prognosis but not histological features. In the case of differentiated thyroid epithelial carcinomas, which include papillary thyroid carcinoma (PTC) and follicular thyroid carcinoma (FTC), considerable information may be gleaned by the inclusion of the clinical hereditary pattern as well as both germline and tumor somatic mutations. In the case of parafollicular C-cell-derived medullary thyroid carcinoma (MTC), genotype-phenotype correlation studies can help determine management strategies for both the timing and extent of surgery, the likelihood of additional components of the MEN syndrome (namely, pheochromocytomas), and potentially the timing to institute small molecule targeted therapy, such as RET (rearranged during transfection) tyrosine kinase inhibitor therapy. Ultimately and regardless of the histological type, the incorporation of molecular medicine into our armamentarium may shift the emphasis from treatment to prevention of disease.

In this chapter, four illustrative clinical cases will be presented. They cover the spectrum of differentiated thyroid carcinoma, integrate clinical challenges and potential solutions nested in molecular biology, and highlight several key points, given below:

1. Common carcinomas (PTC, FTC, and follicular variant of PTC (FV-PTC)) and uncommon carcinomas (MTC, Hürthle-cell thyroid carcinoma (HTC), and anaplastic thyroid carcinoma (ATC)) have different prognoses, within and among histological types, and therefore require tailored levels of therapeutic aggressiveness. For instance, papillary thyroid microcarcinoma (PTMC) has been considered by many to be clinically unimportant, whereas HTC has been considered to be more aggressive than most thyroid malignancies. Yet, some experts suggest that some PTMCs require aggressive management, while some HTCs may require only a hemithyroidectomy. In both types of cancers, size of the tumor and age of the patient at clinical presentation are major pivotal points in deciding on the extent of therapy. It is easy to see how a simple decision tree based on current management strategies may lead to excessive therapy in patients with PTMC (causing treatment-related morbidity) and insufficient therapy in patients with HTC (leading to poorer outcomes). Investigative tools are needed that will predict the likelihood of aggressive tumor behavior, quality of life with certain therapies, and clinical outcomes.
2. There are genetic factors that determine risk for developing certain thyroid malignancies. Though some, but not many, are currently available, one can foresee a time when large panels of genetic factors are available to guide preventive strategies and early decision making for thyroid cancer.
3. Various molecules participate in specific subcellular functions involved in pathogenetic pathways leading to thyroid cancer. Genomic, proteomic, and other systems biology and translational medicine techniques will strengthen the clinical staging strategies currently in use. This will allow the clinician to assign the proper risk of recurrence and death in each type of thyroid cancer and avoid too much or too little treatment. Molecular markers

may be not only used for diagnostic purposes but also exploited strategically for therapeutic purposes. For instance,

a. The *BRAF* mutation is seen in 40% of PTC, and those cancers that show the mutation are aggressive and resistant to RAI. This mutation interferes with tumor suppressor genes and inhibits genes responsible for iodine transport. It also increases the production of vascular endothelial growth factor (VEGF).[1]
b. The forkhead box transcription factor, FOXO3a, is important in inducing cell cycle arrest and apoptosis. FOXO3a accumulates abnormally in thyroid cancers, and this may allow thyroid cancer cells to escape apoptosis and proliferate abnormally.[2]

Following this chapter, many aspects of novel, innovative, and emergent biotechnologies will be presented on thyroid cancer. Salient features of these four cases will weave through the discussions of subsequent chapters for clinical context and ultimately find their way into newly synthesized CPG in the final section of the book.

7.2 CLINICAL CASE 1: FAMILIAL PAPILLARY THYROID CARCINOMA WITH EVOLUTION TO ANAPLASTIC THYROID CARCINOMA

7.2.1 Vignette

The patient was a 72-year-old female presenting with a neck mass. She stated that her physician had recommended investigation one year previously, but since she was asymptomatic, she did not pursue an evaluation. When she could feel the mass herself, she underwent a thyroid US, which revealed a 3 cm irregular hypoechoic nodule in the left lobe with microcalcifications and infiltrating borders. UG-FNAB revealed "numerous cohesive and less cohesive clusters and sheets of follicular cells with nuclear grooves and inclusions, stromal fragments, multinucleated histiocytes, rare Hürthle cells, colloid, and degenerating debris," and was considered compatible with a FV-PTC. A total thyroidectomy and extensive central compartment dissection were performed. The surgical pathology revealed PTC with follicular variant and tall-cell components comprising 90% of the tumor mass (Figure 7.1). There was also ATC composed of spindle and squamoid cells apparently arising from the PTC. RAI scan revealed 1% uptake in the tumor bed and the patient received 200 mCi of ^{131}I. The posttherapy scan did not reveal any distant uptake. She was then treated with external beam radiation and adriamycin, but ultimately succumbed to rapidly progressive ATC eight months after the diagnosis. Both the patient's sister and the sister's son had total thyroidectomies for PTC more than 5 years previously and currently have no demonstrable residual or recurrent disease.

7.2.2 Discussion

This unfortunate woman apparently had anaplastic transformation of a familial PTC (fPTC). It is likely that if she had agreed to evaluation and treatment when the lesion was first discovered, the nidus for the development of the ATC could have been removed and she might have had the same favorable outcome as her sister and nephew. This case emphasizes the importance of genetic markers in the screening and early decision making with thyroid cancer.

Unfortunately, neither germline nor tumor tissue was saved for genetic analysis. This could have been useful in determining whether differences in either germline or somatic molecular features present in the proband, who developed ATC, and her relatives who did not develop ATC could have determined both the risk of the development of PTC and its transformation to ATC. In the future, it is hoped that it may be possible to predict the clinical behavior of PTC and its possible transformation to ATC on the basis of somatic molecular events. If enough germline mutations are found, familial screening of probands with PTC could become a reality. Thus far, germline mutations are rare and do not predict clinicopathologic events in this type of cancer.

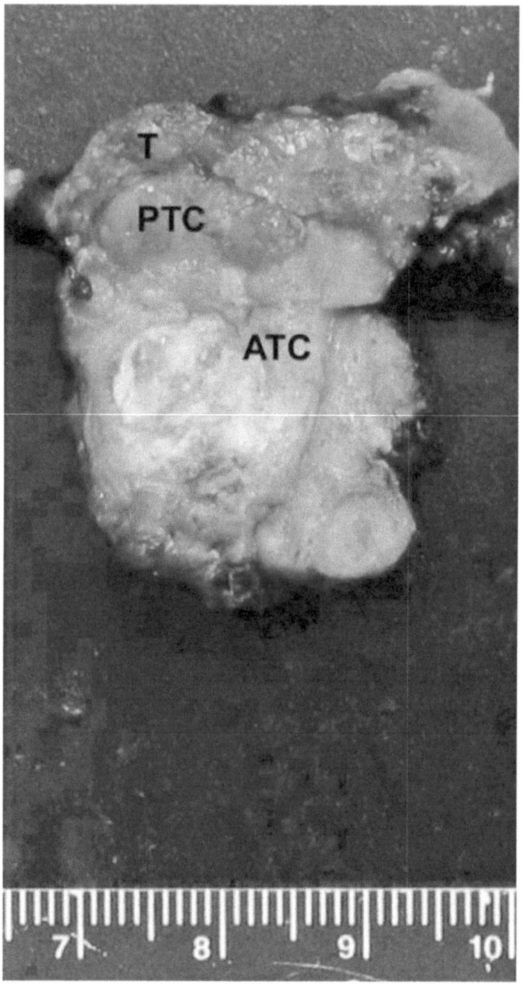

FIGURE 7.1 (See color insert.) Case 1: Gross specimen of PTC with follicular, tall-cell, and anaplastic features. T, normal thyroid tissue; PTC, papillary thyroid carcinoma; ATC, anaplastic thyroid carcinoma.

Familial PTC is rare, but increasingly described. Familial thyroid cancer syndromes are classified into familial MTC (fMTC) and familial nonmedullary thyroid carcinoma (fNMTC). Multifocal PTC is the most frequent presentation of fNMTC. Clinicopathologic findings divide it into two groups. The first includes familial syndromes characterized by a predominance of nonthyroidal tumors, such as familial adenomatous polyposis, phosphatase and tensin homolog on chromosome 10 (PTEN; a known tumor suppressor)–hamartoma tumor (Cowden) syndrome, Carney complex type 1, and Werner syndrome. The second group includes familial syndromes characterized by a predominance of fNMTC, such as pure fPTC with or without oxyphilia, fPTC with papillary renal cell carcinoma, and fPTC with multinodular goiter.[3] In the case of Cowden syndrome and familial adenomatous polyposis, various germline mutations have been identified. However, none have thus far been shown to present an increased risk of transformation to ATC, suggesting that a second hit or somatic mutation may have occurred in the above patient.

At present, there are no peripheral blood signals predictive of either the development of or the clinicopathological behavior of familial thyroid cancer. Studies have focused not only on linkage analysis of candidate genes, but also on the possibility of specific molecular behavior predisposing to carcinogenesis. Although *BRAF* somatic mutations are common in PTC, the T1799A BRAF mutation is not a germline mutation or susceptibility genetic event for fNMTC.[4] Linkage analysis has identified four fNMTC susceptibility loci: *fPTC/PRN* (1p13.2-1q22), *NMTC1* (2q21), *MNG1* (14q32), and *TCO* (19p13.2).[5] Linkage analysis in seven informative families showed no evidence for the involvement of any of the four candidate regions, supporting a genetic heterogeneity for fNMTC. *BRAF-V600E* mutation was observed in 9 out of the 22 PTCs (41%), and H-RAS and N-RAS mutations were detected in 5 out of the 22 PTCs (23%).[5] The data suggest that the four candidate regions are not frequently involved in fNMTC, and that the somatic activation of BRAF and RAS plays a role in fNMTC tumourigenesis.[5]

fPTC displays an imbalance of the telomere-telomerase complex in the peripheral blood, characterized by short telomeres, hTERT gene amplification, and expression.[6] These features may be implicated in the inherited predisposition to develop fPTC.[6]

In contrast to the genomic level, more is known about the molecular basis of carcinogenesis involving signal transduction and gene expression. In thyroid cancer cells, constitutive activation of particular kinases may lead to excessive activation of transduction pathways specific for mitogens or growth factors, resulting in unregulated proliferation. In PTC, the gene for RET—a membrane-located growth factor-receptor with kinase activity—has been shown to undergo at least ten different rearrangements, with the most common being the *RET/PTC3* and *RET/PTC1* rearrangements. Here, the anomalous kinase domain-encoding 3' end of the *RET* gene is aberrantly bound to the 5' end of another gene. The resultant fusion protein activates transduction pathways specific for RET. The nerve growth factor receptor, also with a tyrosine kinase activity *(NTRK)-1* gene, is also rearranged in PTC. However, genes fused to its kinase domain-encoding sequence are different from the ones fused to RET. Other mutations affecting thyroid cancer proteomics and representing potential sites for small molecule targeted therapies include *MET* and triiodothyronine (T3) receptors.[7] Progression from papillary to anaplastic cancer is most possibly caused by the occurrence of additional anomalies within *p53*, *RAS*, *NM23*, β-catenin, and other genes.[7]

Tumor protein p53-induced nuclear protein 1 (TP531NP1) is a stress-induced protein and plays a role in cell cycle arrest and p53-mediated apoptosis. Of 100 PTC, only 4 (4%) expressed high levels of TP531NP1, and the remaining 96 (96%) were classified into the low group.[8] There were no significant relationships between TP531NP1 expression and clinicopathological features of PTC.[8] However, 36 of the 38 (94.7%) ATC expressed high levels of TP531NP1, and the incidence was significantly higher ($p < 0.0001$) compared with other neoplasms.[8] These findings suggest that TP531NP1 may play a significant role in the progression of ATC or contribute to anaplastic transformation from PTC or FTC.[8]

p53 and β-catenin mutations are found with increasing incidence in poorly differentiated thyroid carcinomas and ATC but not in well-differentiated thyroid carcinomas. They may serve as a direct molecular trigger of tumor dedifferentiation. Additional evidence for progression from a preexisting well-differentiated carcinoma to poorly differentiated and ATC is derived from the studies of loss of heterozygosity and comparative genomic hybridization. Molecular studies, although limited by the lack of uniform histological criteria for poorly differentiated carcinomas, revealed no genetic mutations or chromosomal abnormalities that are unique for poorly differentiated carcinoma and not present in well differentiated or ATC. This suggests that poorly differentiated carcinoma, as a group, represents a distinct step in the evolution from well differentiated to ATC, rather than an entirely separate type.[9]

Immunohistochemistry (IHC) analysis of RET/PTC rearrangements revealed no positive staining of RET in any of eight ATC, suggesting that these ATC are not derived from *RET/PTC*- rearranged PTC.[10] In contrast, IHC analysis of *p53* mutation revealed that p53 was detected in the nuclei of five of five *BRAF*-mutated ATC and two of three ATC with wild-type *BRAF*. p53 staining was present only in ATC cells but not in neighboring PTC cells.[10] These results suggest that many ATC with papillary components are derived from *BRAF*-mutated PTC, because of the addition of *p53* mutation.[10]

A number of intracellular growth regulatory proteins have been investigated in PTC and ATC. In PTC, p107, a member of the Rb family, was found to be positive in 84.0% of PTC, regardless of various biological characteristics.[11] FTC less frequently expressed p107 than follicular adenoma.[11] Only 12.5% of ATC were p107 positive, and the incidence was significantly less than that in PTC, suggesting that p107 may play a constitutive role in the progression of PTC and that decreased p107 expression may contribute to the transformation from follicular adenoma to FTC and ATC.[11]

Cyclin G2 is a novel cyclin that negatively regulates the cell cycle progression, contrary to the characteristics of conventional cyclins. Lack of cyclin G2 plays an important role in the malignant transformation of PTC. Also, it may play an adjuvant role in the transformation of follicular adenoma to FTC.[12]

S100A10, a member of the S100 family, forms a heterotetramer with annexin IIH and promotes carcinoma invasion and metastasis by plasminogen activation.[13] Immunohistochemical stains for this protein reveal positive but low level staining in most PTC, without correlation to clinicopathological features.[13] In contrast, high levels are present in most ATC, suggesting that these proteins may contribute to the aggressive characteristics of ATC, while playing a constitutive role in PTC.[13]

7.3 CLINICAL CASE 2: PAPILLARY THYROID MICROCARCINOMA

7.3.1 Vignette

The patient was a 59-year-old woman who previously underwent surgery for primary hyperparathyroidism. A right hemithyroidectomy was performed as part of that parathyroid procedure, in which a 2 mm PTMC was incidentally discovered in the right lobe. Five years later, she presented with bulky cervical lymph node metastases in the left lateral neck. She had a completion thyroidectomy and lymph node dissection. The histology was consistent with metastatic PTC (Figure 7.2). No tumor was found in the remaining lobe. The only primary tumor was the 2 mm PTMC lesion found in the original specimen.

7.3.2 Discussion

Papillary thyroid microcarcinoma: even the name underscores the clinical conundrum. *Micro* suggests that the disease is trivial, requiring little intervention. *Carcinoma* is a reminder that the condition can be occasionally aggressive and associated with considerable morbidity and, at times, lethality.

Thyroid cancer is the fastest growing reported cancer, and most of these are PTMC often incidentally discovered when imaging of the neck is performed for other reasons.[14] Most thyroid cancer staging schemes emphasize size as an important predictor of recurrence. The TNM classification uses 2 cm (or smaller) as a predictor of low risk of recurrence.[15] A scheme described by Tuttle[16] uses risk of recurrence and risk of death as criteria for identifying how aggressive the therapy should be (extent of surgery, need for RAI, and amount of thyroid hormone required). This protocol also emphasizes the indolent nature of PTMC and suggests that less aggressive thyroid surgery, no RAI, and replacement doses of L-thyroxine with the TSH kept at the lower limits of normal (rather than fully suppressed) represent appropriate management.

In a review by Burman,[17] even the need to biopsy small subcentimeter lesions in the thyroid has been questioned. He suggests, based on a careful review of the literature, that small lesions should be aspirated if they are growing or if they have suspicious US characteristics. These suggestions were based partly on the following five articles. A paper by Mazzaferri and Iloos[18] showed that patients with a tumor size of less than 1.5 cm had a 30-year mortality of 0.4%, while those with a size up to 4.4 cm had a mortality of 7%, and those larger than 4.5 cm had a mortality of 18%. In an often quoted observational study, Ito et al.[19] followed 162 patients with biopsy-proven PTC, 1 cm or smaller, and observed that 70% of these tumors remained stable or actually got smaller over eight years. Nodal metastases were noted clinically in only 1.2%.[19] However, of 626 patients with small

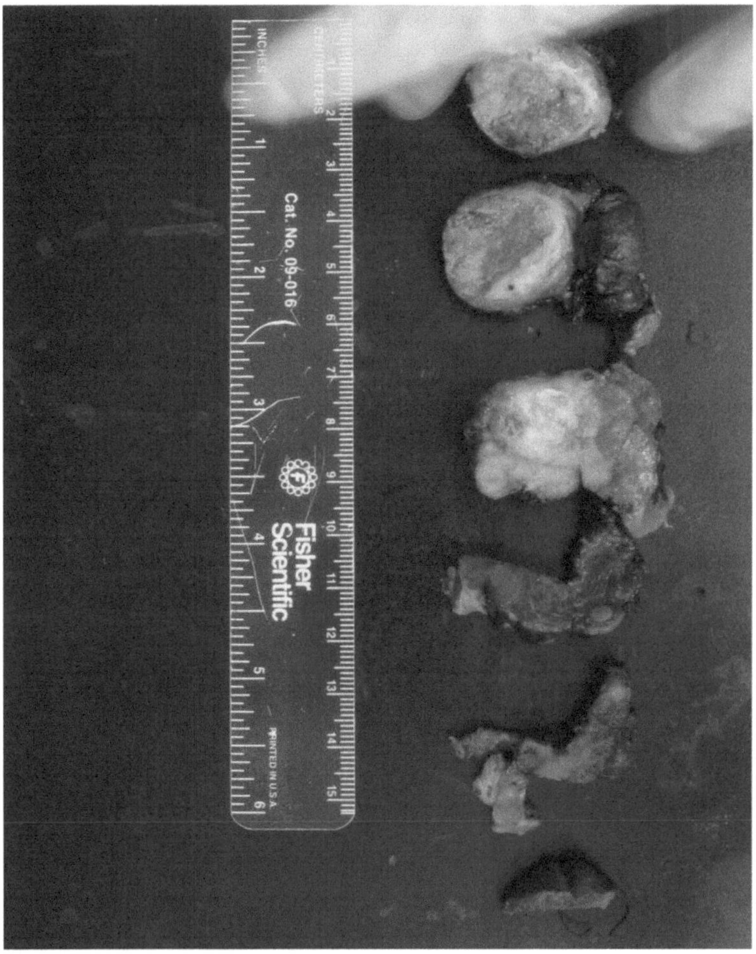

FIGURE 7.2 (See color insert.) Gross specimens from lymph node dissection five years after an incidental 2 mm papillary thyroid microcarcinoma was discovered during parathyroidectomy.

tumors who did elect to have surgery, local metastases were noted in 50.5% and multifocal disease was noted in 42.8%.[19] Recurrence occurred in 5% of patients, 8 years after surgery. These data suggest that PTMC is indolent in most patients when observed clinically but in reality is frequently associated with nodal metastases and multifocality that are not observed clinically.

The American Association of Clinical Endocrinologists and the Associazione Medici Endocrinologi CPG[20] suggested that thyroid nodules should be biopsied based on their suspicious appearance (e.g., microcalcifications and hypervascularity) rather than just size. These CPG suggest that if two or more suspicious US characteristics are found in a nodule, then the sensitivity of detecting a malignancy is between 87 and 93%. Nodule growth is also mentioned as a reason for concern. Cooper et al.[21] suggest that an increase of 15% in nodule volume should warrant a biopsy. On the other hand, Asanuma et al.[22] reported that growth of the nodule did not reliably predict malignancy.

Once a decision is made to do an FNA, then the question becomes: How reliable is this procedure in predicting malignancy? Nikiforov et al.[23] point out that between 10% and 40% of aspirated nodules are classified as indeterminate, meaning that the likelihood of malignancy is uncertain. On the other hand, they remind us of the false negatives (benign cytology in a nodule later proven to be malignant), which result in delay of diagnosis and treatment. This group has looked at the predictive value of chromosomal analysis in thyroid follicular cells. They investigated mutations including

BRAF, *RAS*, *FET/PTC*, and *PAX8-PPARg*. In a series of 470 FNA samples from 328 patients, they found mutations in 32 patients, of which 97% had cancer. In a subset of 37 patients who were classified as indeterminate cytology and negative for mutations, six cancers were found (four PTC and two FTC). In another subset of patients with negative cytology but positive mutations, four out of four were malignant. These data suggest that sending FNA specimens for cytology and detection of chromosomal analysis offers great potential to improve diagnostic accuracy.

Given the conflicting opinions about when a small nodule should be aspirated, the question then becomes: Does early detection really make a difference? The paper by Ito et al.[19] suggests that clinically the patients with small cancers do well, but the paper also shows that many of these patients have microscopic lymph node metastases or multifocal disease.

Roti et al.[24] studied 243 consecutive patients with PTMC over a 9-year period in a retrospective cohort design. There were 52 patients who had surgery for other reasons (such as nodular goiter or Graves' disease) and 191 patients in whom surgery was done because of a suspicious aspiration. They found that 36% of the group with suspicious aspirates had multifocal disease versus 19% of the group where the disease was found incidentally during surgery (NS). Lymph node metastases were found in 16% of the suspicious aspirate group versus 4% of the incidental disease group (NS). All cancers, in both groups, were small, 1 cm or less, and distant metastases only occurred in those patients with carcinomas >8 mm. Since nearly all patients received therapy with total thyroidectomy, RAI, and suppressive doses of thyroid medication, it is not possible to determine whether incidentally discovered PTMCs are really less aggressive than those identified as suspicious prior to surgery. However, these data support the concept that microscopic lymph node metastases and multifocality are not uncommon in patients with PTMC.

If biopsy-proven cervical lymph node metastases are common in PTMC but are generally not clinically apparent, then how important is it to find these nodes at the time of surgery? In other words, do these nodal metastases that are not clinically obvious before surgery change the prognosis, and if they do, should central and lateral neck dissection be done routinely in patients with PTMC? Several studies have addressed these questions.

Zaydfudim et al.[25] reviewed the registry for patients with well-differentiated thyroid cancer. This report described 30,504 patients with PTC and 2584 patients with FTC. In patients with PTC under 45 years of age, lymph node metastases did not influence survival ($p = 0.535$), while in patients 45 years or older, nodal metastases were associated with a 46% increased risk of death ($p < 0.001$). This paper did not look at survival of patients with PTMC nodal metastases separately from all patients with PTC, but it does suggest that clinicians should consider more aggressive therapy in older patients, especially with nodal metastases, with PTMC.

Lim et al.[26] looked at 86 patients with PTMC and a clinically node-negative neck. All patients underwent total thyroidectomy and central compartment neck dissection. Twenty-seven patients (31%) had biopsy-proven metastatic central compartment nodal disease: 18 ipsilateral and 9 bilateral. Male sex and tumor size greater than 5 mm predicted ipsilateral nodal disease. Only the presence of ipsilateral nodal disease predicted the presence of contralateral nodal disease.

Taking another approach, Kwak et al.[27] studied patients with PTMC to see if preoperative thyroid US findings could predict the presence of metastatic nodes and the need for neck dissection. Findings that correlated with surgically proven metastatic nodal disease in the lateral compartment included an upper pole location of the nodule, contact of over 25% of the nodule with the adjacent capsule, and presence of calcification. Presence of central compartment metastatic nodes also predicted the presence of lateral compartment nodes. In this series of 671 PTMC patients, 3.7% were found to have lateral compartment metastatic nodes.

These data still leave some uncertainty about the need for aggressive surgery and RAI ablation in patients with a single PTMC. The data suggest that in patients 45 years old or older, a central node dissection should be done, and in patients with certain thyroid US findings, a lateral neck dissection should be done as well. However, these last comments are based on interpretation of the data and not clear evidence of improved outcomes. Clearly, a better way of identifying those PTMCs destined to be more aggressive is needed, thus setting the stage for molecular medicine.

7.4 CLINICAL CASE 3: HÜRTHLE CELL THYROID CARCINOMA

7.4.1 Vignette

The patient was a 62-year-old woman requesting a second opinion. She had an FNA of a right thyroid nodule suggestive of a Hürthle cell neoplasm. She had a right hemithyroidectomy, and the final surgical pathology demonstrated a well-circumscribed 3.8 × 2.8 × 2.5 cm lesion with focal capsular invasion. The final diagnosis was minimally invasive HTC. A postoperative US identified several "benign cysts" in the remaining left thyroid lobe, the largest of which was 1.4 cm. No abnormal cervical lymph nodes were identified. The question is: Does this patient need a completion thyroidectomy and RAI ablation of any remaining thyroid tissue, or is a hemithyroidectomy in this patient with a well-circumscribed minimally invasive HTC sufficient?

7.4.2 Discussion

HTC is considered a more aggressive form of thyroid carcinoma requiring more aggressive therapy. And yet, this may not always be the case. Hanief et al.[28] emphasized that HTC can be aggressive and lethal or it can run a benign course, and they proposed a way to stratify risk. They suggested that the usual approach to staging all well-differentiated thyroid carcinomas should also apply to HTC. Age over 45, male sex, size of tumor (over 4 cm), degree of invasion, nodal involvement, and recurrence after the initial surgery are some of the factors that predict aggressive behavior. For those with well-localized disease (no capsular invasion or less than 1 mm breaching of the capsule[29]), tumors that are small, and young age, less than total thyroidectomy may be satisfactory. On the other hand, there may be contralateral disease in 40% to 70% of patients with HTC,[30] which may also be less responsive to RAI therapy.[31] This is consistent with the finding that recurrent HTC develops in only 17% of patients who have had a total thyroidectomy, compared with 59% of patients who have had a more limited procedure.[32]

Patients who receive RAI for remnant ablation after total thyroidectomy have been reported to have a lower mortality than patients who received no RAI or who received it to treat residual disease.[33] This supports the idea that clinically occult disease may be present in the remaining thyroid tissue, and that this occult disease may impact clinical outcome. Because of her older age, the patient in this clinical case was advised to have a completion thyroidectomy with remnant ablation. However, the question remains: Was this more aggressive approach really necessary, and could novel molecular markers resolve this issue?

7.5 CLINICAL CASE 4: MEDULLARY THYROID CANCER

7.5.1 Vignette

The patient was a 64-year-old woman who had a near-total thyroidectomy for upper airway obstructive symptoms. She had a large multinodular goiter with chronic thyroiditis on biopsy. Surgical pathology incidentally revealed a 4 mm MTC. On questioning, it was learned that her 86-year-old sister had a nearly identical clinical, surgical, and pathologic diagnosis. In the proband, RET proto-oncogene was positive for an 891 mutation. Her immediate family was tested, and her son and grandson were positive for the mutation. Both underwent total thyroidectomy, with the former having a microscopic MTC, while the latter had C-cell hyperplasia by immunostaining for calcitonin (CTN). The proband's sister was positive for the mutation, while neither of her two children harbored the mutation. The proband's 62-year-old brother was positive for the mutation and had a baseline CTN of 42 ng/L (nl < 10 ng/L) and a negative thyroid US. He had recently undergone surgery for colon cancer and declined any further investigation or surgery. With the exception of the last individual, all affected family members have been regularly screened for the presence of pheochromocytoma and none have been positive. Calcium and parathyroid hormone levels have all been normal.

7.5.2 Discussion

At least 30% of MTC is hereditary, with the remainder being sporadic. With genetic analysis now readily available, all patients with MTC should be evaluated, since some apparently sporadic cases will represent the proband in newly detected families. The ability to detect *RET* proto-oncogene mutations in more than 98% of families with MEN-2 or fMTC has simplified the detection of early disease and the potential for prevention of a lethal cancer. Newly reported genetic information, which correlates the specific mutation type (genotype) with the phenotypic behavior of the thyroid tumor, the presence of other tissue involvement, and the time to development of pheochromocytoma,[34] may yield even more precise practice recommendations.

The *RET* gene is located near the centromere on the long arm of chromosome 10 and includes 21 exons that encode a receptor tyrosine kinase expressed mainly in neural crest and urogential precursor cells. These include adrenal medullary, parasympathetic, sympathetic, colonic ganglia, parathyroid, and thyroid parafollicular C-cells.[35] *RET* is a proto-oncogene, which may undergo activation by DNA rearrangement. The RET protein has four cadherin-like repeats and a cysteine-rich region in the extracellular domain, a transmembrane domain, and two intracellular tyrosine kinase domains. The extracellular cysteine-rich region is important for receptor dimerization, and the intracellular tyrosine kinase domains are involved in signal transduction. Alternative splicing of the 3' region of RET generates three protein isoforms, which differ in the amino acid sequence of the C-terminal tail.[36] Moreover, different isoforms activate different signaling pathways.[37] Four members of the glial cell line–derived neurotrophic factor (GDNF) family have been shown to be RET receptor ligands. RET activation by each ligand is mediated through specific glycosyl-phosphatidylinositol-anchored coreceptors (GFRα1–4).[38] Interaction of one of the RET receptor ligands with its specific GFRα coreceptor induces RET dimerization, which results in tyrosine autophosphorylation of the intracellular domain at specific tyrosine residues. These phosphorylated tyrosines serve as binding sites for a number of transduction molecules to activate downstream signaling pathways. Germline activating point mutations are responsible for the development of MEN-2A, MEN-2B, and fMTC. Since *RET* is a proto-oncogene, a single activating mutation of one allele is sufficient for neoplastic transformation. The mutations in MEN-2 are located mainly in the extracellular cysteine-rich domain corresponding to exons 10 and 11, or the intracellular tyrosine kinase domains corresponding to exons 13 to 16.[39] Cysteine mutations identified in MEN-2A and fMTC families activate RET by inducing ligand-independent, disulfide-linked homodimerization of two RET mutants resulting in constitutive activation.[40] It is thought that replacement of a cysteine with another amino acid due to a fMTC or MEN-2A mutation allows an adjacent cysteine, which would normally be involved in forming an intramolecular disulfide bond, to become free and induce an aberrant intermolecular disulfide bond between two adjacent mutant RET receptors. Mutations in the tyrosine kinase domain induce a conformational change of the catalytic core, which alters substrate specificity and allows RET activation without dimerization.

While it is known that RET mutations in MEN-2 cause constitutive activation by either ligand-independent dimerization in the cysteine-rich domain mutations or conformational changes to the catalytic core of the kinase domain in the tyrosine kinase domain mutations, specific mechanisms in the downstream signaling cascades, which produce different phenotypes, are still uncertain. Mutations may change substrate specificity or alter activation of signaling pathways with variable effects on the timing of development of the tumor and its virulence.

Differences in transforming activity of RET receptors carrying different mutations within the cysteine-rich domain may be due to differences in the ability of the particular cysteine mutation to induce intermolecular disulfide bridge-mediated homodimerization, or to differences in cell surface expression of the mutated RET receptors in various tissues. Differences in susceptibility to the downstream effects of RET constitutive activation may be present in different tissues. For instance, these susceptibilities are more likely in thyroid parafollicular C-cells and less likely in adrenal chromaffin cells. It is possible that not only the position but also the nature of the mutations of the RET proto-oncogene influence the clinical manifestation of MEN-2. In particular, individuals with a Cys634-Arg substitution have a greater risk of developing parathyroid disease,[41] while mutations in other codons are less likely to be associated with hyperparathyroidism and pheochromocytoma.[42]

As new families are reported, it is important to recognize that the prevalence and types of pathogenetic mutations are dependent on the genetic substrate of the population studied.[43,44] It is also important to remember that newly detected mutations may be found due to *de novo* mutations in families or expanded repertoires of germline DNA analysis.[45]

MEN-2B point mutations at codon 914 result in early onset of aggressive MTC, mandating very early thyroidectomy in children harboring the mutation. Associated features include pheochromocytomas in 50% of affected subjects, marfanoid habitus, mucosal neuromas, and ganglioneuromas of the intestine.[46] Within the more common MEN-2A syndrome, both the presence of nonthyroidal tumors (pheochromocytoma and hyperparathyroidism) and the virulence of the MTC have been linked to specific mutations.

In patients with codon 790, 791, 804, or 891 mutations, the mean age is significantly higher, tumor stage at diagnosis more favorable, and clinical course more indolent, compared with patients with the codon 634 mutation.[47] In some series, codon 790/791 mutations have variable penetrance. Therefore, prophylactic thyroidectomy in children is justifiable for codon 790 mutation carriers, while the indication for thyroidectomy might depend on the clinical course of codon 791 carriers.[48]

Likewise, patients with the 804 mutation may display variable penetrance and clinical course. It has been suggested that the V804L mutation causes the low-penetrance MEN-2 syndrome, with a late onset and relatively indolent course, even though some reports of V804L and V804M having an aggressive potential have occurred. It is postulated that the phenotypic heterogeneity in this group of patients is consistent with a model in which codon 804 mutations have low penetrance, with the development of MTC being associated with a second germline or somatic mutation. The activity and (in the case of somatic mutations) timing of these other genetic alterations in the *RET* gene may explain the wide clinical variability associated with germline mutations at codon 804.[49] The clinical implication of this hypothesis is that a very careful analysis of family members with this mutation is necessary due to the relative unpredictability of the disease course.

In this clinical case with the 891 mutation, initial studies of a limited number of families revealed the development only of MCT, without pheochromocytoma, with relatively late onset of clinical disease and a more indolent course. Given this information, the timing of family investigation and the mandate for early thyroidectomy may be tempered. As with all newly described syndromes, however, it is important to recognize that new information may become available. Therefore, the investigation and management of any given family should proceed with the notion that each family is a pioneer in a new field, so that all information should be sought and prevalent literature should not be taken as dogma. For instance, one database reported the absence of pheochromocytomas in families with the codon 891 mutation,[34] yet at nearly the same time, a new family was described linking pheochromocytoma to an 891 mutation.[50] Therefore, the members of the family in this clinical case should be screened regularly for pheochromocytoma even if it is (at present) statistically unlikely.

7.6 CONCLUSION

Using gene mutations to aid in clinical decision making is an exciting next step in transforming the management of thyroid cancer from expert opinion based on histology and clinical features to molecular analysis that can determine the behavior of such lesions. The recognition that gene mutations, transcription factor activity, and other cellular and nuclear abnormalities determine the virulence of thyroid cancers will strengthen clinical staging and management strategies for thyroid cancer. This will allow the clinician to assign the proper risk of recurrence and death for each type of thyroid cancer and deliver appropriate, commensurate treatment protocols.

REFERENCES

1. Sherman S. 2009. Emerging thyroid cancer therapies show promise. *Clin Endocrinol News* 4:1
2. Karger S, Weidinger C, Krause K, et al. 2009. FOXO3a: A novel player in thyroid carcinogenesis? *Endocr Relat Cancer* 16:189–99.
3. Dotto J, Nosé V. 2008. Familial thyroid carcinoma: A diagnostic algorithm. *Adv Anat Pathol* 15(6):332–49.
4. Xing M. 2005. The T1799A BRAF mutation is not a germline mutation in familial nonmedullary thyroid cancer. *Clin Endocrinol* (Oxf) 63(3):263–66.
5. Cavaco BM, Batista PF, Martins C, et al. 2008. Familial non-medullary thyroid carcinoma (FNMTC): Analysis of fPTC/PRN, NMTC1, MNG1 and TCO susceptibility loci and identification of somatic BRAF and RAS mutations. *Endocr Relat Cancer* 15(1):207–15.
6. Capezzone M, Cantara S, Marchisotta S, et al. 2008. Short telomeres, telomerase reverse transcriptase gene amplification, and increased telomerase activity in the blood of familial papillary thyroid cancer patients. *J Clin Endocrinol Metab* 93(10):3950–57.
7. Puzianowska-Kuźnicka M, Pietrzak M. 2005. Genetic factors predisposing to the development of papillary thyroid cancer *Endokrynol Pol* 56(3):339–45.
8. Ito Y, Motoo Y, Yoshida H, Iovanna JL, et al. 2004. High level of tumour protein p53-induced nuclear protein 1 (TP53INP1) expression in anaplastic carcinoma of the thyroid. *Endocr Pathol* 15(4):319–27.
9. Nikiforov YE. 2004. Genetic alterations involved in the transition from well-differentiated to poorly differentiated and anaplastic thyroid carcinomas. *Endocr Pathol* 15(4):319–27.
10. Quiros RM, Ding HG, Gattuso P, et al. 2005. Evidence that one subset of anaplastic thyroid carcinomas are derived from papillary carcinomas due to BRAF and p53 mutations *Cancer* 103(11):2261–68.
11. Ito Y, Yoshida H, Tomoda C, Uruno T, et al. 2003. Decreased expression of p107 is correlated with anaplastic transformation in papillary carcinoma of the thyroid. *Anticancer Res* 23(5A):3819–24.
12. Ito Y, Yoshida H, Uruno T, et al. 2003. Decreased expression of cyclin G2 is significantly linked to the malignant transformation of papillary carcinoma of the thyroid. *Anticancer Res* 23(3B):2335–38.
13. Ito Y, Arai K, Nozawa R, et al. 2007. S100A10 expression in thyroid neoplasms originating from the follicular epithelium: Contribution to the aggressive characteristic of anaplastic carcinoma. *Anticancer Res* 27(4C):2679–83.
14. Davies L, Welch HG. 2006. Increasing incidence of thyroid cancer in the United States, 1973–2002. *JAMA* 295:2164–67.
15. Greene FL, Page DL, Fleming ID, et al. 2002. *AJCC cancer staging handbook: TNM classification of malignant tumors*. 6th ed. New York: Springer-Verlag.
16. Tuttle M. 2008. Risk-adapted management of thyroid cancer. *Endocr Pract* 14:764–74.
17. Burman K. 2006. Micropapillary thyroid cancer: Should we aspirate all nodules regardless of size? *J Clin Endocrinol Metab* 91:2043–46.
18. Mazzaferri EL, Iloos RT. 2001. Clinical review: Current approaches to primary therapy for papillary and follicular thyroid cancer. *J Clin Endocrinol Metab* 86:1447–63.
19. Ito Y, Uruno T, Nakano K, et al. 2003. An observation trial without surgical treatment in patients with papillary microcarcinoma of the thyroid. *Thyroid* 13:381–87.
20. Gharib H, Papini E, Valcavi R, et al. 2006. American Association of Clinical Endocrinologists and Associazione Medici Endocrinologi. Medical guidelines for clinical practice for the diagnosis and management of thyroid nodules. *Endocr Pract* 12:63–102.
21. Cooper DS, Doherty GM, Haugen BR, et al. 2006. Management guidelines for patients with thyroid nodules and differentiated thyroid cancer. *Thyroid* 16:109–41.

22. Asanuma K, Kobayashi S, Shingu K, et al. 2001. The rate of tumour growth does not distinguish between malignant and benign thyroid nodules. *Eur J Surg* 167:102–5.
23. Nikiforov YE, Steward DL, Robinson-Smith TM, et al. 2009. Molecular testing for mutations in improving the fine-needle aspiration diagnosis of thyroid nodules. *J Clin Endocrinol Metab* 94:2092–98.
24. Roti E, Rossi R, Trasforini G, et al. 2006. Clinical and histological characteristics of papillary thyroid microcarcinoma: Results of a retrospective study in 243 patients. *J Clin Endocrinol Metab* 91:2171–78.
25. Zaydfudim V, Feurer ID, Griffin MR, et al. 2008. The impact of lymph node involvement on survival in patients with papillary and follicular thyroid carcinoma. *Surgery* 8:34.
26. Lim YC, Choi EC, Yoon YH, et al. 2009. Central lymph node metastases in unilateral papillary thyroid microcarcinoma. *Br J Surg* 96:253–57.
27. Kwak JY, Kim EK, Kim MJ, et al. Papillary microcarcinoma of the thyroid: Predicting factors of lateral neck node metastasis. *Ann Surg Oncol*, February 2009, 16:1384–1355.
28. Hanief MR, Igali L, Grama D. 2004. Hürthle cell carcinoma: Diagnostic and therapeutic implications. *World J Surg Oncol* 2:27.
29. Genden E, Brett E. 2009. Contemporary management of thyroid carcinoma. Cancer Ther 7:7–20.
30. Gundry SR, Burney RE, Thompson NW, et al. 1983. Total thyroidectomy for Hürthle cell neoplas of the thyroid. *Arch Surg* 118:529–32.
31. Caplan RH, Abellera RM, Kisken WA. 1994. Hurthle cell neoplasms of the thyroid gland: Reassessment of functional capacity. *Thyroid* 4:243–48.
32. McLeod MK, Thompson NW. 1990. Hürthle cell neoplasms of the thyroid. *Otoloryngol Clin North Am* 23:441–52.
33. Lopez-Penabad L, Chiu AC, Hoff AO, et al. 2003. Prognostic factors in patients with Hürthle cell neoplasms of the thyroid. *Cancer* 97:1186–94.
34. Machens A, Brauckhoff M, Holzhausen HJ, et al. 2005. Codon-specific development of pheochromocytoma in multiple endocrine neoplasia type 2. *J Clin Endocrinol Metab* 90:3999–4003.
35. Pachnis V, Mankoo B, Costantini F. 1993. Expression of the c-ret proto-oncogene during mouse embryogenesis. *Development* 119:1005–17.
36. Myers SM, Eng C, Ponder BA, et al. 1995. Characterization of RET proto-oncogene 3' splicing variants and polyadenylation sites: A novel C-terminus for RET. *Oncogene* 11:2039–45.
37. Tsui-Pierchala BA, Ahrens RC, Crowder RJ, et al. 2002. The long and short isoforms of Ret function as independent signaling complexes. *J Biol Chem* 277:34618–25.
38. Airaksinen MS, Saarma M. 2002. The GDNF family: Signalling, biological functions and therapeutic value. *Nat Rev Neurosci* 3:383–94.
39. Akhand AA, Ikeyama T, Akazawa S et al. 2002. Evidence of both extra- and intracellular cysteine targets of protein modification for activation of RET kinase. *Biochem Biophys Res Commun* 292:826–31.
40. Asai N, Iwashita T, Matsuyama M, et al. 1995. Mechanism of activation of the ret proto-oncogene by multiple endocrine neoplasia 2A mutations. *Mol Cell Biol* 15:1613–19.
41. Frank-Raue K, Höppner W, Frilling A, et al. 1996. Mutations of the ret proto-oncogene in German multiple endocrine neoplasia families: Relation between genotype and phenotype. German Medullary Thyroid Carcinoma Study Group. *J Clin Endocrinol Metab* 81:1780–83.
42. Carlomagno F, Salvatore G, Cirafici AM, et al. 1997. The different RET-activating capability of mutations of cysteine 620 or cysteine 634 correlates with the multiple endocrine neoplasia type 2 disease phenotype. *Cancer Res* 57:391–95.
43. Pinna G, Orgiana G, Riola A, et al. 2007. RET proto-oncogene in Sardinia: V804M is the most frequent mutation and may be associated with FMTC/MEN-2A phenotype. *Thyroid* 17:101–4.
44. Chung YJ, Kim HH, Kim HJ, et al. 2004. RET proto-oncogene mutations are restricted to codon 634 and 618 in Korean families with multiple endocrine neoplasia 2A. *Thyroid* 14:813–18.
45. Bethanis S, Koutsodontis G, Palouka T, et al. 2007. Newly detected mutation of the RET protooncogene in exon 8 as a cause of multiple endocrine neoplasia type 2A. *Hormones* (Athens) 6:152–56.
46. Yip L, Cote GJ, Shapiro SE, et al. 2003. Multiple endocrine neoplasia type 2: Evaluation of the genotype–phenotype relationship. *Arch Surg* 138:409–16.
47. Frank-Raue K, Heimbach C, Rondot S, et al. 2003. Hereditary medullary thyroid carcinoma—Genotype-phenotype characterization. *Dtsch Med Wochenschr* 128:1998–2002.
48. Fitze G, Schierz M, Bredow J, et al. 2002. Various penetrance of familial medullary thyroid carcinoma in patients with RET protooncogene codon 790/791 germline mutations. *Ann Surg* 236:570–75.

49. Lesueur F, Cebrian A, Cranston A, et al. 2005. Germline homozygous mutations at codon 804 in the RET protooncogene in medullary thyroid carcinoma/multiple endocrine neoplasia type 2A patients. *J Clin Endocrinol Metab* 90(6):3454–57.
50. Jimenez C, Habra MA, Huang SC, et al. 2004. Pheochromocytoma and medullary thyroid carcinoma: A new genotype-phenotype correlation of the RET protooncogene 891 germline mutation. *J Clin Endocrinol Metab* 89:4142–45.

8 Advances in Thyroid Ultrasound for Detection and Follow-Up of Thyroid Malignancies

Technical and Clinical Aspects

Laurence Leenhardt, François Tranquart, and Jean Tramalloni

CONTENTS

ABSTRACT: Today, thyroid ultrasonography (US) is one of the most frequently used tools in the management of thyroid nodules. In the presence of a thyroid nodule, it is the first morphological examination to be carried out. This examination confirms the diagnosis of the nodule, collects the characteristics suspicious for malignancy, and guides fine-needle aspiration biopsy (FNAB), which today remains the best predictive procedure for malignancy. Several reports, particularly the latest American clinical practice guideline by radiologists, analyze the optimal US characteristics that predict malignancy and prompt ultrasound-guided FNAB (UG-FNAB).

The treatment of thyroid cancer is based on total thyroidectomy completed, if need be, by ^{131}I ablation of any remnants in the case of differentiated thyroid cancer (DTC). Recommendations for the management of DTC have been published recently by different professional societies in Europe, America, and France. The medical imaging procedure in the management of these cancers has changed considerably over the past 20 years, and the ^{131}I whole body scan has been complemented, if not nearly replaced, with cervical US, stimulated thyroglobulin (TG) levels following recombinant human TSH, and other imaging techniques, such as positron emission tomography–computed tomography (PET-CT) with fluorodeoxyglucose (FDG). Due to the widespread use of US over the past few years, and the operator-dependent nature of this examination, it is essential for the clinician to have access to a standardized procedure and criteria, which will enable validation of the quality of the results. The clinician should also be aware of the indications for cervical US and UG-FNAB in the management of these cancers. These issues are discussed in this chapter, which also includes a brief section on future prospects in ultrasonography development.

8.1 INTRODUCTION

Thyroid nodules are present in up to 50% of the adult population, increasing in prevalence with age.[1] The majority of thyroid nodules are benign, although between 5 and 15% are malignant.[2,3] Evaluation typically includes a thyrotropin (TSH) screening, a thyroid ultrasonography (US) examination, and a subsequent fine-needle aspiration biopsy (FNAB), preferably ultrasound-guided (UG-FNAB). US is the most effective method for providing information about location, number, size, echostructure, and echogenecity of thyroid nodules. FNAB of the nodules is the most accurate and cost-effective method for detecting thyroid carcinoma. Several US characteristics have been studied as potential predictors of thyroid malignancy.[4–6] Although there are characteristic features in the US distinction of benign and malignant thyroid nodules, there is also overlap in their appearance. Several publications and numerous societies have attempted to define criteria of malignancy for thyroid nodules. These resources present recommendations based on nodule size, shape, and other characteristics, which merit UG-FNAB.[6,7] The first part of this chapter summarizes the use of US in the preoperative evaluation of thyroid nodules.

Recommendations for the management of DTC have been recently published by different professional societies in Europe,[8] America,[7] and France.[9] After total thyroidectomy and possible radioiodine (RAI) ablation, the goal of the clinician is to detect loco-regional recurrences or distant metastases. The combination of a positive US examination and increased thyroglobulin (TG) following recombinant human TSH stimulation constitutes the best criterion for detecting recurrences. The second part of this chapter analyzes the role of US examination and UG-FNAB of suspected metastatic tissue in the thyroid bed or lymph nodes in patients already diagnosed with thyroid cancer.

8.2 BASICS OF THYROID US TECHNOLOGY

8.2.1 Basic US Knowledge

US applications are based on the interactions between body tissue and acoustic energy. Sound is produced when a wave of alternative compression and decompression travels through matter. The changes in pressure with time generates a sinusoidal waveform (Figure 8.1). Some of the basic sound units of measurement are:[10]

- Wavelength (λ): Distance between two consecutive corresponding points.
- Period (T): Time to complete a single cycle.
- Frequency (f): Number of complete cycles in a unit of time.

Period and frequency are inversely related: **f = 1/T**. The unit of acoustic frequency is the hertz (Hz): 1 Hz = 1 cycle per second, 1 kilohertz (kHz) = 1,000 Hz, and 1 megahertz (MHz) = 10^6 Hz. Ultrasounds (2 to 20 MHz for biological applications) differ from audible sounds (20 to 20,000 Hz) only in their frequency.

8.2.2 Propagation of Sound Wave

Brief pulses of energy are transmitted into the body. Acoustic pressure waves usually travel in the direction of particle displacement (longitudinal waves). The propagation velocity depends on the resistance of the medium to compression. In the body, propagation velocity is assumed to be 1540 m/sec. Propagation velocity (c) is related to frequency and wavelength: $c = f\lambda$. Velocity is critical in determining the distance of a reflecting point to the probe.

US apparatus is based on the detection of reflected sound (echoes). If the medium is homogeneous, then there is no interface that can reflect sound. Therefore, the medium appears anechoic (for example, a cyst or the bladder). Acoustic impedance of a medium (Z) is related to sound velocity and the density of the medium (ρ): $Z = \rho c$.

Sound reflection, necessary for diagnostic US, occurs at the interface of two media with wide acoustic impedance differences. The US beam is also refracted and attenuated by the medium. The attenuation of US is proportional to the frequency: the higher the frequency, the greater the attenuation. For the thyroid gland, high frequencies (10 to 12 MHz) are required, whereas for deeper organ examinations (e.g., the liver), lower frequencies are necessary (3.5 to 6 MHz).

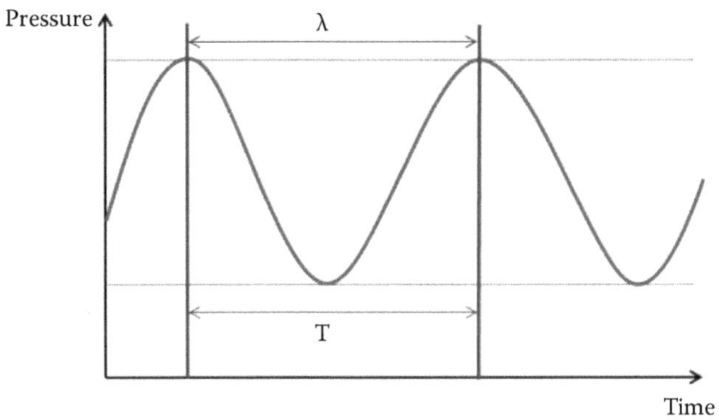

FIGURE 8.1 Sound wave intensity in relation to time.

8.2.3 The Apparatus

US waves are produced by the transducer. The same transducer is also a receiver, which detects backscattered ultrasounds. A processor detects and amplifies the backscattered signal. The US image is then displayed on a screen.

The *transmitter* controls the rate of pulses emitted by the transducer (pulse repetition frequency (PRF)). It produces high amplitude voltage, which is limited in intensity to reduce biological side effects. The *transducer* is a piezoelectric device, which converts the electric energy of the transmitter into acoustic pulses. It also converts the reflected echoes into an electric signal. The piezoelectric device responds to applications of the electric field by changing shape, and it produces electric potentials when compressed. When electrically stimulated, the transducer vibrates. The frequency of vibrations depends on transducer material. The range of frequencies produced by a transducer is its bandwidth. Modern US systems employ broad bandwidth technology. Modern transducers are composed of numerous small piezoelectric units, each with their own electrode.

For real-time B-mode imaging, as many as 60 images must be generated per second. US beams may be steered mechanically or electronically. Mechanical transducers are not in common use today. Electronic beam steering is used in linear or curved or phased array (Figure 8.2). Steering is generated by the precise timing of the firing of each piezoelectric element. There are different types of transducers:

- *Linear array* transducers are commonly used for superficial tissue applications such as thyroid examination. The image field of view is rectangular. Focusing at a selected depth is possible.
- *Curved arrays* are shaped into a convex curve: the field of view is wider in deeper regions. They combine a relatively wide field of view with a sector display format.
- *Phased arrays* look like early mechanical sector probes, but they have no moving parts. They have multiple transducer elements fired in precise sequence under electronic control. The beam can be steered in different directions and a sector image format is produced. Focusing is also possible.

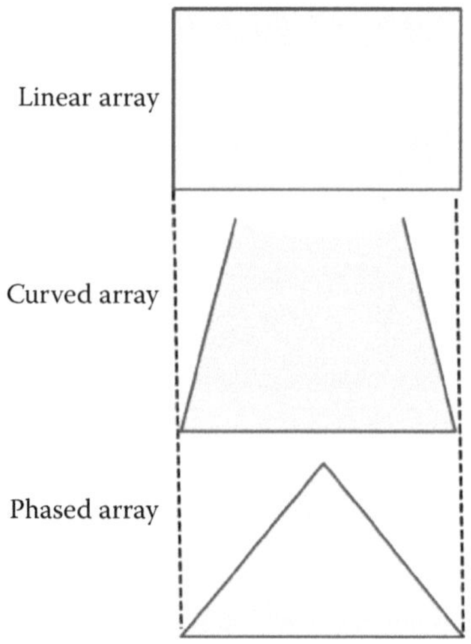

FIGURE 8.2 Field of view obtained respectively with linear, curved, and phased arrays. For the linear array, the size of the field is the size of the transducer. For the curved array, the maximum size of the field is greater than the size of the transducer. Phased arrays allow a large field in depth with a very small transducer.

The *receiver* detects and amplifies the weak potentials produced across the transducer when hit by reflected echoes. The time gain compensation (TGC) compensates the attenuation by different tissue thickness to produce a uniform tissue appearance. It is possible to attenuate strong echoes returning from superficial structures and amplify weak echoes returning from deeper tissues. The receiver also enables the compression of the range of amplitudes returning to the transducer. The ratio of the highest to the lowest amplitudes is the dynamic range, expressed in decibels. This control affects image contrast, which is why it is generally desirable to set as wide a dynamic range as possible.

The *image display* is in real time, gray scale, and B-mode. On each part of the screen, representing the scanned region of the body, corresponding reflected echoes are displayed in white for strongest echoes and in black when no echo is reflected. Intermediate intensity appears in gray scale (usually 256 levels of grays). The final image represents a slice of the scanned region. The impression of motion is produced by generating a series of 2D images at rates from 15 to 60 images per second. The image produced is read on a video display and may be printed or stored on a Picture Archival and Communications System (PACS).

8.2.4 Image Quality Factors

One of the most important factors is the spatial resolution, which is defined as the distance between two closely separated objects that can be distinguished. There are different types of resolutions:

- Separate *axial resolution* (along the axis of an ultrasound's beam propagation)
- *Lateral resolution* (in the plane perpendicular to the beam)
- *Elevation resolution*, which refers to the slice thickness in the plane perpendicular to the beam and the transducer

Axial resolution depends on transducer frequency: the higher the frequency, the higher the axial resolution. The axial resolution varies from 1 mm with a 5 MHz transducer to 0.4 mm with a 15 MHz transducer. Lateral and elevation resolutions are determined by the width of the beam. However, whereas the lateral resolution can be controlled by focusing, the elevation resolution depends on the design of the transducer and cannot be controlled by the user.

8.2.5 Doppler and Color Sonography

When an US-pulsed wave is reflected from an interface, the reflected signal contains amplitude, phase, and frequency information. B-mode imaging uses only amplitude information. When an object is moving (for example, red cells in the blood), the reflected echoes have a different frequency (Doppler-Fizeau effect). This change in frequency is directly proportional to the velocity of the reflecting interface. Doppler sonography enables the study of blood velocity. In color-flow Doppler imaging, velocity information is displayed in the image itself, coded in color in each pixel.

8.2.6 The Operator and His Training

There are no clinical practice guidelines (CPG) regarding this point. Nevertheless, it is strongly recommended that the operator be trained not only for US but also for UG-FNAB proficiency. One of the limits of US examination is the poor reproducibility that has been reported. This is best addressed by having the same operator repeat serial US examinations when following patients preoperatively for thyroid nodules, or postoperatively for thyroid malignancy follow-up.

8.2.7 The Ultrasound Procedure

Two types of probes are recommended to study the thyroid gland or the cervical region after a total thyroidectomy:

- A high-frequency linear probe (10 to 14 MHz). This is an indispensable tool that offers the advantage of high resolution but has the disadvantages of a small field and limited depth. Today, there are high-quality probes on the market with a field of over 40 mm in width. However, small field probes are easy to use on short necks.
- A convex small-radius probe (6 to 8 MHz) used for carrying out UG-FNAB and to study submaxillary and subclavicular regions.

The device must be equipped with at least a sensitive Doppler module capable of recording flows in small vessels: pulsated and color-flow Doppler energy imaging.

Patients are usually scanned in the supine position with the neck slightly hyperextended. Both lobes are scanned individually in the transverse and longitudinal planes. An US examination of the thyroid should always include the entire neck and the lymph node areas, especially the lateral compartment.

8.3 ULTRASOUND AND THYROID NODULES

Ultrasound has become established as the procedure of choice in CPG for the management of thyroid nodules. Professional organizations have provided recommendations for state-of-the-art diagnostic evaluation of thyroid nodules.[6,7] Several US characteristics of nodules (e.g., hypoechogenicity, microcalcifications, blurred nodule margins, and spotted vascularization) are associated with malignancy.

8.3.1 Criteria for Malignancy: Predictive Value of Ultrasound and Color-Flow Doppler Features

The Society of Radiologists in Ultrasound convened a panel of specialists from a variety of medical disciplines to derive a consensus regarding the management of thyroid nodules detected by US and published their report in 2005.[6] For each thyroid nodule, gray-scale and color-flow Doppler US are used to evaluate the US features, which include size, echogenicity (hypoechoic or hyperechoic), and composition (cystic, solid, or mixed), as well as presence or absence of microcalcifications, a halo, irregular margins, and internal, peripheral, or mixed blood flow. Nodule size alone is not predictive of malignancy.

Hypoechogenicity, presence of microcalcifications, irregular margins, absence of halo, solid composition, and intranodular blood flow are characteristics in favor of malignancy (Figure 8.3). However, as shown in Table 8.1, the sensitivities and specificities for these criteria are extremely variable from one study to another.[6] Microcalcifications rank the highest in the positive predictive value (41.8 to 94.2%). However, microcalcifications are only found in 26.1 to 59.1% of cancers (low sensitivity).[6] A combination of factors will improve the positive predictive value of US. For instance, a predominantly solid nodule with microcalcifications has a 31.6% likelihood of being cancer, as compared with a predominantly cystic nodule without calcifications, which has a 1% likelihood of being cancer. Each of these solitary US variables confers an approximate 1.5- to 3-fold increased risk of cancer; when two or more variables are present in combination, the risk of cancer appears substantially higher.[5,11–14]

Color and power Doppler imaging provide useful information regarding the likelihood of malignancy in thyroid nodules. Nevertheless, the positive and negative predictive values of the Doppler examination may not help much with a decision regarding FNAB. Most benign nodules have absent intranodular flow on power Doppler analysis, and most carcinomas have central or mixed flow.

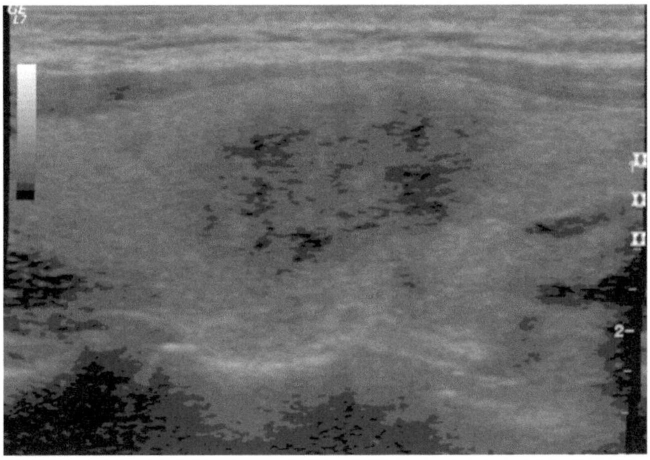

FIGURE 8.3 Hypoechoic thyroid nodule with blurred margins.

TABLE 8.1
US Features that are Associated with Thyroid Malignancies

US Feature	Description	Sensitivity	Specificity	PPV	NPV
		Nodule (Frates et al.[6])			
Cystic/solid	Mostly solid	26–59%	86–95%	24–71%	42–94%
Echogenicity	Decreased	26–87%	43–94%	11–68%	73–94%
Halo/margins	Absent/Irregular	17–77%	39–85%	9–60%	39–98%
Vascularity	Intranodular	54–74%	79–81%	24–42%	86–97%
Tall/Wide	More tall than wide	33%	92%	67%	75%
		Lymphadenopathy (Leboulleux et al.[35])			
AP:transverse ratio	> 0.7–1.0				
Hilum	No echogenic hilar line	100%	29%		
Microcalcification	Punctate	46%	100%		
Crystalization	Present	11%	100%		
Shape	Spherical (long:short axis < 2.5)	46%	64%		
Vascularization	Peripheral and spotted	86%	82%		

Note: Abbreviations: US = ultrasound; PPV = positive predictive value; NPV = negative predictive value.

However, the results of studies about Doppler imaging are mixed: some report that Doppler is helpful, and others that it does not improve diagnostic accuracy. In the Papini et al.[5] study, the diagnostic value of Doppler examination is included in a multivariate analysis and is helpful when combined with other suspicious US characteristics.

8.3.2 Indications for FNAB

FNAB with cytologic evaluation has become the accepted method for screening a thyroid nodule for cancer and, in the hands of an experienced cytologist, is very accurate. Cytologic specimens are typically classified as negative (or benign), suspicious for cancer or follicular neoplasm, positive (or diagnostic for cancer), or nondiagnostic. Of the aspirates read as suspicious for cancer, 30 to 65% will prove to be cancer at surgery. The nondiagnostic rates may be as high as 15 to 20%.

TABLE 8.2
CPG Indications for Thyroid FNAB15

ACT[a]	ATA[b]	AACE	SRU
< 10 mm: if there are clinical risk factors	< 10 mm: if there are clinical risk factors or suspicious US features	< 10 mm: if there are clinical risk factors or suspicious US features	< 10 mm: no recommendation
			> 10 mm: if microcalcifications
5–10 mm: if suspicious US features are present	> 10–15 mm: all nodules	> 10 mm: all nodules	> 15 mm: if solid with coarse calcifications
10–20 mm: may defer if benign US features		> 20 mm: if mixed solid and cystic or just cystic with a mural nodule	
< 20 mm: all nodules			

Note: Abbreviations: CPG = clinical practice guidelines; FNAB = fine needle aspiration biopsy; ACT = Academy of Clinical Thyroidologists; ATA = American Thyroid Association; AACE = American Association of Clinical Endocrinologists; SRU = Society of Radiologists in Ultrasound

[a] Nodule size was not specified as maximum or mean.

[b] Nodule size refers to the maximum dimension.

In patients with multinodular goiter, the US examination guides decision making for selecting the most appropriate lesion(s) for FNAB based on its likelihood of being malignant. Current CPG for UG-FNAB incorporate size (larger or smaller than 1 cm) as a criterion due to the uncertainty as to whether or not a diagnosis of smaller cancers actually improves life expectancy. Nevertheless, for nodules with suspicious US features or clinical criteria associated with a higher risk of cancer (e.g., history of neck irradiation during childhood or family history of thyroid cancer), nodules smaller than 1 cm are appropriate for FNAB. Nevertheless, some discrepancies and areas of uncertainty exist regarding the indications of FNAB, depending on the societies that published them (Table 8.2).[15] They agree upon some criteria regarding the selection of nodules that need not undergo FNAB, such as an entirely cystic nodule or a micronodule <1cm in a low-risk patient without suspicious US features. Nevertheless, there were conspicuous differences of opinion regarding which US findings were more suggestive of malignancy, and to what extent US appearance should take priority over nodule size. Given that there are valid data quantitating specific US findings with cancer risk, why would there still be discrepancies among various CPG recommendations that are presumably evidence based? Moreover, why would the performance of US and UG-FNA not be virtually 100% accurate?

Besides incorporating individual US factors alone to assess thyroid cancer risk, an emerging challenge is one that incorporates many risk factors in a multivariate fashion.[12,13] Thus, several issues regarding the management of thyroid nodules merit future research:

- Constructing strategies for follow-up and for repeated UG-FNAB
- Identifying combinations of US features that optimally direct clinical decision making deriving diagnostic predictions from elastography and other emerging technologies
- Computing cost-effectiveness of new diagnostic approaches

8.4 ULTRASOUND AND THYROID CANCER

It is estimated that there are 4,000 to 5,000 new cases of thyroid cancer discovered every year in France. Recent epidemiological studies report a strong predominance of papillary thyroid carcinoma

(PTC), which accounts for 85–90% of cases, while follicular thyroid carcinoma (FTC) accounts for only 5–8% of cases. The histological subtypes that are not highly differentiated are mainly oncocyte and insular carcinomas and account for 3–5%. The prognosis is good on the whole. Only 5% of patients die from their cancer, and 10–20% of them suffer a recurrence, which in most cases is loco-regional. This relapse sometimes occurs at a much later stage and, in this case, requires a very long monitoring period.

The incidence of small PTC has increased in recent years principally due to more accurate screening. Of those thyroid cancers surgically resected, 40–45% of them measured less than 1 cm. They are often fortuitously discovered during thyroidectomy for other indications. Since 2002, these papillary thyroid microcarcinomas (PTMCs) have been included in the pT1 of the new pTNM classification, which now includes tumors of 2 cm or less. The predominantly lymphatic spread of PTMC with lymph node metastasis further attests to the benefits of performing cervical US scans as the preferred method of diagnosis. More specifically, this supports more comprehensive initial (diagnostic) cervical US that includes a systematic survey of the lateral and central lymph node (LN) compartments in addition to the thyroid gland. This approach is especially true since these small pT1 tumors, when discovered histologically, did not undergo a lymph node neck dissection. Thus, the approach to thyroid cancer will evolve based on parallel advances in molecular pathophysiology and imaging technologies to identify those PTMCs at higher risk for recurrence and possibly mortality.

8.4.1 Goals of Neck US Examination: Detection of Recurrences

Around 5–27% of patients presenting with a differentiated thyroid cancer will relapse,[16–19] including 10.3% in the Frasoldati et al.[20] series and 8.9% of the Peltarri et al.[21] series. These recurrences are generally located in the cervical lymph nodes (60–75%), the thyroid bed (20%), or as invasion of the trachea, cervical muscles, or other neighboring structures (5%).[22–28] The probability of recurrence, on average 10%, depends first and foremost on the stage the disease has reached (based on the pTNM classification).[29] The operational definition of the level of risk used in the French consensus is as follows:[9]

- Very low risk: Intrathyroid unifocal tumor of <1 cm.
- Low risk: Well-differentiated vesicular and papillary T1-T2, N0, and M0 carcinomas.
- High risk: T3-T4 carcinomas, lymph node extension (all T, N1), distant metastasis (all T, all N, M1), and unfavorable histology results: tall-cell, cylindrical-cell, diffuse sclerosing, oncocyte, insular, and poorly differentiated vesicular carcinomas.

The majority of patients have a low risk of recurrence.[21,30] The cervical US scan has been demonstrated to be a reliable method for early testing of recurrences.[31,32] A recurrence is detected using an US scan in 94.1–96% of patients.[17,20] A combination of a positive US scan and increased stimulated TG level constitutes the best criterion for determining recurrence.[20,32,33] In the Pacini et al.[32] study, the diagnostic sensitivity of stimulated TG is 85%, and its negative predictive value is 98.2%. By combining the results of the US scan, the sensitivity associated with these two tests increases to 96.3% and the negative predictive value is 99.5%.[2] The median period for recurrences is 3.5 years.[21] Ninety-three percent of recurrences occur in the first 10 years, and 71% in the first 5 years.[21] Late recurrences (21 years) are possible.[17,26] These recurrences worsen the prognosis and increase the risk of death[17,18] even without distant metastases.

The prognostic factors for local recurrence are age >45 years, male, the size of the tumor, invasion outside of the thyroid, the extent of lymph node extension, follicular histology, partial thyroidectomy, and the absence of RAI ablation. The rate of recurrence is higher in patients presenting with initial lymph node metastases (18.7%) than in those without (N0) (6.5%, $p < 0.001$).[20] In patients with a recurrence, the factors significantly associated with an increased mortality rate in a multivariable analysis (except for age ≥45 years) are vesicular type, capsular invasion, absence of initial RAI ablation, initial presence of distant metastases, and two recurrence characteristics (non-fixing nature and location in the thyroid bed relative to the lymph node location). The relative risk of death is five times higher in cases of a recurrence located in the thyroid space than with a purely lymph node recurrence: relative risk (RR) = 5.05 (95% confidence interval (CI) = 2.62–9.74, $p < 0.05$).[17] Mazzaferri and Kloos[23] reported that the specific mortality rate of cancer at 30 years old in patients with a cervical recurrence in the soft tissue is 30%, twice as high as that observed in patients with a lymph node recurrence.

8.4.2 Cervical Ultrasound Procedure

The actual performance and then interpretation of the US procedure must be standardized in order to limit variability between operators and optimize accuracy. The patient is in the decubitus position with the neck extended. The sonographer must methodically explore the central and lateral compartments. It is easier to study the lateral compartment levels V and II B by rotating the head. The analysis in B-mode should be followed by a color Doppler study.

After a total thyroidectomy, there are special considerations:

- The thyroidectomy spaces are difficult to examine due to edematous changes. It is advisable to wait three months after the surgery to be able to correctly analyze the spaces. Surgical clips can also make the examination difficult.
- The cervical US scan does not reveal retroesophageal, retrotracheal, or upper mediastinal lesions.
- The omohyoid muscle is often sectioned during a total thyroidectomy if lymph node dissections are carried out in the lateral compartment. In this case, it may be difficult to distinguish lateral compartment level III from level IV.
- The jugulocarotid pedicle is close to the lateral side of the trachea and can make the adenopathies in the lateral compartment level VI retract. It may be difficult to distinguish between adenopathies of levels IV and VI (same echographic appearance and same topography).
- Obesity, short-neck morphotypes, and hypertrophic scars alter the quality of the examination and reduce accuracy.

8.4.3 Normal Ultrasound Findings

When performing US of the neck in a patient who has undergone a thyroidectomy, one sees that the carotid artery and jugular vein have migrated medially and are situated close to the trachea. In addition, the thyroid bed has been filled with a varying amount of hyperechoic connective tissue that appears white on US. Normal remnants of thyroid tissue may exist. These normal findings do not need to undergo FNAB. A normal neck contains approximately 300 lymph nodes. They are usually less than 0.5 cm in their short axis and flattened or oval in the transverse view of the neck, with a long axis two or more times the size of the short axis. If they become inflamed or hyperplastic, they enlarge but generally maintain this flattened or oval shape with a white hilar line.

8.4.4 Criteria of Malignancy for Cervical Lymph Nodes

Some criteria for determining malignancy are common to metastatic adenopathies of varying kinds of neoplasia (e.g., head and neck squamous cell cancers and lymphomas), while others are more specific to metastases of thyroid origin. None of the criteria are pathognomonic of malignancy alone, but taken together, they may constitute sufficient justification for presuming malignancy and acting upon this information.[34] The most specific criterion for determining whether an adenopathy is thyroid in origin is the presence of microcalcifications or cystic zones

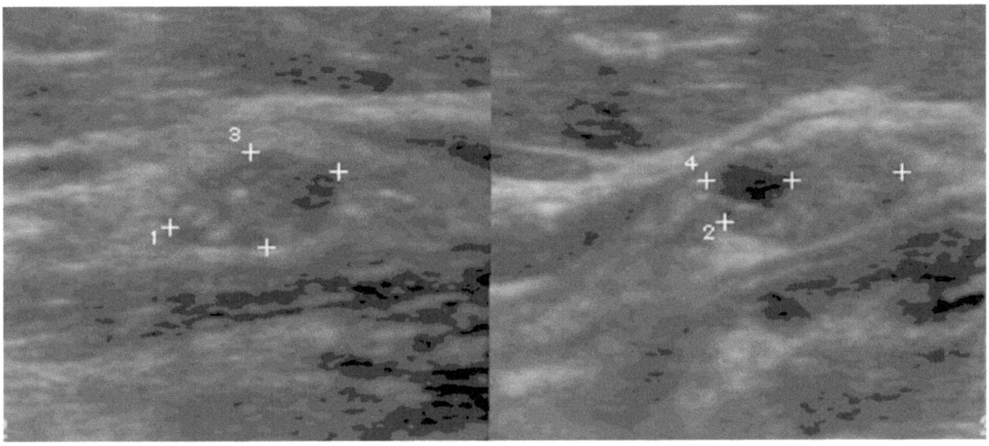

FIGURE 8.4 Suspicious cervical lymph node with cystic component and microcalcifications.

(Figure 8.4).[35] The following seven criteria must be systematically studied in order to characterize lymph node significance: size, form, echogenicity, hilus, microcalcifications, cyst formation, and distribution of vascularization.

The occurrence of the following four echo-Doppler criteria arouse suspicion for a particular lymph node; the presence of at least one of them is enough to recommend a FNAB with an *in situ* TG assay:

- Microcalcification
- The presence of cystic zone(s)
- Peripheral or mixed peripheral and anarchic internal vascularization
- Hyperechogenicity that looks like thyroid tissue (Figure 8.5)

In contrast, the following criteria are highly suggestive of benignity and do not necessarily require a FNAB: hyperechogenic hilus and central hilar vascularization without peripheral vascularization.

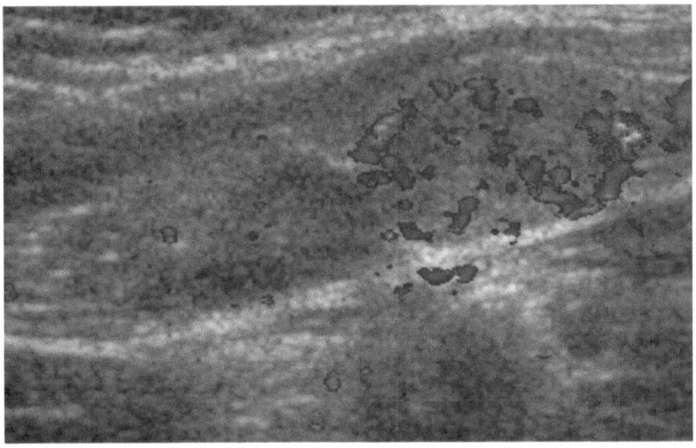

FIGURE 8.5 Hyperechogenic lymph node resembling thyroid tissue with spotted (or mixed) vascularization with color-flow Doppler examination.

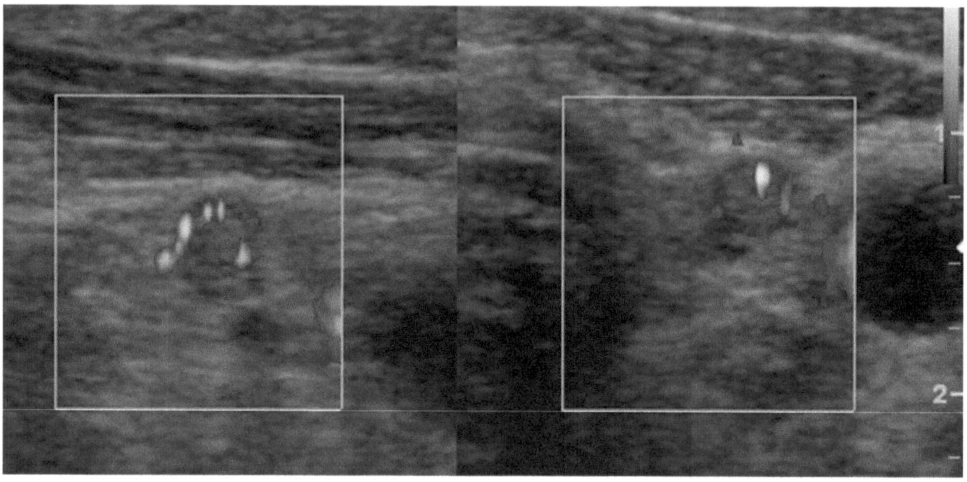

FIGURE 8.6 Thyroid bed recurrence.

8.4.5 Criteria of Malignancy for Thyroid Bed Lesions

A hypoechoic, highly vascularized mass must be considered suspect (Figure 8.6). Other nontumoral masses in the space may be present and commented on, such as hypoechoic glandular remnants, fibrosis, muscle, lymphoceles, and parathyroid gland(s).

8.4.6 Diagnosis of Subcutaneous and Muscular Recurrences

Embedded in the superficial musculocutaneous tissue, subcutaneous recurrences look like solid tissue zones that are highly hypoechoic, presenting varying levels of vascularization. Their anterior topography and hypoechogenic nature should rule out a postoperative granuloma.

8.4.7 Indications for UG-FNAB of Suspicious Lesion with TG Measurement in the FNAB Washout Fluid (TG-FNAB)

8.4.7.1 Technique for UG-FNAB and TG-FNAB

The following points regarding UG-FNAB and TG-FNAB have been gleaned from several publications:[36–39]

- Fine needles (25 to 27 gauge).
- Trained operator.
- To be avoided in the case of blood dyscrasias. If unavoidable, preventive treatment and precautions should be taken in consultation with the hematologist.
- Precautions in the case of anti–vitamin K treatment or platelet aggregation inhibiting drugs with an assessment of the risks and benefits.
- Precautions for aseptic technique.
- Echographic diagram to identify the structure into which the fine needle is to be inserted.
- Smearing, air drying, coloration with May-Grunwald-Giemsa (MGG) or liquid media.
- Rinsing of each aspiration needle in a volume of 1 ml (if several aspirations are carried out on the same lymph node, the rinsing liquid from each needle can be collected in the same tube):
    - Either with a physiological saline solution (0.9% NaCl) or with a buffer of the TG assay or a pool of TG-free serum provided by the laboratory performing the assay.
- Then perform assay of TG using the currently available technique.

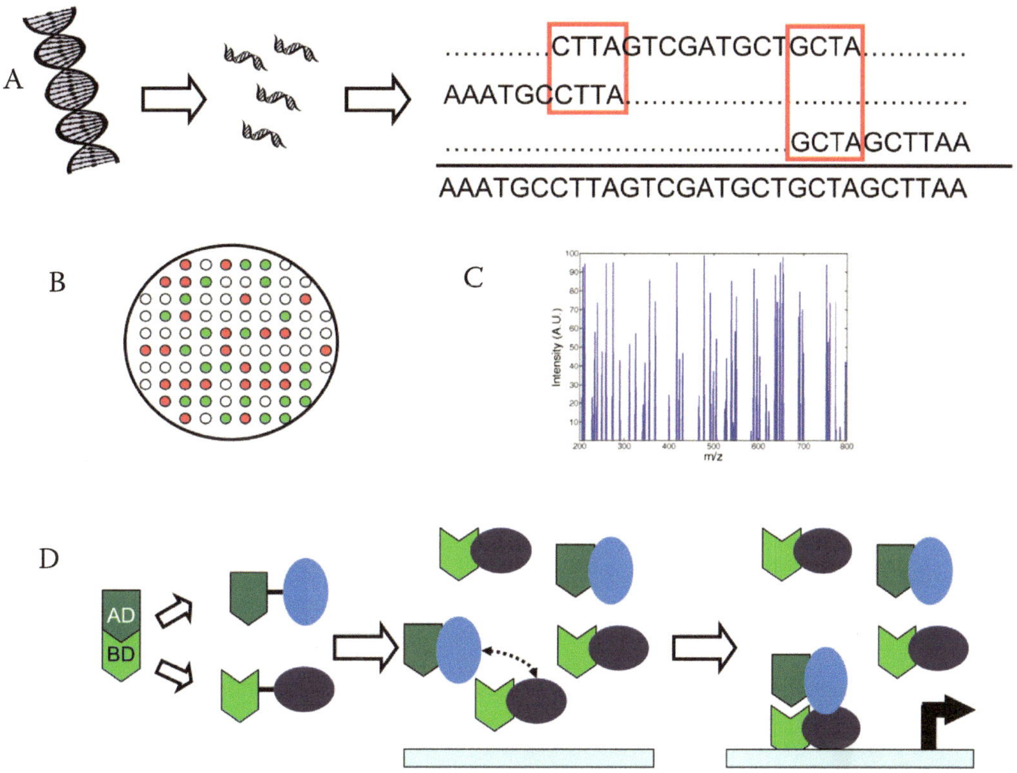

FIGURE 2.1 Schematic summary of various experimental approaches. (a) DNA sequencing: Partially overlapping fragments of DNA are sequenced separately and then merged into a unified sequence. (b) Measuring gene expression profiles: The different colors in microarray indicate the differences in gene expression. (c) Mass spectrometry to obtain the relative abundance of proteins. A typical m/z spectrum obtained from mass spectrometry experiments. High intensity indicates high abundance of the respective component. From the mass, the sequence of the fragment and the protein can be deduced. (d) Measuring protein-protein interactions: yeast 2 hybrid experiments: the activation and binding domains (AD and BD) of a transcription factor are attached to the oval proteins. If there is an interaction between those proteins, then the transcription factor can bind to the promoter and activate it. Increase in gene expression is an indication for the interaction.

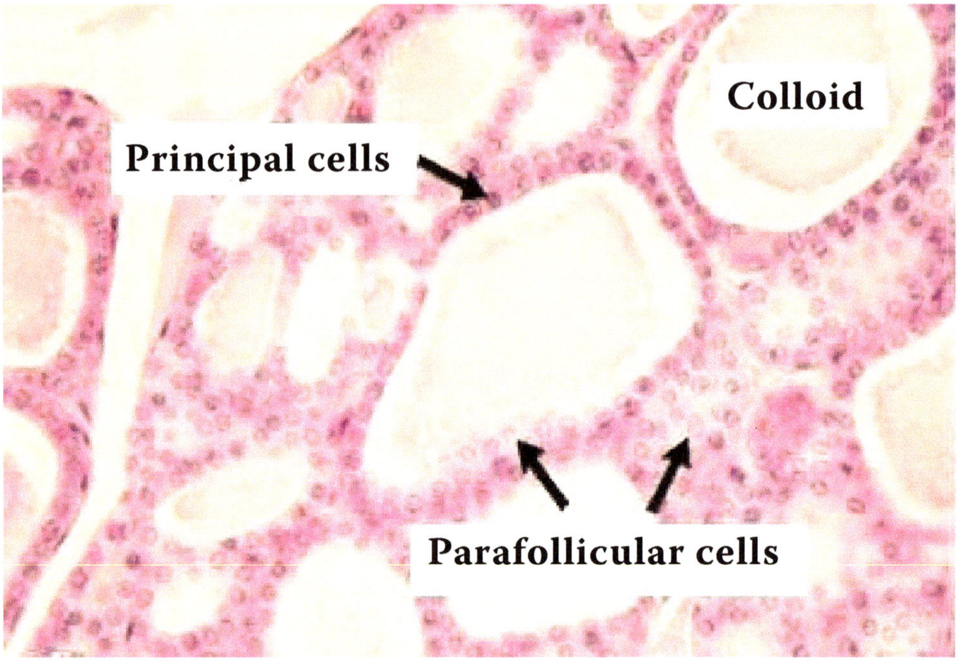

FIGURE 4.1 Thyroid histology. Principal thyroid follicular cells secrete thyroid hormones. Parafollicular cells secrete calcitonin and may be found as single cells in the epithelial lining of the follicle or in groups in the connective tissue between follicles. Parafollicular cells usually appear as large, clear cells since they do not stain well with hematoxylin and eosin. (From http://instruction.cvhs.okstate.edu/.)

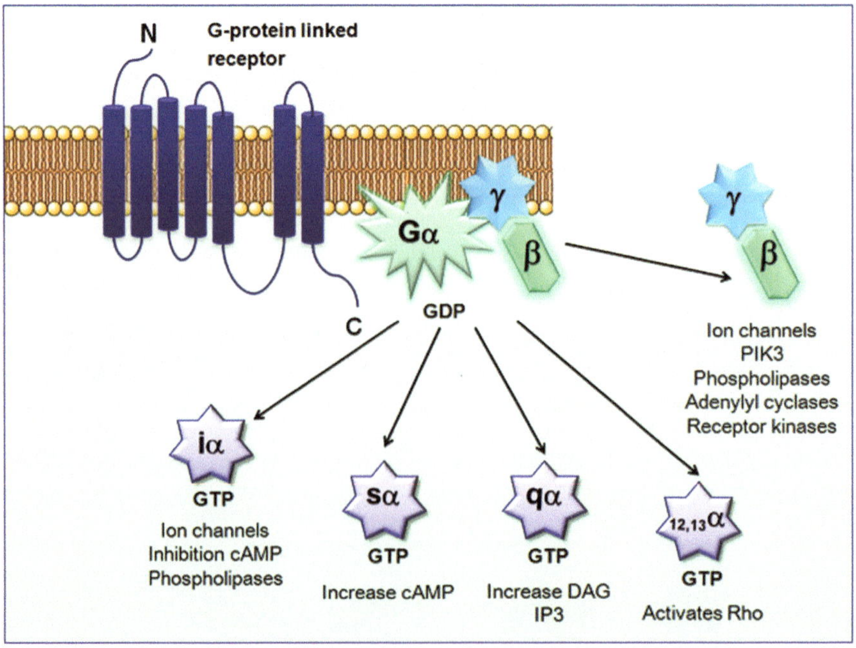

FIGURE 4.3 Calcitonin receptor signal transduction by G-proteins. (From http://medschool.creighton.edu.)

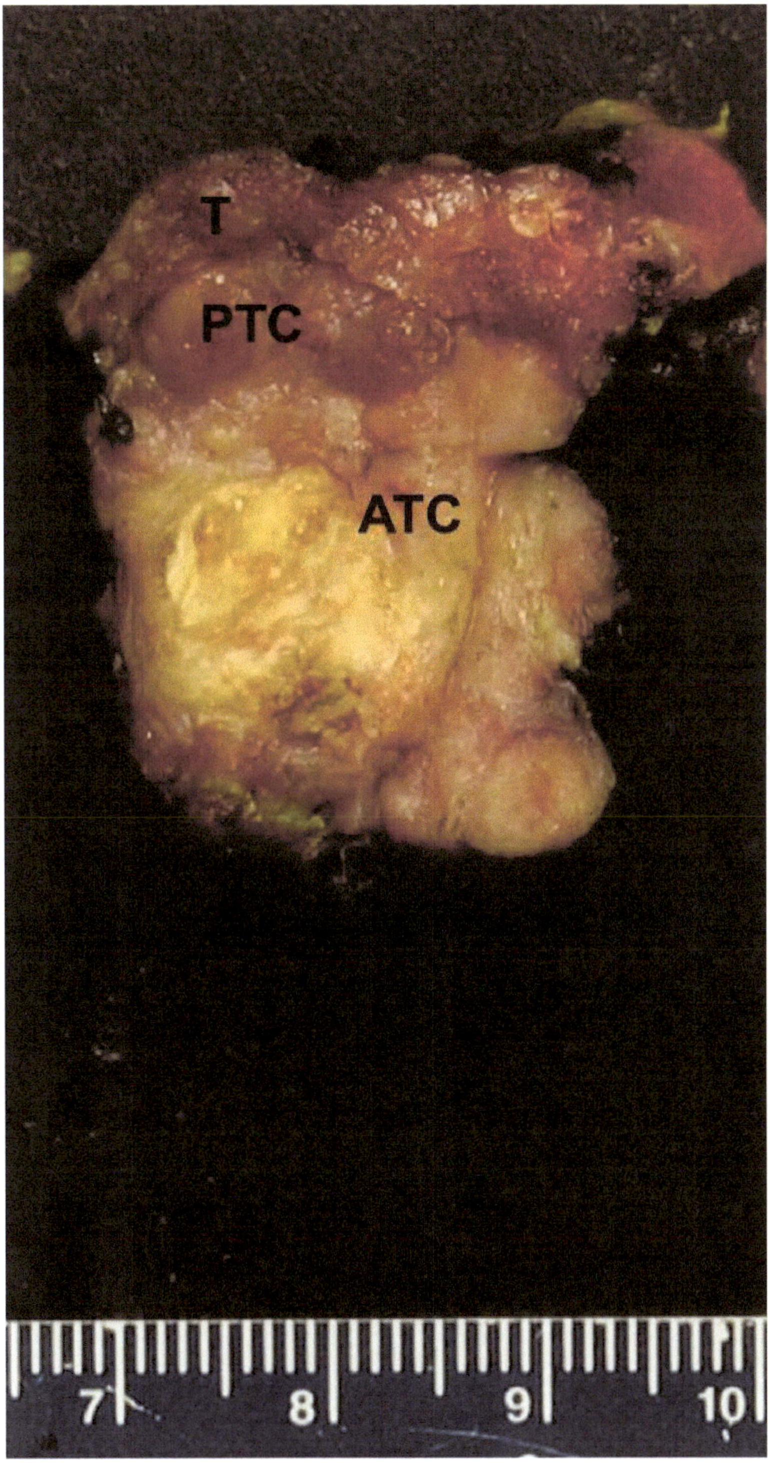

FIGURE 7.1 Case 1: Gross specimen of PTC with follicular, tall-cell, and anaplastic features. T, normal thyroid tissue; PTC, papillary thyroid carcinoma; ATC, anaplastic thyroid carcinoma.

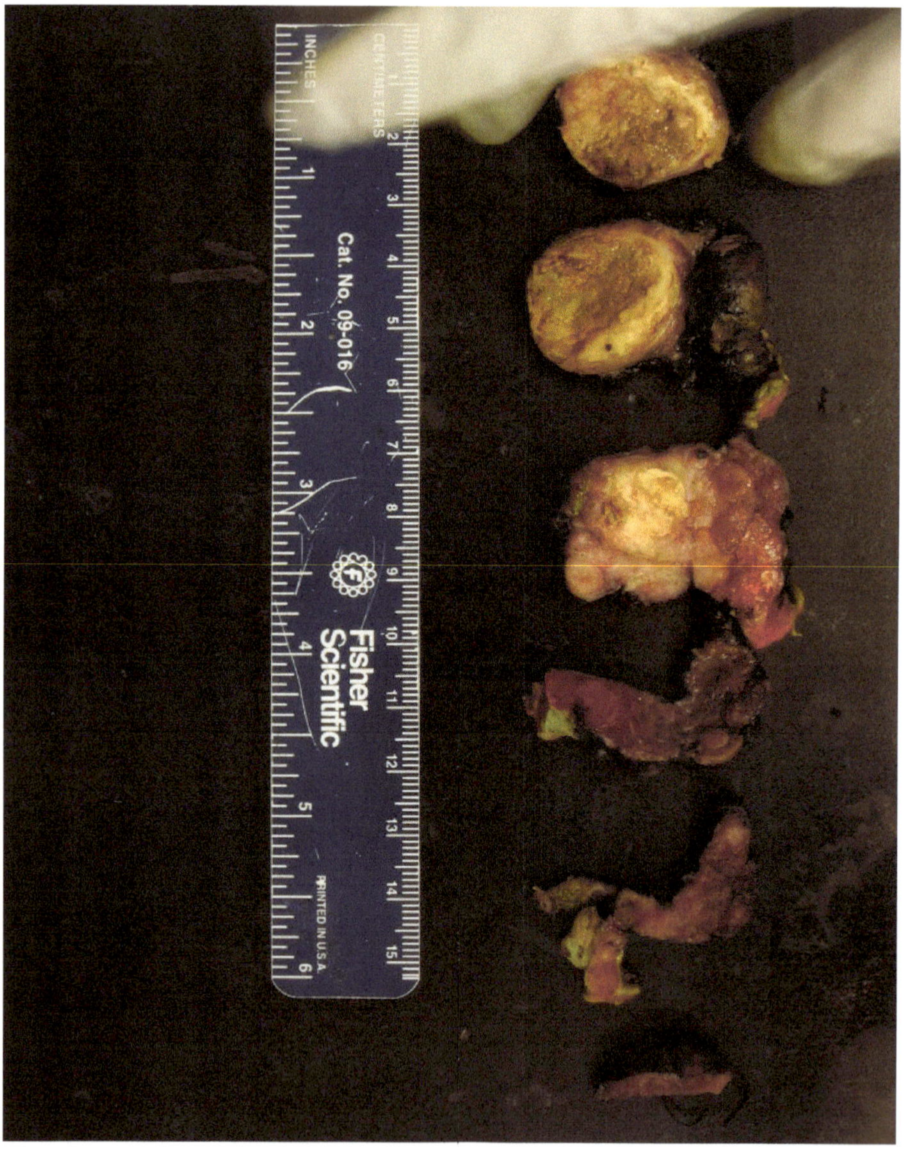

FIGURE 7.2 Gross specimens from lymph node dissection five years after an incidental 2 mm papillary thyroid microcarcinoma was discovered during parathyroidectomy.

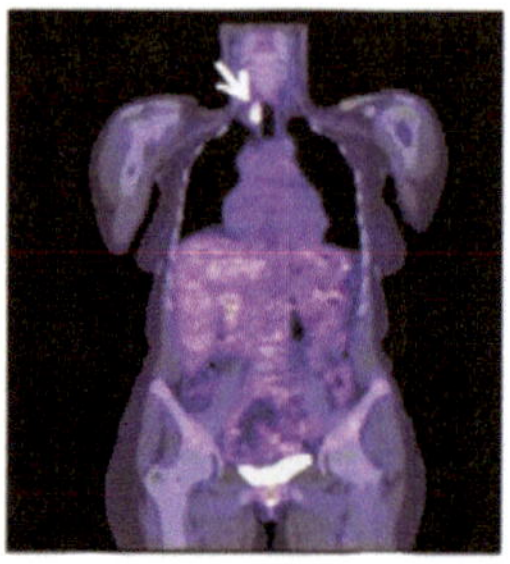

FIGURE 11.1 A 43-year-old woman with negative iodine ^{131}I whole body scan and elevated serum TG level. Combined PET/CT coronal image shows unilateral lymph node metastasis on right (arrow).

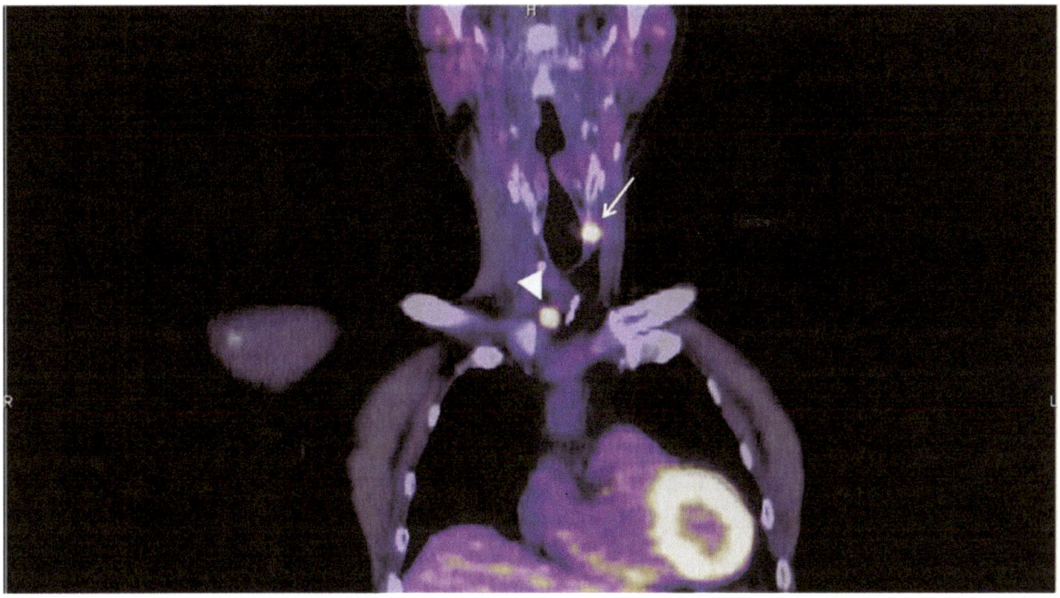

FIGURE 11.2 A 73-year-old man with a sonography suggestive of left lateral compartment lymph nodes metastasis. Combined PET/CT coronal image confirms the left cervical lymph node (arrow) metastasis and reveals a second right supraclavicular tumor focus (large arrowhead).

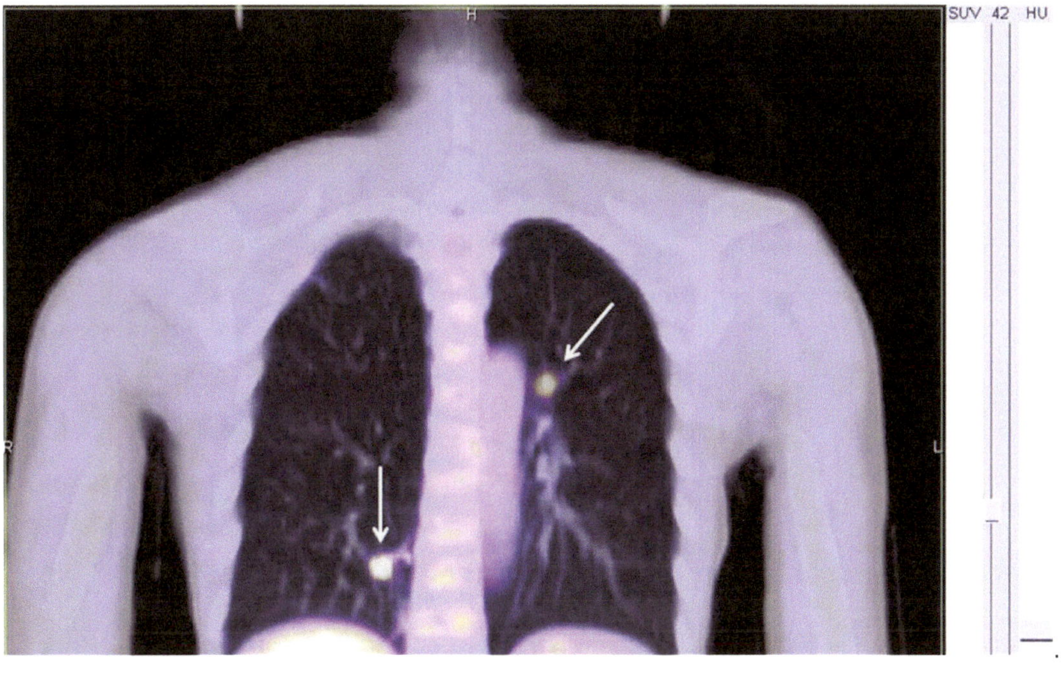

FIGURE 11.3 A 48-year-old woman with lung metastases shown on previous CT examination. Combined PET/CT coronal image confirms the two lung metastases (arrows).

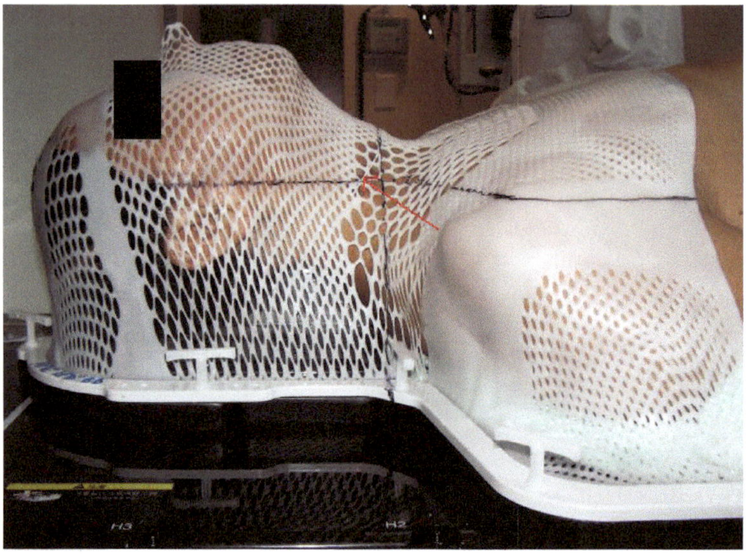

FIGURE 17.1 Patient in treatment position. An aquaplast immobilization mask has been used to ensure reproduction of the head position for daily treatment. Underneath the mask, a metal wire has been placed over the scar in order for visualization on simulation CT (red arrow).

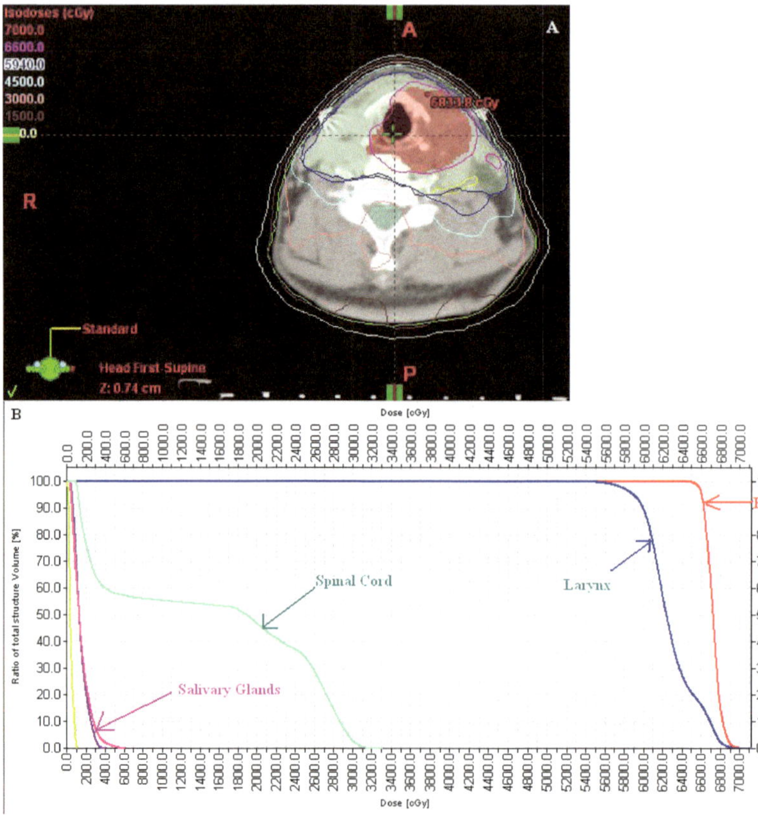

FIGURE 17.2 (A) IMRT plan of a patient with DTC after thyroidectomy. The high-risk PTV (red) is receiving a prescribed dose of 6600 cGy, while the low-risk PTV is simultaneously receiving 5940 cGy, both in 33 fractions. (B) Dose volume histograms assist the radiation oncologist in determining the adequacy of coverage of the PTV and safety of normal tissues irradiated.

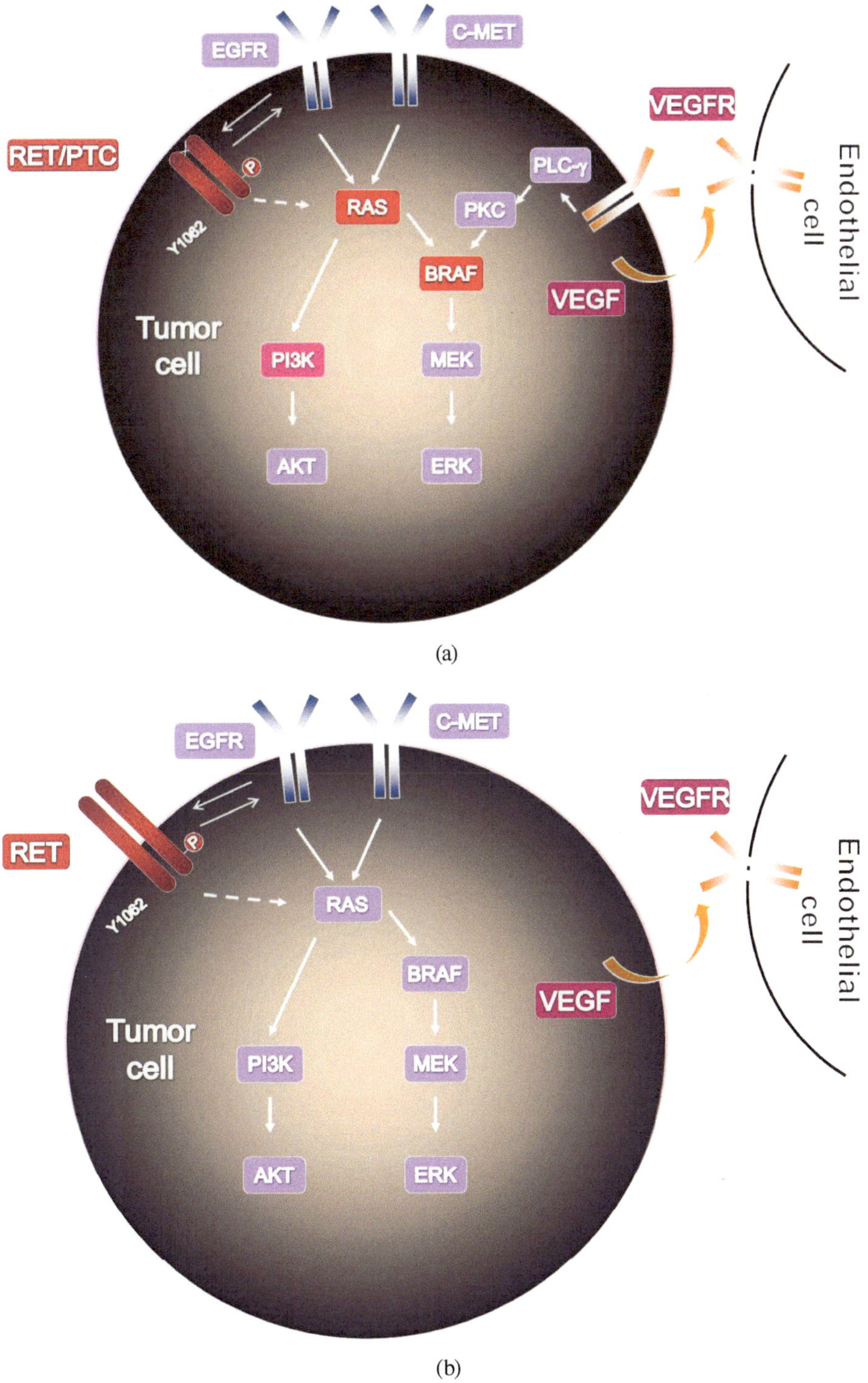

FIGURE 20.1 (a) Kinase signaling pathways relevant to oncogenesis and tumor proliferation of differentiated thyroid carcinoma. (b) Kinase signaling pathways relevant to oncogenesis and tumor proliferation of medullary thyroid carcinoma.

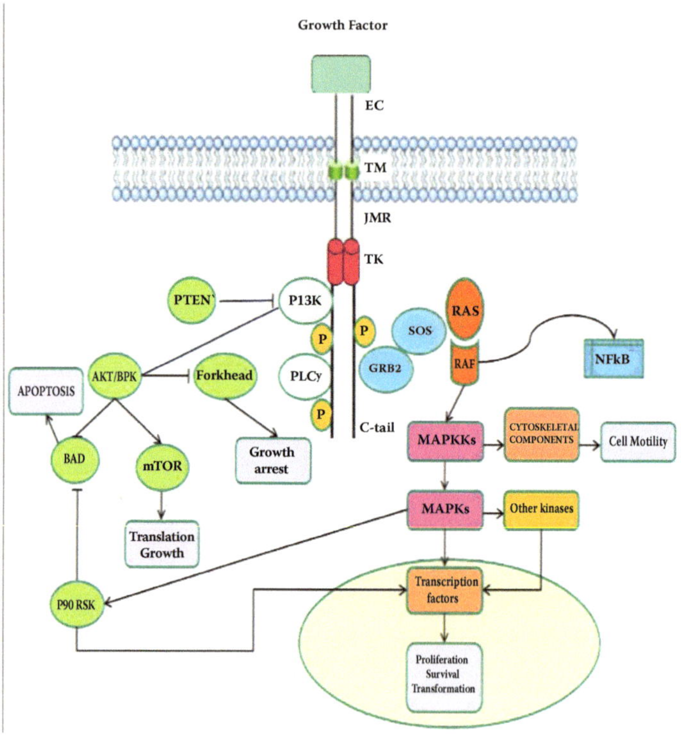

FIGURE 21.1 Receptor tyrosine kinase (RTK) signaling cascade. Signaling mechanism of RTK and major biological outcomes are illustrated. Mutual trans-phosphorylation of tyrosine residues within active RTK dimers recruits intracellular proteins endowed with phosphotyrosine-binding domains. Proximal targets of the RTKs invoke the intracellular signaling cascades RAS-RAF-MAPK (ERK pathway) and the phosphatidylinositol 3-kinase (PI3K) AKT that ultimately lead to diverse biological responses. Abbreviations: Extracellular domain (EC), transmembrane domain (TM), juxtamembrane domain (JMR), and tyrosine kinase domain (TK). (From Castellone, M. D., et al., *Best Pract. Res. Clin. Endocrinol. Metab.*, 22, 1023–38, 2008.)

8.4.7.2 Indications for UG-FNAB and TG-FNAB

For lymph nodes, several Doppler criteria are highly suspicious and indicate the need for a FNAB procedure: the presence of microcalcifications, cyst formation, peripheral vascularization, mixed peripheral and internal anarchic vascularization, or echostructure reminiscent of thyroid tissue (hyperechogenic lymph node). If there are several suspect adenopathies in the same area, a FNAB of only one is sufficient. In cystic lymph nodes where the UG-FNAB result is likely to be inadequate, the TG-FNAB procedure can prove to be very informative. However, in cancers that are not well differentiated, the TG-FNAB assay values may be lower or negative, thus producing false negative results. According to different published series, adding a TG-FNAB measurement to the cytology result improves the sensitivity from 91–100%. For Snozek et al.,[39] the specificity is 96.2% and the positive predictive value is 97.2% at the threshold of 1 μg/L.

8.4.8 Indications for Neck Ultrasonography Examination in the Follow-Up of Patients Presenting with Differentiated Thyroid Carcinoma

These recommendations are derived from American, European, and French CPG.[7–9] Before a thyroidectomy for suspected cancer, an US scan must be performed. If the initial US did not sufficiently delineate the LN chains, particularly in the lateral compartment, then this procedure should be performed preoperatively for all patients with suspicious thyroid nodules. Six to twelve months after a total thyroidectomy, a cervical US scan must be carried out to examine the thyroid bed and the central and lateral lymph node compartments, whether RAI ablation has been carried out or not. In addition, during the monitoring period, a cervical US scan is recommended if the serum TG has increased. In patients at very low risk or low risk, annual US exams are not necessary if the stimulated TG is undetectable and the US scan is satisfactory during the first 6–12 month checkup. In patients at high risk, the frequency of the US monitoring could be adjusted depending on whether there is an initial invasion of the lymph node capsule, which is a risk factor for recurrence. After a lobectomy with the fortuitous discovery of a PTMC, US monitoring of the remaining lobe and the areas around the lymph nodes is suggested, but there is no consensus about the frequency of such an examination.

8.5 INNOVATIONS IN THYROID ULTRASOUND TECHNOLOGY

In recent years, a certain number of developments have shed light to improve the quality of US images, and consequently the quality of the information obtained. Some of these advances are now available on all machines, but others are only integrated into some US machines. Among those that can be cited are speckle reduction imaging, contrast imaging, elastography, and high-intensity focused US.

8.5.1 Speckle Reduction Imaging

Speckles are intrinsic imaging artifacts that appear as a granular texture and can obscure subtle findings in thyroid anatomy.[40] This artifact is frequently referred to as noise and results from complex interference of echo signals when reflectors are closer than the US machine's limit of resolution. There are several technologies available for speckle reduction, including digital filters (available on many currently manufactured US machines today).

8.5.2 Contrast Imaging

The principle of contrast ultrasound scanning is based on examining enhanced microvasculature in tissue after administration of strictly intravascular ultrasound contrast agents of strictly intravascular contrast-enhanced ultrasound (CEUS) agents (micron-sized microbubbles). For the thyroid in particular, the SonoVue™ microbubble product has been less widely used due to the limitations observed in terms of its resonance frequency, which is not well adapted to high-frequency ultrasound. Indeed, this technology is mainly suited to low frequencies below 5–6 MHz, but not high frequencies above 10 MHz. It is therefore necessary to inject more agent to compensate for the lower level of sensitivity while maintaining the mechanical index sufficiently low (<0.3).[41]

It appears, however, that after injection of contrast agents, a benign nodule does not empty quickly over time, whereas a malignant module empties quickly. It can also be used for cervical lymph nodes in order to differentiate between inflammatory lymph nodes (strong enhancement) and metastatic ones (lack of or patchy enhancement).

8.5.3 Ultrasound Elastography

The theoretical principle of elastography imaging[42–47] is based on the analysis of differential tissue characteristics using a tissue rigidity/elasticity modulus (Young's modulus).[43,44] Young's modulus expresses the mathematical relationship between the deformation of a solid and the stress applied. In order to detect differences in tissue behavior, it is necessary to disrupt the medium being explored and at the same time detect the influence of this disruption on the medium. Various methods can be proposed: external or internal palpation or the propagation of a shear wave into the medium.

The first method involves external compression and is based on the fact that a soft region will move and become deformed, while a rigid region will only move. These tissue deformation imaging methods are referred to as static elastography. The deformations caused are recorded in real time in the medium studied so that areas with differing relative behavior can be distinguished, thus revealing any suspected differences in rigidity. These methods cannot be used to quantify the Young modulus because, in fact, they create an image of the compressibility module and not the shear module. They are limited by a certain number of artifacts in the liquid zones or are dependent on the application of external, non-uniform compression.

To determine the Young modulus using US, the shear modulus of the tissue must be measured. This requires simultaneous propagation of a shear wave through the tissue and imaging of the deformation caused by this wave. If the propagation of an external mechanical wave is not implemented for the thyroid, it is then possible to study the propagation of an ultrasound wave or its effect on the tissue. This technology is currently being developed and evaluated by Supersonic Imagine (Aix-en-Provence). A specific module (Acoustic Radiation Force Impulse (ARFI)) is available on a Siemens device. These methods that are currently available in 2D can determine shear modulus, and consequently accurately measure the Young modulus. The absence of artifacts reported with these techniques will undoubtedly improve the quality of the diagnosis.

Elastography can be used to supplement a classic US scan and thereby improve the accuracy of the diagnosis of thyroid tumors, which appear to be harder than the surrounding tissue. In a recent study,[44] tissue found to be hard during a static ultrasound scan is closely correlated with a diagnosis of cancer, while soft nodules are systematically considered benign. These results are similar to those reported by Rago et al.[45] and Lyshchik et al.,[46] who found a high prevalence of malignant nodules in those considered to be hard (with respective sensitivity of 97% and 82% and specificity of 100% and 97%). If no surgery is carried out, the hardness of the tissue will probably lead to a FNAB being carried out without delay, whereas if the tissue is soft, FNAB does not necessarily have to be proposed, even if no response is obtained from cytological tests.[44]

8.5.4 Ultrasound Therapy

The aim of high-intensity focused ultrasound (HIFU) is to maximize the deposit of energy caused by the interaction between high-intensity US and the tissue and to raise the temperature locally in the target area. Two parameters must be adapted: the acoustic intensity and the focus of the US beam. The acoustic intensity should be set at around 100 W/cm^2. This is obtained by using specific transducers together with powerful electronics. Multiple, convergent transducers are used to actively focus the US beam, creating a concentration of energy at a focal point and producing the ablative effect. An effective treatment area of around 5–8 mm in thickness and around 1 mm in diameter is obtained for each sequence of US impulses, which generally lasts about 5 seconds, and is called sonication. For each treatment cycle, around 100 Joules are delivered, heating up the focal point, with a target point locally exceeding 70–75°C in order to destroy the target cells. This method is extremely accurate: tissue 0.3 mm away from the point focal remains intact.

The cervical zone is positioned under the HIFU transducer built into the treatment head. This head also contains a US probe that can locate the zone to be treated during the session. The trachea, the esophagus, and other parts potentially at risk, such as the carotid, must be carefully avoided. The first papers reported satisfactory destruction of the nodule without any noteworthy complications: There was no local damage due to burns, nor any general damage to neighboring thyroid tissue, as reflected by thyroid function testing.[48,49]

This approach might make it possible to treat a certain number of benign or malignant lesions of the thyroid without having to resort to surgery and without disrupting any subsequent treatment. Further research should be carried out regarding the indication of HIFU treatment on local recurrences and metastatic involvement of LN in DTC.

8.5.5 Minimally Invasive Techniques Using Ultrasound

Percutaneous radiofrequency ablation (RFA) and percutaneous ethanol injection using US guidance are relatively new, minimally invasive techniques that have been widely used as an alternative to surgical treatment in patients with hepatocellular carcinoma or liver metastasis from other malignancies. Such techniques with local anesthesia have recently been reported as alternatives to surgery in patients with thyroid nodules or with local recurrence of DTC.[50–58]

8.6 CONCLUSION

Recent advances in US technology have significantly improved diagnoses through improved image quality, especially in terms of providing essential additional information to characterize lesions. Its therapeutic applications are still in their infancy, but it should considerably change the way in which nodules are treated in years to come.

REFERENCES

1. Mazzaferri EL. 1993. Management of a solitary thyroid nodule. *N Engl J Med* 328:553–59.
2. Gharib H. 1994. Fine-needle aspiration biopsy of thyroid nodules: Advantages, limitations, and effect. *Mayo Clin Proc* 69:44–49.
3. Gharib H. 2004. Changing trends in thyroid practice: Understanding nodular thyroid disease. *Endocr Pract* 10:31–39.
4. Kim EK, Park CS, Chung WY, et al. 2002. New sonographic criteria for recommending fine-needle aspiration biopsy of nonpalpable solid nodules of the thyroid. *AJR Am J Roentgenol* 178:687–91.
5. Papini E, Guglielmi R, Bianchini A, et al. 2002. Risk of malignancy in nonpalpable thyroid nodules: Predictive value of ultrasound and color-Doppler features. *J Clin Endocrinol Metab* 87:1941–46.
6. Frates MC, Benson CB, Charboneau JW, et al. 2005. Management of thyroid nodules detected at US: Society of Radiologists in Ultrasound consensus conference statement. *Radiology* 237:794–800.
7. Cooper DS, Doherty GM, Haugen BR, et al. 2006. Management guidelines for patients with thyroid nodules and differentiated thyroid cancer. *Thyroid* 16:109–42.

8. Pacini F, Schlumberger M, Dralle H, Elisei R, Smit JW, Wiersinga W. 2006. European consensus for the management of patients with differentiated thyroid carcinoma of the follicular epithelium. *Eur J Endocrinol* 154:787–803.
9. Borson-Chasot F, Bardet S, Bournaud C, et al. 2008. Guidelines for the management of differentiated thyroid carcinomas of vesicular origin. *Ann Endocrinol* (Paris) 69:472–86.
10. Merrit C. 2005. *Physics of ultrasound in diagnostic ultrasound*, ed. M Jo-Ann, 3–34. Vol. 1. St. Louis, MO: Johnson Elsevier Mosby.
11. Peccin S, de Castsro JA, Furlanetto TW, Furtado AP, Brasil BA, Czepielewski MA. 2002. Ultrasonography: Is it useful in the diagnosis of cancer in thyroid nodules? *J Endocrinol Invest* 25:39–43.
12. Leenhardt L, Menegaux F, Franc B, et al. 2002. Selection of patients with solitary thyroid nodules for operation. *Eur J Surg* 168:236–41.
13. Alexander EK. 2008. Approach to the patient with a cytologically indeterminate thyroid nodule. *J Clin Endocrinol Metab* 93:4175–82.
14. Leenhardt L, Hejblum G, Franc B, et al. 1999. Indications and limits of ultrasound-guided cytology in the management of nonpalpable thyroid nodules. *J Clin Endocrinol Metab* 84:24–28.
15. Cibas ES, Alexander EK, Benson CB, et al. 2008. Indications for thyroid FNA and pre-FNA requirements: A synopsis of the National Cancer Institute Thyroid Fine-Needle Aspiration State of the Science Conference. *Diagn Cytopathol* 36:390–99.
16. Hay ID. 2008. Papillary thyroid carcinoma. *Endocrinol Metab Clin North Am* 19:545–76.
17. Rouxel A, Hejblum G, Bernier MO, et al. 2004. Prognostic factors associated with the survival of patients developing loco-regional recurrences of differentiated thyroid carcinomas. *J Clin Endocrinol Metab* 89:5362–68.
18. Kitamura Y, Shimizu K, Nagahama M, et al. 1999. Immediate causes of death in thyroid carcinoma: Clinicopathological analysis of 161 fatal cases. *J Clin Endocrinol Metab* 84:4043–49.
19. Torlontano M, Crocetti U, Augello G, et al. 2006. Comparative evaluation of recombinant human thyrotropin-stimulated thyroglobulin levels, 131I whole-body scintigraphy, and neck ultrasonography in the follow-up of patients with papillary thyroid microcarcinoma who have not undergone radioiodine therapy. *J Clin Endocrinol Metab* 91:60–63.
20. Frasoldati A, Pesenti M, Gallo M, Caroggio A, Salvo D, Valcavi R. 2003. Diagnosis of neck recurrences in patients with differentiated thyroid carcinoma. *Cancer* 97:90–96.
21. Pelttari H, Laitinen K, Schalin-Jantti C, Valimaki MJ. 2008. Long-term outcome of 495 TNM stage I or II patients with differentiated thyroid carcinoma followed up with neck ultrasonography and thyroglobulin measurements on T4 treatment. *Clin Endocrinol* (Oxf) 69:323–31.
22. DeGroot LJ, Kaplan EL, McCormick M, Straus FH. 1990. Natural history, treatment, and course of papillary thyroid carcinoma. *J Clin Endocrinol Metab* 71:414–24.
23. Mazzaferri EL, Kloos RT. 2001. Clinical review 128: Current approaches to primary therapy for papillary and follicular thyroid cancer. *J Clin Endocrinol Metab* 86:1447–63.
24. Noguchi M, Yagi H, Earashi M, Kinoshita K, Miyazaki I, Mizukami Y. 1995. Recurrence and mortality in patients with differentiated thyroid carcinoma. *Int Surg* 80:162–66.
25. Ortiz S, Rodriguez JM, Parrilla P, et al. 2001. Recurrent papillary thyroid cancer: Analysis of prognostic factors including the histological variant. *Eur J Surg* 167:406–12.
26. Schlumberger M. 1998. Papillary and follicular thyroid carcinoma. *N Engl J Med* 338:297–306.
27. Sugino K, Kure Y, Iwasaki H, et al. 1995. Metastases to the regional lymph nodes, lymph node recurrence, and distant metastases in nonadvanced papillary thyroid carcinoma. *Surg Today* 25:324–28.
28. Mazzaferri EL, Jhiang SM. 1994. Long-term impact of initial surgical and medical therapy on papillary and follicular thyroid cancer. *Am J Med* 97:418–28.
29. Kloos RT. 2008. Approach to the patient with a positive serum thyroglobulin and a negative radioiodine scan after initial therapy for differentiated thyroid cancer. *J Clin Endocrinol Metab* 93:1519–25.
30. Mazzaferri EL. 2003. Changing paradigms in the follow-up of patients with differentiated thyroid cancer: An alternative to [18F]fluorodeoxyglucose positron emission tomographic scanning. *Endocr Pract* 9:324–26.
31. Torlontano M, Attard M, Crocetti U, et al. 2004. Follow-up of low risk patients with papillary thyroid cancer: Role of neck ultrasonography in detecting lymph node metastases. *J Clin Endocrinol Metab* 89:3402–7.
32. Pacini F, Molinaro E, Castagna MG, et al. 2003. Recombinant human thyrotropin-stimulated serum thyroglobulin combined with neck ultrasonography has the highest sensitivity in monitoring differentiated thyroid carcinoma. *J Clin Endocrinol Metab* 88:3668–73.

33. Torlontano M, Crocetti U, D'Aloiso L, et al. 2003. Serum thyroglobulin and 131I whole body scan after recombinant human TSH stimulation in the follow-up of low-risk patients with differentiated thyroid cancer. *Eur J Endocrinol* 148:19–24.
34. Solbiati L, Osti V, Cova L, Tonolini M. 2001. Ultrasound of thyroid, parathyroid glands and neck lymph nodes. *Eur Radiol* 11:2411–24.
35. Leboulleux S, Girard E, Rose M, et al. 2007. Ultrasound criteria of malignancy for cervical lymph nodes in patients followed up for differentiated thyroid cancer. *J Clin Endocrinol Metab* 92:3590–94.
36. Boi F, Baghino G, Atzeni F, Lai ML, Faa G, Mariotti S. 2006. The diagnostic value for differentiated thyroid carcinoma metastases of thyroglobulin (Tg) measurement in washout fluid from fine-needle aspiration biopsy of neck lymph nodes is maintained in the presence of circulating anti-Tg antibodies. *J Clin Endocrinol Metab* 91:1364–69.
37. Lee MJ, Ross DS, Mueller PR, Daniels GH, Dawson SL, Simeone JF. 1993. Fine-needle biopsy of cervical lymph nodes in patients with thyroid cancer: A prospective comparison of cytopathologic and tissue marker analysis. *Radiology* 187:851–54.
38. Pacini F, Fugazzola L, Lippi F, et al. 1992. Detection of thyroglobulin in fine needle aspirates of nonthyroidal neck masses: A clue to the diagnosis of metastatic differentiated thyroid cancer. *J Clin Endocrinol Metab* 74:1401–4.
39. Snozek CL, Chambers EP, Reading CC, et al. 2007. Serum thyroglobulin, high-resolution ultrasound, and lymph node thyroglobulin in diagnosis of differentiated thyroid carcinoma nodal metastases. *J Clin Endocrinol Metab* 92:4278–81.
40. Tsantis S, Dimitropoulos N, Ioannidou M, et al. 2007. Inter-scale wavelet analysis for speckle reduction in thyroid ultrasound images. *Comp Med Imag Graph* 31:117–27.
41. Tranquart , F., Correas, J-M. 2007. *Échographie de contraste, méthodologie et applications cliniques*, Bouakaz, A. (Ed.) Springer, 328.
42. Asteria C, Giovanardi A, Pizzocaro A, et al. 2008. US-elastography in the differential diagnosis of benign and malignant thyroid nodules. *Thyroid* 18:523–31.
43. Bae U, Dighe M, Dubinsky T, Minoshima S, Shamdasani V, Kim Y. 2007. Ultrasound thyroid elastography using carotid artery pulsation: Preliminary study. *J Ultrasound Med* 26:797–805.
44. Tranquart F, Bleuzen A, Pierre-Renoult P, Chabrolle C, Sam Giao M, Lecomte P. 2008. [Elastosonography of thyroid lesions]. *J Radiol* 89:35–39.
45. Rago T, Santini F, Scutari M, Pinchera A, Vitti P. 2007. Elastography: New developments in ultrasound for predicting malignancy in thyroid nodules. *J Clin Endocrinol Metab* 92:2917–22.
46. Lyshchik A, Higashi T, Asato R, et al. 2005. Thyroid gland tumor diagnosis at US elastography. *Radiology* 237:202–11.
47. Rubaltelli L, Corradin S, Dorigo A, et al. 2008. Differential diagnosis of benign and malignant thyroid nodules at elastosonography. *Ultraschall Med.* 2009 Apr. 30(2):175–179. Epub 2008 May 21.
48. Esnault O, Franc B, Monteil JP, Chapelon JY. 2004. High-intensity focused ultrasound for localized thyroid-tissue ablation: Preliminary experimental animal study. *Thyroid* 14:1072–76.
49. Esnault O, Leenhardt L. 2008. High intensity focused ultrasound (HIFU) Ablation therapy for thyroid nodules. In *Thyroid Ultrasound and Ultrasound-Guided FNA*, 2nd ed. HJ Baskin, DS Duick, RA Levine, Eds. 219–36. 2nd ed. Springer.
50. Amabile G, Rotondi M, Chiara GD, et al. 2006. Low-energy interstitial laser photocoagulation for treatment of nonfunctioning thyroid nodules: Therapeutic outcome in relation to pretreatment and treatment parameters. *Thyroid* 16:749–55.
51. Dossing H, Bennedbaek FN, Hegedus L. 2006. Effect of ultrasound-guided interstitial laser photocoagulation on benign solitary solid cold thyroid nodules: One versus three treatments. *Thyroid* 16:763–68.
52. Kim YS, Rhim H, Tae K, Park DW, Kim ST. 2006. Radiofrequency ablation of benign cold thyroid nodules: Initial clinical experience. *Thyroid* 16:361–67.
53. Papini E, Guglielmi R, Bizzarri G, et al. 2007. Treatment of benign cold thyroid nodules: A randomized clinical trial of percutaneous laser ablation versus levothyroxine therapy or follow-up. *Thyroid* 17:229–35.
54. Tarantino L, Francica G, Sordelli I, et al. 2008. Percutaneous ethanol injection of hyperfunctioning thyroid nodules: Long-term follow-up in 125 patients. *AJR Am J Roentgenol* 190:800–8.
55. Monchik JM, Donatini G, Iannuccilli J, Dupuy DE. 2006. Radiofrequency ablation and percutaneous ethanol injection treatment for recurrent local and distant well-differentiated thyroid carcinoma. *Ann Surg* 244:296–304.

56. Dossing H, Bennedbaek FN, Bonnema SJ, Grupe P, Hegedus L. 2007. Randomized prospective study comparing a single radioiodine dose and a single laser therapy session in autonomously functioning thyroid nodules. *Eur J Endocrinol* 157:95–100.
57. Rotondi M, Amabile G, Leporati P, Di Filippo B, Chiovato L. 2009. Repeated laser thermal ablation of a large functioning thyroid nodule restores euthyroidism and ameliorates constrictive symptoms. *J Clin Endocrinol Metab* 94:382–83.
58. Sohn YM, Hong SW, Kim EK, et al. 2009. Complete eradication of metastatic lymph node after percutaneous ethanol injection therapy: Pathologic correlation. *Thyroid* 19:317–19.

9 Multimodality Imaging and Aspiration Biopsy Guidance in the Perioperative Management of Thyroid Carcinoma

Jing Gao, Jonathan K. Kazam, and Elias Kazam

CONTENTS

ABSTRACT: Imaging is a critical tool for the management of thyroid nodules and thyroid cancer. High-resolution thyroid ultrasonography (US) is the principal diagnostic imaging tool preoperatively. Thyroid US is used to determine risk stratification for thyroid nodules in order to guide decision making for surgery. A detailed description of perioperative thyroid US findings is provided with figures. Thyroid US is also used to guide fine-needle aspiration biopsy and may be evaluated in conjunction with other imaging modalities, mainly computed tomography (CT). Thyroid US is also useful to evaluate suspicious lymph nodes preoperatively and suspicious areas of residual or recurrent cancer postoperatively. As thyroid US technology improves, so will comprehensive care of the thyroid cancer patient.

9.1 INTRODUCTION

Recent improvements in sectional imaging modalities have enhanced their usefulness in the diagnosis of thyroid carcinoma and in the assessment of extent of disease. This chapter is based on extensive experience with the use of ultrasonography (ultrasound; US), computed tomography (CT), and ultrasound-guided fine-needle aspiration biopsy (UG-FNAB) in the preoperative and postoperative evaluation of the patient with thyroid carcinoma. Multiple US and CT images are used to illustrate the diagnostic criteria, imaging pitfalls, and relative advantages and limitations of US and CT.

9.2 PREOPERATIVE ASSESSMENT OF THYROID CARCINOMA

9.2.1 Ultrasonography

9.2.1.1 Technique

The patient is scanned in the supine position with the neck extended in order to elevate the thyroid gland as much as possible above the clavicular heads and the manubrium. A high-resolution linear transducer (8–14 MHz) is generally used for thyroid imaging. Lower-frequency transducers are reserved for deeper structures and for UG-FNAB because the needle tip is better visualized at the lower frequencies. Multifocusing is used routinely to improve contrast resolution and enhance the detectability and characterization of thyroid nodules. Single-level focusing is used during UG-FNAB. The dynamic range is varied during the examination. For example, a small thyroid nodule is usually better delineated with a narrower dynamic range, while calcifications and thyroid cysts are more optimally visualized with a wider dynamic range. Curved transducers are useful for improving the measurements of enlarged thyroid lobes, which extend beyond the edges of a linear transducer. Endocavity probes, placed within the suprasternal notch, may be useful for imaging lower cervical or thoracic inlet structures that are otherwise obscured by the sternum or the clavicles.

Color-flow and power Doppler are valuable for assessing the vascularity of the thyroid parenchyma and thyroid nodules. Hypervascular nodules are usually hyperplastic or neoplastic,[1–6] while avascular nodules are usually benign. There are important exceptions to these general rules. For example, an avascular hypoechoic nodule that attenuates the sound beam is usually suspect for papillary thyroid carcinoma (PTC), despite its avascularity. Spectral Doppler provides information on the velocity and resistivity of blood flow as well as possible arteriovenous communication within the nodule. However, spectral Doppler usually does not provide more useful diagnostic information than color Doppler.

9.2.1.2 Sonographic Characteristics of Malignant Thyroid Nodules

The size, shape, contours (margins), echogenicity, vascularity, and calcific content of the thyroid nodule are valuable for ultrasonic characterization. Thyroid nodule size does not correlate with malignancy or benignity, but is valuable in the follow-up of thyroid nodules. Rapid growth of a nodule that is not readily attributable to hemorrhagic liquefaction raises the possibility of a neoplastic process, which may be evaluated with UG-FNAB. For this reason, the length, greatest anteroposterior, and widest diameters of the thyroid nodule are reported. The volume is estimated by multiplying these three measurements by 0.52. Volume estimates that are based on this method can be accurate and precise only if care is taken to obtain the maximal measurements, and are therefore operator dependent. Despite their limitations, such volume estimates provide a degree of quantification that is useful on follow-up studies.

Correlation has been reported between the shape of a thyroid nodule and its malignant potential.[7] "Taller" nodules on the imaging screen (greater anteroposterior than transverse and longitudinal dimensions) have an increased likelihood of malignancy. "Oval" nodules (smaller anteroposterior than transverse and longitudinal dimensions) tend to be benign.

A well-marginated nodule (sharply demarcated border from adjacent thyroid parenchyma) with smooth contours is likely to be benign (Figure 9.1). A poorly marginated nodule (indistinct interface with adjacent thyroid parenchyma) with irregular contours has a potential for malignancy (Figure 9.2).

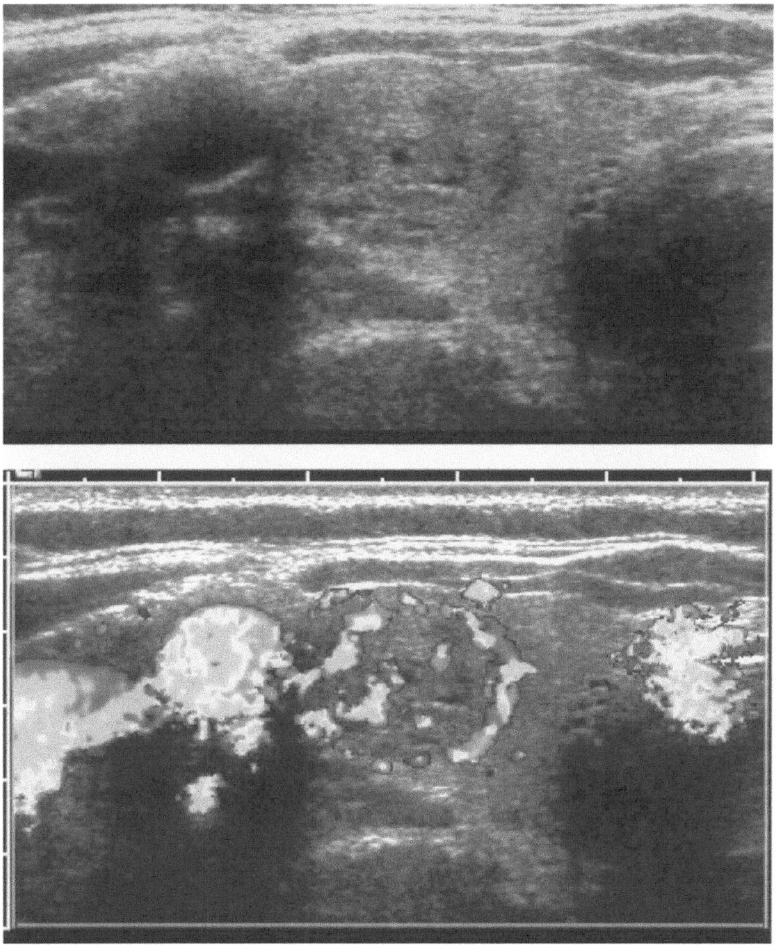

FIGURE 9.1 Ovoid benign hemorrhagic thyroid nodule with colloid. Transverse images without (top) and with (bottom) color Doppler show an echogenic nodule with small cystic foci of probable posthemorrhagic liquefaction, several punctate echogenic foci that may represent posthemorrhagic calcifications, a nonuniform hypoechoic, hypervascular rim, and mild to moderate internal vascularity. The internal echogenicity is consistent with colloid content, and the hypervascularity suggests hyperplastic elements. The US findings suggest a benign hemorrhagic hyperplastic nodule with colloid. UG-FNAB showed a benign colloid nodule.

Echogenicity of the internal content of a thyroid nodule generally correlates with heterogeneous colloid content interspersed with noncolloid hemorrhagic content. This forms multiple acoustic interfaces that lead to increased echogenicity. A uniformly echogenic nodule without hypoechoic content on US may be benign with colloid. However, it may also be a hypercellular follicular lesion that should be biopsied, especially if there is internal vascularity, without cystic internal foci. On the other hand, a hypoechoic nodule may have pure colloid content (especially if it has good through transmission) or may be colloid deficient (especially if there is sound attenuation with irregular contours) (Figure 9.3).

Heterogeneous echotexture (internal content) of thyroid nodules has been classified into cystic, solid, and complex. Most purely cystic thyroid nodules are benign, with the cystic content representing either colloid or posthemorrhagic liquefaction. A complex echotexture with cystic and echogenic components is usually a sign of a benign hemorrhagic nodule with colloid. However, complex cysts can also be follicular lesions (either a follicular adenoma or a follicular carcinoma, which are usually indistinguishable by sonography or even fine-needle aspiration biopsy (FNAB), but are

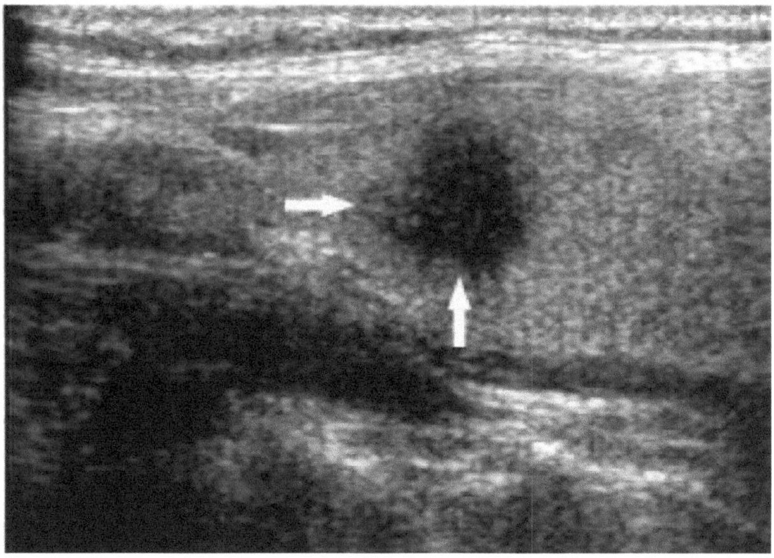

FIGURE 9.2 PTC has irregular superior and posterior interfaces with adjacent thyroid parenchyma (arrows).

diagnosed at surgical pathology, based on vascular or capsular invasion, or nodal metastases) or the follicular variant of PTC. A Hürthle cell neoplasm (either adenoma or carcinoma, distinguishable only at surgical pathology) is characterized by a predominance of eosinophilic Hurthle cells on FNAB (>50%), but usually has a similar US appearance to a follicular lesion or to a hemorrhagic hyperplastic thyroid nodule.

Calcification within a thyroid nodule increases the likelihood of malignancy.[8] Microcalcifications are echogenic foci smaller than 2 mm that are too small to cause posterior acoustic shadowing. A homogeneously hypoechoic noncystic nodule with microcalcifications and hypervascularity is likely to represent a thyroid malignancy, according to the published literature.[9] Microcalcifications, especially

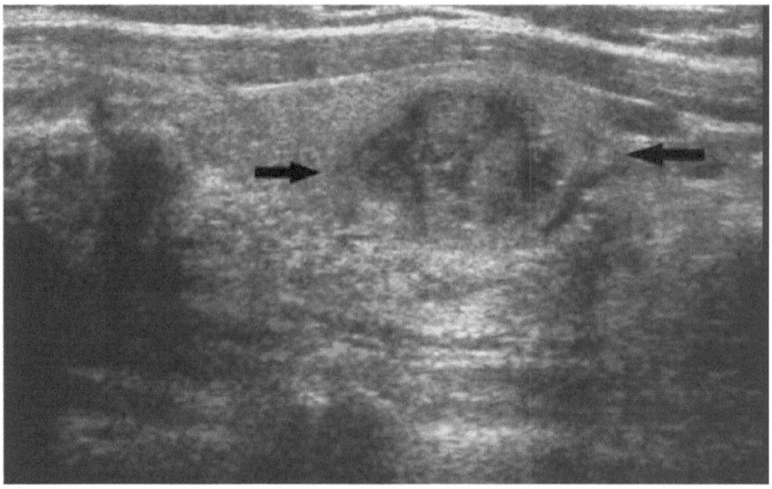

FIGURE 9.3 PTC with colloid content. The echogenic central portion of the nodule is consistent with colloid content. The central hypoechoic foci may represent posthemorrhagic cystic degeneration. The hypoechoic peripheral portion is probably colloid deficient, while the irregular peripheral contours (arrows) are suspicious for a PTC. UG-FNAB confirmed the diagnosis of PTC.

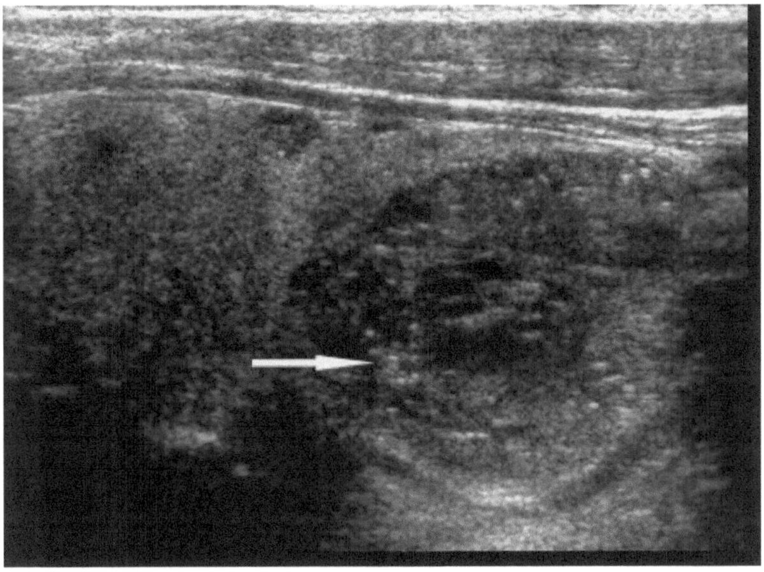

(A)

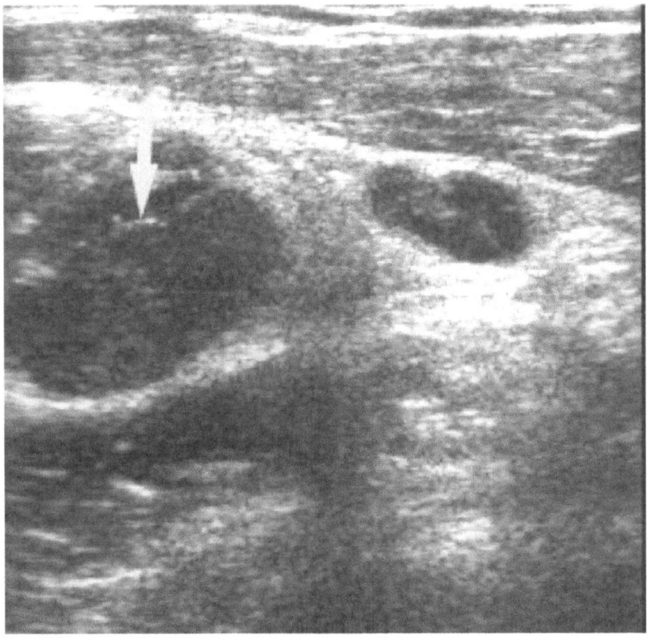

(B)

FIGURE 9.4 Metastatic PTC with microcalcifications. PTC with microcalcifications (A, arrow) associated with metastatic lateral compartment lymph node that contains similar punctate calcifications (B, arrow). *(continued)*

if associated with cystic components in a hypervascular nodule, are indications for FNAB (Figure 9.4), but are more likely to be associated with benign hemorrhagic nodules than malignant neoplasms.

Macrocalcifications, or coarse calcifications, have been classified into three categories. Type 1 macrocalfications are solitary linear or rounded hyperechoic structures >2 mm with or without acoustic shadowing in the middle of the nodule or at its periphery. Type 2 macrocalcifications

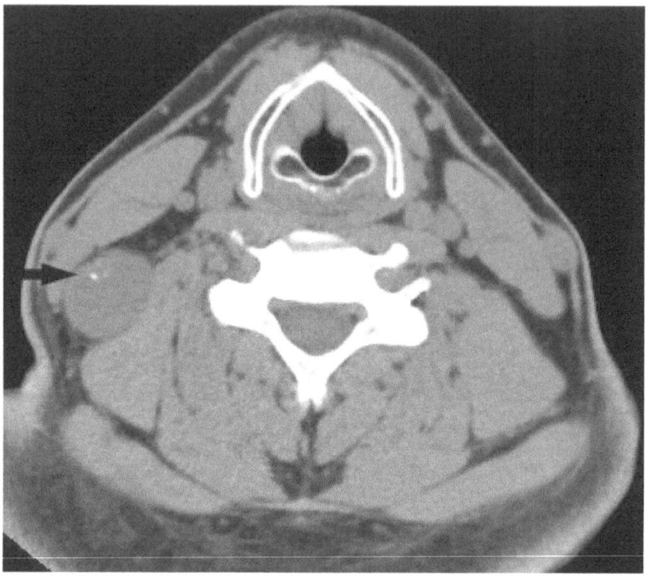

(C)

FIGURE 9.4 (Continued) Metastatic PTC with microcalcifications. CT of metastatic lymph node (C, arrow).

are eggshell calcifications that have curvilinear hyperechoic foci with acoustic shadowing at the margins of the nodule. Type 3 macrocalcifications are all other coarse but nonspecified calcifications.[10] Peripheral coarse rim calcifications (Figure 9.5) and eggshell calcifications (Figure 9.6) with acoustic shadowing are more worrisome for a thyroid malignancy than microcalcifications, but may be seen also with benign posthemorrhagic nodules. There is also an increased incidence

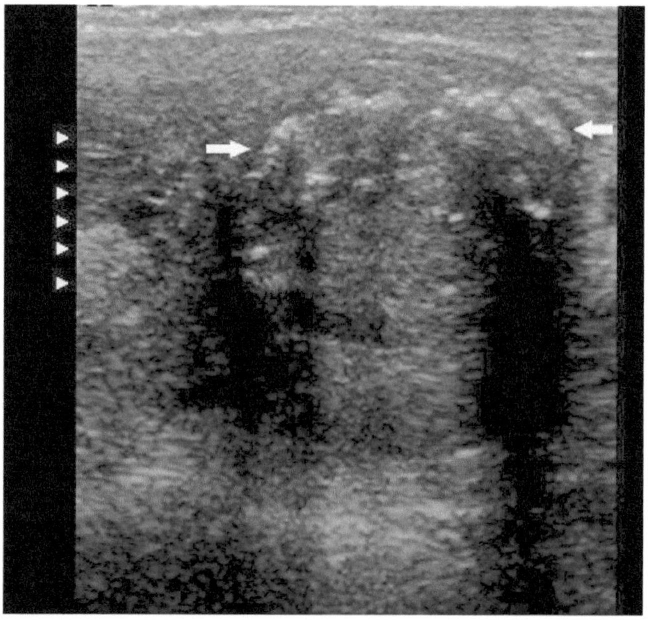

FIGURE 9.5 PTC with coarse peripheral calcifications, associated with acoustic shadowing (arrows). Needle sampling in contiguity to the calcifications is recommended for these lesions.

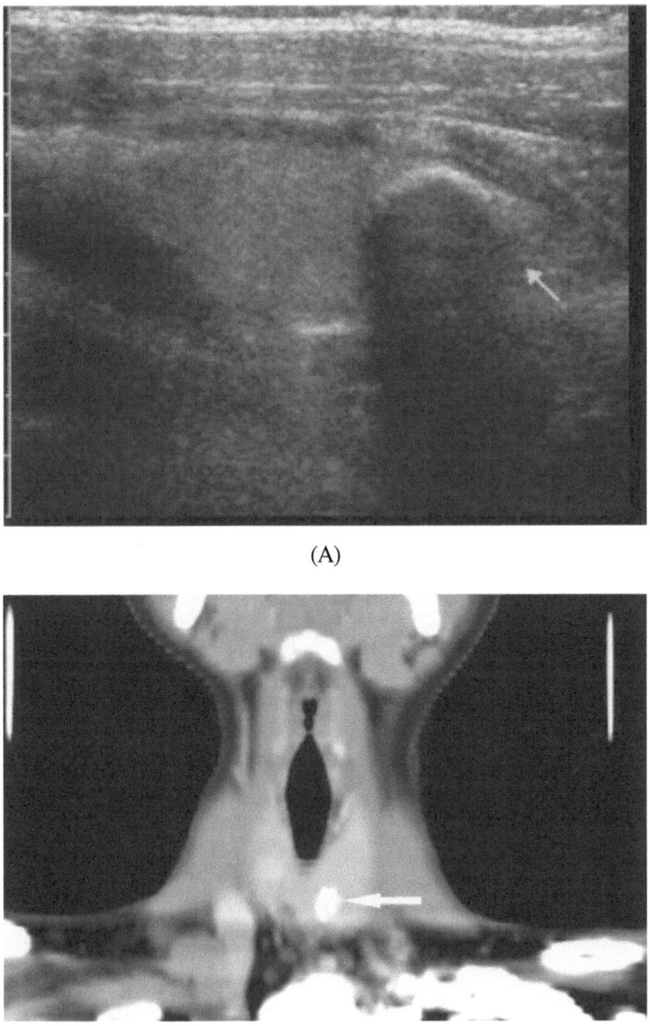

(A)

(B)

FIGURE 9.6 PTC with smoothly outlined eggshell calcification (US—A, arrow; CT—B, arrow), associated with acoustic shadowing. The US-CT findings are indistinguishable from a calcified hemorrhagic nodule.

of coarse calcifications within benign thyroid nodules in patients with hyperparathyroidism and hypercalcemia.

Hypervascularity within a thyroid nodule raises the likelihood of malignancy, but is associated much more often with benignity. Hypervascularity is therefore an important criterion for deciding on whether to biopsy a nodule.[11] A hypoechoic, hypervascular nodule may be malignant, even with good through transmission (Figure 9.7). While a hypoechoic avascular nodule with irregular contours or sound attenuation may be malignant, with or without internal calcifications, a hypoechoic avascular smoothly outlined nodule, without sound attenuation, is most likely benign, even if it contains punctate calcifications. Spectral Doppler evaluation is useful for differentiating true internal vascularity from color Doppler artifacts, but it is not generally used routinely to characterize thyroid nodules.

Overall, hypoechogenicity, internal calcifications, sound attenuation, hypervascularity, and irregular or lobulated contours are generally used as criteria for selecting suspicious thyroid nodules for UG-FNAB. In a multinodular goiter, all nodules that meet these criteria need to be aspirated,

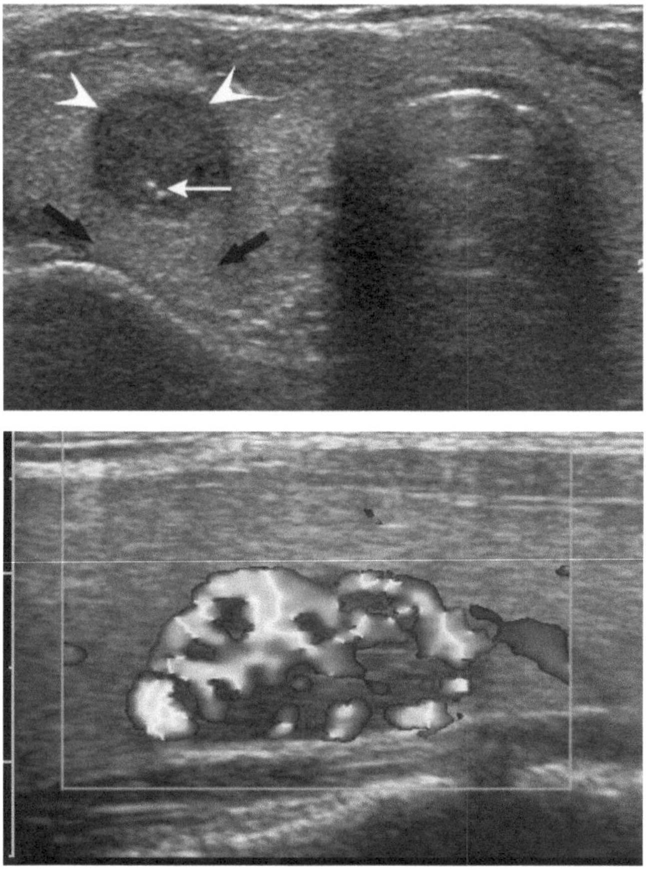

FIGURE 9.7 Hypervascular medullary thyroid carcinoma with good through transmission of US. Transverse image (top) shows a hypoechoic nodule (two arrowheads) with two punctate calcifications (white arrow) and good through transmission (black arrows). The nodule is moderately hypervascular on color Doppler examination (sagittal image, bottom).

even if it means sampling five or six nodules during one biopsy procedure. Moreover, equivocal nodules are more likely to be aspirated in high-risk patients, such as those with a history of external beam radiation,[11] Hashimoto's thyroiditis, or family history of thyroid carcinoma.[12]

Ultrasound elastography is an emerging imaging technique that can assist in the differential diagnosis of malignant from benign thyroid nodules.[13] Three-dimensional ultrasound may also prove to be useful in the future. This technology provides a geographical display of the internal echotexture of a nodule. Contrast-enhanced ultrasound has also demonstrated some promise, but has been approved only for echocardiographic applications in the United States. It may be useful in differentiating malignant from benign cervical lymph nodes (LNs).[14]

9.2.2 Computed Tomography (CT)

CT is most useful if it is performed with thin sections (preferably 2–2.5 mm collimation), with and without intravenous (IV) contrast enhancement. Malignant thyroid nodules and metastatic cervical lymph nodes tend to be iodine deficient on CT[15] (relatively lucent on pre-contrast scans; Figure 9.8) and may contain calcifications.[15,16] Coarse calcifications are more often associated with primary thyroid malignancy than tiny punctate calcifications. The primary and metastatic lesions are usually hypervascular on CT. Capillary phase hypervascularity on CT, an important finding

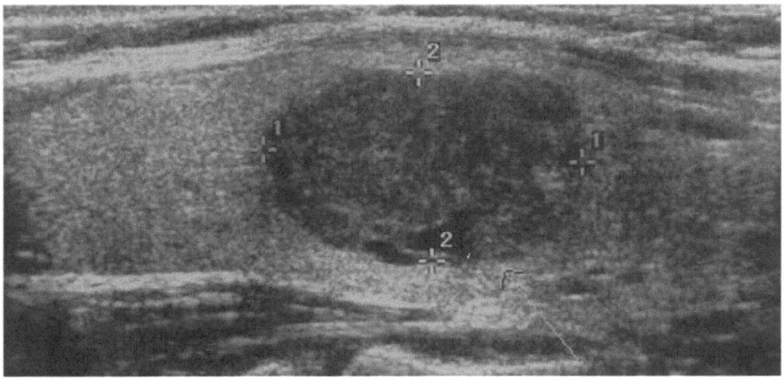

(A)

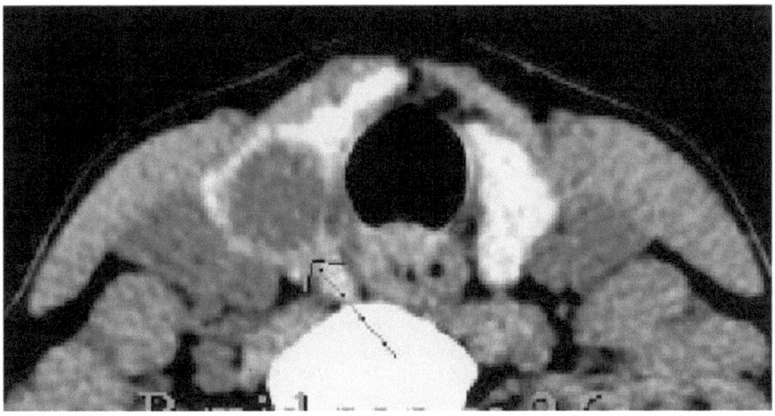

(B)

FIGURE 9.8 Right PTC on US (A, arrow) and iodine deficient on pre-contrast CT (B, arrow). Colloid content is associated with partial hyperdensity of the nodule (iodine content) on pre-contrast CT. The echogenic content of this nodule on sonography is due to microcalcifications and tumor necrosis (consistent with iodine deficiency on CT).

in some malignant lesions, may be undetectable with Doppler US. Low-lying or posterior cervical lymph node metastases that may be obscured on US can be clearly delineated with CT (Figure 9.9). CT may be used in combination with US in order to assess the extent of disease in preoperative or postoperative thyroid carcinoma patients.[17,18]

9.2.3 Radionuclide Scanning

Nearly all malignant thyroid nodules are photopenic, or "cold," on thyroid scans. On the other hand, an isotope-avid, or "hot," thyroid nodule is unlikely to be malignant and probably should not be aspirated. Notwithstanding the above, the high diagnostic accuracy of US and FNAB obviates the need for thyroid scanning except in cases of borderline TSH levels where demonstration of nodular functionality needs to be excluded.

9.2.4 Positron Emission Tomography (PET)/CT Scanning

PET/CT scanning is not routinely performed for the preoperative evaluation of thyroid nodules. A fluorodeoxyglucose (FDG)-avid thyroid nodule may be detected incidentally on a PET/CT scan that

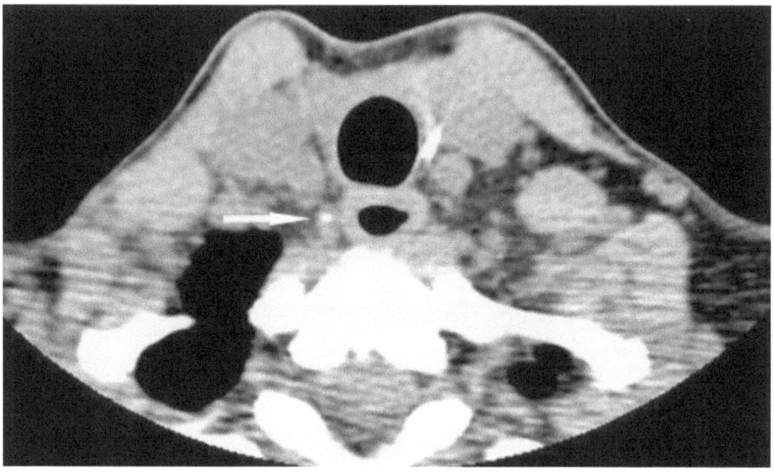

(A)

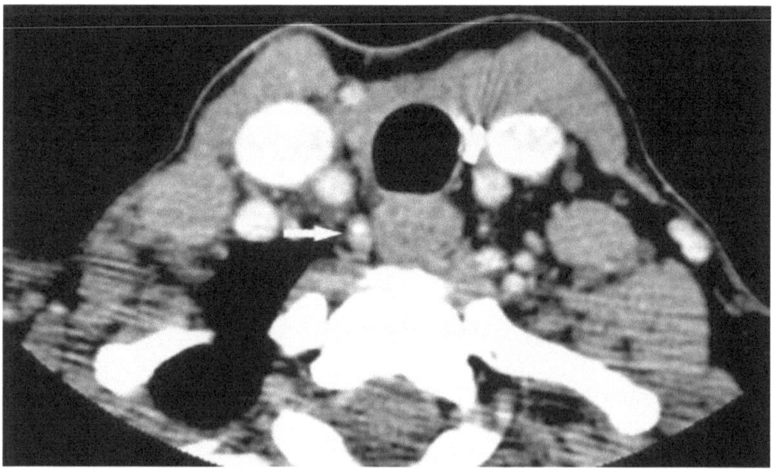

(B)

FIGURE 9.9 Metastatic PTC to a right posterior central compartment lymph node. The metastatic node contains a punctate calcification on pre-contrast CT (A, arrow) and is mildly to moderately hypervascular on the contrast-enhanced CT scan (B).

is performed to evaluate other malignancies, such as lymphoma or carcinoma. These nodules have a 14–50% chance of malignancy and should be aspirated.[11] An FDG-avid cervical lymph node in a patient with thyroid carcinoma is very likely to be metastatic, with more FDG uptake expected for less differentiated thyroid metastases.[19–22] Sestamibi scanning is also not used routinely for thyroid imaging, but a thyroid nodule with sestamibi uptake has a 22–66% chance of malignancy and should be aspirated.[11]

9.2.5 Magnetic Resonance Imaging (MRI)

Incidental discovery of thyroid nodules may be made on MRI of the cervical spine or magnetic resonance angiogram (MRA) of the carotid arteries. MRI is not commonly employed in the assessment of extent of disease in the thyroid carcinoma patient. However, brain and spine metastases from thyroid carcinoma may be well visualized on MRI.

9.3 CERVICAL LYMPH NODE METASTASES

Cervical lymph node metastases occur in 30–80% of patients with PTC,[23] but in less than 10% of follicular thyroid carcinomas (FTCs).[24] In a 1992 study using CT or MRI, the presence of a unilateral metastatic cervical lymph node reduces the 5-year survival rate to 50%, while bilateral metastatic cervical nodal metastases are associated with a further reduction in the 5-year survival rate to 25%.[25]

The metastatic cervical nodes may be centrally or laterally located, with lateral nodal metastases usually associated with central metastases (i.e., skip lesions with positive lateral but negative central compartment metastases are rare). The central compartment extends from the hyoid bone superiorly to the innominate vein inferiorly and is delimited laterally by the carotid sheaths and posteriorly by the prevertebral fascia. The lateral compartment extends from the hypoglossal nerve superiorly down to the subclavian vein; it is delimited laterally by the trapezius muscles and medially by carotid sheath.[26,27]

Pathological cervical lymph nodes may be located within one of six anatomical regions (Figure 9.10):

- Level 1: Submental and submandibular.
- Level 2: Jugular digastric chain, just above and at the carotid bifurcation.
- Level 3: Mid-cervical or mid-jugular chain, from just below the carotid bifurcation to the cricothyroid notch or omohyoid muscle.
- Level 4: Deep cervical chain, just below the cricothyroid notch or omohyoid muscle and above the clavicles.
- Level 5: Posterior triangle behind the sternocleidomastoid muscle, along the lower half of the spinal accessory nerve and transverse cervical artery.
- Level 6: Anterior mid-neck.

We use these anatomical levels when reporting the location of lymph nodes on US, but prefer more precise localization relative to anatomical landmarks, such as the cricoid cartilage or the suprasternal notch, when localizing lymph nodes on CT.

Sonographic lymph node characteristics that are associated with an increased likelihood for metastatic cervical lymphadenopathy[28–30] include:

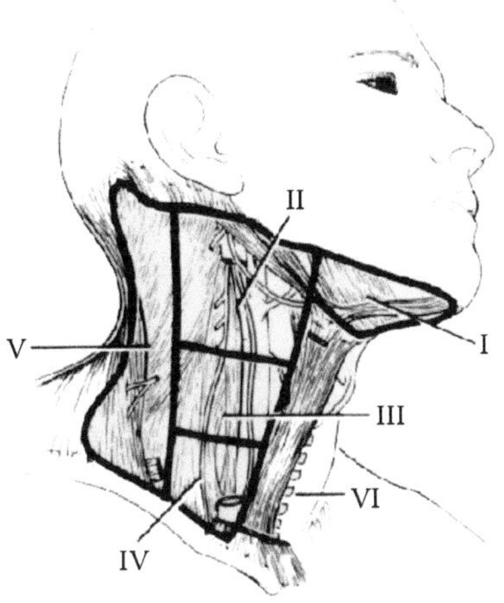

FIGURE 9.10 Anatomical levels for pathological lymph nodes in the neck. Levels II to V are in the lateral compartment; level VI is in the central compartment. (From http://www.bcm.edu/oto/studs/anat/neck.html, accessed January 2, 2010.)

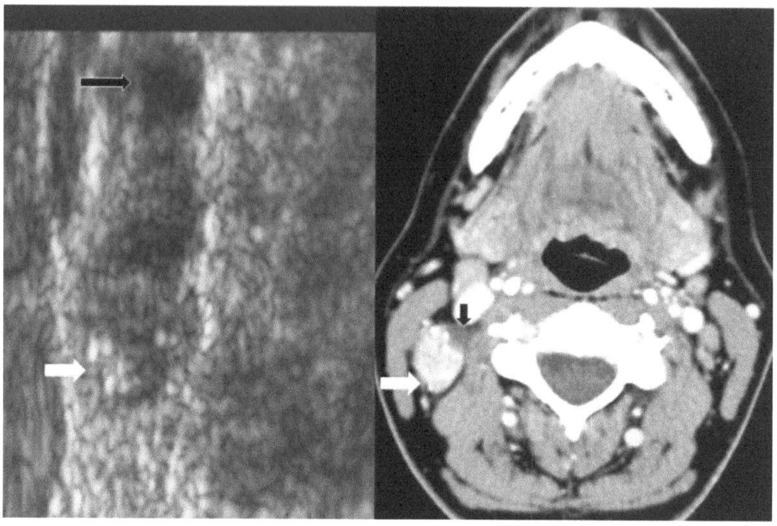

FIGURE 9.11 Metastatic PTC to a right retrojugular lymph node. Sagittal US (left) shows a heterogeneous lymph node with punctate calcifications (white arrow) and cystic foci (black arrow). Contrast-enhanced CT (right) shows the metastatic lymph node to be moderately hypervascular. The posterior calcification was demonstrated on the pre-contrast CT scan (not shown here).

- Size increase up to 25% during a short follow-up interval
- A rounded rather than ovoid contour (a short/long-axis diameter ratio of 0.5 or more)
- Heterogeneous internal echogenicity (loss of the normal echogenic hilum), echogenic or mixed echogenic/hypoechoic echotexture
- Cystic foci[31]
- Intranodal calcifications[32] (Figure 9.11)
- Hypervascularity on color Doppler examination

Partially cystic cervical lymph node metastases may also be seen with metastatic squamous cell carcinoma, mimicking metastatic thyroid carcinoma.

9.4 POSTTHYROIDECTOMY SURGICAL BED

Hypoechoic, hypervascular nodules within the postthyroidectomy surgical bed, or in the adjacent strap muscles, are strongly suspicious for recurrent thyroid carcinoma by US,[33] but can occasionally represent prominent parathyroid glands. Thin-section CT (2 or 2.5 mm collimation), without and with IV contrast, is exquisitely sensitive for tiny subcentimeter foci of recurrent neoplasm within the thyroid surgical bed. These are usually relatively lucent pre-contrast and at least moderately hypervascular on the contrast-enhanced CT scans. Prominent parathyroid glands may have a similar CT appearance. Many postthyroidectomy patients are evaluated with combined US-CT imaging. CT is more sensitive to hypervascularity, especially in the capillary phase, than Doppler sonography (Figures 9.11 and 9.12). CT may also demonstrate small foci of recurrence that are not initially apparent on US, because of their small sizes or relatively deep locations. Small foci of recurrent carcinoma in the thyroid surgical bed may remain stable for several years without treatment. Postoperative ganglia (small nodules that contain tiny nerve branches, surrounded by postoperative fibrosis) are usually hypoechoic and hypovascular or mildly hypervascular on sonography, but may be moderately or markedly hypervascular on CT (capillary phase enhancement). UG-FNAB may be used for suspected focal recurrence of thyroid surgical bed neoplasm on CT, even if the nodule is not delineated on the initial diagnostic US, and sent for cytological studies as well as TG, and even

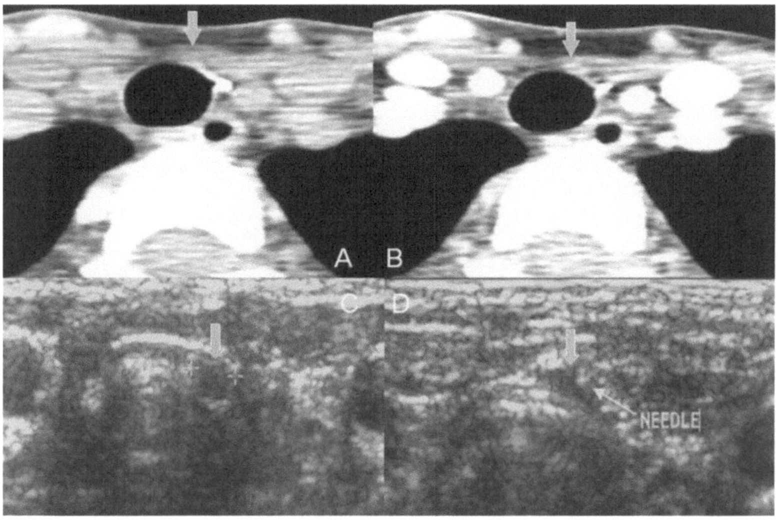

FIGURE 9.12 Recurrent PTC (postthyroidectomy), hypervascular on CT but not on US. CT without (A) and with (B) IV contrast shows a hypervascular left pretracheal nodule (arrows), contiguous to the left strap muscles. The nodule was overlooked on initial US, but was shown to be hypoechoic and avascular (C) after correlation with the CT scan. The apparent avascularity on sonography may be due to the location of the nodule in the relatively insensitive near field of the transducer, or to the relative insensitivity of color/power Doppler to capillary phase hypervascularity. Nevertheless, sonography was valuable for UG-FNAB (D), which showed recurrent PTC in this patient with undetectable TG levels.

parathyroid hormone (PTH) if needed for clarification, from the aspirate sample. Diagnostic criteria for evaluation of metastatic cervical lymph nodes in the postthyroidectomy patient are similar to the criteria used for the preoperative evaluation.

9.5 ULTRASOUND-GUIDED FINE-NEEDLE ASPIRATION BIOPSY (UG-FNAB)

FNAB allows precise sampling of thyroid nodules or suspicious cervical lymph nodes and has become an invaluable tool in the preoperative and postoperative evaluation of the patient with thyroid carcinoma.[34–36] The patient is usually asked to discontinue aspirin, nonsteroidal anti-inflammatory drugs (NSAIDS), antiplatelet agents, and anticoagulants at least 3 days prior to FNAB and for three days afterwards. If clinically necessary, however, FNAB may be performed safely in anticoagulated patients, with prolonged postprocedure compression of the puncture site.

FNAB is preferably performed under sterile conditions, using a linear transducer (at 8–14 MHz for imaging and at 5 MHz for needle guidance to improve visualization of the needle tip), a handheld 1.5-inch (3.8 cm) 25-gauge needle, and a biopsy needle pathway that forms a 30–40° angle with the long axis of the transducer (Figure 9.13). Lidocaine 2% is administered for local anesthesia, along the course of the biopsy needle(s), from the skin to the thyroid capsule, or from the skin to the near surface of the biopsied lymph node. If necessary, syringe suction is applied through a sterile tube that is connected to the biopsy needle. This method allows precise sampling of nodules as small as 3 mm.

Initially, an attempt is made to avoid intranodular cystic foci, which can be aspirated subsequently, if necessary. Samples are obtained from solid, hypervascular, or calcified regions within the nodule, using a slight rotation an in/out motion of the biopsy needle within the nodule. A nodule that measures 1–2 cm is usually sampled at least twice, while a 2–3 cm nodule is usually sampled at least three times. The fine-needle aspirate enters the biopsy needle by the capillary effect. In a minority of cases (<10%, e.g., rim-calcified or suspected fibrotic nodules), syringe suction is applied

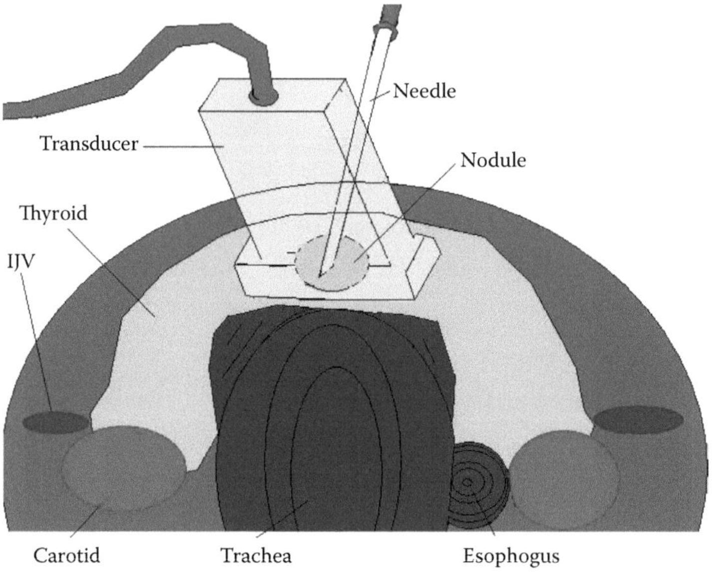

FIGURE 9.13 Schematic diagram shows short-axis approach to UG-FNAB. After the thyroid nodule is localized with an initial US, the skin is cleansed with a germicidal solution, and lidocaine 2% is injected slowly under sterile conditions in the overlying subcutaneous tissues, down to the level of the strap muscles or thyroid capsule. Lidocaine is also used to provide an acoustic interface between the transducer and the skin, as ultrasonic gel may cause artifacts under the microscope. An 8 MHz linear transducer is held at a nearly 90° angle to the overlying skin, while the handheld 25-gauge, nearly 3.5 cm long biopsy needle enters directly above it, and courses at a nearly 30 to 40° angle with the transducer. Additional minor adjustments in the course of the biopsy needle are then made when the needle tip is at approximately the level of the strap muscles. The transducer may be moved slightly inferiorly, to allow for additional cephalad angulation of the biopsy needle, if necessary.

via a sterile tube that is attached to the biopsy needle. This allows for fine movements of the biopsy needle that may be lost if the syringe is attach directly to the needle.

The availability of an expert cytologist at the time of FNAB can provide tremendous benefit. Two sets of slides are usually prepared, with one set air dried and stained with a quick stain for examination at the time of biopsy, and another set fixed in alcohol for subsequent cytological evaluation. Additional FNABs are performed if the initial quick stain shows an insufficient number of cells. As a result, the incidence of indeterminate biopsy is <1% (personal experience). Post-FNAB images are obtained routinely. Post-FNAB hematomas (<1% incidence in more than 2000 patients with UG-FNAB; personal experience) are compressed under US guidance, usually for 2–3 minutes, and reevaluated to ascertain stability or regression prior to the patient's discharge.

9.6 SUMMARY

High-resolution US with color Doppler supplementation is an excellent imaging modality for delineating thyroid nodules, selecting potentially malignant thyroid nodules for FNAB, and guiding the FNAB needle. US, especially when combined with pre-contrast and contrast-enhanced thin-section CT scanning, is a powerful tool for the preoperative evaluation and postoperative follow-up of the patient with thyroid carcinoma.

ACKNOWLEDGMENTS

We thank Madeline Vazquez, MD (formerly clinical professor of pathology and radiology at Weill Cornell Medical College), for her invaluable cytological expertise, Juan Espinosa, RDMS, for his invaluable technical assistance and Figure 9.13, and Jian-chu Li, MD, at Peking Union Medical University Hospital for providing Figure 9.5 and Figure 9.7.

REFERENCES

1. Watters DA, Ahuja AT, Evans RM. 1992. Role of ultrasound in the management of thyroid nodules. *Am J Surg* 164:654–57.
2. Papini E, Guglielmi R, Bianchini A, Crescenzi A, Taccogna S, Nardi F, Panunzi C, Rinaldi R, Toscano V, Pacella CM. 2002. Risk of malignancy in nonpalpable thyroid nodules: Predictive value of ultrasound and color-Doppler features. *J Clin Endocrinol Metab* 87(5):1941–46.
3. Jun P, Chow LC, Jeffrey RB. 2005. The sonographic features of papillary thyroid carcinomas: Pictorial essay. *Ultrasound Q* 21:39–45.
4. Chan BK, Desser TS, McDougall IR, Weigel RJ, Jeffrey RB Jr. 2003. Common and uncommon sonographic features of papillary thyroid carcinoma. *J Ultrasound Med* 22(10):1083–90.
5. Wienke JR, Chong WK, Fielding JR, Zou KH, Mittelstaedt CA. 2003. Sonographic features of benign thyroid nodules: Interobserver reliability and overlap with malignancy. *J Ultrasound Med* 22(10):1027–31.
6. Frates MC, Benson CB, Doubilet PM, Cibas ES, Marqusee E. 2003. Can color Doppler sonography aid in the prediction of thyroid nodules? *J Ultrasound Med* 22(2):127–31.
7. Moon WJ, Jung SL, Lee JH, Na DG, Baek JH, Lee YH, Kim J, Kim HS, Byun JS, Lee DH. 2008. Benign and malignant thyroid nodules: US differentiation—multicenter retrospective study. *Radiology* 247(3):762–70.
8. Frates MC, Benson CB, Charboneau W, Cibas ES, Clark OH, Coleman BG, et al. 2005. Management of thyroid nodules detected on ultrasound: Society of Radiologists in Ultrasound consensus conference statement. *Radiology* 237:794–800.
9. Taki S, Terahata S, Yamashita R, Kinuya K, Nobata K, Kakuda K, Kodama Y, Yamamoto I. 2004. Thyroid calcifications: Sonographic patterns and incidence of cancer. *Clin Imaging* 28(5):368–71.
10. Kim MJ, Kim EK, Kwak JY, Park CS, Chung WY, Nam KH, Youk JH. 2008. Differentiation of thyroid nodules with macrocalcifications: Role of suspicious sonographic findings. *J Ultrasound Med* 27:1179–84.
11. Baloch ZW, Cibas ES, Clark DD, et al. 2008. The National Cancer Institute thyroid fine-needle aspiration state of the science conference: A summation. *CytoJournal* 5: 6.
12. Ott RA, Calandra DB, Shah KH, Lawrence AM, Paloyan E. 1987. The incidence of thyroid carcinoma in Hashimoto's thyroiditis. *Am Surg* 53(8):442–45.
13. Lyshchik A, Higashi T, Asato R, Tanaka S, Ito J, Mai JJ, et al. 2005. Thyroid gland tumor diagnosis at US elastography. *Radiology* 237:202–11.
14. Moritz JD, Ludwig A, Oestmann JW. 2000. Contrast-enhanced color Doppler sonography for evaluation of enlarged cervical lymph nodes in head and neck tumors. *AJR* 174:1279–84.
15. Yoon DH, Chang SK, Choi CS, Yun EJ, Seo YL, Nam ES, Cho SJ, Rho YS, Ahn HY. 2008. The prevalence and significance of incidental thyroid nodules identified on computed tomography. *J Comput Assist Tomogr* 23(5):810–15.
16. Shetty SK, Maher MM, Hahn PF, et al. 2006. Significance of incidental thyroid lesions detected on CT: Correlation among CT, sonography, and pathology. *AJR* 187:1349–56.
17. Kim E, Park JS, Son KR, Kim JH, Seon SJ, Na DG. 2008. Preoperative diagnosis of cervical metastatic lymph nodes in papillary thyroid carcinoma: Comparison of ultrasound, computed tomography, and combined ultrasound with computed tomography. *Thyroid* 18(4):411–18.
18. Aygun N. 2008. Imaging of recurrent thyroid cancer. *Otolaryngol Clin N Am* 41:1095–106.
19. Helal BO, Merlet P, Toubert ME, Franc B, Schvartz C, Gauthier-Koelesnikov H, Prigent A, Syrota A. 2001. Clinical Impact of ^{18}F-FDG PET in thyroid carcinoma patients with elevated thyroglobulin levels and negative ^{131}I scanning results after therapy. *J Nucl Med* 42(10):1461–69.
20. Zuijdwijk MD, et al. 2008. Utility of fluorodeoxyglucose-PET in patients with differentiated thyroid carcinoma. *Nucl Med Commun* 29(7):636–41.
21. Sundram FX. 2006. Clinical use of PET/CT in thyroid cancer diagnosis and management. *Biomed Imaging Intervention J.*

22. Nahas Z, Goldenberg D, Fakhry C, Ewertz M, Zeiger M, Ladenson PW, Wahl R, Tufano RP. 2005. The role of positron emission tomography/computed tomography in the management of recurrent papillary thyroid carcinoma. *Laryngoscope* 115(2):237–43.
23. Mazzaferri EL. 1993. Management of a solitary thyroid nodule. *N Engl J Med* 328:553–59.
24. Alfalah H, Cranshaw I, Jany T, Arnalsteen L, Leteurtre E, Cardot C, Pattou F, Carnaille B. 2008. Risk factors for lateral cervical lymph node involvement in follicular thyroid carcinoma. *World J Surg* 32:2623–26.
25. Som PM. 1992. Detection of metastasis in cervical lymph nodes: CT and MR criteria and differential diagnosis. *AJR* 158:961–69.
26. Wada N, Duh QY, Iwasaki KS, Kameyama KK, Mimura T, Takami KH, Takanashi Y. 2003. Lymph node metastasis from 259 papillary thyroid microcarcinomas. *Ann Surg* 237(93):399–407.
27. Ahn JE, Lee JH, Yi JS, Shong YK, Hong SJ, Lee DH, Choi CG, Kim SJ. 2008. Diagnostic accuracy of CT and ultrasonography for evaluating metastatic cervical lymph nodes in patients with thyroid cancer. *World J Surg* 32:1552–58.
28. Roh JL, Kim JM, Park CI. 2008. Lateral cervical lymph node metastases from papillary thyroid carcinoma: Pattern of nodal metastases and optimal strategy for neck dissection. *Ann Surg Oncol* 15(9):1177–82.
29. Roh JL, Kim JM, Park CI. 2008. Central cervical nodal metastases from papillary thyroid microcarcinoma: Pattern and factors predictive of nodal metastasis. *Ann Surg Oncol* 15(9):2482–86.
30. Chuang YS, Kim JY, Bae JS, Song BJ, Kim JS, Jeon HM, Jeong SS, Kim EK, Park WC. 2009. Lateral lymph node metastasis in papillary thyroid carcinoma: Results of therapeutic lymph node dissection. *Thyroid* 19(3):241–46.
31. Wunderbaldinger P, Harisinghani MG, Hahn PF, et al. 2002. Cystic lymph node metastases in papillary thyroid carcinoma. *AJR* 178:693–97.
32. Ahuja AT, Chow L, Chick W, King W, Metreweli C. 1995. Metastatic cervical nodes in papillary carcinoma of the thyroid: Ultrasound and histological correlation. *Clin Radiol* 50:229–31.
33. Frates MC. 2008. Ultrasound in recurrent thyroid disease. *Otolaryngol Clin N Am* 41:1107–16.
34. Iannuccilli JD, Cronan JJ, Monchik JM. 2004. Risk for malignancy of thyroid nodules assessed by sonographic criteria: The need for biopsy. *J Ultrasound Med* 23(11):1455–64.
35. Tollin SR, Mery GM, Jelveh N, Fallon EF, Mikhail M, Blumenfeld W, Perlmutter S. 2000. The use of fine-needle aspiration biopsy under ultrasound guidance to assess the risk of malignancy in patients with a multinodular goiter. *Thyroid* 10:235–41.
36. Kim EK, Park CS, Chung WY. 2002. New sonographic criteria for recommending fine-needle aspiration biopsy of nonpalpable solid nodules of the thyroid. *AJR* 178(3):687–91.

10 The Use of Diagnostic Radiotracers in Thyroid Cancer

Josef Machac

CONTENTS

ABSTRACT: This review of the diagnostic uses of radiotracers in thyroid cancer begins with radionuclide imaging of the intact thyroid gland for distinguishing malignant from benign thyroid nodules, in both intentional and incidental diagnostic settings. This is followed by a discussion of whole body radioiodine imaging after thyroidectomy and in the follow-up of thyroid cancer. Patient preparation and imaging techniques are reviewed. Dosimetry of thyroid remnants, cancer metastases, and whole organs such as the lungs and bone marrow are discussed. Diagnostic imaging of well-differentiated, poorly differentiated, or anaplastic thyroid carcinoma with nonradioiodine tracers is explored, using such tracers as F-18 FDG, Thallium-201, Technetium-99m sestamibi, and Tc-99m tetrofosmine. Radionuclide imaging in medullary thyroid carcinoma is also reviewed, using somatostatin analogs such as indium-111 octreotide or tracers concentrated by neurosecretory granules, such as iodine-123 MIBG or iodine-131 MIBG. Work on new approaches is aimed to fill in gaps among the above diagnostic methods.

10.1 INTRODUCTION

Radioactive iodine (RAI) imaging of thyroid cancer utilizes to a large extent the retained ability of the cells to trap and organify iodine for the production of thyroid hormone. There are other cells interspersed throughout the gland, the parafollicular C-cells, which secrete the peptide hormone calcitonin, lymphocytes, and other migratory cells. Papillary (PTC) and follicular (FTC) thyroid carcinomas may begin with variable degrees of differentiation and in the course of their career, undergo variable degrees of dedifferentiation, with loss of ability to respond to thyroid-stimulating hormone (TSH) stimulation, and loss of ability to take up RAI. Anaplastic thyroid carcinoma (ATC) is the most dangerous of the thyroid carcinomas and is essentially undifferentiated with no RAI concentrating ability. Because of these variations in RAI concentrating abilities, diagnostic imaging and treatment of thyroid carcinoma can utilize a variety of different radioisotope tracers.

Medullary thyroid carcinoma (MTC) is derived from the parafollicular C-cells. These tumors secrete calcitonin (CTN) and carcinoembryonic antigen (CEA), which serve as tumor markers. Since the only chance for cure is early surgical resection of tumor, with few patients responding to radiotherapy and chemotherapy, initial accurate staging is important. The imaging and therapy of tumors derived from these cells utilize retained properties stemming from the neuroendocrine origin of these cells, such as presence of somatostatin receptors or the storage of amine precursors in neurosecretory granules.

10.2 DIAGNOSTIC IMAGING OF THE THYROID

10.2.1 Use of RAI and Its Analogs

The use of radioactive isotopes of iodine-127, the naturally occurring iodine isotope, begins with a 1934 suggestion by Enrico Fermi about their potential use in medicine, consequent to his group's production of new radioisotopes by neutron bombardment of natural elements.[1] The first radioactive isotope of iodine, iodine-128, was produced by Robert Evans at the Massachusetts Institute of Technology.[2] Herz et al.[3] demonstrated the concentration of I-128 in the thyroid glands of rabbits and suggested that RAI could be used for studying the physiology of the thyroid gland and for therapy. Iodine-123 was discovered by I. Pearlman at the Crocker Medical Cyclotron at Berkeley in 1949.[4] At that time, I-123, with a half-life of 13.3 days, I-130, with a half-life of 12 hours, and I-131, with a half-life of 8 days, were produced in very small amounts. It was only in the mid-1940s, when the U.S. Atomic Energy Commission was first able to provide a plentiful supply of I-131, that the first human subject to receive RAI did so at MIT in 1946.[2] The number of patients receiving I-131 or I-123 sodium iodide has increased to 430,000 on a yearly basis.[5]

10.2.2 The Molecular Basis of RAI Imaging

Radioactive iodine follows the same physiologic and biochemical pathways as nonradioactive iodide in the diet. Iodide is consumed by mouth and absorbed through the intestinal tract,[3] from where it is transported in the blood to the rest of the body. Within the blood, iodine is preferentially carried by the RBCs (60%), compared to 40% in plasma.[6] From the bloodstream, iodide freely diffuses into the interstitial space. Besides the thyroid, other organs involved in the clearance of iodide from the blood are the kidneys, salivary glands, gastric mucosa, sweat glands, and mammary glands. The salivary glands take up radioiodide, although most is secreted into saliva and is again reabsorbed from the gastrointestinal tract. In the stomach, the gastric juice-to-plasma ratio is as high as 40:1. RAI is also excreted in sweat, about 4 to 8% of the total dose. RAI is cleared by the mammary glands, especially during lactation, with the milk-to-plasma ratio being as high as 39:2.[7] The concentration in milk is such that as little as 25 μCi of I-131 may cause adverse effects to the infant's thyroid gland.[8] There is placental transport of RAI.[9] While the human thyroid does not take up RAI in the first 12 weeks of gestation, there is an increasing amount of RAI uptake after the 14th week.

In the thyroid gland, iodide is trapped by the epithelial cells. The ability of thyroid tissue and thyroid cancers to concentrate RAI depends partially upon the expression and functional integrity of the sodium iodide symporter (NIS), a membrane protein transport mechanism. Some differentiated thyroid cancer (DTC) and most ATC have lost this ability, and thus do not show uptake of RAI on diagnostic imaging and are insensitive to RAI therapy. Castro et al.[10] studied the expression of the NIS and RAI uptake in the primary tumor in 60 patients with metastatic thyroid cancer. They found that of 43 patients with positive RAI uptake in their metastases, 37 (90%) also had positive NIS in their primary tumors, while 10 of the 17 patients with negative RAI scans also had negative NIS.

Once trapped, the iodine becomes activated by oxidation and then organified through binding with the tyrosine group of the thyroglobulin protein to form monoiodotyrosine and diiodotyrosine. These moieties undergo a coupling reaction to form the hormones triiodothyronine (T3) and thyroxine (T4), which are then stored in colloid form and released into the systemic circulation. RAI is incorporated into thyroid tissue, stored, and released from the thyroid in the same way as nonradioactive iodine, and thus is a useful tool to study thyroid gland physiology.[6]

The biologic half-life of iodine in the thyroid gland appears to be related to uptake. A thyroid uptake of 5% is associated with a biologic half-life of 52 days, a 15% uptake is associated with a biologic half-life of 57 days, and an uptake of 25% is associated with a biological half-life of 65 days.[11]

For RAI, the effective half-life is equal to the inverse reciprocal sum of the biologic half-life of iodide and the physical half-life of the iodine isotope. For example, for a physical half-life of 8 days and a biological half-life of 50 days, the effective half-life is 6.9 days. Uptake in the thyroid peaks at about 2 days after ingestion in a euthyroid subject.[8] The excretion of RAI is mainly by glomerular filtration, with a mean clearance of 35.6 ml/min/1.73 m^2 in the euthyroid subject, while hypothyroid patients demonstrate a lowered mean glomerular filtration rate of 26.8 ml/min/m.[2,12] The fraction of iodide excreted is inversely related to thyroid uptake.

10.2.3 Clinically Useful Radioisotopes of Iodine

Iodine-127 is the stable, naturally found isotope of iodine. Iodine-123 is produced in an accelerator or a cyclotron, and decays to stable tellurium-123 by electron capture, with a physical half-life of 13.3 hours. Its principal radiation emissions (83.4%) are gamma photons with a mean energy of 159 keV (Table 10.1). The electron volt is the amount of energy acquired by an electron accelerated through an electrical potential of 1 volt. A keV is 1000 eV. I-123 is available in liquid or capsule form for oral administration. The suggested oral dose for measurement of thyroid function with I-123 of the intact thyroid is 100 to 200 μCi. For intact thyroid imaging, activity of 200 to 400 μCi is recommended. Figure 10.1 shows a normal iodine-123 scan. The radiation absorbed dose to the thyroid gland ranges from 150 to 260 rads for a 200 μCi dose, and a whole body exposure of 0.08 to 0.14 rad, depending on the thyroid uptake.[13] The rad is a unit of absorbed radiation energy. One rad is equal to 100 ergs of energy per gram of tissue. Iodine-123 is also used in some centers for whole body imaging of thyroid cancer. For that purpose, a dose of several mCi is used. The whole body radiation exposure is approximately 0.24 rem/mCi, about one-twentieth the exposure from a similar amount of iodine-131.

Iodine-131 is a product of nuclear fission or neutron activation of tellurium-130, although it can also be produced in cyclotron. I-131 has a physical half-life of 8.04 days and decays by beta and

TABLE 10.1
Properties of Radionuclides Used for Radionuclide Studies of Thyroid

			Principal Photons		
Radionuclide	**Physical Half-Life**	**Mode of Decay**	**Type**	**Energy (keV)**	**Abundance (%)**
Iodine-123	13.3 hours	Electron capture	Gamma	159	83.4
Iodine-131	8.04 days	Beta minus	Gamma	364, 637, 285	81.2, 7.3, 6.0
Iodine-124	4.2 days	Beta plus (23%), electron capture (74%)	Gamma	603, 723, 1,691	70
Technetium-99m	6.02 hours	Isomeric transition	Gamma	140	88.9
Thallium-201	73.1 hours	Electron capture	Mercury x-rays	69–83	94.1
			Gamma	135, 167	2.7, 10.0
Fluorine-18-fluorogeoxyglucose	109 minutes	Beta plus			

Source: Adapted from D. C. Kocher, Radioactive decay data table. *DOE/TIC*, 11026, 133, 1981.

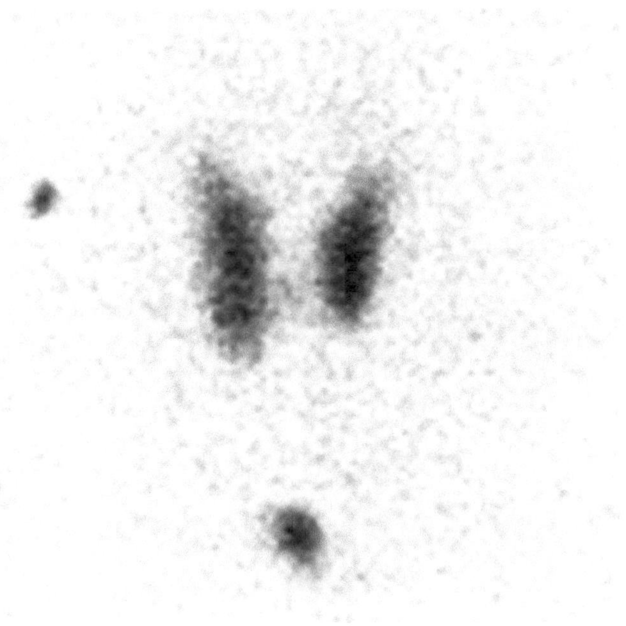

FIGURE 10.1 A normal anterior pinhole iodine-123 scan, with markers, of a 70-year-old woman, showing slight asymmetry in size but uniform distribution.

gamma emission to stable xenon-131; 81.2% of the gamma decay has a mean energy of 364 keV, 7.3% of 637 keV, and 6.05% of 285 keV. The beta decay occurs principally with a mean energy of 193 keV[14] (Table 10.1). It is the production of high-energy beta particles with diagnostic dosing that produces the high ionization potential of iodine-131, which, together with the longer physical half-life, leads to higher exposure to the thyroid gland and the whole body than iodine-123. This is an unwanted side effect, but necessary for desired therapeutic purposes in hyperthyroidism or in thyroid cancer.

I-131 is available either in liquid or in capsule form for oral administration. The liquid form of I-131 must be handled with care in a ventilated hood to avoid absorption of volatilized isotope during dilution by personnel handling it. The suggested oral dose of I-131 for measurement of intact thyroid gland function is 6 to 10 μCi. Iodine-131 is rarely used today for imaging of the intact thyroid gland due to the high radiation exposure to the gland from a dose such as 100 μCi, and also due to its inferior imaging characteristics compared to I-123 or Tc-99m pertechnetate. Iodine-131 imaging is sometimes used today for characterization of suspected substernal thyroid gland, due to the need to allow several days for tracer clearance from the mediastinal and cardiac blood pool. Iodine-131 continues to be used for whole body imaging for thyroid cancer, using activities of 1 to 5 mCi, where successful detection of chest and peripheral metastatic disease is enhanced by allowing several days of blood pool and background activity clearance, and where radiation exposure to the thyroid gland is no longer a concern. The use of iodine-131 for whole body imaging does expose the subject to some risk from radiation, which is approximately 4 rem effective dose per mCi, or 8 rem for a 2 mCi dose, or 16 rem for a 4 mCi dose.[15] This compares witha 0.62 rem average yearly exposure from all sources in the United States, and a 5 rem maximal permissible exposure limit set by the Nuclear Regulatory Commission for workers with occupational radiation exposure.

Iodine-124 is produced in a cyclotron and has a half-life of 4.2 days, and decay with either electron capture, resulting in a high-energy gamma ray emission (70%), or positron, with an abundance of about 23%. It decays to tellurium-124. It has not been widely used because of a complicated decay scheme that includes several high-energy gamma rays, which offer poor imaging characteristics with conventional gamma cameras, and a high radiation exposure due to its positron emission

(Table 10.1). However, iodine-124 lends itself to PET imaging, which would offer enhanced resolution and quantification capability, which has been investigated for lesion dosimetry estimation in thyroid cancer.[16]

10.2.4 Iodine Allergy

The common usage of the term *iodine allergy* covers a wide array of compounds that may contain iodine, such as allergy to contrast dye or allergy to shellfish. Those who have so-called iodine allergy are almost never allergic to iodide or radioactive iodide, a condition that is so extremely rare that it is not tested for routinely. The rarity of real allergy to iodide, which is essential for the production of thyroid hormone, is attested to by its wide abundance in table salt and other foods.

10.2.5 Technetium-99m Pertechnetate

Technetium-99m is produced by decay of the isotope molybdenum-99. It decays by nuclear isomeric transition with a half-life of 6.02 hours into technetium-99m, giving off the principal gamma rays with an energy of 140 keV. Tc-99m-pertechnetate, a group VII anionic form of technetium-99m with a +7 valence, is the form of technetium-99m eluted from a molybdenum-99/Tc-99m generator. Combined with four oxygen atoms with a total valence of –8, the net valence of the Tc-99m pertechnetate ion is –1, the same as iodide. Technetium-99m pertechnetate was introduced as a thyroid imaging agent with the introduction of the Mo-99/Tc-99m generator.[17–19] To some extent, Tc-99m pertechnetate behaves biologically similarly to iodide during its early distribution following administration. It is concentrated in the thyroid gland in the same manner as iodide, due to similar anionic size, shape, and charge.[20] Unlike iodide, however, once pertechnetate is trapped by the thyroid epithelial cells, it does not undergo further metabolism, or organification. Other competing anions, such as perchlorate, can readily cause the discharge of pertechnetate that has accumulated in the gland, since pertechnetate remains unchanged in the epithelial cells.[21–23] Thyroid uptake within the first 1–2 hours after injection ranges from 0.5 and 4% in euthyroid patients.[21] Imaging of the thyroid gland is done usually at the peak uptake concentration, which occurs at 20–30 minutes after injection.[24] The usual administered dose of Tc-99m pertechnetate for thyroid imaging is 1 to 10 mCi.

Pertechnetate is also concentrated in the choroid plexus, salivary and thyroid glands, and gastric mucosa.[21] Tc-99m pertechnetate also crosses the placental barrier and delivers radiation to the fetus.[25,26] Tc-99m pertechnetate is excreted in human breast milk.[27] Recommendations for the length of time to interrupt nursing range from 12–48 hours, but 24 hours (4 half-lives) is considered sufficient.[28] Tc-99m pertechnetate is eliminated in urine and feces.[21]

10.2.6 Thyroid Imaging

In previous clinical practice guidelines, functional radionuclide thyroid imaging would be used to differentiate between functional (generally benign) and nonfunctional (potentially malignant) nodules. More recently, patients with thyroid nodules and an indeterminate fine-needle aspiration (FNA) cytological finding and low-normal TSH level would undergo radionuclide thyroid imaging to assess risk for malignancy (low if functional).[29] Patients are also referred for radionuclide thyroid imaging because of hyperthyroidism, where a significant hypofunctional nodule may be coincidentally detected. Thus, the role of radionuclide thyroid imaging in the evaluation of a thyroid nodule is primarily to determine if the nodule is functional ("warm"; Figure 10.2), hyperfunctional ("hot"), or hypofunctional ("cold"; Figure 10.3). In the presence of normal or elevated TSH, the thyroid scan describes the functional status in relation to the surrounding parenchyma. Solitary hypofunctional nodules have an incidence of 20–35% being malignant (Figure 10.4), while the risk of malignancy in functional or hyperfunctional nodules is very low.[30–32] Patients with a history of

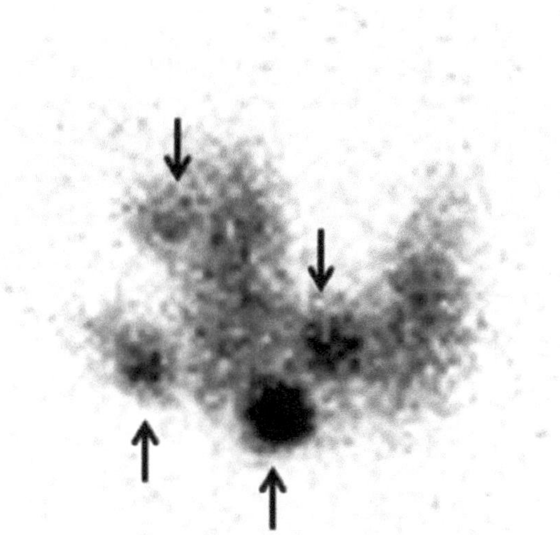

FIGURE 10.2 Four markers (arrows) outline a palpable nodule in the lower portion of the right lobe demonstrating a functional nodule.

childhood irradiation of the neck are particularly at risk of malignancy,[33] as well as those of male gender, extreme in age, and with a multinodular goiter with a predominant hypofunctional nodule or a hypofunctional nodule that is increasing in size (Figure 10.5). In the presence of two or more thyroid nodules larger than 1–1.5 cm, the rate of malignant nodules was similar to the rate of 8–10% in a contemporaneous group of patients with solitary cold nodules.[34]

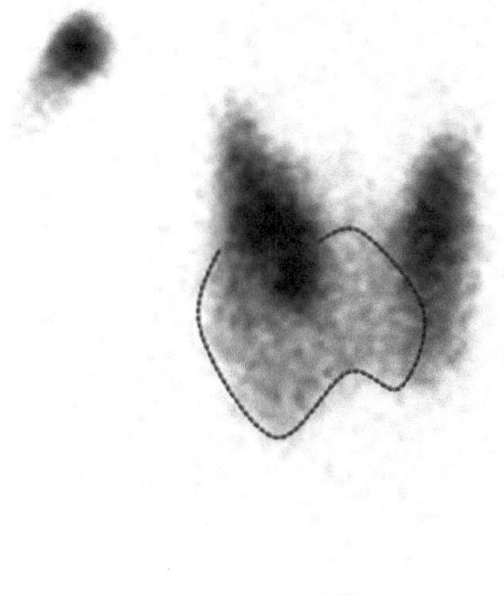

FIGURE 10.3 Anterior views with markers of an iodine-123 scan of a 53-year-old woman presenting with a nodule in the inferior portion of the right lobe of the thyroid, corresponding to a hypofunctional area on the scan (dotted line), subsequently documented to be a thyroid carcinoma.

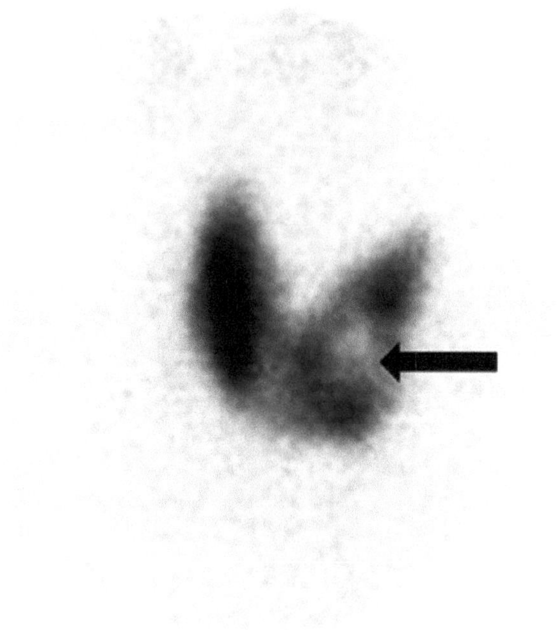

FIGURE 10.4 Tc-99m pertechnetate left anterior oblique (LAO) view of the neck in a 60-year-old woman referred for RAI therapy for Graves' disease, demonstrating a hypofunctional nodule (arrow) in the mid-lateral portion of the left lobe, corresponding to a 1 cm nodule, which was documented by FNA to be a PTC. Instead of RAI treatment, the patient underwent a thyroidectomy.

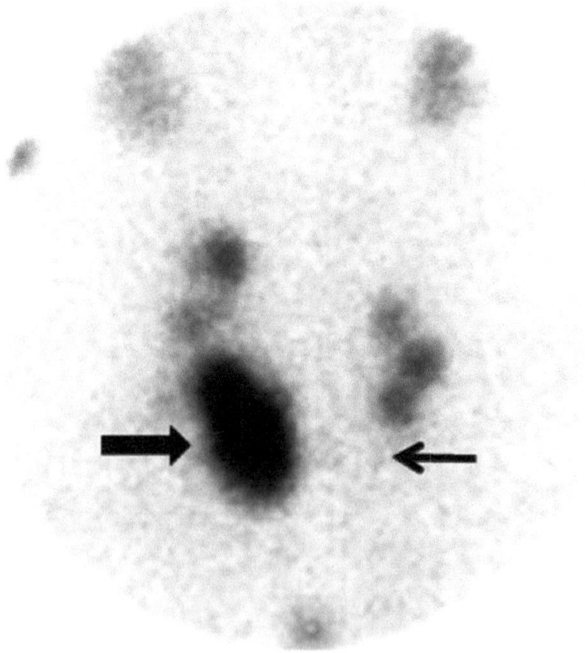

FIGURE 10.5 Tc-99m pertechnetate anterior pinhole view of the neck of a 66-year-old woman with a multi-nodular goiter demonstrating both hyperfunctioning (large arrow) and hypofunctioning (small arrow) regions.

10.2.7 Thyroid Imaging in Children

Although thyroid nodules occur less commonly in children than in adults,[35] some studies suggest that thyroid malignancy is more common in children who have them, with a rate of 15–20%,[36–38] while others suggest it is similar to the frequency found in adults.[39,40] Therefore, the recommended diagnostic approach to pediatric thyroid nodules is the same as in adults.[29]

10.2.8 Thyroid Nodules in Pregnant Women

The evaluation of thyroid nodules in pregnant women with normal or elevated TSH levels is the same as for nonpregnant women, with the exception that radionuclide imaging is contraindicated. In pregnant women with a suppressed TSH and detectable nodule(s), radionuclide imaging and FNA of nonfunctional nodule(s) can generally be deferred until after delivery.[29]

10.2.9 Thyroid Imaging Technique

Thyroid imaging is best performed with a gamma camera equipped with a pinhole collimator, which yields better resolution, compared to imaging with parallel hole collimators. Diagnostic imaging is performed either with Tc-99m pertechnetate or with I-123. I-131 is not used in routine imaging of the intact gland because of high radiation exposure. Anterior, LAO oblique, and RAO oblique views are obtained. The gland is assessed for size, which is best performed by palpation or by ultrasound (US), since the apparent size of the gland on the scan performed with a pinhole collimator is prone to error. This is because thyroid size is a function of the distance of the gland from the focal point of the collimator. Imaging should be performed with a landmark marker, such as the sternal notch and the angle of the neck. If a nodule is palpated on physical examination, or if an area on the thyroid scan shows decreased or increased uptake, an effort should be made to correlate it with palpation with the help of hot or cold markers, respectively. This requires skill and practice since parallax error with the pinhole collimator can misidentify the nodule location outside of the gland or in a different part of the gland, such that the functional status of the nodule is improperly assigned.

10.2.10 Choice of Radiotracer

The use of Tc-99m pertechnetate is more convenient, more readily available, and less expensive than I-123 for thyroid imaging. It results in a relatively low radiation exposure, and the higher administered activity provides a higher photon flux, thus providing good quality images, albeit with a slightly lower contrast than for I-123. Historically, there has been a preference for I-123 imaging when used for the evaluation of the thyroid nodule, because while Tc-99m pertechnetate is trapped in normal thyroid tissue and functional nodules, it does not undergo organification, accounting for potential mismatch between the two tracers. RAI is usually imaged at 4 hours after dosing. Kusic et al.[41] compared Tc-99m pertechnetate and I-123 imaging in 316 patients. In a few cases, there was a significant difference between the two scans. The difference involving thyroid carcinoma was rare. Thus, many laboratories currently use Tc-99m pertechnetate to image the thyroid gland for evaluation of thyroid nodules.

10.3 WHOLE BODY RAI IMAGING

10.3.1 Indications

Whole body RAI imaging is used following total or near-total thryoidectomy prior to ablation of thyroid remnants and treatment of residual disease, in posttherapy imaging, as 6–12 months' follow-up after ablation therapy, as part of surveillance, and in patients with known or suspected metastatic disease.

10.3.2 Patient Preparation

Patients are prepared by either the withdrawal from thyroid hormone therapy or the use of recombinant human TSH (rhTSH). In addition, patients should undergo two weeks of low-iodine diet before having a RAI whole body scan. If it is determined that the patient needs I-131 treatment, the low-iodine diet should be continued until after the treatment is completed. Care must be taken to determine if a woman is pregnant prior to the administration of RAI for either diagnostic or therapeutic purposes, as it has been determined that the I-131 concentration is 6–7 times more concentrated in the fetal thyroid than in maternal thyroid.[42]

10.3.3 Iodine Intake Management

The rationale for dietary management is to minimize the amount of iodine in the blood that would compete with the RAI for uptake into thyroid tissue or thyroid cancer, thus decreasing the diagnostic and therapeutic efficacy, while exposing the rest of the body to unnecessary radiation. The target is to decrease the amount of intake (and excretion) to below 50 mcg/day, preferably 30–50 mcg/day from an average that is several-fold higher.[43,44] A recommended low-iodine diet is shown in Table 10.2, with a listing of proscribed foods rich in iodine and a list of allowed foods low in iodine. The low-iodine diet should be followed for 2 weeks prior to the diagnostic or therapeutic procedure. Noniodized salt should be used during this time, and it is easy to find in the grocery store or supermarket. Kosher salt has traditionally been iodine free, although recently an iodized kosher salt has been introduced as well. More elaborate and detailed lists are available on numerous Internet sites, with some being opaque and confusing, often arising from an attempt to exclude absolutely all sources of iodide in the diet. A pragmatic approach does not require a zero-iodine diet, but merely a low-iodine diet. The low-iodine diet should be continued for several days after the RAI dose administration. There is evidence that uptake in thyroid tissue peaks at 24–48 hours after administration, which is not displaced by nonradioactive iodine.[45]

Drugs that interfere with RAI uptake are numerous and should be avoided if possible. These include competing anions (perchlorate, thiocyanate, and pertechnetate), inorganic iodide and

TABLE 10.2
Low-Iodine Diet: Proscribed and Allowed Foods

Proscribed Foods
Seafood (fish, shellfish, shrimp, oysters, lobster, kelp, seaweed)
Iodized salt (contained in processed foods with added salt, especially salted crackers, potato chips, pretzels, salted nuts)
Minimal milk or milk products
Minimal commercially baked products (to avoid added salt, preservatives, and conditioners)
Artificially dyed foods of red color
Allowed Foods
Fresh fruits
Fresh or frozen vegetables (without added sauce)
Fresh or frozen meats without seasoning or marinades, including red meats, chicken, or turkey
Potatoes, pasta, and rice (cooked with no salt or with noniodized salt)
Noniodized salt
Raw spices (avoiding commercially prepared seasonings, which usually contain salt)

organic iodine-containing preparations (Lugol's solution, supersaturated potassium iodide (SSKI), vitamin and mineral supplements, kelp, iodide-containing antitussives, topical iodine tincture, amiodarone, and various contrast dyes), and antithyroid drugs that result in decreased iodine uptake.[46]

10.3.4 Monitoring of Urinary Iodide

Routine testing of urinary iodine is usually not required. However, if the patient has had excessive iodine intake, orally or parenterally, or if compliance with a low-iodine diet is unknown, a 24-hour urinary iodide collection may be performed. The excreted iodide reflects intake in a steady state. A value of greater than 50 mcg/day indicates insufficient preparation for RAI procedures.[1,44]

10.3.5 TSH Stimulation

In order to be able to visualize thyroid remnant tissue and loco-regional or distant metastatic thyroid cancer with RAI, it is necessary to provide RAI uptake stimulation either with conventional thyroid hormone withdrawal to stimulate endogenous TSH elevation or by directly injecting rhTSH.

10.3.5.1 Conventional Thyroid Hormone Withdrawal

The traditional method of achieving TSH stimulation is thyroid hormone withdrawal of sufficient duration and hypothyroidism so that endogenous pituitary TSH levels reach or exceed 30 mcU/L.[47,48] This requires approximately 4–6 weeks of withdrawal from levo-thyroxine (T4) replacement therapy, or lack of supplementation altogether after total or subtotal thyroidectomy. The relatively long time to achieve this is due to the slow turnover of endogenous or exogenous T4. After the withdrawal period, it is recommended to check the serum TSH level to confirm levels above 30 mcU/L before proceeding with diagnostic imaging and, if indicated, therapy with RAI. However, if a large amount of remnant thyroid tissue is still present, there may be a sufficient amount of thyroid hormone produced that would not allow the TSH to rise sufficiently. In this instance, I-131 diagnostic imaging may proceed, with the demonstration of the thyroid remnants, which can then be eliminated through RAI ablation therapy. One may repeat the procedure in 6–12 months, this time after the thyroid remnants have been eliminated or reduced, to allow optimal imaging for thyroid cancer metastases. Following completion of RAI imaging or, if indicated, after RAI therapy is administered, the patient is placed back on thyroid hormone therapy again. It takes again about 4–6 weeks to restore the proper therapeutic steady state. Thus the patient may be clinically hypothyroid for a total of 2 months. Some patients may hardly report any symptoms, while other patients may become incapacitated. One should also keep in mind that the metabolism or excretion of patients' nonthyroid medications may be altered.

10.3.5.1.1 Levo-Thyroxine/Triiodothyronine Combination

In order to shorten the period of hypothyroidism, many physicians choose to substitute, during the first 2–4 weeks of T4 withdrawal, the short-acting thyroid hormone levo-triiodothyronine (T3). The usual dose is 25 mcg, two to three times a day, depending on body mass. T3 therapy is discontinued 10 days to 2 weeks prior to the RAI scan. This approach reduces the amount of time that the patient is clinically hypothyroid. Thyroid hormone withdrawal, with or without interim use of T3, can achieve adequate TSH levels in 90% of patients.[49–58] These two methods have not been compared with each other for imaging or ablation effectiveness.

10.3.5.1.2 Side Effects of Hypothyroidism

For patients who are prepared for RAI imaging and therapy with thyroid hormone withdrawal, the most significant complaints are due to the prolonged hypothyroid state, though many patients tolerate this quite well. These complaints include fatigue, depressed mood, cold intolerance, dry skin, weight increase, constipation, hoarseness, numbness/tingling, and decreased sweating. Common clinical signs include coarse/dry skin, periorbital puffiness, delayed deep tendon reflexes, and bradycardia.[59] These adverse effects can be avoided if one uses stimulation with rhTSH, since T4 therapy may be continued.

10.3.5.1.3 Restarting Thyroid Hormone

If the patient is to receive RAI treatment right after the diagnostic whole body scan, the patient is maintained without thyroid hormone. If the treatment must be postponed by a few weeks or more, then the patient can be placed back on T3 treatment, which is then discontinued 10 days to 2 weeks prior to the therapy. Following the imaging procedure, or if the patient does not undergo therapy, or if therapy has been administered, the patient is restarted on T4 therapy at least 24 hours after the RAI administration and the completion of the scan or therapy. There is evidence that maximal RAI uptake occurs at 24 hours after RAI administration, with no displacement thereafter other than hormone excretion.[45] T3 is usually not given when restarting thyroid hormone, in order to avoid a rapid rise in the thyroid hormone level, unless one is young, healthy, and significantly symptomatic from the hypothyroidism associated with thyroid hormone withdrawal.

10.3.5.2 Use of Recombinant Human TSH

There have been attempts to avoid the clinical hypothyroidism that generally attends the requisite TSH elevation for RAI imaging/treatment preparation. Recent advances in biotechnology have introduced the use of rhTSH. Single-dose rhTSH studies demonstrate maximal iodide uptake at TSH levels between 51 and 82 mU/l.[60,61] A preliminary study by Meier et al.[62] compared the efficacy of one or two doses of rhTSH with thyroid hormone withdrawal. Two injections of rhTSH were given 24 hours apart, with a diagnostic RAI dose given 24 hours after the second rhTSH dose, followed by whole body scans. Blinded readings of the imaging results with rhTSH stimulation were compared with images obtained after thyroid hormone withdrawal. Nineteen paired images, interpreted in both paired and unpaired fashion, produced similar results. A subsequent larger phase III study in 152 patients used 0.9 mg rhTSH doses injected 24 hours apart; dosing with 2–4 mCi of I-131 on the third day and imaging on the fifth day were compared with 4–6 weeks of thyroid hormone withdrawal.[63,64] The mean serum levels of TSH were 132 mU/L 24 hours after the second injection of rhTSH, compared to 101 mU/L after thyroid hormone withdrawal. The findings on scans were concordant in 66% of the patients with positive scans. Among the 34% judged to be discordant, the postwithdrawal scan was superior in 29% of the studies, and the post-rhTSH studies were judged superior in 5% of the studies. The uptake in fractional uptake of I-131 was measured in the same patients after rhTSH stimulation and thyroid hormone withdrawal.[63] The uptake after rhTSH stimulation was not as high as that after hormone withdrawal. At the same time, whole body retention was decreased after rhTSH, suggesting a higher rate of excretion of I-131. Maximal TG levels were achieved 48–72 hours after the second rhTSH injection.[63] The sensitivity and specificity of TG at a level of 3 ng/mL were 72% and 95%, respectively, after rhTSH injection, similar to 71% and 100%, respectively, after thyroid hormone withdrawal. Thus, the combination of RAI imaging and TG level measurement after rhTSH was equivalent to withdrawal from thyroid hormone.

The resulting recommended protocol for diagnosis consists of IM injections of 0.9 mg of rhTSH on days 1 and 2. The patient receives a diagnostic dose of I-131 on day 3 and is imaged on day 5. Blood samples for TG levels are obtained on day 1, just prior to the rhTSH injection, and again on day 5.

In comparisons of the sensitivity of RAI imaging with either thyroid hormone withdrawal or rhTSH stimulation, multiple factors play a role. Thyroid hormone withdrawal provides sustained stimulation for thyroid remnant and thyroid cancer tissues, favoring greater uptake, compared to the much shorter duration of stimulation. On the other hand, the measured TSH levels resulting from serial rhTSH injections are higher, resulting in more intense stimulation, favoring rhTSH. Blood clearance of RAI is faster in the euthyroid state than in the hypothyroid state, favoring a higher target-to-background ratio with the use of rhTSH. This set of complex influences can only be evaluated empirically.

In a phase III study, 229 patients with differentiated thyroid cancer were studied with TG level measurements and whole body I-131 scans, after administration of rhTSH and again after thyroid hormone withdrawal. Whole body scans were concordant between rhTSH and withdrawal stimulation in 89 patients. Of the discordant scans, 9 (4%) had superior scans after rhTSH and 17 (8%) after thyroid hormone withdrawal, a difference that was not statistically significant. Using a threshold of 2 ng/ml or greater for serum TG measurements, thyroid tissue or cancer was detected in 22% while on thyroid hormone therapy, 52% after rhTSH stimulation, and 56% after thyroid hormone withdrawal when disease was limited to the thyroid bed.[65] This is in contrast to 80% while on thyroid hormone therapy, 100% after rhTSH stimulation, and 100% of patients after thyroid hormone withdrawal with metastatic disease.[65] This, plus a second phase III study, showed that rhTSH scans failed to detect remnant or cancer localized to the thyroid bed in 16% (20/124) of patients in whom it was detected by a scan after thyroid hormone withdrawal. In addition, the rhTSH scan failed to detect metastatic disease in 24% (9/38) of patients in whom it was detected after thyroid hormone withdrawal. In the same patients, metastatic disease was confirmed by posttreatment scan or lymph node biopsy in 35 patients. Post-rhTSH stimulation TG levels greater than 2.5 ng/ml detected metastatic disease in all 35 patients, compared to 79% still on thyroxine suppressive therapy.[59]

Side effects of rhTSH include nausea (10.5%), vomiting (2.1%), headache (7.3%), asthenia (3.4%), chills (1.0%), fever (1.0%), flu symptoms (1.0%), dizziness (1.6%), and paresthesias (1.6%).[59]

10.3.6 Postthyroidectomy Whole Body RAI Imaging

The purpose of postthyroidectomy RAI imaging is to assist in initial staging for prognostication and follow-up management strategy. Imaging supplements operative staging with information about regional and distant metastatic disease, as well as providing guidance for RAI treatment. Rosario et al.[66] demonstrated the usefulness of diagnostic I-131 imaging by finding that thyroid remnant uptake and ablation efficacy are inversely related.

The patient who has undergone a thyroidectomy usually has a small amount of thyroid remnant tissue (Figure 10.6). In the presence of a large thyroid remnant, the area of the neck on the scan may be overwhelmed by uptake within the remnant (Figure 10.7), making the detection of nearby uptake within loco-regional lymph nodes difficult.[67] Moreover, the TSH level may remain depressed even after hormone withdrawal.[66] If the thyroidectomy is known to be incomplete, one may image the thyroid with Tc-99m pertechnetate to assess the size of the remnant, possibly with a small (7 μCi) dose of I-131 to measure the uptake, and then plan to administer a small ablation dose (15 to 30 mCi) of I-131. Alternatively, one may administer 200 μCi of I-123 where both size and uptake can be assessed, before low-dose RAI ablation. In patients presenting for the first time after a thyroidectomy, some experts advocate performing a Tc-99m pertechnetate scan to estimate the volume of the thyroid remnant. This is followed by a small dose (0.200–0.400 mCi) of I-131, and then measuring the uptake at 24, 48, and 72 hours to calculate the retention time and dosimetry for an ablative dose of I-131. This is then followed by a whole body scan performed after the RAI therapy. If the patient shows metastatic disease, the patient is reevaluated at 6–12 months with the intent of treating the metastatic disease.

If the thyroidectomy is known to be thorough, since one almost always finds a thyroid remnant, one would use a 1 to 5 mCi dose of I-131. A whole body anterior and posterior scan is performed at 48 hours, which allows sufficient clearance of blood and soft tissue background activity. The scan should extend from the head to the knees or mid-thighs, plus high-count spot views of the neck and chest, using a high-energy collimator. A high-energy pinhole collimator of the neck helps to resolve the individual neck remnant foci, which are often multiple. If it is suspected that metastatic disease is present, imaging can be done with single-photon-emission computed tomography (SPECT)–computed tomography (CT) imaging, which, if available, allows 3D anatomic

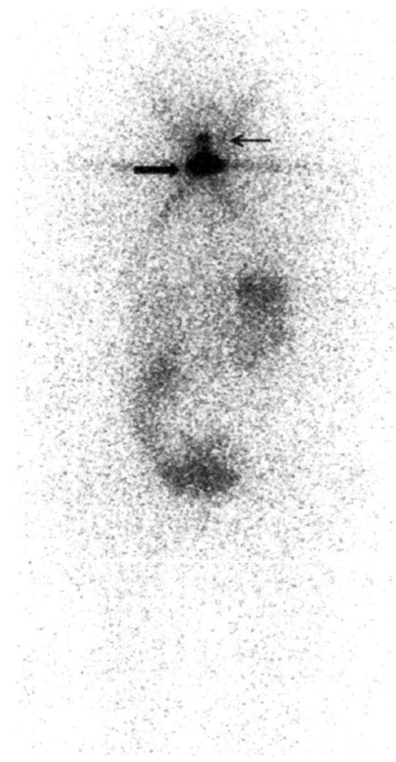

FIGURE 10.6 Anterior whole body image of a postthyroidectomy RAI imaging study with 2.0 mCi of I-131 at 48 hours in a 37-year-old female diagnosed with PTC. The images show two foci in the thyroid bed (large arrow) and one in the mid-laryngeal region (small arrow), consistent with residual thyroid remnants. Residual remnants of a pyramidal lobe are common.

localization for documentation, future follow-up, and to guide potential surgical or external beam radiation therapy (Figure 10.8). For imaging of metastases in the chest and the rest of the body, one should take care that the time allowed for imaging is high enough to achieve good counting statistics. In instances of equivocal results, one may image again the next day and after a longer period of time to clarify the results.

The uptake in the neck, and any other area of uptake, is calculated from the counts in a region of interest (ROI) drawn over the neck, minus background activity obtained from a ROI over the thigh, divided by the total administered activity, adjusted for decay, and then converted into counts with the help of a scanned calibration source.

The dose used for diagnostic whole body I-131 imaging has undergone considerable evolution. Some 30 years ago, the standard dose was 1.0 mCi. Because the sensitivity of detection of metastatic disease was shown to improve with increasing dose, the commonly used dose was then increased to 5 mCi, and then even to 10 mCi, about 15 years ago. Due to rising concern about "stunning" effects and the availability of accurate TG assays, the commonly used dose was then revised downwards. Doses of 1 to 5 mCi are therefore commonly used at this time.

10.3.7 Stunning

At the time when high doses (5 to 10 mCi) of I-131 were used for diagnostic whole body RAI imaging, it was noted that the uptake of I-131 in the thyroid remnant from the therapy dose was lower than the uptake in the same remnant on the diagnostic scan; this was relevant for doses between 5–10 mCi, but not for lower doses of 1–3 mCi.[68–70] Murat et al.[71] studied the therapeutic

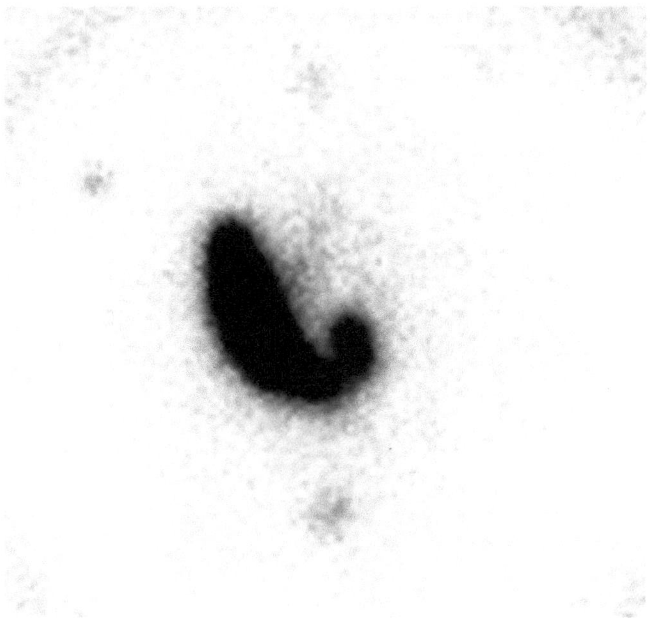

FIGURE 10.7 Anterior neck pinhole views with markers of a 75-year-old man status post surgery for acute airway obstruction, demonstrating a residual right lobe and isthmus of thyroid tissue, with 48-hour uptake of 19%, but no extrathyroidal uptake. The patient had a 4.5 cm poorly differentiated left-sided thyroid cancer with extensive local and lymphovascular invasion, which was resected, but leaving the right lobe. The patient was not able to undergo completion thyroidectomy and was treated with an initial 30 mCi RAI ablation dose. The patient then returned for 180 mCi RAI therapy of loco-regional disease.

outcome of scanning doses of 1 versus 3 mCi (111 vs. 37 MBq) of I-131, with a fixed ablative dose of 100 mCi (3.7 GBq) 9 days later. They observed 76% success with 1 mCi and 50% with 3 mCi MBq.[72] On the other hand, Dam et al.[73] showed no evidence of stunning with 5 mCi (185 MBq) doses in 81% of 135 patients. No significant difference was noted in treatment success in the two groups, either in ablation success or in treatment of local lymph node metastases. The stunning

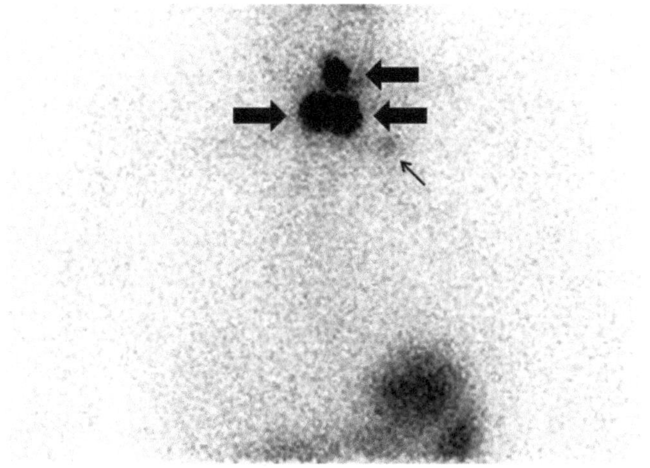

FIGURE 10.8 Anterior neck and chest view of a rhTSH-stimulated diagnostic I-131 scan of a 30-year-old man with multifocal PTC with local tumor extension and multiple positive lymph nodes status post total thyroidectomy and lymph node dissection. This scan demonstrates three likely thyroid bed remnants (large arrows), plus a faint left supraclavicular nodal metastasis (small arrow).

effect was also noted with increasing number of days between the administered diagnostic and therapeutic doses.[74] On the other hand, Cholewinski et al.[75] observed that when ablation therapy was performed on the same day as whole body imaging, at 72 hours after a 5 mCi (185 MBq) of I-131, there was no evidence of stunning. Nonetheless, the stunning phenomenon finding resulted in the tendency to reduce the diagnostic dose. Thus, one can argue that the choice of dose of I-131 should be based on the purpose of the study. A low diagnostic dose of I-131 is appropriate where the chance of peripheral metastasis is relatively low and where the determination of the presence and size of thyroid remnant is the sole purpose of preablation imaging. In patients with increased risk for metastases, a larger dose is justified. There has been a tendency in some centers to omit the diagnostic imaging study prior to initial ablation altogether, in order to avoid potential stunning of the remnants and extant cancer.[76]

10.3.8 Use of Iodine-123 for Diagnostic Whole Body Imaging

An alternative to I-131 is the use of iodine-123 for diagnostic whole body imaging. A 5 to 10 mCi dose of I-123 orally is followed by whole body scan approximately 4 hours later, with a low-energy collimator. The advantages include a lower dose of radiation absorbed by the patient, a lower likelihood of stunning, as well as improved image quality due to more counts and higher resolution stemming from the use of a low-energy collimator. The disadvantage of I-123 is its higher cost. Although comparisons show good correlation between I-123 and I-131 for tumor detection, optimal I-123 activity and time of imaging after I-123 have not yet been established.[77] Mandel et al.[78] compared 1.3 to 1.5 mCi of I-123 followed by imaging 5 hours later and 3.0 mCi of I-131 imaged 42–44 hours later. They found the same number of foci, but the I-123 scans appeared to be superior in quality. In 177 pairs of pretherapy I-123 scans and posttherapy I-131 scans, Alzahrani et al.[79] showed complete concordance in 94% pairs. With a second I-131 therapy, they showed complete concordance in 82% of 34 pairs. Sarkar et al.[80] found that I-123 was adequate for imaging of residual tissue, but it appeared less sensitive than I-131 for imaging of thyroid cancer metastases. Gerard et al.[81] found only a 70% sensitivity with pretherapy I-123 scans for detection of 37 positive sites on posttherapy I-131 scans, and 86% compared to early posttherapy I-131 scans obtained at 24–48 hours (as opposed to the commonly used 3–7 days). In spite of the better quality of I-123 images, a factor in scan sensitivity is the longer amount of time available for blood pool and background tracer clearance with the longer half-life of I-131. On the other hand, Thomas et al.[82] found concordance in only 26% of 53 patients between pretherapy I-123 scans with posttherapy I-131 scans: 35% of total lesions were detected by I-123 only, and 15% of total lesions were detected by I-131 posttherapy scans only. The lower rate of I-131 detection may be due to the 7-day waiting period for the posttherapy scans.

An alternative approach in patients who have undergone a thorough thyroidectomy involves omitting diagnostic pretherapy imaging altogether and administering an empirical ablative dose of 50 to 150 mCi of I-131, depending on risk, followed by a diagnostic scan 2 to 7 days after the therapy. The argument for this approach is that in patients with a low likelihood of peripheral metastases, the aim of therapy is to perform ablation, plus treatment of local disease, if any, with the post-therapy scan used for staging.

Salvatori et al.[83] examined the risk of omitting the pre-ablation diagnostic scan. In patients without anti-TG antibodies, distant metastases, or macroscopic tumors at surgery, thyroid remnants were found in 94% of patients by post-therapy scans. Of these, 91% had detectable TG. The most significant predictive factors for the absence of thyroid remnants or metastases were an RAI uptake of less than 1% and undetectable TG levels off thyroxine. They argued that a diagnostic whole body scan can be omitted before therapy in the majority of patients. Nevertheless, 6% of their patients had no thyroid remnants (Figure 10.9), TG elevation, or RAI uptake, and therefore would not have needed the ablation therapy.

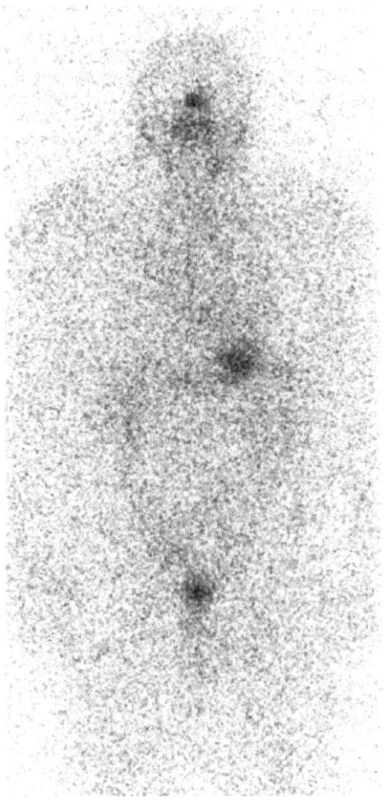

FIGURE 10.9 Anterior whole body views of a diagnostic I-131 scan showing absence of any abnormal uptake in a 60-year-old man with PTC following total thyroidectomy.

10.3.9 Detection of Local and Distal Metastases

Haugen et al.[84] compared the clinical effectiveness of I-131 whole body imaging and serum TG measurement following stimulation with rhTSH in 83 patients following thyroidectomy. Ten (12%) of the patients had positive scans, 8 of which had findings limited to the thyroid bed. Of the 9 patients with a TG of 5 ng/ml or more and a negative scan, 7 were assessed further and had disease on US. Thus, whole body scans and TG measurements are complementary, with TG levels being more sensitive for disease.

Whole body RAI scans are specific for detecting functioning thyroid tissue or metastases, but not 100%, since there are also normal structures that also take up RAI avidly, such as salivary glands and stomach, and other structures that demonstrate excreted activity, such as colon, kidneys, and urinary bladder. Activity may also be seen in the esophagus as a result of swallowed saliva (Figure 10.10A,B), which is cleared by drinking liquids and eating a piece of solid food (Figure 10.10C,D). Aspirated saliva can occasionally be seen in the carina, verified by SPECT-CT imaging and resolution on repeat imaging on successive days. There are sometimes tissues that take up RAI as artifact, particularly inflammatory lesions. Superficial contamination from saliva or urine may cause diagnostic confusion or degradation of image quality (Figure 10.11). One should be wary of superficial posttraumatic dermal scabs, a common finding as a source of potential false positive RAI uptake.[85]

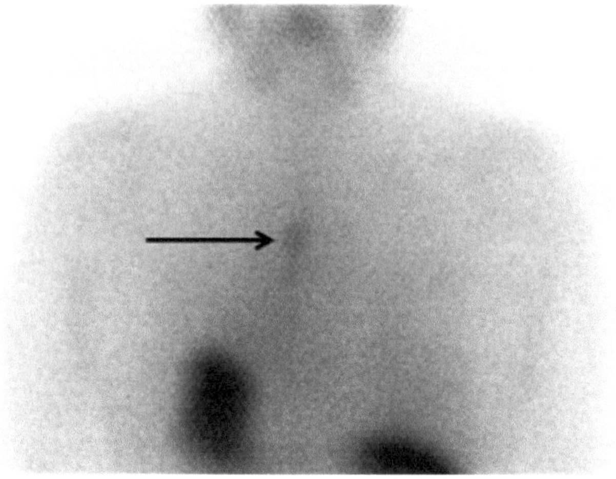

FIGURE 10.10 Anterior I-131 view of the chest showing activity (arrow) in the mid-chest due to esophageal retained saliva. This activity disappeared following drinking water and eating a piece of bread.

10.3.10 Postablation Therapy Imaging

The larger activity of I-131, 20 to 50 times the activity compared to the pretherapy diagnostic RAI imaging, significantly better image quality, and improved contrast due to enhanced clearance of background activity, because of the 5–8-day interval between dosing and imaging, all contribute to increased lesion detectability (Figure 10.12). Scanning at 5 days allows SPECT-CT imaging with

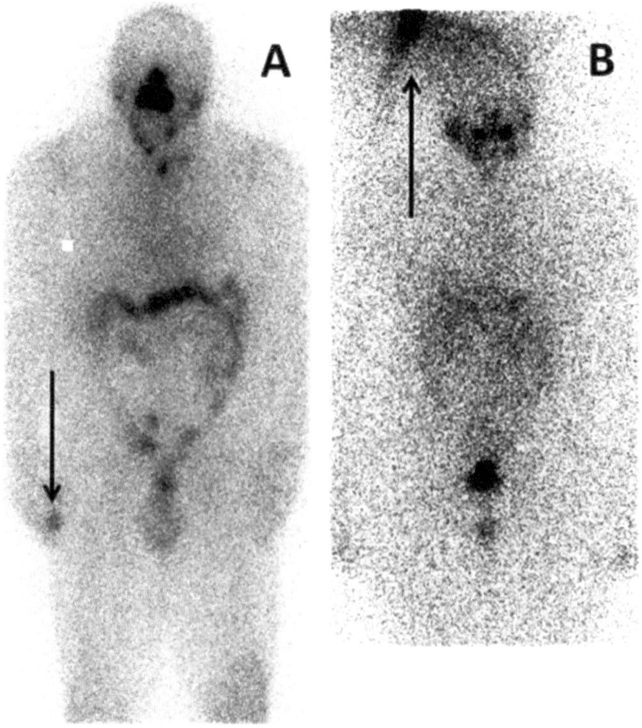

FIGURE 10.11 Anterior whole body I-131 scan in a 60-year-old man showing contamination of the right hand (A), and in a 37-year-old woman showing contamination of activity in her hair (B).

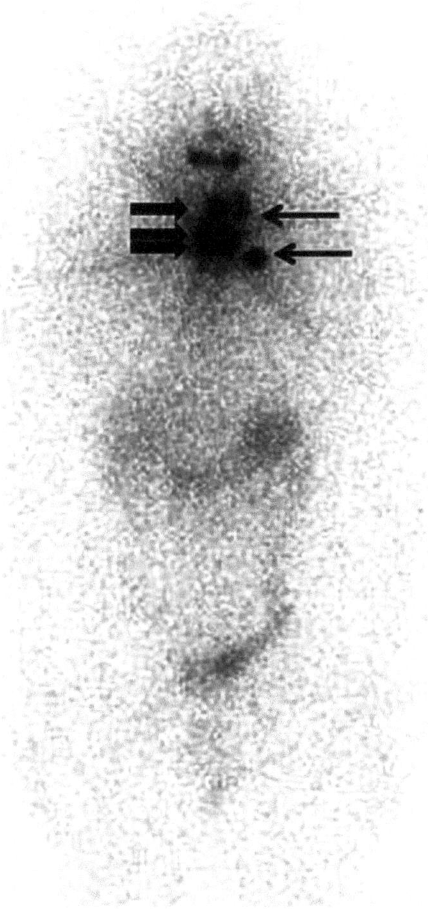

FIGURE 10.12 Post-therapy RAI images of the same patient as in Figure 10.8, five days after receiving 150 mCi of I-131. Anterior whole body views showing previously seen bright thyroid remnants (large arrows) and much more easily detectable metastases in the left supraclavicular region (level IV) and in the left mid-cervical region (small arrows).

good quality. Waiting until 7 or 8 days after administration may increase signal–noise contrast, but the decreasing count density can compromise SPECT-CT usefulness. Additional metastatic foci have been detected in 10–20% of patients imaged after high-dose RAI treatment compared to the diagnostic scan.[86,87] New findings on the posttherapy scan in the neck, lungs, and mediastinum have been shown to alter the staging in approximately 10% of patients, affecting clinical management in 9–15% of patients[86–88] (Figures 10.13 and 10.14).

10.3.11 Postablation Follow-Up Imaging

Patients who have undergone ablation therapy are usually evaluated 6–12 months later to assess the efficacy of ablation (Figure 10.15) and the presence of metastatic disease, with the intent of treating any metastatic disease. If the patient has a low risk of tumor recurrence, negative thyroid hormone withdrawal or TSH-stimulated TG levels, and negative neck US, further follow-up diagnostic RAI scans are not generally necessary. In patients with higher risks of recurrence, one may monitor the patient with sequential scans and TG level measurements for 3–10 years, or more. Patients with highly suspected or known metastases are often reexamined at 4–6 months after ablation. Patients with known new or previously treated metastatic disease may return for blood dosimetry

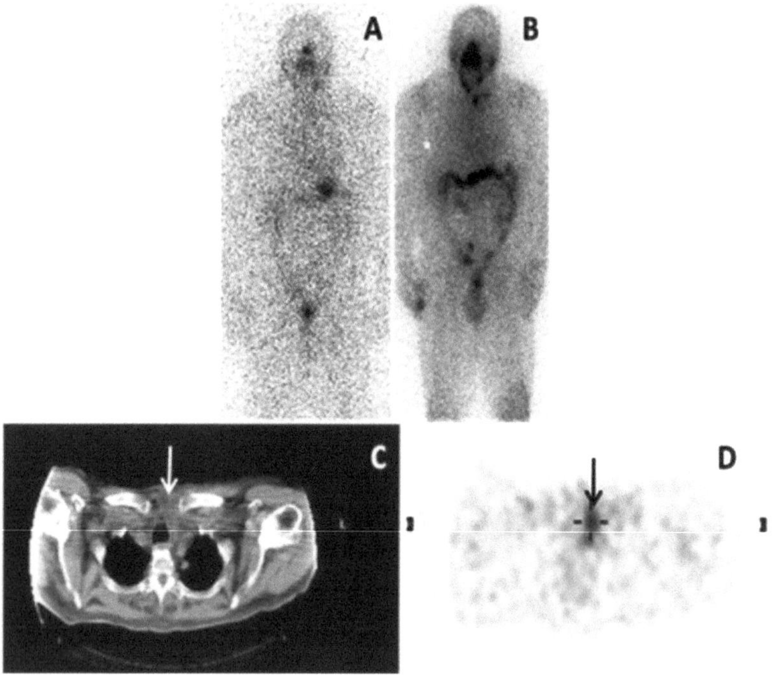

FIGURE 10.13 Anterior whole body scan (A) after a diagnostic dose of I-131 in a 60-year-old man with invasive PTC s/p thyroidectomy, showing no evidence of thyroid remnant or metastasis. Anterior whole body post-therapy RAI scan (B) and SPECT-CT transaxial image of the lower neck (C) and corresponding RAI uptake (D) following a 150 mCi ablation dose, demonstrating evidence of local metastases (arrows).

measurement and imaging with a low diagnostic dose of I-131 therapy, with a high dose or maximal tolerated dose of I-131, and then repeat imaging 5–7 days after therapy (Figures 10.16 to 10.19).

10.3.12 Follow-Up Monitoring

Patients are followed with TG testing, both at baseline and during rhTSH or withdrawal stimulation. A rising TG level, or the conversion of a negative TG level to an elevated TG level, or the presence of anti-TG antibody, especially an increasing titer, raises the suspicion of recurrence or persistence of thyroid cancer, which is an indication for RAI imaging. If the test is negative while on T4 therapy and the anti-TG antibody is also negative, then a case can be made that one can omit follow-up RAI whole body scans in previously treated patients, although continued follow-up with TG level measurements is still recommended.[89] Approximately 20% of patients who are clinically free of disease with serum TG levels less than 1 ng/ml during thyroid hormone suppression will have serum TG levels greater than 2 ng/ml after rhTSH stimulation or thyroid hormone withdrawal. A concomitant RAI scan will be positive in about one-third of such patients.[90] If the RAI scan is positive for persistent or recurrent tumor (Figures 10.18 and 10.19), then the patient is considered for RAI retreatment. If the RAI scan is negative but the TG is positive, then one may still consider treatment, as discussed below.

10.3.13 Use of I-123 for Follow-Up Whole Body Scans

DeGeus-Oei et al.[91] compared the sensitivities of serum TG and I-123 scintigraphy in the detection of thyroid cancer recurrence or metastasis. They studied 55 patients with recurrent (36) or

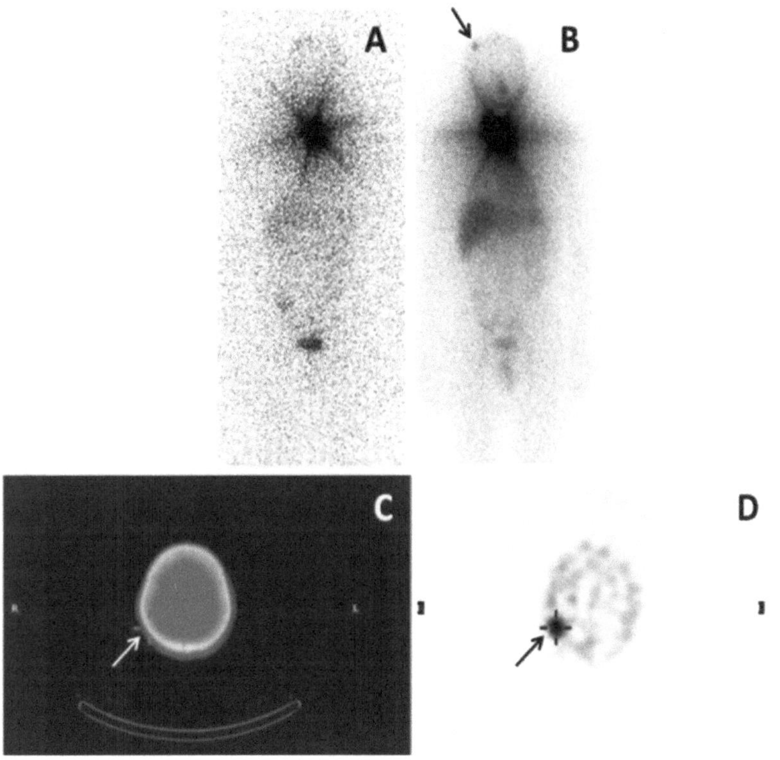

FIGURE 10.14 Anterior whole body diagnostic I-131 scan (A) of a 63-year-old man s/p total thyroidectomy for a 1.5 cm multifocal PTC that was invasive into the skeletal muscle without positive regional nodes. The scan shows local remnant thyroid tissue and possible local neck metastases but no apparent peripheral metastases. After receiving a 150 mCi dose of I-131, the post-therapy anterior whole body scan (B) shows a metastasis in the right calvarium (arrow), confirmed on the SPECT-CT image (C, arrows), changing the staging markedly and consequently necessitating further resection surgery.

metastatic cancer (19) during 3–6 months' follow-up, with 3–8 mCi (111–360 MBq) of I-123 following conventional withdrawal, I-131 therapy, and posttherapy scanning. They found that the I-123 scan had a sensitivity of 75% compared to 82% for TG measurement during thyroxine therapy or 98% for TG measurement after thyroid hormone withdrawal. Siddiqi et al.[92] investigated the use of I-123 for follow-up whole body imaging with the aim of avoidance of stunning prior to subsequent I-131 treatment. They studied 12 patients with elevated TG levels and negative diagnostic I-131 whole body scans prior to therapy with I-131. They found nearly complete concordance between I-123 scans and the posttherapeutic I-131 scans in 11 of 12 patients at the first follow-up scan, in 4 patients who received I-123 at their second follow-up scan, and in 1 patient who received I-123 at a third follow-up scan. The authors suggest that I-123 is not likely to cause a stunning effect, and argue the use of I-123 as the preferred imaging agent prior to possible RAI treatment for metastases.

10.3.14 Sensitization Methods for RAI Imaging

A significant proportion (30–50%) of thyroid cancers are not RAI avid, either at the onset of treatment or in the course of therapy. A number of methods have been tried to produce or enhance RAI sensitivity in thyroid cancer metastases. It has been determined that lithium inhibits iodine release from thyroid tissues, without inhibiting the intake of iodine. The overall effect is greater I-131 retention in normal thyroid tissue and thyroid tumors.[93] It has been estimated that the I-131 radiation dose in metastatic tumors was doubled, primarily in tumors with rapid clearance of iodine.[94] However,

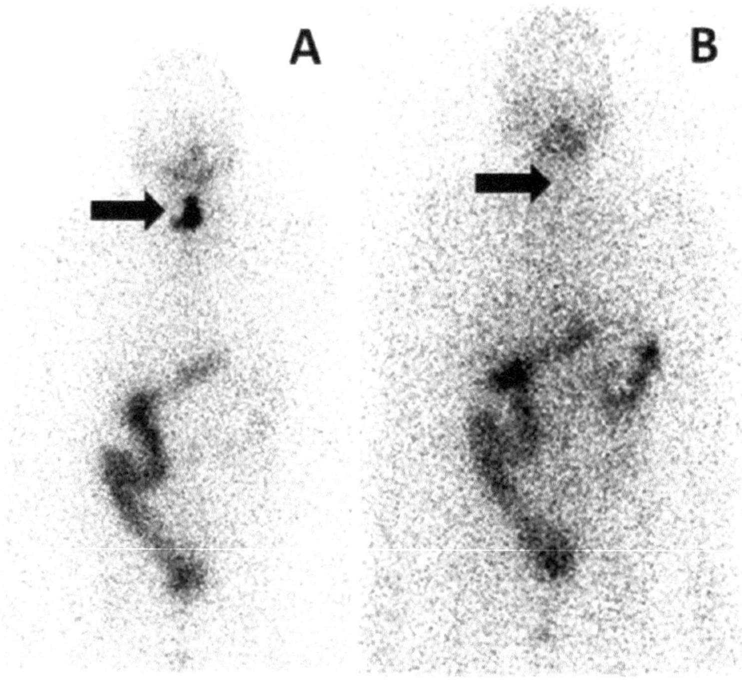

FIGURE 10.15 Anterior whole body (A) view of a diagnostic I-131 scan of a 60-year-old woman status post-thyroidectomy demonstrates thyroid remnants in the thyroid bed with an uptake at 48 hours of 1.16%. After ablation therapy with 100 mCi of I-131, one year later, anterior whole body (B) view of a diagnostic I-131 scan demonstrates successful complete ablation of the thyroid remnants.

it has been used sporadically, and randomized data in support of lithium treatment as an adjuvant leading to a better outcome are lacking.

Retinoids have been demonstrated to inhibit thyroid tumor growth and induce RAI uptake. The antiproliferative effect appears to be via antiangiogenesis and reduced VEGF accumulation.[95] Handkiewicz-Junak et al.[96] treated 53 patients with RAI nonavid metastatic disease (45 patients) or elevated TG levels and negative RAI scans (8) with 13-cis-retinoic acid (1.0 mg/kg/day over the first week, and then with 1.5 mg/kg/day) for 6 weeks prior to rhTSH-stimulated I-131 treatment. They observed RAI uptake in 9 patients (17%). In 41 patients, response was monitored with TG levels before and after treatment. In these patients, there was a significant increase in median TG level from 60–90 ng/ml. Interestingly, there was no difference in TG response between RAI uptake responders and nonresponders. In 9 responders who underwent repeat therapy with the same protocol, only 4/9 (44%) showed tumor progression. A prevalent side effect was lip dryness (98%). Fernandez et al.[97] also assessed 13-cis-retinoic acid treatment (0.66–1.5 mg/kg for 5–12 weeks), followed by a therapeutic I-131 dose of 100–200 mCi (3700–7400 MBq) in 27 patients with RAI negative metastatic disease. In 9 out of 27 (33%) patients, a positive RAI scan was observed after the treatment, while 10 patients had minimal and 8 patients had no RAI uptake. In 17 of the 19 patients with good or minimal RAI uptake that had follow-up, 7 patients had a decrease or stabilization of tumor, while tumor mass increased in the remaining 10 patients. The treatments were well tolerated. Kim et al.[98] studied 11 patients treated with 1.5 mg/kg daily for 5 weeks, followed by 200 mCi of I-131 therapy. Six of eleven (54%) patients had significantly increased TG levels just after the 13-cis-retinoic acid therapy. I-131 uptake showed marginal improvement in only 2 patients. TG levels continued to increase in 7 patients as evidence of progression. Dry skin and lips were the chief side effects. By contrast, Courbon et al.[99] demonstrated the absence of efficacy of I-131 irradiation combined with 13-cis-retinoic acid for the treatment of patients with aggressive, rapidly growing

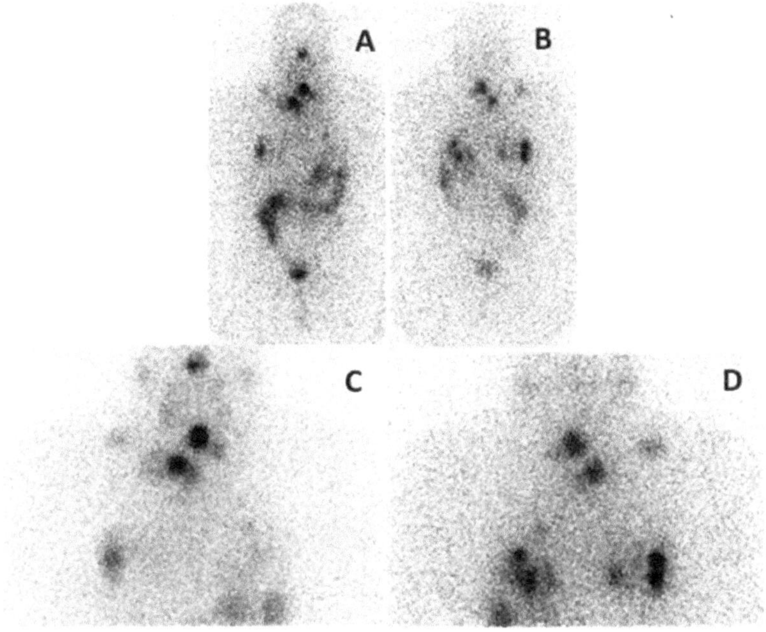

FIGURE 10.16 Anterior (A) and posterior (B) whole body views and anterior (C) and posterior (D) chest views on a diagnostic I-131 whole body scan in a 47-year-old man diagnosed with thyroid cancer following a pathological fracture of the right clavicle due to a large tumor. The patient underwent a total thyroidectomy, right neck dissection, removal of tumor from the mediastinum, and surgical resection of the right clavicular tumor. The scan shows several foci in the thyroid bed, faint activity in the right clavicle region, and multiple foci of activity in the chest, consistent with metastatic disease.

metastatic thyroid cancer. Thus, patients with highly aggressive disease, rapidly growing in a short period from 2–6 months, should not be considered for 13-cis-retinoic acid therapy. Overall, the benefit of 13-cis-retinoic acid in these trials was modest.[100–102] Nonetheless, 13-cis retinoic acid may be considered in patients with non-RAI-avid metastatic disease.[103]

Abnormal DNA methylation may be an early event in thyroid tumorigenesis, and methylation of the sodium iodide symporter (NIS) may play a role in the loss of iodine concentration in thyroid tumors. Inhibitors of methylation (5-azacytidine, phenylacetate, and sodium butyrate) have been shown to increase NIS expression and iodine uptake in cell culture models. Proteins that cause histone deacetylation inhibit gene transcription and differentiated function. Inhibitors of histone deacetylation (depsipeptide and trichostatin A) have been shown to increase NIS expression and iodine uptake in poorly differentiated and undifferentiated cell lines. Phase II human trials are currently under way for depsipeptide. Finally, commonly used agents such as thiazolidinediones and HMG-CoA reductase inhibitors have shown promise in preliminary *in vitro* studies in advanced thyroid cancer cell lines.[104] In addition, the development of tyrosine kinase inhibitors has raised hopes in providing alternative therapy for bone metastases, especially in older age groups with poorly differentiated tumors with evidence of tumor but no uptake of RAI.[103]

10.3.15 Dosimetry Calculation

Dosimetry calculation is the estimation of the expected delivered radiation dose to a target tissue from a therapeutic dose, using measurements based on a tracer dose of the radionuclide. Dosimetry can be performed for the thyroid remnant(s), tumor metastases, and other tissues, such as the salivary

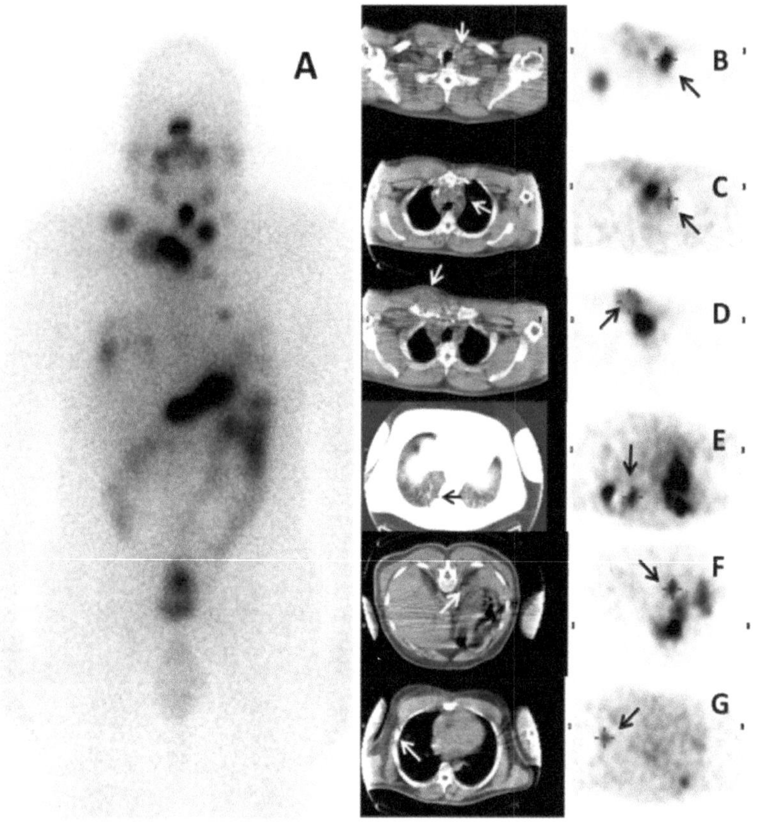

FIGURE 10.17 Post-therapy RAI images of the same patient as in Figure 10.16, who underwent high-dose I-131 therapy with 318 mCi of I-131. Anterior (A) whole body image and SPECT-CT images of the neck and shoulder (B), upper mediastinum (C), right clavicle (D), lower chest (E), left adrenal gland (F), and ribs (G) demonstrating multiple metastases and clarifying their anatomical location.

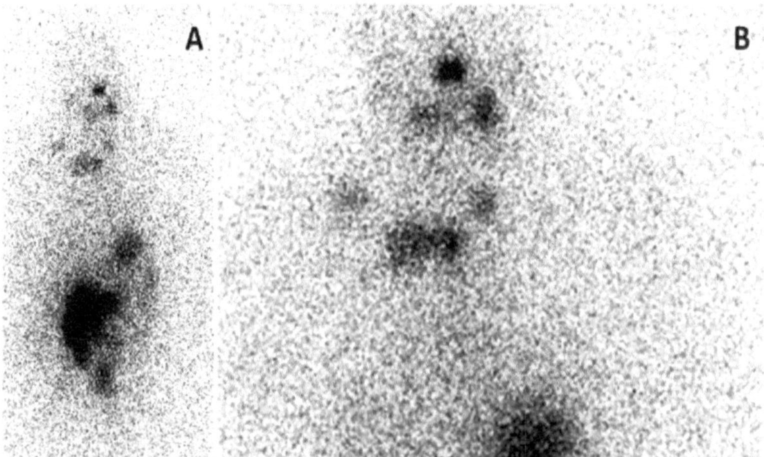

FIGURE 10.18 Six months later, the patient in Figures 10.16 and 10.17 underwent diagnostic I-131 imaging. Anterior whole body (A) and chest (B) images reveal resolution of the foci in the thyroid bed and chest with improvement in the foci in the upper mediastinum and base of neck.

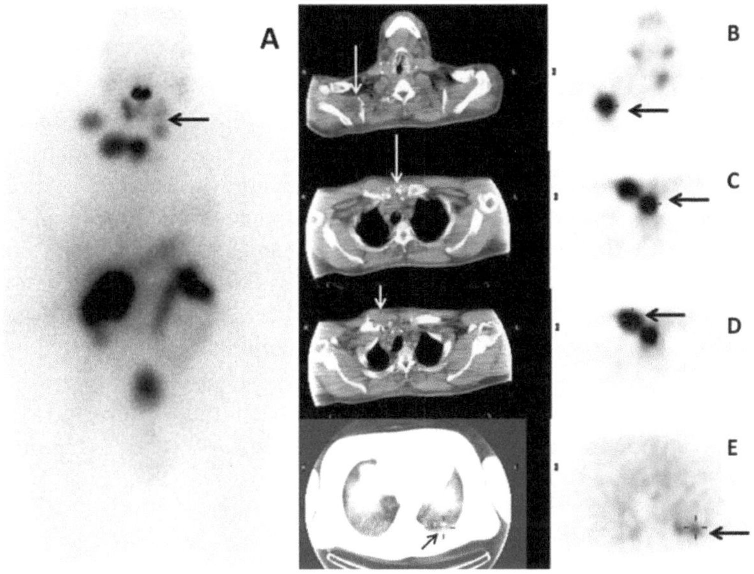

FIGURE 10.19 Six months later and following another dose of 399 mCi of I-131, anterior (A) whole body RAI image demonstrates regression of uptake in the thyroid bed (arrow), and SPECT-CT images (see arrows) demonstrate a stable lesion in the right scapula (B), two stable lesions at the base of the neck and the upper mediastinum (C), a recurrence of tumor in the right clavicle (D), and regression of uptake in the lungs (E). The lesion in the right clavicle, which was painful, was treated with EBRT.

glands, the lungs, and the bone marrow. Ideally, dosimetry should be performed in all instances of radionuclide therapy. Practically, dosimetry in thyroid cancer therapy is performed to a varied degree in selected settings only.

10.3.15.1 Dosimetry Calculation to Thyroid Remnants

The varying approaches fall into several categories: treatment with an empirical fixed dose for all patients, starting with an empirical dose and then modifying it based on the tracer uptake and mass in the target tissue, and patient-specific dosimetry, which recognizes the effects not only of uptake and mass of RAI in the target, but also of the tissue biologic turnover rate. The latter approach is the most rational, although it is the most time-consuming and costly.

Maxon et al.[53] have demonstrated the utility of dosimetry calculations to the thyroid remnants in deciding on the ablation therapy dose in ensuring treatment efficacy, minimizing exposure to the patient, minimizing costs of radioisotope and ancillary costs of hospitalization, and observing radiation safety guidelines. First, the size of the remnant is estimated by US, CT, or MRI, whichever is available. Alternatively, one may utilize Tc-99m pertechnetate imaging to estimate the remnant volume as a percentage of normal volume. Second, a small dose of I-131 is administered, ranging from 50 μCi to 1–3 mCi in order to measure the 24-, 48-, and 72-hour uptake and estimate the area under the curve, which allows one to then estimate the absorbed dose per mCi of I-131.[53]

10.3.15.2 Dosimetry to Metastatic Lesions

Theoretically, dosimetry estimation of the dose to metastatic lesions can be carried out in the same manner as for thyroid remnants, as long as the metastases take up RAI. It is even more difficult to carry out in practice, due to the difficulty in estimating lesion volume, unless it can be defined on ultrasound, CT, or MRI. The resolution with I-131 is too poor to accurately measure volume, and only slightly better with I-123. If RAI uptake is only faint, then the error in RAI uptake measurement and the turnover measurement is large. The use of I-124 with PET imaging has shown greater promise.

10.3.15.3 Iodine-124 Use for Dosimetry of Metastatic Lesions

The use of I-124 with PET imaging allows uptake to be quantified and volumes estimated with greater accuracy than with conventional imaging. Eschmann et al.[105] studied both phantom models and patients, in whom they administered oral therapeutic I-131 and diagnostic I-124 doses. They used the I-124 measurements to calculate patient dosimetry and compared the results with intratherapeutic probe measurements and clinical outcomes. They found the accuracy of measured activity to be within 10%, and volumetry accuracy within 20%, for spheres exceeding 13 mm, and an accuracy of 50% in activity and volumetry for spheres 10 mm in diameter. (For smaller lesions, the accuracy would be expected to be even worse.) They were able to estimate thyroid remnants to be receiving doses of 350–1420 Gy, and tumor metastases of 70–170 Gy. Kolbert et al.[106] also used I-124 PET imaging over 2–8 days prior to I-131 therapy for the estimation of dosimetry to normal organs, such as the brain, salivary glands, lungs, heart, liver, kidneys, spleen, and pancreas. This work demonstrates that tumor dosimetry is feasible with this approach, thus offering a rational approach to tumor therapy. A similar approach was investigated by others as well[107,108] (Figure 10.20).

10.3.15.4 Dosimetry to the Lungs with Metastases

With pulmonary metastases (Figure 10.21), the delivered dose to the lungs needs to be assessed to avoid the rare but possible pulmonary pneumonitis and eventual fibrosis due to high-dose I-131 therapy.[109] If the RAI uptake is distributed uniformly, one can estimate the dosimetry to the lungs by obtaining the (1) volume of the lungs by CT or MRI, (2) estimated percent of solid tissue (or use an estimated constant mass of solid tissue, e.g., 3 kg), and (3) serial tracer dose measurements of uptake to compute the uptake and residence time. The recognized upper limit of radiation tolerance to the lungs is 30 Gy.[110,111] If the lung uptake is nodular, the method would revert to one discussed above for discrete metastases. In practice, most centers use an easy empirical method to prevent excessive lung exposure by measuring the retained activity. That is, a limit of 80 mCi whole body retention at 48 hours should be observed in patients with diffuse pulmonary I-131 uptake.[112]

10.3.15.5 Dosimetry to the Bone Marrow

The calculation of dosimetry to the bone marrow assumes absence of direct uptake of iodine by the bone marrow. Instead, the exposure derives from the beta radiation from the blood bathing the bone marrow, and the gamma radiation from the rest of the body. The activity in the blood is sampled at the time of administration of the diagnostic tracer dose of I-131, which can be as little as 100 μCi at 24, 48, and 72 hours. By graphic or analytic methods, the mean half-life and area under the curve are found. The area under the curve, the maximal uptake, and the administered activity yield the radiation dose to the blood, which is assumed to equal the dose to the bone marrow. The dose to the bone marrow can be estimated by a similar calculation using a method developed by Benua et al.[112] Alternatively, the dose of radiation can also be determined from the estimated blood volume, dose to the blood, and retained activity in the body from sequential whole body scans.[113,114] The result is then used to calculate the activity of I-131 required to deliver a 200 rad (cGy) dose. The radiation dose absorbed by the bone marrow from the blood is actually believed to range from 30–60% of the dose to the blood,[53] thus providing a margin of extra safety.

10.4 NONRADIOIODINE IMAGING

Inasmuch as RAI imaging is only 60–70% sensitive for the detection of thyroid cancer, other radiotracers may be used for tumor detection, staging, and follow-up. Some thyroid malignancies may be poorly differentiated, tall-cell variants, or composed of Hürthle cells, which are often less RAI avid. In addition, many patients with metastatic well-differentiated PTC or FTC may initially be I-131 avid, but then become, over time with multiple I-131 treatments, I-131 less or nonavid. General tumor-avid agents like thallium-201, Tc-99m sestamibi, F-18 fluorodeoxyglucose, and

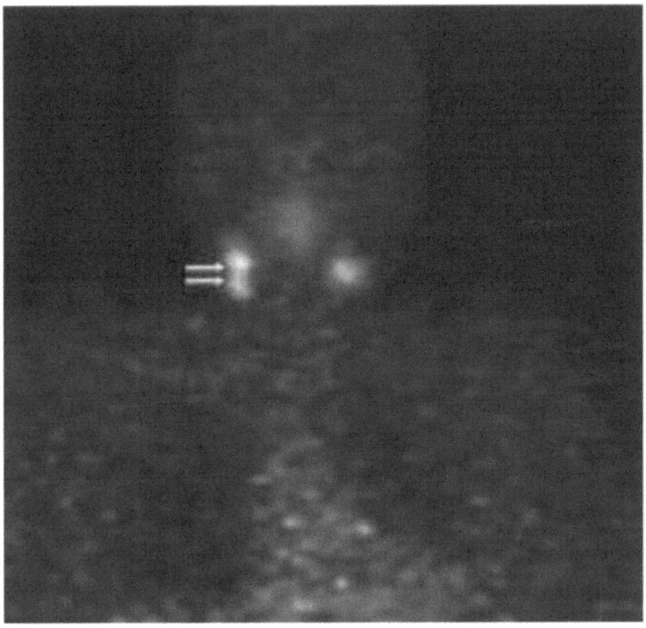

A

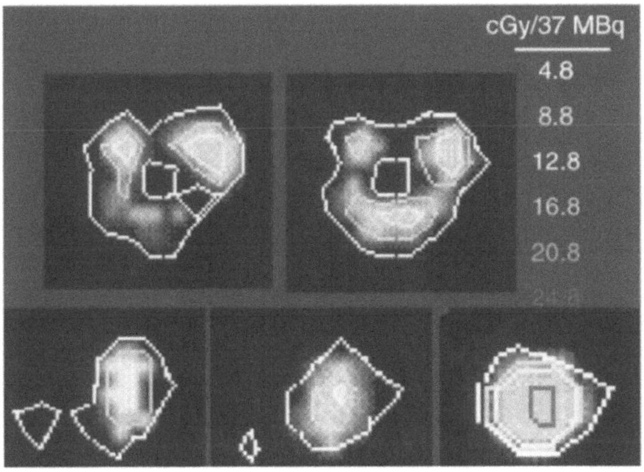

B

FIGURE 10.20 (A) Iodine-124 PET image of a patient with recurrent differentiated thyroid carcinoma with two foci on the right and one focus on the left side of the neck. (Reproduced from L. S. Freudenberg et al., *Eur. J. Nucl. Med. Mol. Imaging*, 35, 950–57, 2008. With permission.) (B) Calculated absorbed dose distribution from planned I-131 in individual thyroid cancer lesions based on sequential I-124 activity obtained with serial PET imaging. (Reproduced from Y. E. Erdi et al., *Clin. Positron Imaging*, 2, 41–46, 1999. With permission.)

gallium-67 have, in those cases of less to no I-131 avidity, been useful. F-18 FDG, in particular, has become a modality of choice when the tumor has decreased I-131 avidity. These agents may be used when the patient is euthyroid and does not need to be on a low-iodine diet. The detection and localization of disease influence decision making regarding therapy, whether it is surgical resection, local external beam radiotherapy (EBRT), or even a trial of I-131 after diligent patient preparation with a low-iodine diet and proper TSH stimulation.

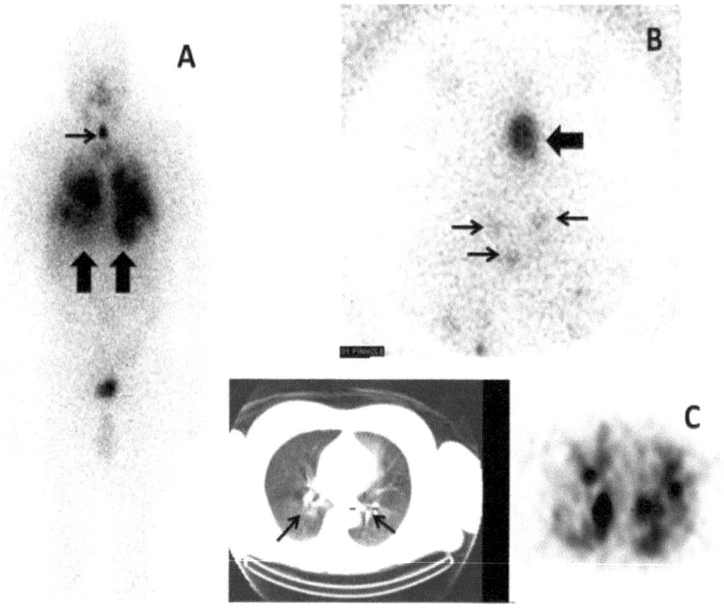

FIGURE 10.21 Anterior (A) whole body view of a diagnostic I-131 scan of a 56-year-old man s/p total thyroidectomy for a 4 cm PTC with lymphovascular invasion and one positive lymph node. The scan demonstrates a thyroid remnant in the neck (small arrow) and both diffuse and multifocal pulmonary and hilar uptake (large arrows). The 48-hour uptake in the lungs was 1.1%, and whole body retention was 2.8%. A pinhole view of the neck (B) demonstrates the thyroid remnant (large arrow) as well as several small faint foci in the lower anterior neck (small arrows). A SPECT-CT transaxial view of the chest (C) clarifies the anatomical locations of the pulmonary and hilar involvement (arrows).

10.4.1 Bone Scintigraphy

Bone imaging is useful when there are bone metastases of thyroid cancer. Bone imaging is performed with Tc-99m-labeled diphosphonates, such as methylene diphosphonate (MDP) or hydroxymethylene diplosphonate (HDP). No special patient preparation is required, other than good hydration. Several hours after radiotracer injection, one performs whole body imaging from head to toes. If needed, additional spot views are obtained (Figure 10.22). SPECT and especially SPECT-CT imaging are particularly helpful when imaging suspected lesions of the skull, spine, and pelvis. Thyroid cancer metastases are typically osteolytic, that is, appear photopenic, with a small osteopblastic rim on the edges. Small lesions simply appear as foci of increased uptake. Multiple studies reveal bone imaging to have a sensitivity ranging from 45–100% for bone lesions.[115,116] Castillo et al.[116] found that 60% of 39 differentiated thyroid carcinoma bone metastases in 8 patients were negative in the bone scans. Another 20% of the bony metastases showed only minimal increased uptake of the bone imaging tracer. The authors concluded that bone imaging has a relatively low sensitivity to thyroid cancer metastases.

10.4.2 Thallium-201 Imaging

Thallium-201 thallous chloride was suggested as a tumor imaging agent in the late 1970s, particularly for the study of lesions in the neck and chest, including tumors of the thyroid.[117–119] Thallium behaves similarly to potassium in biologic systems, with cellular uptake occurring via the sodium-potassium ATPase-dependent pump. The intracellular concentration of thallous ion is apparently greater than that of potassium.[120,121] The radiotracer was subsequently investigated primarily for its usefulness as a myocardial perfusion imaging agent. A small portion of Tl-201 concentrates in the

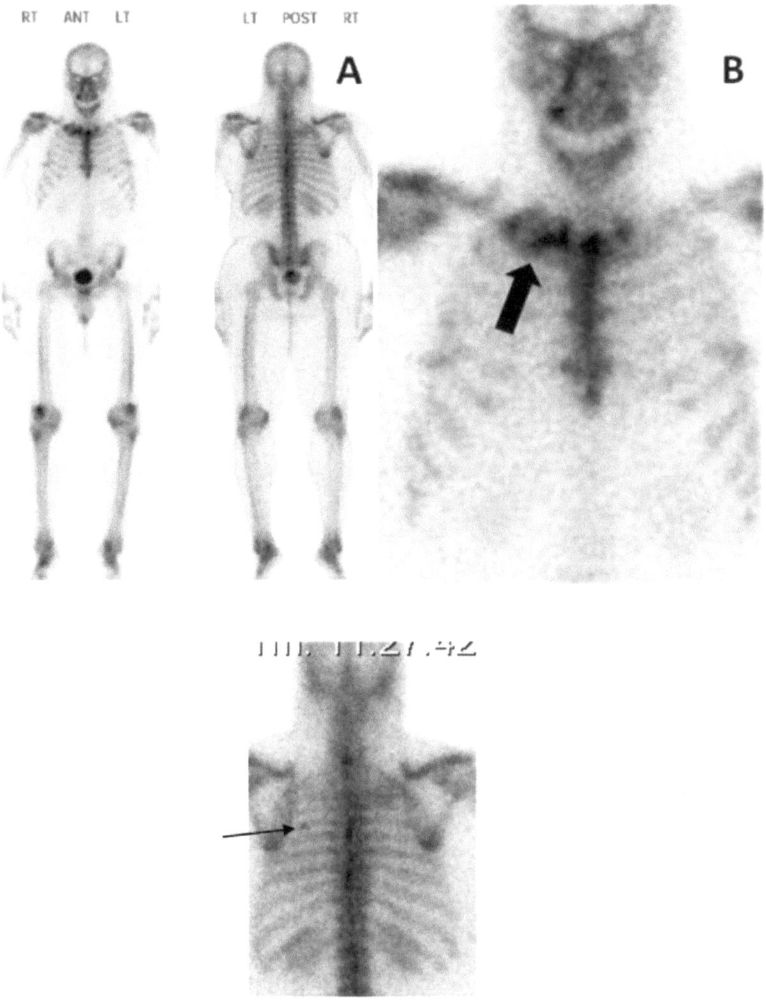

FIGURE 10.22 Bone scan with Tc-99m MDP of the same patient in Figure 10.15, demonstrating a lesion in the right clavicle (large arrow) and a small metastasis in a left fifth posterior rib (small arrow).

thyroid, and this activity washes out faster than that from the myocardium.[122] Nevertheless, because of the even faster clearance of Tl-201 from the blood, the ratio of Tl-201 uptake in the thyroid to activity in the blood is high.[123] Tl-201 was observed to accumulate in tumors even more than normal tissue, leading to investigations concerning the value of Tl-201 imaging of RAI nonavid nodules.[123–128] Since uptake in tissues is related to the fact that thallium is metabolically similar to potassium, it distributes to tissues in proportion to blood flow. The greater accumulation of thallium-201 in various tumors relative to normal tissue is partially due to increased metabolic activity and blood flow. Increased membrane permeability may be another factor.[117,118,124] Thallium-201 imaging was used for some time in the imaging of parathyroid adenomas,[129,130] and for the imaging of local and distant thyroid metastases with low avidity for RAI.

After intravenous injection, Tl-201 clears rapidly from the blood. Ninety-two percent of the injected dose is cleared with a half-life of 5 minutes, and the remaining 8% with a half-life of 40 hours, due to its distribution in the body's large intracellular potassium pool.[122] Only 0.2% of the injected dose of Tl-201 is found in the normal thyroid gland. This activity leaves the thyroid within the first 24 hours of injection. Mruck et al.[131] have found that Tl-201 in uptake in thyroid cell cultures is enhanced by TSH.

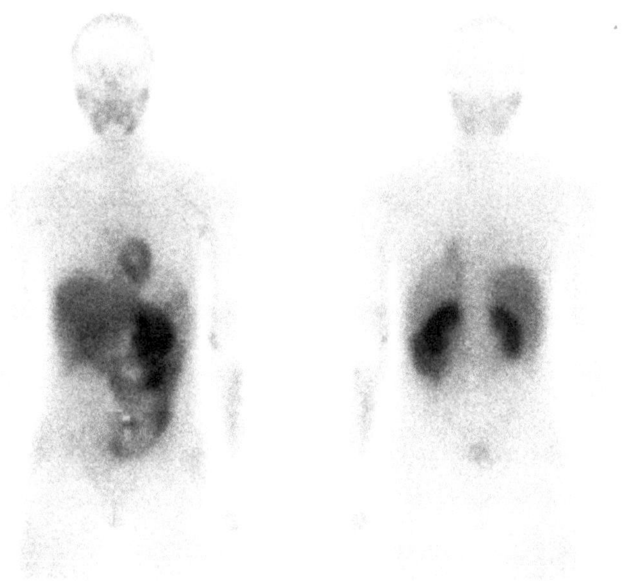

FIGURE 10.23 (A) Normal follow-up posterior and anterior images with thallium-201 in a 37-year-old female patient with history of PTC, the same as in Figure 10.17, who had undergone RAI ablation therapy.

Because of its long half-life and the long biological residence time, the radiation exposure to the body is high for a tracer that decays by electron capture, with a whole body dose of 0.65 rad/mCi, and the kidney being the critical organ, receiving 1.2 rad/mCi.[132] Therefore, the dose is limited to 2.0–5.0 mCi. Thallium-201 is excreted in maternal milk, which requires cessation of breast feeding for 2–3 weeks. This tracer is not generally given to pregnant women.

There is no special preparation for Tl-201 imaging. A half-hour after Tl-201 injection, one performs a whole body scan from head to toes or at least down to the thighs (Figure 10.23). This is usually accompanied by spot imaging, with high counts of the neck and chest, or any other organ with suspected or past involvement. SPECT or SPECT-CT imaging of the neck, chest, or any other suspected areas is often helpful.

The efficacy of Tl-201 imaging is variable. Lorberboym et al.[133] studied 50 patients with differentiated thyroid carcinoma after thyroidectomy with Tl-201 and I-131. Fifteen pairs were obtained before ablative RAI therapy, 30 pairs after ablative RAI therapy, and 15 pairs after RAI therapy for metastatic disease. Thirty-six I-131 whole body scans in 34 patients showed residual uptake in the neck, but only 6 (17%) Tl-201 studies showed uptake in the neck. Fourteen scans in nine patients identified multiple metastatic lesions, whereas the Tl-201 scans were either negative, nonspecific, or showed fewer lesions. In four study pairs, the Tl-201 scans showed solitary lesions that were not detected on the RAI scans. In 16 scans, the Tl-201 studies gave false positive results. Thus, Tl-201 imaging was less sensitive and less specific. Alam et al.[115] report a 71% sensitivity in a small sample of 14 patients. Nakada et al.[134] found that a high Tl-201 uptake in metastatic lesions indicates a greater incidence of abnormal DNA content and aggressive metastatic tumor behavior. One group used Tl-201 whole body imaging in patients who have had previous ablation therapy presenting with persistently elevated TG levels despite the presence of anti-TG antibodies.[135] Thus, Tl-201 scintigraphy is not very useful in patients with RAI-avid tumor, but can be helpful in characterizing metastatic thyroid carcinoma in patients with RAI nonavid tumor (Figure 10.24), and identifying patients with poorer prognosis, although this role has largely been supplanted by FDG PET imaging.

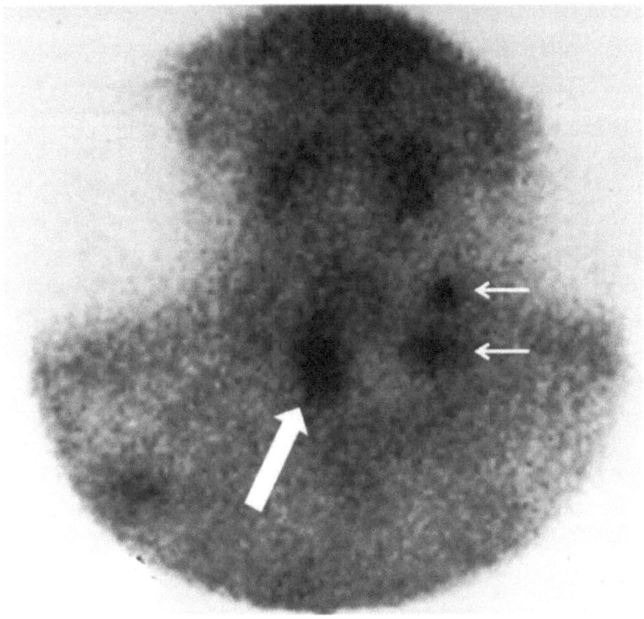

FIGURE 10.24 Thallium-201 scan of the neck of a patient with FTC with a rising serum TG level despite negative I-131 scans, showing several foci of abnormal uptake in the anterior lower neck (large arrow) and in the left supraclavicular region (small arrows). (Reproduced with permission from Fogelman I, Maisay MN, and Clarke SEM, *An Atlas of Clinical Nuclear Medicine*, 2nd ed., slide collection, Martin Dunitz, Ltd., 1995, Figure 2.76.)

10.4.3 Tc-99m Sestamibi/Tc-99m Tetrofosmine Imaging

Several lipophilic cationic Tc-99m radiopharmaceuticals (sestamibi, tetrofosmin) were developed for myocardial perfusion imaging. In the course of its use, Tc-99m sestamibi localized in lung cancers and a number of other tumors, such as parathyroid adenomas, osteosarcoma, and tumors of the brain, breast, lung, and thyroid.[136] Although initial cell membrane transport involves only passive diffusion that is nonsaturable,[137] intracellular binding of Tc-99m sestamibi is mainly associated with mitochondria,[138] related to mitochondrial metabolism and the negative inner membrane potential of the mitochondria.[139] While ouabain, a cell membrane Na+/K+ ATPase inhibitor, either had no effect or increased the cellular uptake of sestamibi, the cellular uptake of Tc-99m tetrofosmin was reduced significantly, which suggests that 20–30% of the uptake of tetrofosmin is mediated by an active transport process.[140] Nigericin, a drug that causes disruption of the cell membrane, significantly decreased the cellular uptake of Tl-201 and increased the accumulation of both sestamibi and tetrofosmin. Using precarbonyl cyanide m-chlorophenylhydrazine preincubation, an uncoupler of oxidative phosphorylation that depolarizes the mitochondrial membrane potential, it was found that 90% of sestamibi uptake is associated with mitochondria, whereas most of tetrofosmin accumulates in the cytosolic fraction.[141] The mitochondrial retention of Tc-99m sestamibi is not organ specific or tumor specific and appears to be common to most tissues.

10.4.3.1 Imaging Methods

After 20–25 mCi of Tc-99m sestamibi or Tc-99m tetrafosmine, one can perform imaging within 10 minutes. One performs a whole body scan from head to at least the thighs, with selected spot views of the neck and chest, and any other areas of interest. SPECT and SPECT-CT imaging of suspected areas help localize and identify the organ or tissue involved.

10.4.3.2 Applications

In patients studied 5 weeks after thyroidectomy with Tc-99m sestamibi and then I-131 imaging, 82 of 259 patients with low TG levels had positive Tc-99m sestamibi scans, and 113 had positive I-131 scans. Of the patients with positive I-131 scans, 72% also had positive sestamibi scans, while in patients with positive sestamibi scans, 99% also had positive I-131 scans. Thus, sestamibi showed a lower sensitivity for thyroid malignancies. In patients with elevated TG levels, of whom 69% had lymph node disease or metastases, I-131 scans still revealed more thyroid remnants and lung metastases than Tc-99m setamibi scans.[142] Another study compared I-131 imaging, CT, FDG PET, and tc-99m furifosmin, which is similar to Tc-99m tetrofosmin, in thyroid cancer with cervical, mediastinal, pulmonary, and bone metastases. Tc-99m furifosmin was found to accumulate quickly in the tumor sites, with significant washout over the next 4 hours. The visual contrast was relatively low, about 1.2 to 1. However, tumor was localized in some patients, whose results were negative by other methods of imaging. The sensitivity on a lesion-by-lesion basis was 34%, compared with 91% for FDG PET.[143]

Of interest, on Tl-201, Tc-99m sestamibi, or Tc-99m, tetrofosmine planar studies have been reported by Kim et al.[144] to frequently show a small focus of increased uptake in the right mid-parasternal region (40–49%) or in the lower mid-chest at the level of the lower sternum (20–39%). This frequent finding on planar images has been determined on SPECT and SPECT-CT studies to represent myocardial uptake in the superior right atrium and the superoanterior right ventricular wall, respectively.

10.4.4 Comparison of Tl-201, Tc-99m Sestamibi, and I-131

Tc-99m sestamibi and Tc-99m tetrofosmine demonstrate improved resolution owing to the optimal 140 keV gamma photons of Tc-99m and the higher activity that can be administered with a lower radiation dose due to the shorter (6-hour) half-life of Tc-99m. Ugar et al.[145] reported a 70% agreement between I-131, Tl-201, and Tc-99m sestamibi in patients with well-differentiated thyroid carcinoma. Both Tl-201 and Tc-99m sestamibi showed false negative results in cases where the tumor was I-131 avid. In some studies, Tc-99m tetrofosmine showed similar results as Tl-201, and both showed variable results compared to I-131.[146] In 19 patients subsequently found to have metastatic disease following thyroidectomy and I-131 ablation,[147] Tc-99mb tetrofosmine imaging identified 11 of these patients with a sensitivity of 58%. In another study, Tc-99m tetrofosmine was reported to be superior to either Tl-201 or Tc-99m sestamibi.[148] Thus, Tc-99m setamibi and Tc-99m tetrofosmine imaging are options in regions where FDG PET imaging is not available.

10.4.4.1 Differentiation of Malignant from Benign Thyroid Nodules

Tc-99m sestamibi and Tl-201 were investigated in patients with thyroid nodules, in an attempt to differentiate benign from malignant nodules. Delayed image retention of Tc-99m sestamibi and Tl-201 was better than the early uptake for detecting malignant nodules (Figure 10.25), with a sensitivity of 90%, specificity of 95%, and accuracy of 92%.[149] Benign lesions tend to show less uptake. This is illustrated by a case report by Murthy et al.[150] of a patient referred for thyroid imaging with I-123 for evaluation of a thyroid nodule that showed uptake of I-123 with suppression of the rest of the gland. Subsequent Tl-201 imaging revealed no Tl-201 uptake in the nodule, with good uptake in the rest of the thyroid gland. Surgical histopathology demonstrated a follicular adenoma.

10.4.4.2 Multidrug Resistance and P-Glycoprotein

The net uptake and retention of lipophilic Tc-99m cationic agents like Tc-99m sestamibi or Tc-99m tetrafosmine by tumor cells appear to be related to the efflux of the tracer from the cell. The efflux of these tracers is mediated by a P-glycoprotein (Pgp),[151] controlled by the human multidrug resistance (MDR) gene.[152] Mutations of p53 suppressor and p21 ras oncogenes result in increased expression

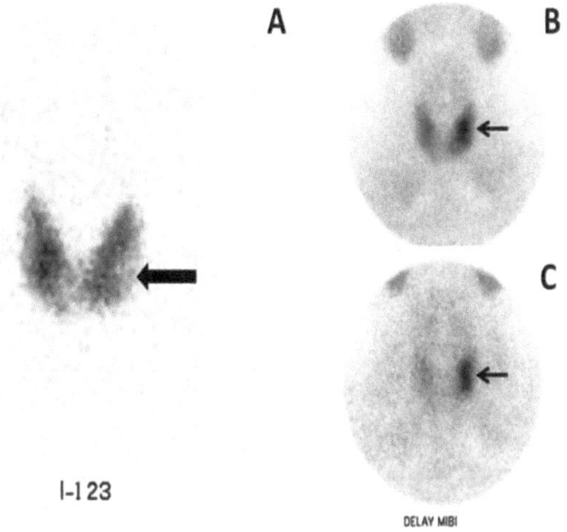

FIGURE 10.25 Anterior pinhole iodine-123 (A), and early (B) and delayed (C) Tc-99m sestamibi images in the same patient as in Figure 10.13 prior to thyroidectomy for thyroid cancer, demonstrating a hypofunctional lesion (large arrow) in the mid-lateral aspect of the left lobe on the I-123 study, with increased uptake on the early Tc-99m sestamibi image (small arrow) and enhanced uptake on the delayed Tc-99m sestamibi image (small image).

of the MDR1 gene. Increased levels of Pgp result in the ability to pump cationic and lipiphilic cytotoxic and harmful agents out of the cells by an ATP-dependent process, resulting in relative resistance to such chemotherapy agents as anthracyclines, taxanes, and vinca alkaloids. Tc-99m sestamibi has been demonstrated to be a substrate for the Pgp. Drugs such a cyclosporine A and verapamil, which inhibit Pgp function, enhance Tc-99m sestamibi accumulation. Tetrofosmin has also been found to be a substrate for Pgp.[153] Tetrofosmin is also a substrate for Pgp, although only at 30 to 70% of the level seen with Tc-99 sestamibi.[154] Thus, while the role for Tc-99m sestamibi and Tc-99m tetrafosmine as tumor imaging agents is limited with the widespread availability of FDG PET imaging, they have a potential and emerging role in the assessment of the level of expression of MDR genes, and thus likely effectiveness of certain types of pharmacological therapies.

10.4.5 F-18 Fluorodeoxyglucose (FDG) PET Imaging

A number of imaging agents take advantage of the fact that malignant tumors have a higher cell proliferation rate and, consequently, a higher metabolic rate. FDG, an analog of glucose, is transported into living cells by facilitated diffusion and is a substrate for the enzyme hexokinase. The resultant FDG-6-phosphate is trapped in the cell and not able to enter into further glycolytic or synthetic pathways. It can be dephosphorylated back into FDG by glucose-6-phosphatase, thereby leaving the cell, but that enzyme is in very low concentrations in most cancer tissues, so for all practical purposes, FDG accumulation is an indication of glucose utilization. FDG was demonstrated to accumulate in a variety of tumors. However, since all cells metabolize glucose to some degree, FDG is not entirely tumor specific. Moreover, tumor uptake is not closely correlated with the proliferation rates of all tumors. FDG also accumulates in granulomatous tissue and in macrophages at sites of infection/inflammation, and a significant percentage of FDG tumor uptake is in macrophages frequently seen in a large number in tumors.[155]

10.4.5.1 Methods

The method of FDG PET imaging for thyroid cancer is similar to methods used for other malignancies. The patient is instructed to fast for 4–6 hours. Patients with diabetes should be well controlled prior to the study, as elevated glucose competes with FDG uptake. The use of insulin just prior to the study should be avoided, as insulin drives the FDG into skeletal muscle, which interferes with interpretation and may interfere with FDG tumor uptake. The patient is injected with 10 mCi of FDG and is instructed to rest for 45–60 minutes. The patient then undergoes imaging from the base of the brain down to the thighs. The use of CT along with PET, as PET/CT imaging, has significantly improved the value of FDG PET imaging in localization of FDG uptake, improving specificity, and in guiding surgical resection and EBRT.[156]

The uptake of FDG in tissues and lesions is commonly measured in standard uptake value (SUV) units obtained by placing a region of interest (ROI) of volume of interest (VOI) over a lesion or the portion of a lesion with the most intense FDG uptake. The highest activity in the smallest imaging volume (the pixel, in mCi or megabequerels/g) is divided by the injected total FDG dose (in mCi or megabequerels), corrected for decay between injection and imaging, and divided by the total weight in grams, to yield the SUV_{max}. Thus, if the SUV were to be uniformly distributed throughout the body, then the SUV would be 1.0. An SUV_{max} of 4.0 means the maximal uptake in the lesion is four times greater than the average activity in the whole body. The SUV measurement is sensitive to the blood glucose levels, the metabolic state of the body, and the time between injection of FDG and imaging, so that careful attention to standardized preparation in terms of fasting blood sugar and the time between injection and imaging must be adhered to.

10.4.5.2 Applications of FDG PET

The uptake of FDG complements the uptake of RAI in thyroid cancer. In some patients with thyroid cancer, there is uptake of both RAI and FDG; in other patients, only RAI uptake is positive; while in still other patients, only FDG scanning is positive. Together, however, the two agents achieve a high sensitivity of detection.[156–158] Chung et al.[159] demonstrated that FDG PET accurately localized metastatic sites with a sensitivity of 94% and specificity of 95%, compared to a sensitivity of 54% and specificity of 76% for TG, and sensitivity of 64% and specificity of 56% for I-131 imaging. Since some metastases were found in normal-sized lymph nodes and half of those patients had normal serum TG levels, FDG PET was useful in planning the patients' surgical management. Wang et al.[157] noted that FDG uptake correlated with increased aggressiveness of thyroid cancer, also correlating with decreased uptake of I-131. In 37 patients with elevated TG levels, FDG PET localized disease in 71% patients. These findings resulted in altered management in 19 of the 37 patients, with a positive predictive value of 92%. In patients with low TG levels, FDG PET had a negative predictive value of 93%. Although none of the FDG PET scans were positive with stage I disease, they were positive in all patients with stage IV disease and increased TG levels. At the same time, the methodology was not sensitive enough to detect minimal residual disease in the cervical nodes.[160] Another study in 22 patients showed FDG PET had a 80% sensitivity and 83% specificity in detecting cervical lymph node metastases of all sizes, with a sensitivity of 96% for malignant lymph nodes greater than 1.0 cm in size, and a sensitivity of 70% for normal-sized lymph nodes 1.0 cm or smaller. This contrasted with TG, which has a sensitivity of only 55% in detecting cervical lymph node metastases of all sizes in a subset of 12 patients.[161]

Practically, FDG PET imaging is indicated in patients with evidence of thyroid cancer where I-131 imaging is negative[162] (Figure 10.26). Alnafisi et al.[163] demonstrated positive FDG PET scans in 7 of 11 patients with elevated TG levels, but negative I-131 scans and negative CT, US, or MRI scans. The identification of biopsy sites resulted in treatment changes in most patients. Frilling et al.[164] studied 24 patients with negative I-131 diagnostic scans and elevated TG levels, resulting in the identification of 38 foci with FDG uptake, and a sensitivity of 94.6%, specificity of 25%, and accuracy of 88%, compared to surgical findings, with three false positives, in benign cervical lymph nodes. Helal et al.[165] evaluated FDG PET in 37 patients with increased TG levels and negative post-high-dose I-131 scans. In patients with positive conventional imaging findings, FDG PET identified 17 of 18 known tumor sites and 11 additional sites. Among patients with negative conventional imaging studies, FDG PET identified positive sites in 19 of 27 patients who had not been identified to have tumor sites.

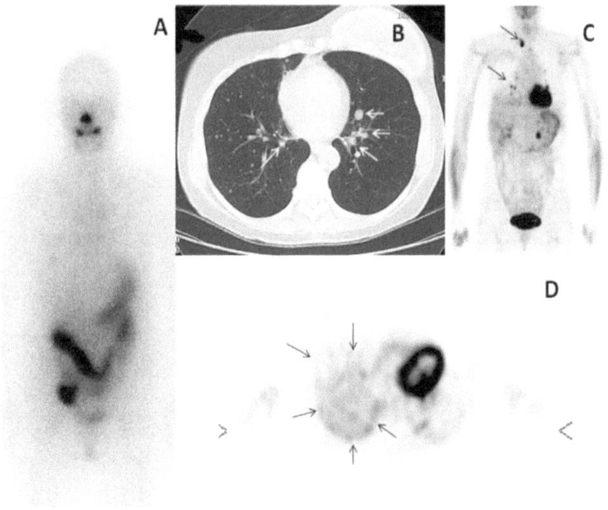

FIGURE 10.26 Sixty-year-old female who had undergone a total thyroidectomy for locally invasive PTC. Pre-ablation RAI scan demonstrated nodal involvement. She received high-dose RAI therapy with 200 mCi of I-131 with successful ablation of all tumor uptake. Three years later, she received 260 mCi RAI because of a rising serum TG level. The anterior whole body post-therapy I-131 scan was negative (A) despite an abnormal CT scan of the chest (B) showing new pulmonary nodules (arrows). A FDG PET whole body scan (C) demonstrates hypermetabolic lymph nodes in the neck and hila with multiple foci in the lung fields (arrows), as well as mild diffuse FDG uptake in the basal lung parenchyma (D, arrows), corresponding to the multiple small lung nodules on the CT.

PET imaging resulted in a change in treatment in 29 of the 37 patients, including further resection. Bertagna et al.[166] also found FDG PET/CT to be a useful diagnostic tool in such patients. The highest accuracy reached was with TG levels higher than 21 ng/ml. FDG PET imaging may also direct the physician to specific surgical interventions, or EBRT if the disease is still limited and localized.

FDG PET imaging is also of value in providing prognostic information in thyroid cancer. Robbins et al.[167] studied 400 patients with a median follow-up of 7.9 years. While age, initial stage, histology, TG, RAI uptake, and PET uptake all correlated with survival by univariate analysis, only age and PET results continued to be strong independent predictors of survival using multivariate analysis. The combination of FDG negative and either RAI positive or negative results predicted a good prognosis, while FDG positive uptake with either RAI positive or negative results predicted a much worse prognosis.

TSH levels affect the sensitivity of FDG PET. Two studies showed that an increase in TSH levels did not increase the ability of FDG PET to detect lesions.[168,169] On the other hand, in 10 subjects with thyroid carcinoma studied by Moog et al.,[170] 15 of 17 lesions with increased FDG uptake had an average target-to-background ratio (TBR) during TSH suppression of 3.85, but a TBR during TSH stimulation of 5.84, representing a 52% increase. Of these 17 lesions, I-131 uptake was absent in 12. Leboulleux et al.[171] found that the performance of FDG PET during TSH stimulation using thyroid hormone withdrawal was either superior or equal to the performance of FDG PET during thyroid hormone suppression, but never inferior, leading to a recommendation to perform FDG PET imaging during TSH stimulation.

10.4.5.3 Comparison of FDG PET with Tl-201, Tc-99m Sestamibi, and Tc-99m Tetrofosmin

Shiga et al.[172] studied FDG PET and Tl-201 in 32 patients with differentiated thyroid cancer also undergoing I-131, CT, and MRI after total thyroidectomy. Out of 47 total lesions, all scintigraphic methods detected 87% of the lesions. FDG was concordant with I-131 for only 38% of the lesions. In 94% of the lesions, FDG uptake was concordant with Tl-201 uptake. Of the three patients that were discordant, two were positive with PET and one was positive with Tl-201. This supports the earlier comment that Tl-201 is a reasonable alternative to FDG PET where PET imaging is not available.

10.4.5.4 Use of FDG PET in MTC and ATC

The challenge in both MTC and ATC is early diagnosis, complete surgical resection, and early diagnosis of recurrence. These represent the key factors that can lead to improved outcomes, primarily due to the fact that conventional imaging and biochemical testing have poor sensitivity and RAI has no role, since these tumors are not iodine avid. While MTC comprises only about 10% of thyroid cancers, it has a relatively poor prognosis, though not nearly as poor as ATC.[173,174] Imaging of occult MTC includes FDG PET imaging, CT, or MRI, and if FDG PET imaging is not available, Tc-99m DMSA, In-111 octreotide scintigraphy, and I-123 MIBG.

Brandt-Mainz et al.[175] studied 20 MTC patients with suspected disease due to increased CTN levels or sonographic abnormalities in the neck and found 13 of 17 patients positive on FDG PET imaging, validated by CT and histology, while 5 were completely negative, including 4 false negatives. DeGroot et al.[176] studied 26 patients with increased serum markers with FDG PET imaging after undergoing total thyroidectomy, central compartment dissection, and further neck dissection as needed. The results were compared to other conventional anatomical imaging and bone scintigraphy. Compared to 20% positive findings by Tc-99m DMSA, 21% by In-111 octreotide imaging, and anatomic imaging in 40%, FDG PET imaging was superior and identified positive foci in 50% of the patients. The use of FDG PET imaging resulted in 35% of patients requiring further surgery. Thus, FDG PET appears to be valuable for detecting and localizing metastatic lesions in MTC at the time of initial staging and during follow-up. Although a review of the literature reveals that no large study of these cancers with FDG PET imaging exists, it does suggest a sensitivity of 66–100% and specificity of 79–90%, better than conventional anatomic imaging, and suggesting an important role of FDG PET in these tumors.[177]

10.4.5.5 FDG PET Imaging of the Thyroid Preoperatively

Incidental findings in the thyroid gland on US or CT are common, ranging from 19–46%; 1.5–10% prove to be carcinoma. In patients with multinodular goiters, screening for thyroid cancer is difficult. Kresnik et al.[178] studied 16 patients with thyroid carcinomas, 23 patients with thyroid adenomas, and 4 patients with degenerative goiter, and found FDG PET to have 100% sensitivity, 63% specificity, and 100% negative predictive value for carcinoma. Mitchell et al.[179] performed a similar study in 31 patients with 48 thyroid lesions, of which 15 lesions were malignant (of which 9 showed increased FDG uptake corresponding to a 60% sensitivity). However, 30 of the 33 lesions that were benign were cold on FDG imaging, for a specificity of 91%, positive predictive value of 75%, and negative predictive value of 83%. De Gues-Oei et al.[180] studied 44 patients with inconclusive fine-needle aspiration procedures. Histopathological examination revealed 7 well-differentiated thyroid carcinomas in 6 patients, all accumulating FDG with a negative predictive value of 100%. FDG accumulated in 13 of 38 benign nodules. The authors argue that after FDG PET, the number of unnecessary hemithyroidectomies would be only 30%, compared to 86% without PET. These results underscore the fact that many well-differentiated thyroid cancers do not take up FDG, so that FDG PET is unlikely to be used routinely for evaluation of thyroid nodules. This is of interest mainly in incidental findings, as the use of FDG PET for this situation is not indicated or reimbursed.

Occasionally, one finds focal FDG uptake in the thyroid on whole body FDG PET scans performed for nonthyroid oncological staging or follow-up. This is to be distinguished from diffuse FDG uptake in both thyroid lobes, which usually indicates thyroiditis (subacute thyroiditis, Hashimoto's thyroiditis, or Graves' disease). Ramos et al.[181] found that out of 4600 whole body scans, 4 patients had focal thyroid uptake, which was later found to represent thyroid carcinoma (SUV values ranging from 2.3 to 53). Davis et al.[182] studied 1285 patients and found focal uptake in the thyroid in 5 patients, which showed malignant thyroid carcinoma by FNA in all 5 patients. Eloy et al.[183] studied 630 FDG PET scans performed for evaluation of cancers and found 30 patients with incidental FDG PET uptake in the thyroid, 18 of which had focal uptake. Five of these eighteen patients (28%) were demonstrated to have well-differentiated PTC. The mean SUV was 3.0 (range 1.1–7.4), with no difference in uptake SUV between the malignant and benign nodules. Van den Bruel et al.[184] investigated 8 patients with findings of incidental thyroid "hot spots" on FDG PET scans performed for other indications, and then studied with thyroid US and FNA; surgical histology was obtained in 7. They found that 5 had malignancies, 2 of which were MTC, one of which was locally invasive, and 3 were PTC, two of which were locally invasive. Two patients showed benign follicular adenomas. Thus, malignancy was high among these findings. Chu et al.[185] reviewed 6241 PET scans and found focal thyroid uptake in 76 patients (1.2%). Only 14 of these 76 (18%) patients underwent biopsy. Of these 14, 4 (28.6%) had PTC, 7 (50%) had hyperplasia, and one each had thyroiditis, nodular goiter, or follicular neoplasm. Cohen et al.[186] reviewed 4525 whole body FDG PET examinations and found a 2.3% incidence of thyroid incidentaloma. Of those that underwent biopsy, 47% were malignant. In addition, malignant lesions had higher SUV values than benign lesions. Thus, PET scan results can be useful in revealing unsuspected thyroid cancer in patients being evaluated for other malignancies.

10.4.6 (I-131 or I-123)-Metaiodobenzylguanidine (MIBG) and C-11 Hydroxyephedrine

Tumors of tissues arising from the neural crest share the characteristic of amine precursor uptake and decarboxylation (APUD), and contain secretory granules containing either precursors or products of catecholamines. C-cells of the thyroid gland give rise to MTC and are an example of these types of APUD cells.

MIBG is similar in structure to norepinephrine, one of the circulating catecholamines and a sympathetic nerve neurotransmitter, and guanethidine, a neuronal blocking drug sometimes used in hypertension. Hydroxyephedrine is an analog of ephedrine, a sympathomimetic drug.[187] MIBG and hydroxyephedrine are taken up in APUD-like neuroendocrine cells and their tumors in neurosecretory granules. After a 0.5 mCi dose of I-131 MIBG, one images the whole body and spot views of the neck and chest 3–7 days after injection (Figure 10.27). After an injection of 3 to 10 mCi of I-123 MIBG, one images the same regions at 24 hours, and as late as 48 hours, with superior image quality compared to I-131 MIBG imaging. Potentially, one could use I-124 MIBG with PET imaging, although the latter is still an investigational tracer. More experience has been obtained with imaging of C-11 hydroxyephedrine, 15–60 minutes after injection in neuroendocrine tumor imaging, although it, too, is investigational. While initial reports of imaging of MTC were encouraging, the technique was found to have limited utility for diagnosis, mainly because of a low sensitivity of 40–50%.[188–191] One possible use of MIBG is in selecting those patients that may respond to therapy with high-dose I-131 MIBG.

10.4.7 Indium-111 Octreotide Imaging

Many types of endocrine cells are regulated by chemical messenger molecules that bind to specific protein receptors and initiate physiological responses. Medullary cells of the thyroid often have receptors for somatostatin (SST) and vasoactive intestinal peptide (VIP).[192–194] Five subtypes of SST receptors have been identified, with SST receptor type-2 (SSTR2) being the most common. The expression of SSTR subtypes varies among different tumor types, including SSTR3 and SSTR5.[195] The diverse biologic effects of SST are mediated through G-protein-coupled receptors.[196] Somatostatin exerts an antiproliferative effect on neuroendocrine tumors.[197] A number of analogs, such as octreotide, seglitide, and lanreotide, have been developed that have greater stability in plasma than somatostatin, which itself has a very short half-life. While somatostatin binds to all five SSTR subtypes with comparable affinity, octreotide binds with higher affinity to SSTR2, STTR3,

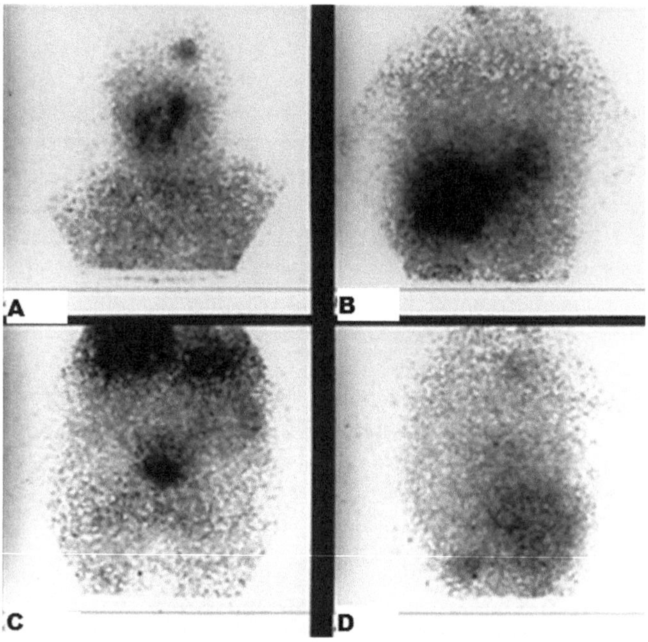

FIGURE 10.27 Anterior head and neck (A), anterior chest (B), anterior abdomen and pelvis (C), and posterior chest (D) I-131 MIBG scan views in a 64-year-old man with MTC. They show MIBG uptake in the skull (A, arrow), right paralumbar region (B and D, arrows), and left iliac crest (C, arrow). The patient received high-dose I-131 MBI therapy with a satisfactory palliative response. (Reproduced from I., Fogelman et al., *An Atlas of Clinical Nuclear Medicine*, 2nd ed., Martin Dunitz, Ltd., 1995, slide collection, Figure 2.82. With permission.)

and SSTR5, whereas lanreotide binds to SSTR1 through SSTR4 with comparable affinity. A number of receptor-binding radiotracers have been developed to target these receptors in tumor tissue. Indium-111-DTPA-D-Phel-pentetreotide (Octreoscan, Mallincrodt, St. Louis) became widely available for the imaging of neuroendocrine tumors. A number of other radiolabeled SSTR-binding peptides have been investigated as potential imaging agents.[198–200] Labeling with positron-emitting radionuclides such as Cu-64, Ga-68, F-18, and Y-86 has been investigated as a potential diagnostic imaging agent,[201–203] with the potential for quantification of biodistribution and pharmacokinetic properties suitable for therapeutic use of labeled somatostatin analog peptides.

10.4.7.1 Methods

One injects usually 5.0 mCi of In-111 octreotide intravenously. After injection, In-111 octreotide is rapidly cleared from the circulation with a half-life of about 5 hours. Whole body imaging is performed at 24 hours, starting with whole body imaging, followed by spot imaging of the neck or chest, or other areas of interest, such as the liver. SPECT and SPECT-CT imaging are particularly helpful in both localization and identifying the specificity of the findings.

10.4.7.2 Applications of In-111 Octreotide

The normal thyroid has a low level of uptake of In-111 octreotide, thought to be due to uptake in lymphoid tissue present in the normal thyroid, since lymphocytes have receptors for somatostatin. One may find asymmetrical uptake in the thyroid gland, if one of the lobes is larger than the other, or in the case of a nodular gland or goiter. The uptake of In-111 octreotide has been reported in both I-131-avid and I-131 negative thyroid tumors. Positive In-111 octreotide images were obtained in 75% of 16 I-131 negative patients with thyroid cancer, while octreotide scans were positive in

89% of 9 patients with positive I-131 scans.[204] Another study reported positive octreotide scans in 63% of 8 patients.[205] Stokkel et al.[206] reported the diagnostic value of In-111 octreotide monitoring of differentiated thyroid carcinoma in 23 patients with increasing TG levels unresponsive to I-131 treatment. The overall sensitivity of In-111 octreotide was 74%, with a 82% sensitivity in patients who had no abnormal uptake on I-131 scintigraphy. Patients with low or minimal In-111 octreotide uptake in thyroid metastases had a good (100%) 10-year prognosis, while survival was only 33% in patients with moderate to intense uptake. Sarlis et al.[207] found that In-111 octreotide had a 50% sensitivity for metastatic thyroid cancer, compared with 67% for FDG PET. However, some lesions were detected that were negative by RAI or FDG PET. This imaging modality is an alternative to I-131 in regions that do not have access to FDG PET imaging, or in the rare case of lesions that have negative FDG PET uptake. The mechanism of uptake of In-111 octreotide in differentiated thyroid cancer is unclear, whether it is through the exhibition of SST receptors by the cancer, or due to the presence of lymphocytes mixed in with the tumor. In-111 octreotide imaging is not so much of interest as another diagnostic modality, but more so as a predictor for the utility of therapeutic octreotide (Sandostatin), high-dose In-111 octreotide for therapy in patients refractory to I-131, or other analogs, such as Y90-labeled somatostatin analogs in octreotide positive patients.

10.4.7.3 Use of In-111 Octreotide in MTC

Arslan et al.[208] compared In-111 octreotide, Tc-99m DMSA, CT, MRI, and US in 14 patients with histologically verified MTC. In-111 octreotide showed a sensitivity of 78.5%, similar to MRI and CT, compared to 57% for Tc-99m DMSA on a patient basis, and a sensitivity of 44% and 30% for individual lesions, respectively. In-111 octreotide and Tc-99m DMSA together showed a sensitivity of 85.7% on a patient basis, whereas the combined use of CT and MRI showed the best sensitivity rate of 81% on a lesion basis. For MTC, it appears that multiple scan modalities should be used to achieve best results (Figure 10.28).

10.4.8 Imaging with Radiolabeled Antibodies

Because of the high prevalence of CEA antigen expression on the surface membranes of MTC, imaging with radiolabeled anti-CEA monoclocal antibodies has been investigated.[209–213] Pretargeted immunoscintigraphy using an affinity enhancement system (AES) allowed the identification of MTC metastases or recurrences. Radioimmunoguided surgery was made possible by this approach,

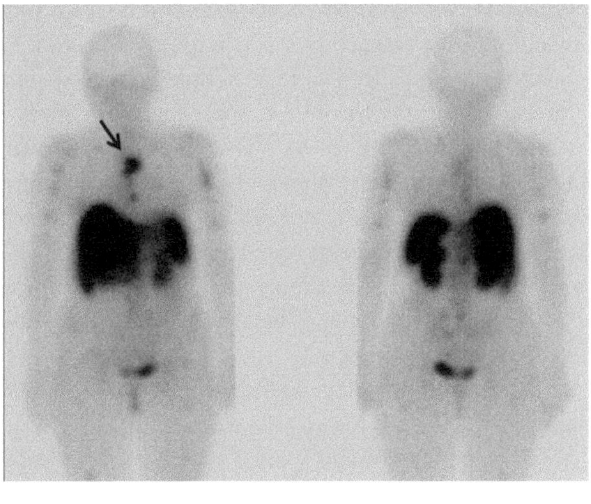

FIGURE 10.28 Indium-111 octreotide anterior and posterior whole body images in a woman with MTC metastatic to the skeleton (small arrow) and to the liver.

revealing metastases ranging from 5–15 mg, for a total of 34 metastases, 22 of which were missed by other methods, for an accuracy of 86%, followed by a return of CTN levels to normal in some patients.[214] MTC immunoscintigraphy may improve predictions of response to therapy with high-dose radiolabeled antibodies.

10.5 ANATOMICAL IMAGING

Ultrasound, CT, and MR examinations of the neck are important adjunct imaging modalities with radiotracer imaging. For lung parenchyma, one does not require CT contrast. However, in CT imaging of the neck and mediastinum, CT contrast is very helpful in distinguishing lymph nodes from normal structures and blood vessels. One should carefully time such an exam with RAI imaging, to avoid loading the patient with iodinated dye, which would delay imaging or therapy. MRI, with or without gadolinium contrast, is another alternative. Their advantage is that they do not interfere with RAI imaging or therapy. While SPECT-MRI or PET-MRI scanners are not currently available, development work is under way in that direction, augmenting the capabilities in anatomical-functional hybrid imaging.

Hybrid imaging incorporating both functional imaging, whether it be RAI imaging, In-111 octreotide imaging of I-123 MIBG, or F-18 FDG PET imaging, and CT imaging in a single unit allows precise image coregistration. This markedly facilitates the correlation between the functional imaging and anatomical imaging. Radionuclide imaging often lacks anatomical landmarks for precise localization; plus the resolution in planar or SPECT imaging is limited. Since the introduction of hybrid SPECT-CT and PET-CT imaging units, CT imaging helps localize the abnormal lesion and, when of good quality, provides important morphological information in its own right, such as the size of the lymph node or residual tumor, and helps differentiate metastatic disease from false positive findings. The combined hybrid imaging also helps localize disease for possible surgical or EBRT approaches.

10.6 CONCLUSION

Radionuclide imaging of the intact thyroid gland has been largely supplanted by US imaging and fine-needle aspiration for the initial diagnosis of thyroid cancer. Nonetheless, thyroid radionuclide imaging continues to play an important role when these approaches do not confirm a diagnosis, or when suspicious thyroid nodules are detected as incidental findings on radionuclide imaging studies done for other unrelated indications, such as for hyperthyroidism or for parathyroid imaging, or for other oncological conditions with F-18 FDG PET. Whole body RAI imaging with I-131 or I-123 continues to play a major role in the staging of thyroid cancer, particularly in patients at intermediate or high risk of metastatic disease, and in the preparation for RAI therapy, and the evaluation of such therapy on follow-up and in the assessment of suspected recurrence. The use of iodine-131 or even iodine-124 imaging for dosimetry is not always performed routinely, but the therapeutic thought process should always include concepts based on dosimetric studies. Patients receiving high doses of iodine-131 therapy, particularly the elderly or those with compromised renal function, or those with a large mass of functional pulmonary metastases, should routinely undergo preliminary dosimetric measurements before I-131 therapy.

Since the value of RAI imaging in patients with dedifferentiated thyroid cancer, ATC, or MTC is limited, diagnostic imaging with non-RAI tracers, including FDG, Tl-201, Tc-99m sestamibi, and Tc-99m tetrofosmine, SST analogs, and MIBG and its analogs, plays an important role. Because even these tracers have a limited sensitivity, the use of multiple tracers may sometimes be necessary. Ongoing work is needed to augment the sensitivity of RAI and to develop new and hopefully more efficacious agents for the detection and localization of these more difficult subsets of thyroid cancer.

REFERENCES

1. Fermi E. 1934. Radioactivity induced by neutron bombardment. *Nature* 133:757.
2. Chapman E. 1983. History of the discovery and early use of radioactive iodine. *JAMA* 250:2042.
3. Herz S, Roberts A, Evans RD. 1983. Radioactive iodine as an indicator in the study of thyroid physiology. *Proc Soc Exp Biol Med* 38:510–14.
4. Ernest LO. 1939. The medical cyclotron of the Crocker Radiation Laboratory. *Science* 90:407.
5. Parker TW, Mettler FA, Christie JH, et al. 1984. Radionuclide thyroid studies: A survey of practice in the United State in 1981. *Radiology* 150:547.
6. Riggs DS. 1952. Iodine metabolism in man. *Pharmacol Rev* 4:285.
7. Honor AJ, Myant NB, Rowlands EN. 1952. Secretion of radioiodine in digestive juices and milk in man. *Clin Sci* 11:447.
8. Miller H, Weetch RS, Glasq MB. 1955. The excretion of radioactive iodine in human milk. *Lancet* 2:1013.
9. Chapman EM, Corner GW Jr, Robinson D, et al. 1984. The collection of radioactive iodine by the human fetal thyroid. *J Clin Endocrinol* 8:717.
10. Castro MR, Bergert ER, Goellner JR, et al. 2001. Immunohistochemical analysis of sodium iodide symporter expression in metastatic differentiated thyroid cancer: Correlation with radioiodine uptake. *J Clin Endocrinol Metab* 86:5627–32.
11. MIRD Dose Estimate Report No. 5. 1975. Summary of current radiation dose estimates to humans from I-123, I-124, I-125, I-126, I-130, I-131 as sodium iodide. *J Nucl Med* 16:857.
12. Berson SA, Yalow RS, Sorrentino J, et al. 1952. The determination of thyroidal and renal plasma I-131 clearance rates as a routine diagnostic test of thyroid dysfunction. *J Clin Invest* 31:141.
13. Package insert for sodium iodine I-123. Revised July 1995. Mallincrodt Medical, Inc.
14. Kocher DC. 1981. Radioactive decay data table. *DOE/TIC* 11026:133.
15. Package insert for sodium I-131 iodide.
16. Pentlow KS, Graham MC, Lambrecht RM, et al. 1996. Quantitative imaging of iodine-124 with PET. *J Nucl Med* 37:1557–62.
17. Harper PV, Beck R, Charleston D, et al. 1964. Optimization of a scanning method using Tc-99m. *Nucleonics* 22:50.
18. Andros G, Harper PV, Lathrop KA, et al. 1965. Pertechnetate-99m localization in man with application to thyroid scanning and the study of thyroid physiology. *J Clin Endocrinol* 25:1067.
19. Harper PV, Lathrop KA, Jimenez F, et al. 1965. Technetium-99m as a scanning agent. *Radiology* 85:101.
20. Wolff J, Maurey JR. 1963. Thyroidal iodide transport IV. The role of ion size. *Biochim Biophys Acta* 69:48.
21. Lathrop KA, Harper PV. 1972. Biologic behavior of Tc-99m from Tc-99m pertechnetate ion. *Prog Nucl Med* 1:145.
22. Heck L, Lathrop k, Gottschalk A, et al. 1968. Perchlorate washout of pertechnetate from the thyroid gland. *J Nucl Med* 9:323.
23. Shimmins JG, Harden RM, Alexander WD. 1969. Loss of pertechnetate from the human thyroid. *J Nucl Med* 10:637.
24. Atkins HL, Richards P. 1968. Assessment of thyroid function and anatomy with technetium-99m as pertechnetate. *J Nucl Med* 9:7.
25. Wegst A. Goin J, Robinson R. 1983. Cumulated activities determined from biodistribution data in pregnant rats ranging from 13 to 21 days gestation. I. Tc-99m pertechnetate. *Med Phys* 10:841.
26. Hahn V, Brod K, Wolf R. 1980. The radiation dose to the fetus during isotope investigations of the mother. *Fortschr Rontgenstr* 132:326.
27. Wyburn J. 1973. Human breast milk excretion of radionuclides following administration of radiopharmaceuticals. *J Nucl Med* 14:115.
28. Romney BM, Nickoloff EL, Esser PD, et al. 1986. Radionuclide administration to nursing mothers: Mathematically derived guidelines. *Radiology* 160:549.
29. Ridgeway EC. 1992. Clinician evaluation of a solitary thyroid nodule. *J Clin Endocrinol Metab* 8:215–24.
30. Christensen SB, Bondeson L, Ericsson UB, et al. 1984. Prediction of malignancy in the solitary thyroid nodule by physical examination, thyroid scan, fine-needle biopsy and serum thyroglobulin: A prospective study of 100 surgically treated patients. *Acta Chir Scand* 150:433–40.
31. Belflore A, LaRosa GL, LaPorta et al. 1992. Cancer risk in patients with cold thyroid nodules: Relevance of iodine uptake, sex, age, and multinodularity. *Am J Med* 93:363–69.
32. Mazzaferri EL. 1993. Management of a solitary thyroid nodule. *N Engl J Med* 328:553–58.

33. Schneider AB, Shore-Freedman E, Ryo UY, et al. 1985. Radiation-induced tumors of the head and neck following childhood irradiation: Prospective studies. *Medicine* 64:1–15.
34. Sachmechi I, Miller E, Varatharajah R, et al. 2000. Thyroid carcinoma in single cold nodules and in cold nodules of multinodular goiters. *Endocr Pract* 6:5–7.
35. Rallison ML, Dobyns BM, Keating FR Jr, et al. 1975. Thyroid nodularity in children. *JAMA* 233:1069–72.
36. Raab SS, Silverman JF, Elsheikh TM, et al. 1995. Pediatric thyroid nodules: Disease demographics and clinical management as determined by fine needle aspiration biopsy. *Pediatrics* 95:446–49.
37. Corrias A, Einaudi S, Chiorboli E, et al. 2001. Accuracy of fine needle aspiration biopsy of thyroid nodules in detecting malignancy in childhood: Comparison with conventional clinical, laboratory, and imaging approaches. *J Clin Endocrinol Metab* 86:4644–48.
38. Hung W. 1999. Solitary thyroid nodules in 93 children and adolescents, a 35-years experience. *Horm Res* 52:15–18.
39. Gharib H, Zimmerman D, Goellner Jr, et al. 1995. Fine-needle aspiration biopsy: Use in diagnosis and management of pediatric thyroid diseases. *Endocr Pract* 1:9–13.
40. Arda IS, Yildirim S, Demirhan B, et al. 2001. Fine needle aspiration biopsy of thyroid nodules. *Arch Dis Child* 85:313–17.
41. Kusic Z, Becker DV, Saenger EL, et al. 1990. Comparison of technetium-99m and iodine-123 imaging of thyroid nodules: Correlation with pathologic findings. *J Nucl Med* 31:393–99.
42. Hodges RE, Evans TC, Bradbury JT, et al. 1955. The accumulation of radioactive iodine by human fetal thyroids. *J Clin Endocrinol Metab* 6:661.
43. Maxon HR, Thomas SR, Boehringer A, et al. 1983. Low iodine diet in I-131 ablation of thyroid remnants. *Clin Nucl Med* 8:123–26.
44. Pluijmen MJ, Eustatia-Rutten C, Goslings BM, et al. 2003. Effects of low-iodide diet on postsurgical radioiodide ablation therapy in patients with differentiated thyroid carcinoma. *Clin Endocrinol* (Oxf) 58:428–35.
45. Becker D. 1983. Physiological basis for the use of potassium iodide as thyroid blocking agent: Logistic issues in its distribution. *Bull NY Acad Med* 59:1004.
46. Hladik WB, Nigg KK, Rhodes BA. 1982. Drug-induced changes in the biologic distribution of radiopharmaceuticals. *Semin Nucl Med* 12:184.
47. Edmonds CJ, Hayes S, Kermode JC, et al. 1977. Measurement of serum TSH and thyroid hormones in the management of treatment of thyroid carcinoma with radioiodine. *Br J Radiol* 50:799–807.
48. Hilts SV, Hellman D, Anderson J, Woolfenden J, Van AntwerpJ, Patton D. 1979 Serial TSH determination after T3 withdrawal or thyroidectomy in the therapy of thyroid carcinoma. *J Nucl Med* 20:928–932..
49. Martin ND. 1978. Endogenous serum TSH levels and metastatic survey scans in thyroid cancer patients using triodothyronine withdrawal. *Clin Nucl Med* 3:401–3.
50. Hilts SV, Hellman D, Anderson J, et al. 1979. Serial TSH determination after T3 withdrawal or thyroidectomy in the therapy of thyroid carcinoma. *J Nucl Med* 20:928–32.
51. Goldman JM, Line BR, Aamodt RL, et al. 1980. Influence of triiodothyronine withdrawal time on I131 uptake postthyroidectomy for thyroid cancer. *J Clin Endocrinol Metab* 50:734–39.
52. Schneider AB, Line B, Goldman JM, et al. 1981. Sequential serum thyroglobulin determinations, I-131 scans, and I-131 uptakes after triodothyronine withdrawal in patients with thyroid cancer. *J Clin Endocrinol Metab* 53:1199–206.
53. Maxon HR, Thomas SR, Hetzberg VS, et al. 1983. Relation between effective radiation dose and outcome of radioiodine therapy for thyroid cancer. *N Engl J Med* 309:937–41.
54. Liel Y. 2002. Preparation for radioactive iodine administration in differentiated thyroid cancer patients. *Clin Endocrinol* 57:523–27.
55. Sanchez R, Espinosa-de-los-Monteros AL, Mendoza V, et al. 2002. Adequate thyroid-stimulating hormone levels after levothyroxine discontinuation in the follow-up of patients with well-differentiated thyroid carcinoma. *Arch Med Res* 33:478–81.
56. Grigsby PW, Siegel BA, Bekker S, et al. 2004. Preparation of patients with thyroid cancer for I-131 scintigraphy or therapy by 1–3 weeks of thyroxine discontinuation. *J Nucl Med* 45:567–70.
57. Serhal DI, Nasrallah MP, Arafah BM. 2004. Rapid rise in serum thyrotropin concentrations after thyroidectomy or withdrawal of suppressive thyroxine therapy in preparation for radioactive iodine administration to patients with differentiated thyroid cancer. *J Clin Endocrinol Metab* 89:3285–89.
58. Guimaraes V, DeGroot LJ. 1996. Moderate hypothyroidism in preparation for whole body I-131 scintiscans and thyroglobulin testing. *Thyroid* 6:69–73.
59. Thyrogen package insert. November 1998. Genzyme Corporation.

60. Torres MS, Ramirez L, Simkin PH, et al. 2001. Effect of various doses of recombinant human thyrotropin on the thyroid radioactive iodine uptake and serum levels of thyroid hormones and thyroglobulin in normal subjects. *J Clin Endocrinol Metab* 86:1660–64.
61. Hershman JM, Edwards CL. 1972. Serum thyrotropin (TSH) levels after thyroid ablation compared with TSH levels after exogenous bovine TSH: Implications for 131-I treatment of thyroid carcinoma. *J Clin Endocrinol Metab* 34:814–818.
62. Meier CA, Braverman LE, Ebner SA, et al. 1994. Diagnostic use of recombinant human thyrotropin versus thyroid hormone withdrawal in patients with thyroid carcinoma (phase I/II) study. *J Clin Endocrinol Metab* 78:188–96.
63. Ladenson PW, Braverman LE, Mazzaferri EL, et al. 1997. Comparison of administration of recombinant human thyrotropin with withdrawal of thyroid hormone for radioactive iodine scanning in patients with thyroid carcinoma. *N Engl J Med* 337:888–96.
64. Ladenson PW. 1999. Recombinant human thyrotropin symposium: Strategies for thyrotropin use to monitor patients with treated thyroid carcinoma. *Thyroid* 9:429–33.
65. Haugen RB, Pacini F, Reiners C, et al. 1999. A comparison of recombinant human thyrotropin and thyroid hormone withdrawal for the detection of thyroid remnant or cancer. *J Clin Endocrinol Metab* 84:3877–85.
66. Rosario PWS, Ribeiro Maia FF, Cardoso LD, et al. 2004. Correlation between cervical uptake and results of post-surgical radioiodine ablation in patients with thyroid carcinoma. *Clin Nucl Med* 29:358–61.
67. Carril JM, Quirce R, Serrano J, et al. 1997. Total-body scintigraphy with thallium-201 and iodine-131 in the follow-up of differentiated thyroid cancer. *J Nucl Med* 38:686–92.
68. Park H. 1992. Stunned thyroid after high-dose I-131 imaging. *Clin Nucl Med* 17:501–2.
69. Jeevanram RK, Shah DH, Sharma SM, et al. 1986. Influence of initial large dose on subsequent uptake of therapeutic radioiodine in thyroid cancer patients. *Nucl Med Biol* 13:277–79.
70. Hilditch TE, Dempsey MF, Bolster AA, et al. 2002. Self-stunning in thyroid ablation: Evidence from comparative studies of diagnostic I-131 and I-123. *Eur J Nucl Med Mol Imaging* 29:783–88.
71. Murat J-P, Daver A, Minier J-F, et al. 1998. Influence of scanning doses of iodine-131 on subsequent first ablative treatment outcome in patients operated on for differentiated thyroid carcinomas. *J Nucl Med* 39:1546–50.
72. Murat J-P, Daver A, Minier J-F, et al. 1998. Influence of scanning doses of iodine-131 on subsequent first ablative treatment outcome in patients operated on for differentiated thyroid carcinomas. *J Nucl Med* 39:1546–50.
73. Dam HQ, Kim SM, Kin HC, et al. 2004. I-131 therapeutic efficacy is not influenced by stunning after diagnostic whole-body scanning. *Radiology* 2332:527–33.
74. Lassmann M, Luster M, Hanscheid H, et al. 2004. Impact of I-131 diagnostic activities on the biokinetics of thyroid remnants. *J Nucl Med* 45:619–25.
75. Cholewinski SP, Yoo KS, Klieger PS, et al. 2000. Absence of thyroid stunning after diagnostic whole-body scanning with 185 MBq I-131. *J Nucl Med* 41:1198–202.
76. Muratet JP, Giraud P, Daver A, et al. 1997. Predicting the efficacy of first iodine-131 treatment in differentiated thyroid carcinoma. *J Nucl Med* 38:1362–68.
77. Anderson GS, Fish S, Nakhoda K, et al. 2003. Comparison of I-123 and I-131 for whole-body imaging after stimulation by recombinant human thyrotropin: A preliminary report. *Clin Nucl Med* 28:93–96.
78. Mandel SJ, Shankar LK, Benard F, et al. 2001. Superiority of iodine-123 compared with iodine-131 scanning for thyroid remnants in patients with differentiated thyroid cancer. *Clin Nucl Med* 26:6–9.
79. Alzahrani AS, Bakheet S, Mandil MA, et al. 2001. I-123 isotope as a diagnostic agent in the follow-up of patients with differentiated thyroid cancer: Comparison with post I-131 therapy whole body scanning. *J Clin Endocrinol Metab* 86:5294–300.
80. Sarkar SD, Kalapparambath TP, Palestro CJ, et al. 2002. Comparison of I-123 and I-131 for whole-body imaging in thyroid cancer. *J Nucl Med* 43:632–34.
81. Gerard SK, Cavalieri RR. 2002. I-123 diagnostic thyroid tumor whole-body scanning with imaging at 6, 24, and 48 hours. *Clin Nucl Med* 27:1–8.
82. Thomas DL, Menda Y, Bushnell D, et al. 2009. A comparison between diagnostic I-123 and post-therapy I-131 scans in the detection of remnant and locoregional thyroid disease. *Clin Nucl Med* 34:745–48.
83. Salvatori M, Perotti G, Rufini V, et al. 2004. Are there disadvantages in administering I-131 ablation therapy in patients with differentiated thyroid carcinoma without a pre-ablative diagnostic I-131 whole body scan? *Clin Endocrinol* (Oxf) 61:704–10.
84. Haugen BR, Ridgway EC, Mclaughlin BA, et al. 2002. Clinical comparison of whole-body radioiodine scan and serum thyroglobulin after stimulation with recombinant human thyrotropin. *Thyroid* 12:37–43.

85. Regalbuto C, Buscema M, Arena S, et al. 2002. False-positive findings on I-131 whole-body scans because of post-traumatic superficial scabs. *J Nucl Med* 43:207–9.
86. Fatourechi V, Hay ID, Mullan BP, et al. 2000. Are post-therapy radioiodine scans informative and do they influence subsequent therapy of patients with differentiated thyroid cancer? *Thyroid* 10:573–77.
87. Sherman SI, Tielens ET, Sostre S, et al. 1994. Clinical utility of posttreatment radioiodine scans in the management of patients with thyroid carcinoma. *J Clin Endocrinol Metab* 78:629–34.
88. Souza Rosario PW, Barroso AL, Rezende LL, et al. 2004. Post-131 therapy scanning in patients with thyroid carcinoma metastases: An unnecessary cost or a relevant contribution? *Clin Nucl Med* 29:795–98.
89. Pacini F, Capezzone M, Elisei R, et al. 2002. Diagnostic 131-iodine whole body scan may be avoided in thyroid cancer patients who have undetectable stimulated TG levels after initial treatment. *J Clin Endocirnol Metab* 87:1499–501.
90. David A, Blotta A, Bondanelli M, et al. 2001. Serum thyroglobulin concentrations and I-131 whole body scan results in patients with differentiated thyroid carcinoma after administration of recombinant human thyroid stimulating hormone. *J Nucl Med* 42:1470–75.
91. DeGeus-Oei L-F, Oei H-Y, Hennemann G, et al. 2002. Sensitivity of I-12 whole body scan and thyroglobulin in the detection of metastases or recurrent differentiated thyroid cancer. *Eur J Nucl Med* 29:768–74.
92. Siddiqi A, Foley RR, Britton KE, et al. 2001. The role of I-123 diagnostic imaging in the follow-up of patients with differentiated thyroid carcinoma as compared to I-131 scanning: Avoidance of negative therapeutic uptake due to stunning. *Clin Endocrinol* 55:515–21.
93. Pons F, Carrio I, Estorch M, et al. 1987. Lithium as an adjuvant of iodine-131 uptake when treating patients with well-differentiated thyroid carcinoma. *Clin Nucl Med* 12:644–47.
94. Koong SS, Reynolds JC, Movius EG, et al. 1999. Lithium as a potential adjuvant to I-131 therapy of metastatic, well-differentiated thyroid carcinoma. *J Clin Endocrinol Metab* 84:912–16.
95. Hoffman S, Rockenstein A, Ramaswamy A, et al. 2007. Retinoic acid inhibits angiogenesis and tumor growth of thyroid cancer cells. *Mol Cell Endocrinol* 264:74–81.
96. Handkiewicz-Junak D, Roskosz J, Hasse-Lazar K, et al. 2009. 13-Cis-retinoic acid re-differentiation therapy and recombinant human thyrotropin-aided radioiodine treatment of non-functional metastatic thyroid cancer: A single-center, 53-patient phase 2 study. *Thyroid Res* 2:8.
97. Fernandez CA, Puig-Domingo M, Lomena F, et al. 2009. Effectiveness of retinoic acid treatment for redifferentiation of thyroid cancer in relation to recovery of radioiodine uptake. *J Endocrinol Invest* 32:228–33.
98. Kim WG, Kim EY, Kim TY, et al. 2009. Redifferentiation therapy with 13-cis retinoic acid in radioiodine-resistant thyroid cancer. *Endocr J* 56:105–12.
99. Courbon F, Zerdoud S, Bastie D, et al. 2006. Defective efficacy of retinoic acid treatment in patients with metastatic thyroid carcinoma. *Thyroid* 16:1025–31.
100. Adamczewski Z, Makarexicz J, Mikosinski S, et al. 2006. Application of 13-cis-retinoic acid in patients with 131I scintigraphically-negative metastases of differentiated thyroid carcinoma. *Encokrynol Pol* 57:403–6.
101. Short SC, Suovuori A, Cook G, et al. 2004. A phase II study using retinoids as redifferentiation agents to increase iodine. *Clin Oncol (R Coll Radiol)* 16:569–74.
102. Coelho SM, Corbo R, Buescu A, et al. 2004. Retinoic acid in patients with radioiodine non-responsive thyroid carcinoma. *J Endocrinol Invest* 27:334–39.
103. Toubert ME, Hindie E, Rampin L, et al. 2007. Distant metastases of differentiated thyroid cancer: Diagnosis, treatment and outcome. *Nucl Med Rev Cent East Eur* 10:106–9.
104. Haugen BR. 2004. Redifferentiation therapy in advanced thyroid cancer. *Curr Drug Targets Immune Endocr Metabol Disord* 4:174–80.
105. Eschmann SM, Reischl G, Bilger K, et al. 2002. Evaluation of dosimetry of radioiodine therapy in benign and malignant thyroid disorders of iodine-124 and PET. *Eur J Nucl Med* 29:760–67.
106. Kolbert KS, Pentlow KS, Pearson JR, et al. 2007. Prediction of absorbed dose to normal organs in thyroid cancer patients treated with I-131 by use of I-124 PET and 3-dimensional internal dosimetry software. *J Nucl Med* 48:143–49.
107. Erdi YE, Macapinlac H, Larson SM, et al. 1999. Radiation dose assessment for I-131 therapy of thyroid cancer using I-124 PET imaging. *Clin Pos Imaging* 2:41–46.
108. Sgouros G, Kolbert KS, Sheikh A, et al. 2004. Patient-specific dosimetry for I-131 thyroid cancer therapy using I-124 PET and 3-dimensional-internal dosimetry (3D-ID) software. *J Nucl Med* 45:1366–72.
109. Beierwalters WH. 1978. The treatment of thyroid carcinoma with radioactive iodine. *Semin Nucl Med* 8:79.
110. Williams JR, Thwaites UI, eds. 1993. *Radiotherapy physics in practice.* Oxford: Oxford Medical Publishers, Oxford University Press.

111. Stabin MG. 1996. MIRDOSE: Personal computer software for internal dose assessment in nuclear medicine. *J Nucl Med* 37:538–46.
112. Benua RS, Cicale NR, Sonenberg M, et al. 1962. The relation of radioiodine dosimetry to results and complications in the treatment of metastatic thyroid cancer. *AJR* 87:171.
113. Maxon MR, Smith HS. 1990. Radioiodine-131 in the diagnosis and treatment of metastatic well-differentiated thyroid cancer. *Endocrinol Metab Clin North Am* 19:685–719.
114. Wahl RL, Kroll S, Zasadny KR. 1998. Patient-specific whole-body dosimetry: Principles and a simplified method for clinical implementation. *J Nucl Med* 39(Suppl):14S–20S.
115. Alam MS, Takeuchi L, Kasagi K, et al. 1997. Value of combined technetium-99m hydroxymethylene diphosphanate and thallium-201 imaging in detecting bone metastases from thyroid carcinoma. *Thyroid* 7:705–12.
116. Castillo LA, Yeh SD, Leeper RD, et al. 1980. Bone scans in metastases from functioning thyroid carcinoma. *Clin Nucl Med* 5:200–9.
117. Tonami N, Hisada K. 1977. Clinical experience of tumor imaging with Tl-201 chloride. *Clin Nucl Med* 2:75.
118. Hisada K, Tonami N, Miyamae T, et al. 1978. Clinical evaluation of tumor imaging with Tl-201 chloride. *Nucl Med* 129:497.
119. Salvatore M, Carrati L, Bazzicalupo L, et al. 1976. The use of Tl-201 thallium chloride as a positive tumor indicator. *Eur J Nucl Med* 1:15.
120. Gehring PJ, Hammond PB. 1964. The uptake of thallium by rabbit erythrocytes. *J Pharmacol Exp Ther* 145:215.
121. Gerhing PJ, Hammond PB. 1967. The interrelationship between thallium and potassium in animals. *J Pharmacol Exp Ther* 55:187.
122. Atkins HL, Budinger TF, Lebowitz E, et al. 1977. Thallium-201 for medical use. Part 3. Human distribution and physical imaging properties. *J Nucl Med* 18:133.
123. Palermo F, Bruniera F, Caldato L, et al. 1979. Tl-201 in the scintigraphic evaluation of the "cold" thyroid areas. *Eur J Nucl Med* 4:43.
124. Harada T, Ito Y, Shimaoka K, et al. 1980. Clinical evaluation of 201-thallium chloride scan for thyroid nodule. *Eur J Nucl Med* 5:125.
125. Pavoni P, Mango L. 1981. Clinical evaluation of 2021-thallium chloride scan for thyroid nodule. *Eur J Nucl Med* 6:47.
126. Tonami N, Bunko H, Michigishi T, et al. 1978. Clinical application of Tl-201 scintigraphy in patients with cold nodule. *Clin Nucl Med* 3:217.
127. Tennvall J, Plamer J, Cederquist E, et al. 1981. Scintigraphic evaluation and dynamic studies with thallium-201 in thyroid lesions with suspected cancer. *Eur J Nucl Med* 6:295.
128. Piers DA, Sluiter WJ, Willemse PHB, et al. 1982. Scintigraphy with Tl-201 for detection of thyroid cancer metastases. *Eur J Nucl Med* 7:515.
129. Fukunaga M, Morita R, Yonekura Y, et al. 1979. Accumulation of Tl-201 chloride in a parathyroid adenoma. *Clin Nucl Med* 4:229.
130. Makiuschi M, Miyakawa M, Sugenoya A, et al. 1981. Diagnostic usefulness of Tl-201 chloride scintigraphy for preoperative localization of parathyroid tumor. *Jpn J Surg* 3:162.
131. Mruck S, Pfahlberg A, Papadopoulos T, et al. 2002. Uptake of Tl-201 into primary cell cultures from human thyroid tissue is multiplied by TSH. *J Nucl Med* 43:145–52.
132. Chilton HM, Callaran, RJ, Thrall JH. 1990. Radiopharmaceutical for cardiac imaging: Myocardial infarction, perfusion, metabolism, and ventricular function. In *Pharmaceutical in medical imaging*, ed. DP Swanson, HM Chilton, JH Thrall, 419. New York: Macmillan.
133. Lorberboym M, Murthy S, Mechanick JI, et al. 1996. Thallium-201 and iodine-131 scintigraphy in differentiated thyroid carcinoma. *J Nucl Med* 37:1487–91.
134. Nakada K, Katoh C, Morita K, et al. 1999. Relationship among Tl-201 uptake, nuclear DNA content and clinical behavior in metastatic thyroid carcinoma. *J Nucl Med* 40:963–67.
135. Oyen WJG, Verhagen C, Saris E, et al. 2000. Follow-up regimen of differentiated thyroid carcinoma in thyroidectomized patients after thyroid hormone withdrawal. *J Nucl Med* 41:643–46.
136. Muller S, Guth-Tougelidis B, Creutzig H. 1987. Imaging of malignant tumors with Tc-99m-MIBI SPECT (abstract). *J Nucl Med* 28:562.
137. Delmon-Moingeon LI, Piwnica-Worms D, Van den Abbeele AD, et al. 1990. Uptake of the cation hexakis 2-methoxyisobutylisonitrile-technetium-99m by human carcinoma cells lines *in vitro*. *Cancer Res* 50:2198–202.
138. Crane P, Laliberte R, Hemingway S, et al. 1993. Effect of mitochondrial viability and metabolism on technesium-99m-sestamibi myocardial retention. *Eur J Nucl Med* 20:20–25.

139. Piwnica-Worms D, Kronauge JF, Chiu ML. 1990. Uptake and retention of hexakis (2-methoxyisobutyl isonitrile) technetium in cultured chick myocardial cells. Mitochondrial and plasma membrane potential dependence. *Circulation* 82:1826–38.
140. Maublant JC, Moins N, Gachon P, et al. 1993. Uptake of technetium-99m teboroxime in cultured myocardial cells: Comparison with thallium-201 and technetium-99m-setamibi. *J Nucl Med* 34:255–59.
141. Arbab AS, Koizumi K, Toyama K, et al. 1996. Uptake of technetium-99m-tetrofosmin, technetium-99m-MIBI and thallium-201 in tumor cell lines. *J Nucl Med* 37:1551–56.
142. NG DCE, Sundram FX, Sin AE. 2000. Tc-99m sestamibi and I-131 whole-body scintigraphy and initial serum thyroglobulin in the management of differentiated thyroid carcinoma. *J Nucl Med* 41:631–35.
143. Brandt-Mainz K, Muller SP, Sonnenschein W, et al. 1998. Technetium-99m-furifosmin in the follow-up of differentiated thyroid carcinoma. *J Nucl Med* 39:1536–41.
144. Kim CK, Jung E, Yun M, et al. 2001. A normal variant on tl-201 and Tc-99m MIBI whole-body imaging: The superior right atrial wall (auricle) and superoanterior right ventricular wall are often seen as mediastinal lesions. *Clin Nucl Med* 26:412–18.
145. Ugar O, Kostakoglu L, Caner B, et al. 1996. Comparison of Tl-201, Tc-99m MIBI and I-131 imaging in the follow-up of patients with well-differentiated thyroid carcinoma. *Nucl Med Commun* 17:373–77.
146. Unal S, Menda Y, Adalet I, et al. 1998. Thallium-201, technetium-99m-tetrofosmin and iodine-131 in detecting differentiated thyroid carcinoma metastases. *J Nucl Med* 39:1897–902.
147. Wu H-S, Liu F-Y, Huang W-S, et al. 2003. Technetium-99m tetrofosmin single photon emission computed tomography to detect metastatic papillary thyroid carcinoma in patients with elevated human serum thyroglobulin levels but negative I-131 whole body scan. *Clin Radiol* 58:787–90.
148. Lind P, Gallowitsch HJ, Landsteger W, et al. 1997. Technetium-99m tetrofosmin whole body scintigraphy in the follow-up of differentiated thyroid carcinoma. *J Nucl Med* 38:348–52.
149. Erdil TY, Kabasakal L, et al. 2000. Correlation of technetium-99m MIBI and thallium-201 retention in solitary cold thyroid nodules with postoperative histopathology. *Eur J Nucl Med* 27:713–20.
150. Murthy S, Mechanick JI, Chau P, et al. 1993. Unusual scintigraphic findings in a thyroid adenoma. *J Nucl Med* 34:465–66.
151. Piwnica-Worms D, Chiu ML, Budding M, et al. 1993. Functional imaging of multidrug-resistant P-glycoprotein with an organotechnetium complex. *Cancer Res* 53:977–84.
152. Gottesman MM, Pastan I. 1993. Biochemistry of multidrug resistance mediated by the multidrug transporter. *Ann Rev Biochem* 6:385–427.
153. Ballinger JR, Bannerman J, Boxen I, et al. 1996. Technetium-99m-tetrofosmin as a substrate for P-glycoprotein *in vitro* studies in multidrug-resistant breast tumor cells. *J Nucl Med* 37:1578–82.
154. Bernard BF, Krenning EP, Breeman WAP, et al. 1998. Tc-99m MIBI, Tc-99m tetrofosmin and Tc-99m-q12 *in vitro* and *in vivo*. *Nucl Med Biol* 25:233–40.
155. Zimmer LA, McCook B, Meltzer C, et al. 2003. Combined positron emission tomography/computed tomography imaging of recurrent thyroid cancer. *Otolaryngol Head Neck Surg* 128:178–84.
156. Feine V, Lietzenmayer R, Hanke JP, et al. 1996. Fluorine-18-FDG and iodine-131 uptake in thyroid cancer. *J Nucl Med* 37:1468–72.
157. Wang W, Macapinlac H, Finn R, et al. 1999. PET scanning with F-18-fluorodeoxyglucose can localize differentiated thyroid carcinoma in patients with negative I-131 whole body scans. *J Clin Endocrinol Metab* 84:2291–302.
158. Wang W, Larson SM, Fazzari, et al. 2000. Prognostic value of F-18-fluorodeoxyglucose positron emission tomography scanning in patients with thyroid carcinoma. *J Clin Endocrinol Metab* 85:1107–13.
159. Chung J-K, So Y, Lee JS, et al. 1999. Value of FDG PET in papillary thyroid carcinoma with negative I-131 whole-body scan. *J Nucl Med* 40:986–92.
160. Wang W, Macapinlac H, Larson SM, et al. 1999. F-18 fluoro-2-deoxyglocuse positron emission tomography localizes residual thyroid cancer in patients with negative diagnostic I-131 whole body scans and elevated serum thyroglobulin levels. *J Clin Endocrinol Metab* 84:2291–302.
161. Yeo JS, Chung J-K, So Y, et al. 2001. F-18-fluorodeoxyglucose positron emission tomography as a presurgical evaluation modality for I-131 scan-negative thyroid carcinoma patients with local recurrence in cervical lymph nodes. *Head Neck* 23:94–103.
162. Mechanick JI, Kim CK, Krynyckyi BR, et al. 2000. Multiple papillary thyroid carcinoma metastases revealed on positron emission tomography scan in a patient with negative ^{131}I scan. *Thyroid* 10(10):929.
163. Alnafisi NS, Drieger AA, Coates G, et al. 2000. FDG PET of recurrent or metastatic I-131 negative papillary thyroid carcinoma. *J Nucl Med* 41:1010–15.

164. Frilling A, Tecklenborg K, Gorges R, et al. 2001. Preoperative value of F-18 fluorodeoxyglucose positron emission tomography in patients with radioiodine-negative recurrent well-differentiated thyroid carcinoma. *Ann Surg* 234:804–11.
165. Helal BPO, merlet P, Toubert M-E, et al. 2001. Clinical impact of F-18-FDG PET in thyroid carcinoma in patients with elevated thyroglobulin levels and negative I-131 scanning results after therapy. *J Nucl Med* 42:1464–69.
166. Bertagna F, Bosio G, Biasiotto G, et al. 2009. F-18 FDG-PET/CT evaluation of patients with differentiated thyroid cancer with negative I-131 total body scan and high thyroglobulin level. *Clin Nucl Med* 34:756–61.
167. Robbins RJ, Wan W, Grewal RK, et al. 2006. Real-time prognosis for metastatic thyroid carcinoma based on 2-[F-18] fluoro-2-deoxy-d-glucose-positron emission tomography scanning. *J Clin Endocrinol Metab* 91:498–505.
168. Wang W, Macapinlac H, Larson SM, et al. 1999. F-18 fluoro-2-deoxyglocuse positron emission tomography localizes residual thyroid cancer in patients with negative diagnostic I-131 whole body scans and elevated serum thyroglobulin levels. *J Clin Endocrinol Metab* 84:2291–302.
169. Bertagna F, Bosio G, Biasiotto G, et al. 2009. F-18 FDG-PET/CT evaluation of patients with differentiated thyroid cancer with negative I-131 total body scan and high thyroglobulin level. *Clin Nucl Med* 34:756–61.
170. Moog F, Linke R, Manthey N, et al. 2000. Influence of thyroid-stimulating hormone levels on uptake of FDG in recurrent and metastatic differentiated thyroid carcinoma. *J Nucl Med* 41:1989–95.
171. Leboulleux S, Schroeder PR, Busaidy NL, et al. Assessment of the incremental value of recombinant TSH stimulation before FDG PET=CT imaging to localize residual differentiated thyroid cancer. *J Clin Endocrinol Metab* 94:1310–1316.
172. Shiga T, Tsukamoto E. Nakada K, et al. 2001. Comparison of F-18 FDG, I-131-Na, Tl-201 in diagnosis of recurrent or metastatic thyroid carcinoma. *J Nucl Med* 42:414–19.
173. Robbins J, Merino MJ, Boice JD, et al. 1991. NIH conference. Thyroid cancer: A lethal endocrine neoplasm. *Ann Intern Med* 115:133–47.
174. Norton JA, Levin B, Jensen R. 1989. Cancer of the endocrine system. In *Cancer: Principles and practice of oncology*, ed. VT DeVita, S Hellman, SA Rosen, 1333–435. New York: JB Lippincott Co.
175. Brandt-Mainz K, et al. 2000. The value of fluorine-18 fluorodeoxyglucose PET in patients with medullary thyroid cancer. *Eur J Nucl Med* 27:490–96.
176. de Groot JWB, Links TP, Jager PL, et al. 2004. Impact of F-18-fluoro-2-deoxyglucose positron emission tomography (FDG-PET) in patients with biochemical evidence of recurrent or residual medullary thyroid cancer. *Ann Surg Oncol* 11:786–94.
177. Khan N, Oriuchi N, Higuchi T, et al. 2005. Review of fluorine-18–2-fluoro-2-deoxyglucose positron emission tomography (FDG-PET) in the followup of medullary and anaplastic thyroid carcinomas. *Cancer Control* 12:254–60.
178. Kresnik E, Gallowitsch HJ, Mikosch P, et al. 2003. Flurine-18-fluorodeoxyglucose positron emission tomography in the preoperative assessment of thyroid nodules in an endemic goiter area. *Surgery* 133:294–99.
179. Mitchell JC, Grant F, Evenson AR, et al. 2005. Preoperative evaluation of thyroid nodules with F-18 FDG PET/CT. *Surgery* 138:1166–75.
180. De Gues-Oei L-F, Pieters GFFM, Bonekamp JJ, et al. 2006. F-18-FDG PET reduces unnecessary hemithyroidectomies for thyroid nodules with inconclusive cytologic results. *J Nucl Med* 47:770–75.
181. Ramos CD, Chisin R, Yeung HWD, et al. 2001. Incidental focal thyroid uptake on FDG positron emission tomographic scans may represent a second primary tumor. *Clin Nucl Med* 26:193–97.
182. Davis PW, Perrier ND, Adler L, et al. 2001. Incidental thyroid carcinoma identified by positron emission tomography scanning obtained for metastatic evaluation. *Am Surg* 67:582–84.
183. Eloy JA, Brett EM, Fatterpekar GM, et al. 2009. The significance and management of incidental F-18 fluorodeoxyglucose-positron emission tomography uptake in the thyroid gland in patients with cancer. *Am J Neuroradiol* 30:1431–34.
184. Van den Bruel A, Maes A, De Potter T, et al. 2002. Clinical relevance of thyroid fluorodeoxyglucose-whole body positron emission tomography incidentaloma. *J Clin Endocrinol Metab* 87:1517–20.
185. Chu QD, Connor MS, lilien DL, et al. 2006. Positron emission tomography (PET) positive thyroid incidentaloma: The risk of malignancy observed in a tertiary referral center. *Am Surg* 72:272–75.
186. Cohen MS, Arslan N, Dehdashti, et al. 2001. Risk of malignancy in thyroid incidentalomas identified by fluorodeoxyglucose-positron emission tomography. *Surgery* 130:941–46.
187. Shulkin BL, Wieland DM, Schwaiger M, et al. 1992. PET scanning with hydroxyephedrine: A new approach to the localization of pheochromocytoma. *J Nucl Med* 33:1125–31.

188. Endo K, Shiomi K, Kasagi K, et al. 1984. Imaging of medullary thyroid cancer with I-131 MIBG. *Lancet* 2:233.
189. Hoefnagel CA, Vaute PA, de Kraker J, et al. 1987. Radionuclide diagnosis and therapy of neural crest tumors using iodine-131 meta-iodobenzylguanidine. *J Nucl Med* 28:308–14.
190. Ohta H, Yamamoto K, Endo K. 1984. A new imaging agent for medullary carcinoma of the thyroid. *J Nucl Med* 25:323–25.
191. Van Moll L, McEnan AJ, Shapiro B, et al. 1987. Iodine-131 MIBG scintigraphy of neuroendocrine tumors other than pheochromocytoma and neuroblastoma. *J Nucl Med* 28:979–88.
192. Reubi JC, Laissue J, Krenning EP, et al. 1992. Somatostatin receptors in human cancer: Incidence, characteristics, functional correlates and clinical implications. *J Steroid Biochem Mol Biol* 43:27–35.
193. Reubi JC. 1995. *In vitro* identification of vasoactive intestinal peptide receptors in human tumors: Implications for tumor imaging. *J Nucl Med* 36:1846–53.
194. Krenning EP, Kwekkeboom DJ, Bakker WH, et al. 1993. Somatostatin receptor scintigraphy with In-111-DTPA-D-Phe1- and I-123-Tyr3-octreotide: The Rotterdam experience with more than 1000 patients. *Eur J Nucl Med* 20:716–31.
195. Virgolini I, Pangerl T, Bischof C, et al. 1997. Somatostatin receptor subtype expression in human tissues: A prediction for diagnosis and treatment of cancer? *Eur J Clin Invest* 27:645–47.
196. Patel YC, Greenwood MT, Panetta R, et al. 1995. Minireview: The somatostatin receptor family. *Life Sci* 57:1249–65.
197. Gomez-Pan A, Rodriguez-Arnao MD. 1983. Somatostatin and growth hormone-releasing factor: Synthesis, location, metabolism and function. *Clin Endocrinol Metab* 12:469–509.
198. Beeman WAP, Hofland LJ, Bakker WH. 1993. Radioiodinated somatostatin analogue RC-160: Preparation, biological activity, *in vivo* application in rats and comparison with [I-123-Tyr3] octreotide. *Eur J Nucl Med* 20:1089–94.
199. Breeman WAP, van Hagen PM, Kwekkeboom DJ. 1998. Somatostatin receptor scintigraphy with [In-111-DTPA]-RC-160 in humans: A comparison with [In-111-DTPA] octreotide. *Eur J Nucl Med* 25:182–86.
200. Thakur ML, Kolan, Li J, et al. 1997. Radiolabeled somatostatin analogs in prostate cancer. *Nucl Med Biol* 24:105–13.
201. Anderson CJ, Pajeau TS, Edwards WB, et al. 1995. *In vitro* and *in vivo* evaluation of copper-64 octreotide conjugates. *J Nucl Med* 36:2315–2325.
202. Smith-Jones PM, Stolz B, Bruns C, et al. 1994. Gallium-67/gallium-68-[DFO]-octreotide: A potential radiopharmaceutical for PET imaging of somatostatin receptor-positive tumors: Synthesis and radiolabeling, *in vitro* and preliminary *in vivo* studies. *J Nucl Med* 35:317–25.
203. Wester HJ, Brockmann J, Rosch F, et al. 1997. PET-pharmacokinetics of F-18-octreotide: A comparison with Ga-67-DFO and Y-86-DTPA-octreotide. *Nucl Med Biol* 24:275–86.
204. Baudin E, Schlumberger M, Lumbroso J, et al. 1996. Octreotide scintigraphy in patients with differentiated thyroid carcinoma: Contribution for patients with negative radioiodine scan. *J Clin Endocrinol Metab* 81:2541–44.
205. Krenning E, Kwekkeboom DJ, Bakker WH, et al. 1993. Somatostatin receptor scintigraphy with [In-111-DTPA-D-Phe]- and [I-123-Tyr-3]-octreotide: The Rotterdam experience with more than 1000 patients. *Eur J Nucl Med* 20:716–31.
206. Stokkel MPM, Verkooijen R, Smit JWA, et al. 2004. Indium-111 octreotide scintigraphy for the detection of non-functioning metastases from differentiated thyroid cancer: Diagnostic and prognostic value. *Eur J Nucl Med Mol Imaging* 31:950–57.
207. Sarlis NJ, Gourgiotis L, Guthrie LC, et al. 2003. In-111 DTPA octreotide scintigraphy for disease detection in metastatic thyroid cancer: Comparison with F-18 FDG positron emission tomography and extensive conventional radiographic imaging. *Clin Nucl Med* 28:208–17.
208. Arslan N, Ilgan S, Yuksel D, et al. 2001. Comparison of In-111 octreotide and Tc-99m DMSA scintigraphy in the detection of medullary thyroid tumor foci in patients with elevated levels of tumor markers after surgery. *Clin Nucl Med* 26:683–88.
209. Berche C, Mach JP, Lumbrusco JD, et al. 1982. Tomoscintigraphy for detecting gastrointenstinal and medullary thyroid cancer: First clinical results using radiolabeled monoclonal antibodies against carcinoembryonic antigen. *Br Med J* 285:1447–51.
210. Reiners CH, Eilles CH, Spiegel W, et al. 1986. Immunoscintigraphy in medullary thyroid cancer using an I-123- or In-111-labelled monoclonal anti-CEA antibody fragment. *Nucl Med* 25:227–31.
211. Edington HD, Watson CG, Levine, et al. 1988. Radioimmunoimaging of metastatic medullary carcinoma of the thyroid gland using an indium-111-labeled monoclonal antibody to CEA. *Surgery* 104:1004–11.

212. Vuillez JP, Peltier P, Caravel JP, et al. 1992. Immunoscintigraphy using In-111-labeled F(ab')2 fragments of anticarcinoembryonic antigen monoclonal antibody for detecting recurrences of medullary thyroid carcinoma. *J Clin Endocrinol Metab* 74:157–63.
213. Juweid M, Sharkey RM, Behr T, et al. 1996. Improved detection of medullary thyroid carcinoma with radiolabeled antibodies to carcinoembryonic antigen. *J Clin Oncol* 74:1209–17.
214. de Labriolle-Vaylet C, Cattan P, Sarfati E, et al. 2000. Successful surgical removal of occult metastases of medullary thyroid carcinoma recurrences with the help of immunoscintigrpahy and radioimmunoguided surgery. *Clin Cancer Res* 6:363–71.
215. Freudenberg LS, Antoch G, Frilling, et al. 2008. Combined metabolic and morphologic imaging in thyroid carcinoma patients with elevated serum thyroglobulin and negative ultrasonography: Role of I124-PET/CT and FDG-PET. *Eur J Nucl Med Mol Imaging* 35:950–57.
216. Erdi YE, Macapinlac H, Larson SM, et al. 1999. Radiation dose assessment for I-131 therapy of thyroid cancer using I-124 PET imaging. *Clin Positron Imaging* 2:41–46.
217. Fogelman I, Maisay MN, and Clarke SEM. 1995. *An atlas of clinical nuclear medicine*. 2nd ed. Martin Dunitz, London.

11 PET-CT Scanning in the Management of Thyroid Cancer

Experience with a Cohort of 100 Patients

Dario Casara, Faise Al Bunni, Maria Rosa Pelizzo, and Fabio Pomerri

CONTENTS

ABSTRACT: Differentiated thyroid carcinoma (DTC) has an excellent prognosis with a 10-year survival rate higher than 90%. However, persistent or recurrent disease occurs in 5 to 24% of patients. Patients with increasing serum thyroglobulin (TG) values and negative diagnostic ^{131}I (RAI) whole body scan (WBS) results should undergo appropriate imaging modalities to localize the disease and plan adequate treatment. Recently, ^{18}FDG positron emission tomography/computed tomography (PET/CT) has been proposed as a useful imaging modality to study these patients. The aim of this report is to evaluate the rational use of PET/CT as a diagnostic and therapeutic tool for patients with DTC recurrence. One hundred DTC patients underwent thyroidectomy and postoperative thyroid remnant ablation with ^{131}I and were studied using PET/CT because of rising basal or recombinant thyroid-stimulating hormone (TSH)-stimulated serum TG levels. The patients were divided into three groups based on the indication for PET/CT: diagnostic indication for patients with RAI negative WBS and increased TG levels (group 1), surgical planning for patients with known recurrence (group 2), and prognostic or therapeutic efficacy evaluation for patients with advanced disease (group 3). PET/CT positive foci were found in 53 of 100 studies. In group 1 (N = 33), PET/CT localized recurrent disease in 42% of patients. In group 2 (N = 45), PET/CT scan predicted loco-regional resectable disease in 42% of patients. In group 3 (N = 22), PET/CT confirmed sites of known metastatic disease in 59% of cases. PET/CT results led to surgical treatment with radical intent in 19 patients (complete remission in 8 patients) and palliative intent in 6 cases. In addition,

PET/CT results led to high-dose RAI treatment for 11 patients and external beam radiotherapy for 3 patients. The remaining 14 patients underwent only clinical follow-up. The study emphasizes how PET/CT indications are changing from a purely diagnostic role in localizing RAI negative DTC recurrences to designing appropriate DTC surgical or combined therapeutic strategies.

11.1 INTRODUCTION

Thyroid cancer accounts for more than 90% of endocrine tumors, but only 1 to 2% of all cancers. The National Cancer Institute Surveillance Epidemiology End Results (SEER)[1] reports an incidence of 9.1 cases per 100,000 inhabitants based on the population adjusted for age, with 13.4 cases for women and 4.6 cases for men. This study also reports an increased incidence of this disease with an increased trend of 6.4%, during the years 2001 to 2005, compared with 5.8%, during the years 1996 to 2005. The prognosis of thyroid cancer is favorable,[1] with a 5-year survival at 96.9% and 10-year survival at 95.6%. These rates are due to the efficacy of complementary treatment of total or near-total thyroidectomy and radioiodine (RAI) therapy with ^{131}I. RAI therapy is used for remnant ablation after thyroid surgery and in more advanced stages to treat lymph node and distant metastases. However, a small percentage of patients, approximately 5 to 24% of cases, show persistence or progression of disease or metastases during clinical follow-up, despite the use of repeated cycles of RAI.[2] The persistence or recurrence of disease can be suspected early on the basis of increasing levels of circulating thyroglobulin (TG), a sensitive marker for differentiated thyroid carcinoma (DTC).[3] Thus, patients with increasing basal or recombinant human (rh) TSH-stimulated TG and negative diagnostic RAI whole body scans (WBSs) require accurate imaging modalities for the detection of disease recurrence. The rationale for utilizing novel functional imaging techniques is predicated on the potential efficacy of radical surgery to control loco-regional disease. Ultrasonography (US) and US-guided fine-needle aspiration (UG-FNA) biopsy are highly useful for monitoring progression of disease, but they cannot evaluate anatomy beyond the neck. Alternatively, ^{18}fluorodeoxyglucose positron emission tomography (^{18}FDG PET) is useful for staging disease and evaluating the effectiveness of systemic therapies.[4,5]

^{18}FDG PET is an imaging technique that can detect metabolic changes, even before structural alterations can be found. However, the specificity of ^{18}FDG PET is compromised by the absence of anatomical details. However, this deficiency can be addressed by anatomical studies, such as diagnostic computed tomography (CT), which can identify false positives, such as nonneoplastic tissue showing high metabolic activity.[6] The hybridized imaging technique of ^{18}FDG PET/CT (hereafter named PET/CT) provides a single complementary image, both anatomical and functional, performed in the same scanning session, that exceeds the accuracy of the two methods taken individually and facilitates appreciation of detail, which is difficult with each technique alone.[6]

Recently, ^{18}FDG PET and PET/CT have been proposed as effective investigative studies of DTC patients with increased TG and negative RAI WBS.[7–12] The first studies that introduced the use of ^{18}FDG PET in the diagnosis of thyroid cancer metastases date back to the observations of Joensuu and Ahonen,[13] later confirmed by Feine et al.,[14] which showed how some metastatic lesions that do not concentrate RAI were able to capture FDG, and were thus detected by ^{18}FDG PET. Feine et al.[14] and Grunwald et al.[4] described the reciprocal relationship, called flipflop characteristic, between FDG avidity and RAI avidity: well-differentiated thyroid cancers show RAI concentration but not FDG, while poorly differentiated or anaplastic thyroid cancers are FDG avid but do not concentrate RAI.

Tumor cells exhibit increased glucose uptake in part due to the presence of an increased number of glucose transporters (GLUT) in their cell membranes.[15] Well-differentiated thyroid cancer cells are rich in Na^+/I^- symporter (NIS) molecules, but not GLUT molecules. Poorly differentiated thyroid cancer cells are rich in GLUT molecules, but not NIS molecules. These events were investigated in a paper from our unit (Mian et al.[16]) using tissue specimens from patients who underwent reintervention for disease recurrence that was not RAI avid. We found that GLUT-1 gene expression is higher in metastases than primary tumors, and the loss of RAI may not depend on a drop in NIS gene expression alone, but also on modifications in enzymes regulating intracellular iodine metabolism. Therefore, the sensitivity of PET/CT to detect occult metastases is correlated with the presence of GLUT receptors (metabolic characterization) and the degree of tissue differentiation.

In a study conducted on a cohort of 125 subjects by Wang et al.,[17] univariate analysis of survival was correlated with (1) the presence of distant metastases, (2) age older than 45 years, (3) positive ^{18}FDG PET imaging, (4) maximum standardized uptake value (SUV), and (5) the total volume of ^{18}FDG PET disease. Multivariate analysis of survival demonstrated that the volume of ^{18}FDG PET positive lesions was the most significant factor related to survival. It is interesting to note that a similar analysis was reported by Casara et al.[18] on a cohort of thyroid cancer patients with distant metastases. The survival rate of these patients studied by univariate analysis was correlated to (1) the size of metastases, (2) patient's age, and (3) RAI uptake, but multivariate analysis demonstrated that the main factor connected to survival rate was RAI avidity.

It is therefore possible that the result of PET/CT examination could be used to refine the prognosis of the patient once metastatic lesions are identified and characterized, providing valuable support to the endocrinologist, oncologist, nuclear medicine specialist, and surgeon, in order to implement the most appropriate therapeutic strategy. We report here an original study on a cohort of DTC patients with the purpose of assessing the accuracy of PET/CT in evaluating and managing DTC recurrence.

11.2 MATERIAL AND METHODS

11.2.1 Patients

A group of 100 patients enrolled at Istituto Oncologico Veneto of Padua underwent total thyroidectomy with at least one cycle of RAI remnant ablation or treatment of metastatic disease. All patients gave their informed consent to take part in the study, which was approved by the local institutional review board. Out of these 100 patients, 62 were female (median age, 59 years; range, 18–88) and 38 males (median age, 59 years; range, 30–82). Of the 100 patients enrolled, 46 underwent total thyroidectomy, and 54 thyroidectomy plus lymphoadenectomy. Eleven patients had already undergone a reintervention, 9 for nodal recurrence, 1 for lung metastasis, and 1 for bone metastases before PET/CT scanning.

The surgical pathological stage at initial diagnosis using the AJCC/UICC classification was as follows: stage I, 23 patients; stage II, 6 patients; stage III, 44 patients; and stage IV, 27 patients. In 87 patients, the histological type was papillary thyroid carcinoma, and in the remaining 13 patients it was follicular thyroid carcinoma.

The 100 patients had an increased TG level both during suppressive hormonal therapy (mean, 23.6 ± 80.4; median, 3.2 ng/ml; range, 1–634 ng/ml) or with stimulated TSH, following thyroid hormone withdrawal, or using rhTSH (mean, 137.2 ± 207.5; median, 37.6 ng/ml; range, 2.2–831 ng/ml). Thirty-nine of these patients underwent additional cycles of RAI for local recurrences (N = 15), lymph node metastases (N = 15), or distant metastases (N = 9). The RAI dose for remnant ablation was an average of 3.7 GBq (range, 1.7–5.5); treatment of lymph node metastases, 7.3 GBq (range, 5.5–11.0 GBq); and distant metastases, 9.8 GBq (range, 7.4–11.0). The period considered free of disease (based on levels of circulating TG) revealed an average of 3 years (range, 0.5–10 years).

For the statistical analysis the Mann-Whitney test was used with a p-value of 0.05 considered significant.

11.2.2 ^{18}FDG PET-CT Protocol

All patients underwent PET/CT during TSH suppressive therapy with levothyroxine. This methodology was chosen because there is still no consensus in literature for the use of PET with TSH stimulation in patients with DTC.[19,20] PET/CT images were obtained from 2004 to the end of 2007 using a GE Discovery ST tomograph, while from January 2008 on, using a Siemens CT Biograph 16 HR tomograph. The metabolic tracer was ^{18}F-fluorodeoxyglucose (^{18}FDG) for all patients, ranging from 180 to 400 MBq (activity calculated on 3.7 MBq/kg/BW).

The ^{18}FDG was administered during rest and fasting for 4–6 hours and washing the vein by means of 40–50 ml of saline infusion. The acquisition was obtained at 60 minutes after tracer administration. Patients were asked to drink freely before scanning to both stimulate diuresis and obtain an adequate distension of the intestinal walls.

11.3 RESULTS

11.3.1 Clinical Indications for PET/CT

The overall results of PET/CT scan and treatment characteristics are presented in Table 11.1. The PET/CT confirmed the presence of disease in one or more sites in 53 subjects. Table 11.2 summarizes the outcome of PET/CT positive patients subdivided on therapeutic management.

In 33 of 100 patients, the indication for the use of PET/CT was based on an increased TG value with a negative diagnostic RAI WBS during clinical follow-up. These patients were included in a classification henceforth called group 1 (diagnostic indication) (Table 11.3).

TABLE 11.1
Number of Patients Subdivided by PET/CT Results and Therapeutic Management

Management	PET/CT +	PET/CT −	Total
Surgery	25	3	28
RAI	11	12	23
EBRT	3	0	3
F/U	14	32	46
Total	53	47	100

Note: RAI, radioactive ^{131}I therapy; EBRT, external beam radiotherapy; F/U, follow-up (observation only).

TABLE 11.2
Number of PET/CT Positive Patients Subdivided by Therapeutic Management and Outcome

Outcome of PET/CT + Patients	Surgery	RAI	EBRT	F/U	Total
Remission	8	2	0	1	11
Progression	6	0	2	2	10
Unchanged	5	9	1	11	26
n.a.	6	0	0	0	6
Total	25	11	3	14	53

Note: RAI, radioactive ^{131}I therapy; EBRT, external beam radiotherapy; F/U, follow-up (observation only); n.a., not available.

TABLE 11.3
Number of Patients with Positive and Negative PET/CT Results Subdivided by Groups

Group	PET/CT +	PET/CT –	Total
1	14	19	33
2	26	19	45
3	13	9	22
Total	53	47	100

In 45 of 100 patients the field of PET/CT application was based on an assessment of a demonstrated or highly suspected disease recurrence on the basis of US and FNA cytology in order to plan a surgical reintervention (group 2; surgical indication) (Table 11.3). The reason for this approach stemmed from the consideration that further treatments with RAI were not able to resolve the recurrence, while a radical surgical approach could potentially achieve some control of the disease when the patient had a single neck recurrence.

In 22 of 100 patients the indication for PET/CT was based on the assessment of progression of known metastatic disease or effectiveness of a systemic therapy (group 3; post-treatment evaluation) (Table 11.3).

11.3.2 Results of PET/CT in Group 1

Of 33 patients with negative RAI diagnostic WBS and increased TG, PET/CT was positive for the presence of disease in at least one location in 14 patients (42%) (Table 11.4). The average value of basal TG was 83.8 ± 189.4 ng/ml (median, 9.0 ng/ml; range, 1.4–634 ng/ml). In 6 of the 14 patients, PET/CT was positive in one site, including 2 patients with lateral compartment lymph nodes (Figure 11.1), 1 with a retroclavicular lymph node, and 3 with soft tissue structure involvement neighboring the thyroid bed. In 2 other patients PET/CT was positive in cervical lymph nodes bilaterally or in mediastinal lymph nodes, and in the remaining 6 patients, PET/CT was positive for lung or bone metastases.

In 6 of the patients with RAI negative WBS and evidence of disease recurrence in a single site by PET/CT images, 3 were referred to surgery, 1 with tracheo-esophegeal invasion underwent both surgery and external beam radiotherapy (EBRT), 1 was treated with high-dose RAI, and 1 was monitored only by clinical follow-up.

Two patients with evidence of recurrence of disease bilaterally or in a retrosternal site, and therefore not easily curable with surgery, were referred for high-dose RAI therapy (7.4 GBq). Among 6 patients with positive PET/CT for secondary lung lesions, 1 patient was referred for palliative

TABLE 11.4
Number of Patients in Group 1 (Diagnostic Indication) Subdivided by Disease Localization and Therapeutic Management

Group 1 PET/CT+	Surgery	RAI	EBRT	F/U	Total
Localized disease	4[a]	1	0	1	6
Bilateral or retrosternal lymph nodes	0	2	0	0	2
Distant metastasis	1	2	2	1	6
Total	5	5	2	2	14

Note: RAI, radioactive ^{131}I therapy; EBRT, external beam radiotherapy; F/U, follow-up (observation only).
[a] One patient underwent both surgery and EBRT.

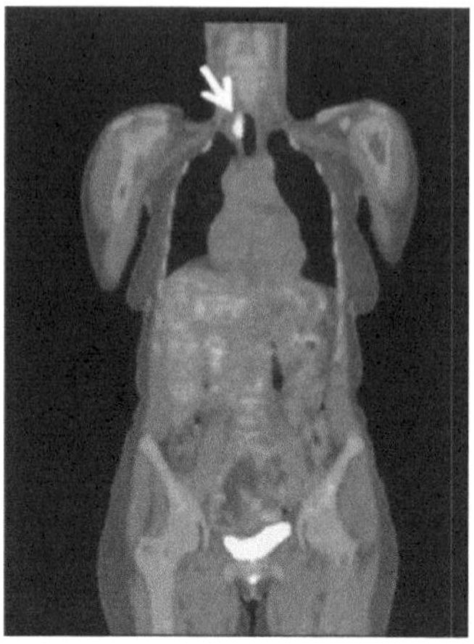

FIGURE 11.1 (See color insert.) A 43-year-old woman with negative iodine ^{131}I whole body scan and elevated serum TG level. Combined PET/CT coronal image shows unilateral lymph node metastasis on right (arrow).

surgery to reduce the presence of significant disease in a regional site, 2 patients were referred for high-dose RAI therapy (7.4 GBq), 2 patients were referred to EBRT for palliation due to the presence of lung and painful bone metastases, and 1 patient was referred for clinical follow-up of secondary lung lesions.

In one patient with low TG levels, the PET/CT positive lung lesions were thought to be due to breast cancer, which was operated on before a diagnosis of thyroid cancer was made. Among 19 patients in whom the PET/CT was negative, 13 patients just had clinical follow-up and 6 patients had high-dose RAI therapy. It is interesting to note that in these patients with negative ^{18}FDG PET uptake, the average basal TG value was 6.8 ± 5.6 ng/ml (median, 4.8 ng/ml; range, 1 to 16 ng/ml) and significantly lower in comparison with the ^{18}FDG PET positive group of patients ($p < 0.05$).

Additionally, among 19 patients with negative PET/CT, 1 patient was discovered to have an occult primary breast cancer (external outer quadrant of left breast), and another patient was discovered to have an occult tumor in the left maxillary sinus (oncocytic papilloma). The patient with the breast cancer underwent a quadrantectomy plus sentinel node biopsy and hormonal therapy. The patient with oncocytic papilloma in the left maxillary sinus underwent surgical excision.

11.3.3 Results of PET/CT in Group 2

Among the 45 patients undergoing PET/CT for preoperative planning, it was positive in 26 (Table 11.5). In 19 of these 26 patients, the PET/CT confirmed the presence of disease located in a single neck site, and in 7 patients the PET/CT was positive for bilateral lesions or lung metastases (Figure 11.2). Average values of basal TG were 20.7 ± 84.8 ng/ml (median, 2.0 ng/ml; range, 1–156 ng/ml) in the positive PET/CT patients.

Out of the 19 patients in which the PET/CT confirmed the presence of loco-regional cervical disease, 15 were referred for radical surgery, 1 for high-dose RAI therapy (7.4 GBq) due to excessive extension of the disease into neighboring tissues, and 3 for close follow-up since they were judged unfit for an immediate surgery.

Of the 15 patients referred for radical surgery, the subsequent clinical follow-up was available in 10 patients: 6 of these 10 were later found to be in complete remission on the basis of serum TG values and negative US; in 1 of these 6, a complete remission followed a subsequent, postoperative, course of EBRT. One of the patients not considered to be in complete remission developed a clear rise in anti-TG antibodies precluding reliable interpretation of TG values.

TABLE 11.5
Number of Patients in Group 2 (Surgical Indication) Subdivided by Disease Localization and Therapeutic Management

Group 2 PET/CT+	Surgery	RAI	EBRT	F/U	Total
Localized disease	15	1	0	3	19
Bilateral lymph nodes	4	0	0	0	4
Distant metastasis	1	2	0	0	3
Total	20	3	0	3	26

Note: RAI, radioactive ^{131}I therapy; EBRT, external beam radiotherapy; F/U, follow-up (observation only).

Among the 7 patients with evidence of bilateral loco-regional recurrence or distant metastatic disease, 5 underwent palliative interventions to reduce the size of local lesions and 2 underwent high-dose RAI therapy.

In 19 of the 45 patients in group 2, there were negative PET/CT results. The average basal TG value was 7.6 ± 19.5 ng/ml (median, 1.4 ng/ml; range, 0.2–80 ng/ml). Three of these 19 patients underwent subsequent surgery based on the opinion of the surgeon involved, and all had complete remissions, 5 were treated with high-dose RAI therapy, and the remaining 11 patients were subjected to follow-up.

11.3.4 Results of PET/CT in Group 3

Twenty-two subjects with previously known metastatic disease made up this group. Eighteen patients had been treated with RAI or EBRT, and the grounds for the application of PET/CT was based on assessment of the development or progression of disease already known from previous diagnostic investigations. In 13 of the 22 patients in group 3, the PET/CT produced a positive result (Table 11.6) and the PET/CT results matched with the previously known metastatic sites (2 cases were conducted postoperative to identify the presence of residual metastatic disease). The average basal TG value was 20.4 ± 32.2 ng/ml (median, 3.1 ng/ml; range, 1–89 ng/ml). In 9 cases, PET/CT did not show foci of disease.

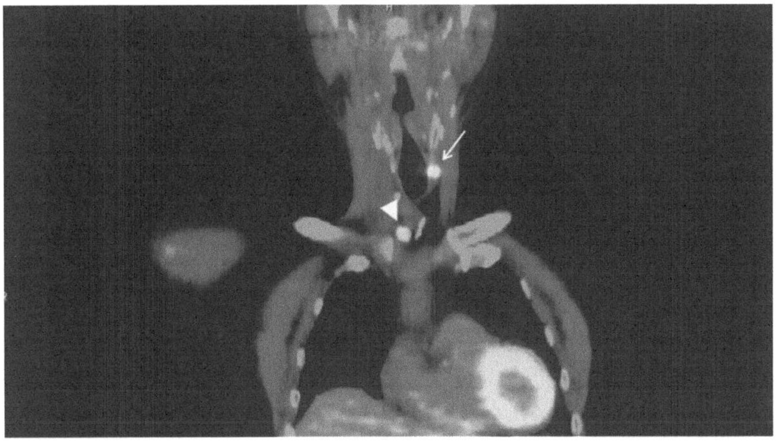

FIGURE 11.2 (See color insert.) A 73-year-old man with a sonography suggestive of left lateral compartment lymph nodes metastasis. Combined PET/CT coronal image confirms the left cervical lymph node (arrow) metastasis and reveals a second right supraclavicular tumor focus (large arrowhead).

TABLE 11.6
Number of Patients in Group 3 (Posttreatment Indication) Subdivided by Therapeutic Management

Group 3 PET/CT+	Surgery	RAI	EBRT	F/U	Total
Distant metastasis	0	3	1	7	11
Postoperative	0	0	0	2	2
Total	0	3	1	9	13

Note: RAI, radioactive ^{131}I therapy; EBRT, external beam radiotherapy; F/U, follow-up (observation only).

In 4 of the 13 patients with a positive result, further treatment was planned: 3 patients with lung metastases had additional high-dose RAI (Figure 11.3) and one patient with vertebral bone metastasis had EBRT of the lesion.

The remaining 9 other patients, of the 13 with a positive result, were followed by clinical follow-up. Significantly, in one patient with clearly positive PET/CT for pulmonary lesions, a lung adenocarcinoma was diagnosed by CT-guided FNA. In another patient of this group, the PET/CT scan showed the presence of a localized rectal tumor. This patient underwent surgical removal of the rectal cancer.

11.4 DISCUSSION

PET/CT is a diagnostic tool now largely accepted in the clinical follow-up of patients with DTC. However, clinical practice guidelines (CPG) generally recognize as a unique indication the negative diagnostic RAI WBS with positive TG levels.[21] For example, the leading American health care agencies recognize only this claim for reimbursement of PET (*Medicare National Coverage Determinations Manual*[22]). Our cancer center, Istituto Oncologico Veneto, has one of the largest series among the Italian institutions that deal with cancer (approximately 6000 cases of thyroid carcinoma enrolled from 1972 to 2008), so it can be considered a reference center for the treatment

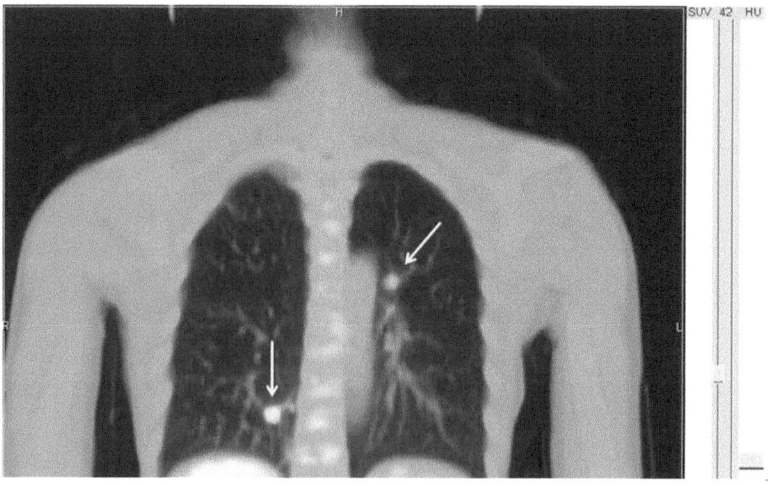

FIGURE 11.3 (See color insert.) A 48-year-old woman with lung metastases shown on previous CT examination. Combined PET/CT coronal image confirms the two lung metastases (arrows).

of thyroid cancer. This study was developed with the aim of evaluating the impact of PET/CT in the management of DTC patients.

The first general observation concerns the rationale that prompted endocrinologists, surgeons, and oncologists to require a PET/CT scan. In our study, the purely diagnostic indication for PET/CT (group 1) accounted for 33% of the requests, while the remaining 67% had a rationale to plan a targeted intervention or evaluate a systemic therapy.

The results in group 1 show that PET/CT detected the presence of previously unknown disease in 14 out of 33 patients (42.3%), while in the remaining 19 patients (57.7%) the PET/CT was negative. The absence of FDG tracer uptake indicates the possible presence of indolent disease, capable of being controlled with suppressive levothyroxine therapy and close follow-up. In contrast, a positive PET result should prompt the clinician to a more aggressive therapeutic decision: surgery if possible or other specific therapies.

Among positive cases in group 1 for one or more sites of disease relapse, 5 underwent surgery: 3 with radical intent and 2 with palliative intent. Five other patients were referred for high-dose RAI and 2 for EBRT. Thus, 12 out of a total of 33 patients (36.3%) in group 1 underwent therapeutic strategy based on the PET/CT findings. As reported in the literature,[23] PET/CT can play an important role in modifying the therapeutic decision of patients who otherwise would have been empirically treated with high doses of ^{131}I.

Group 2 accounts for 45 patients. This group of patients is generally not considered by CPG. The PET/CT results confirmed the presence of resectable disease in 19 patients (42.2%). Here, PET/CT also influenced therapeutic decision making. Specifically, radical surgery could be planned for localized and potentially curable disease.[2] Significantly, 15 of the 19 patients were referred for radical surgery, and 6 of these 15 had documented disease remission. In 5 cases the surgical result was not valuable because of a too short clinical follow-up. Nineteen out of forty-five patients had a negative PET/CT result, which had a prognostic impact. PET/CT characterized by absent uptake or low standardized uptake value (SUV) measurement correlates with less aggressive disease and leads to a noninvasive therapeutic approach.[17] In only 2 patients with negative PET/CT, there was histological confirmation of disease after surgery. This false negative PET/CT result is in agreement with data reported in literature[24] as representing the presence of small nodal involvement by foci of well-differentiated disease.

In group 3, we assessed the increasing role of PET/CT as a tool for evaluation of patients undergoing specific treatments, especially in patients with loss of RAI avidity, and therefore not evaluable with ^{131}I scan. Again, therapeutic management could be directed by the PET/CT findings.

All 100 cases in this study were evaluated by PET/CT as potential carriers of disease recurrences. The PET/CT confirmed the presence of disease in one or more sites in 53 subjects. However, even a negative result or a result associated with a low SUV was able to direct management. As shown by studies of Wang et al.[17] and Robbins et al.,[24,25] negative results are associated with a less aggressive disease, and therefore could, in some cases, be managed with thyroid hormone suppressive therapy and close clinical monitoring.

REFERENCES

1. Horner MJ, Ries LAG, Krapcho M, et al. (eds). 2009. *SEER Cancer statistics review, 1975–2006*. Bethesda, MD: National Cancer Institute.
2. Mirallié E, Guillan T, Bridji B et al. 2007. Therapeutic impact of ^{18}FDG PET/CT in the management of iodine-negative recurrence of differentiated thyroid carcinoma. *Surgery* 142:952–58.
3. Ozata M, Suzuki S, Miyamoto T, et al. 1994. Serum thyroglobulin in the follow up of patient with treated differentiated thyroid cancer. *J Clin Endocrinol Metab* 79:98–105.
4. Grunwald F, Schomburg A, Bender H, et al. 1996. Fluorine-18 fluorodeoxyglucose positron emission tomography in the follow-up of differentiated thyroid cancer. *Eur J Nucl Med* 23:312–19.
5. Grunwald F, Menzel C, Bender H, et al. 1997. Comparison of F-18 FDG-PET with iodine-131 and Tc-99m-sestamibi scintigraphy in differentiated thyroid cancer. *Thyroid* 7:327–35.

6. Wechalear K, Sharma B, Cook G. 2005. PET/CT in oncology—A major advance. *Clin Radiol* 60:1143–55.
7. Alnafisi NS, Driedger AA, Coates G, et al. 2000. FDG PET of recurrent or metastatic ^{131}I-negative papillary thyroid carcinoma. *J Nucl Med* 41:1010–15.
8. Helal BO, Merlet P, Toubert ME, et al. 2001. Clinical impact of (18)F-FDG PET in thyroid carcinoma patients with elevated thyroglobulin levels and negative (131)I scanning results after therapy. *J Nucl Med* 42:1464–69.
9. Palmedo H, Bucerius J, Joe A, et al. 2006. Integrated PET/CT in differentiated thyroid cancer: Diagnostic accuracy and impact on patient management. *J Nucl Med* 47:616–24.
10. Finkelstein SE, Grigsby BW, Siegel BA, et al. 2008. Combined [18F]fluorodeoxyglucose positron emission tomography and computed tomography (FDG-PET/CT) for detection of recurrent, 131I-negative thyroid cancer. *Ann Surg Oncol* 15:286–92.
11. Shammas A, Degirmenci B, Mountz JM, et al. 2007. 18F-FDG PET/CT in patients with suspected recurrent or metastatic well-differentiated thyroid cancer. *J Nucl Med* 48:221–26.
12. Freudenberg LS, Frilling A, Kuhl H, et al. 2007. Dual-modality FDG-PET/CT in follow-up of patients with recurrent iodine-negative differentiated thyroid cancer. *Eur Radiol* 17:3139–47.
13. Joensuu H, Ahonen A. 1987. Imaging of metastases of thyroid carcinoma with fluorine-18 fluorodeoxyglucose *J Nucl Med* 28:910–14.
14. Feine U, Lietzenmayer R, Hanke JP, et al. 1996. Fluorine-18-FDG and iodine-131-iodide uptake in thyroid cancer. *Eur J Nucl Med* 37:1468–72.
15. Warburg O, Wind F, Negleis E. 1930. On the metabolism of tumors in body. In *Metabolism of tumors*, ed. O Warburg, 254–70. London: Costable.
16. Mian C, Barollo S, Pennelli GM, et al. 2008. Molecular characteristics in papillary thyroid cancers (PTCs) with no 131-I uptake. *Clin Endocrinol* 68:108–16.
17. Wang W, Larson SM, Fazzari M, et al. 2000. Prognostic value of [18F]fluorodeoxyglucose positron emission tomographic scanning in patients with thyroid cancer. *J Clin Endocrinol Metab* 85:1107–13.
18. Casara D, Rubello D, Saladini G, et al. 1993. Different features of pulmonary metastases in differentiated thyroid cancer: Natural history and multivariate statistical analysis of prognostic variables. *J Nucl Med* 10:626–31.
19. Saab G, Driedger AA, Pavlosky W, et al. 2006. Thyroid-stimulating hormone-stimulated fused positron tomography/computed tomography in the evaluation of recurrence in 131I-negative papillary thyroid carcinoma. *Thyroid* 16:267–72.
20. Chin BB, Patel P, Cohade C, et al. 2004. Recombinant human thyrotropin stimulation of fluoro-D-glucose positron emission tomography uptake in well differentiated thyroid carcinoma. *J Clin Endocrinol Metab* 89:91–95.
21. Boellaard R, O'Doherty M, Weber W, et al. 2010. FDG PET and PET/CT: EANM procedure guidelines for tumour PET imaging: Version 1.0. *Eur J Nucl Med Mol Imaging* 37:181–200.
22. http://www.cms.hhs.gov/manuals/downloads/ncd103c1_part4.pdf.
23. Freudenberg LS, Frilling A, Kuhl H, et al. 2007. Dual-modality FDG-PET/CT in follow-up of patients with recurrent iodine-negative differentiated thyroid cancer. *Eur Radiol* 17:3139–47.
24. Robbins RJ, Larson SM. 2008. The value of positron emission tomography (PET) in the management of patients with thyroid cancer. *Best Pract Res Clin Endocrinol Metab* 6:1047–59.
25. Robbins RJ, Wan Q, Grewal RK, et al. 2006. Real time prognosis for metastatic thyroid carcinoma based on 2(18F)fluoro-2-deoxy-D-glucose-positron emission tomography scanning. *J Clin Endocrinol Metab* 91:498–505.

12 The Emerging Role of Percutaneous Needle Biopsy and Molecular Tumor Marker Analysis for the Preoperative Selection of Thyroid Nodules

Angelo Carpi, Jeffrey I. Mechanick, and Andrea Nicolini

CONTENTS

ABSTRACT: There is significant heterogeneity of fine-needle aspiration biopsy (FNAB) experience and performance reflected in the medical literature. Part of this heterogeneity results from indeterminate FNAB diagnoses, especially in normal, hyperplastic, or other benign tissue. The number of indeterminate diagnoses in the many centers that do not report their results can be much higher than the 20–30% reported in the literature. As a consequence, the number of the nodules that are excised still appear excessive. Large-needle aspiration biopsy (LNAB) can reduce the number of surgical operations for benign nodules because it diagnoses as benign about 50% of the nodules with inadequate or indeterminate follicular FNAB findings. These effects are explained by the larger needle size (LNAB vs. FNAB), which is more adequate for the investigation of these nodules. Many molecular and genetic tests have been recently developed and proposed for improving diagnostic accuracy of preoperative FNAB cytology. Among the tissue molecular markers, galectin-3 is the most widely investigated for improving accuracy of preoperative diagnosis of thyroid nodules. LNAB provides a much more abundant and adequate substrate for galectin-3 immunodetection than FNAB. Decision trees and clinical algorithms that assist the evaluation of thyroid nodules with inadequate or indeterminate follicular FNAB findings can include LNAB and galectin-3 immunodetection on LNAB.

12.1 INTRODUCTION

Thyroid nodules are a frequent clinical finding in adults, especially in women. The principal goal in the management of thyroid nodules is to differentiate the relatively infrequent (≤10%) malignant nodules that require excision from the relatively frequent (≥90%) benign nodules that can generally be observed.[1,2] Despite the widespread application of high-resolution ultrasonography (US) and US-guided fine-needle aspiration biopsy (UG-FNAB) for cytological analysis, recent studies still report the proportion of excised nodules higher than 20%.[3] These figures are based on diagnostic algorithms centered on FNAB cytology, which is considered the most accurate technique available for preoperative thyroid nodule diagnosis.[1] Emergent technologies should improve the performance of preoperative selection of thyroid nodules in a safe manner. This chapter will focus on the use of one such emergent technology: large-needle percutaneous aspiration biopsy (LNAB) with molecular analysis, particularly galectin-3 (gal-3) immunodetection, of the sampled material. The premise is that as sufficient evidence is manifest, then this and other procedures will be incorporated into decision trees, clinical algorithms, and eventually clinical practice guidelines (CPG).

12.2 FNAB DIAGNOSIS OF BENIGN THYROID NODULES IS NOT SUFFICIENTLY ACCURATE

The excess of excised nodules reflects the relative inaccuracy of FNAB cytology in diagnosing benign nodules. In fact, FNAB cytology will classify some nodules with follicular structure as indeterminate. The FNAB finding of follicular indeterminate nodule shows a relatively high probability of malignancy (~30%) at postoperative examination; therefore surgical excision cannot be avoided. However the majority of these nodules are actually benign at postoperative histology.[4] It is a smaller group of nodules with follicular components at FNAB that are suspicious for malignancy and are postoperatively confirmed in about 50%. The most frequent cytological term used is *follicular proliferation*, which means an indeterminate cytology report.[5] Indeed, there is a broad range of reported frequencies of postoperatively confirmed malignancies in patients with indeterminate preoperative cytology.[2,5–10]

The most common morphologic feature in cytology specimens is represented by variously arranged microfollicles with scanty or absent colloid.[11] These nodules are better characterized by distinguishing between hyperplastic nodules and follicular neoplasms (follicular adenoma and follicular thyroid carcinoma [FTC]).[12] Some authors have also given importance to the architectural follicular pattern[11] or to the nuclear features.[14] The majority of clinical reports suggest that high cellularity, colloid, and microfollicles are the principal features to be considered for the FNAB diagnosis of a follicular neoplasm.[11,13,15] However, some reports indicate that "the presence of nuclear atypia is necessary to define a follicular neoplasia."[16] When these FNAB-derived features were compared among postoperatively defined hyperplastic nodules, follicular adenomas, and papillary thyroid carcinomas (PTCs), "none of the features considered was of sufficient statistical frequency either alone or in combination to be useful in distinguishing neoplastic from non-neoplastic lesions."[17] In a report examining 339 cytological diagnoses of hyperplastic/adenomatoid nodule or follicular neoplasm and 120 cytological diagnoses of follicular-derived neoplasm with focal nuclear features suspicious for PTC, the postoperative malignancy rate was 22 and 72%, respectively.[18]

Some studies suggest that the FNAB diagnosis of indeterminate follicular nodule is overused in the current clinical practice. Segev et al.[19] reported a study on 181 indeterminate and suspicious FNABs carried out at Johns Hopkins Medical institutions that had available histopathologic diagnoses. These authors separated the indeterminate FNABs into a suspected follicular neoplasm group (66% of cohort) and a suspicious lesion group (34% of cohort) and found postoperatively that 37 and 59%, respectively, were eventually diagnosed as cancer.[19] They added that "although the classic diagnostic dilemma is the differentiation of FTC from benign follicular adenoma, it has been our experience that indeterminate FNAB diagnoses can result from any benign or malignant thyroid nodule."[19]

Mowschenson et al.[20] showed the results of FNAB performed from the grossly normal contralateral lobe during thyroidectomy in 42 patients. Cytopathologists examined the slides without knowing the source of the tissue. FNAB of grossly normal thyroid tissue was adequate for interpretation in 32 of 42 patients, and in nine of 42 cases it was interpreted as unremarkable.[20] However, the remaining specimens were classified as microfollicular lesions (18), mixed macromicrofollicular lesions (3), Hürthle cell lesion (1), and PTC (1).[20] FNAB of grossly normal thyroid tissue suggested a microfollicular lesion in 18 (56%) patients, a result that would raise the possibility of a FTC and often lead to the recommendation for operation.[20] Carpi et al.[21] reported a series of 320 FNAB diagnoses of indeterminate follicular nodules from small local hospitals that were reevaluated in a university setting. The revised diagnosis was benign in 112, while inadequate in 84 cases and unchanged in 124 cases. This suggests that the frequency of indeterminate FNAB diagnosis is higher in small centers with presumably lower experience.[21]

There may also be interobserver variability in the cytological interpretation of the preoperative FNAB or the postoperative histology. In one study, the predictive performance of FNAB cytological diagnoses for surgical outcome was lower for nonneoplastic diagnoses (53–74%) than for follicular neoplasms (77–90%).[22] There was also little correlation among observers regarding the FNAB diagnosis of follicular lesion vs. follicular neoplasm,[23] as well as regarding postoperative histologic diagnosis of adenomatous goiters vs. follicular adenoma.[24] These data highlight the diagnostic pitfalls when evaluating hyperplastic goiters, especially by a cytopathologist who deliberately errs on the side of minimizing false negative results. This is consistent with previous studies comparing preoperative FNAB cytology with postoperative histology of follicular neoplasms, which showed that the positive predictive value of this diagnostic technique varied from 0.06–0.5.[10,25,26] It was commented that the low predictive value of an FNAB diagnosis of follicular neoplasm is largely accounted for by the confusion encountered between hyperplastic nodular goiter and follicular adenoma. This error is generally accepted as unavoidable because of cytomorphological similarities between the pathologies and the subjective aversion to missing a neoplastic process requiring surgery.[7,10]

Another factor that leads to surgical excision of benign nodules is the occurrence of inadequate material for cytological analysis from FNAB.[1,27] Nodules with a significant cystic component are usually associated with this finding.[1] Even in "good hands" the rate of inadequate samples is rarely lower than 10%.[1,2,28] In a 2002 survey of clinical endocrinologists, 14% of the respondents had nondiagnostic rates (due to inadequate samples) greater than 21%, and 4% of the respondents had nondiagnostic rates greater than 30%.[29] If physicians performing the aspirations have insufficient experience, these proportions can rise to 70%.[2,30,31]

12.2.1 Summary

- There is significant heterogeneity of FNAB experience and performance reflected in the medical literature.
- Part of this heterogeneity results from indeterminate FNAB diagnoses, especially in normal or hyperplastic benign tissue.
- The number of indeterminate diagnoses in the many centers that do not report their results can be much higher than the 20 to 30% reported in the literature.
- Limited experience and technical skill are major factors that are responsible for this heterogeneity.

12.3 MULTINATIONAL VARIATIONS IN PHYSICIAN CONFIDENCE IN FNAB CYTOLOGICAL ANALYSIS SET THE STAGE FOR NEW TECHNIQUES

Uncertainties in the contemporary management of thyroid nodules jeopardize clinicians' confidence in the FNAB technique. As a result of these uncertainties, there has been a relative excess of excised benign thyroid nodules. Even in the age of CPG promoting the primacy of FNAB in the evaluation of a thyroid nodule, large multicenter reports in Europe and the United States have shown that FNAB was only used as the initial procedure in about half of thyroid nodules found to be malignant.[32–34] One of these studies[34] reported that among 448 FNABs from patients with histologically proven PTC, only 36% of the cytological diagnoses had been confirmed or suggested PTC (16% were reported as follicular neoplasia, 30% as normal cytologic findings, and approximately 15% as nondiagnostic). In this nationwide study involving approximately 100 laboratories, only 41% of the thyroid carcinoma cases were detected cytologically.[34] In approximately 33% of the 890 patients with thyroid carcinoma, FNAB was not even attempted.[21,34] These findings reinforce the claim that the same technique (FNAB cytology) that can provide excellent results in "the best hands" can appear scarcely accurate and unreliable when tested in "many hands." Ultimately these studies demonstrated low clinicians' confidence in FNAB, so that "it did not appear to be a diagnostic test that they could trust or would routinely order."[35]

A large survey on thyroid nodule management conducted by the European Thyroid Association in 1988 showed that FNAB was indicated as the first diagnostic procedure by 41% of the responders if they had to choose only one diagnostic test, and by only 25% when more than one test could be performed.[36] A subsequent European survey published in 1999 showed that FNAB was routinely used by 99% of the responders as a diagnostic test in the case of a palpable nodule.[37] However, 23% of the responders preferred surgical treatment to medical follow-up in spite of the benign FNAB diagnosis.[37] A further survey among American thyroidologists in 2000 showed that for the same index case (palpable nodule with benign FNAB diagnosis and without any clinical suspicion) as in the European survey, surgery was advocated by only 1.3%.[38] However, a contemporaneous study in a U.S. endocrine clinic reported that a third of all study patients with a solitary thyroid nodule did not have an FNAB.[39]

Quite different were the results of a survey among Danish endocrinologists: 65% of the responders suggested surgical excision of a palpable solitary thyroid nodule regardless of a benign FNAB cytology and absence of any clinical suspicions.[40] Finally, a recent survey of the German Society of Endocrinology reported in 2005 that only 18% of the responders suggested performance of FNAB routinely for a thyroid nodule.[41] So the question remains: Despite the presence of evidence-based CPG by credentialed thyroidologists from large professional medical societies,[42,43] why is FNAB utilization inconsistent?

12.3.1 Summary

- Clinicians' confidence in thyroid FNAB is low in some geographic areas.

- Inconsistent utilization of FNAB results from inherent uncertainties associated with the procedure despite published evidence-based recommendations.

12.4 LARGE-NEEDLE ASPIRATION BIOPSY (LNAB) AND MOLECULAR TUMOR MARKERS ARE EMERGENT TECHNOLOGIES THAT MAY IMPROVE THE PREOPERATIVE DIAGNOSIS OF THYROID NODULES

Several adjuvant techniques have been reported to improve the performance of thyroid FNAB.[44] Molecular tumor markers and LNAB histology are prime examples and can be complementary.[21,44]

12.4.1 Percutaneous LNAB

12.4.1.1 General Aspects

In recent years, a number of reports on large-needle biopsy techniques have been published. The American Association of Clinical Endocrinologists CPG for thyroid nodule evaluation[42] recently reported that "large-needle biopsy does not have a higher diagnostic accuracy than FNAB, is cumbersome and is associated with pain and, occasionally, severe bleeding. Currently, the use of large-needle biopsy in the routine management of thyroid nodules is not advised."[42] In contrast, Carpi et al.[44] have found the routine use of LNAB to be safe with the patient usually accepting combined LNAB and FNAB evaluations.

The general term *large-needle biopsy* (LNB) describes techniques that provide tissue cylinders or fragments for conventional histological evaluation. These techniques can be grouped into two methods defined as large-needle cutting biopsy (LNCB) or large-needle aspiration biopsy (LNAB).

The LNCB techniques (also called core-, coarse-, or cutting-needle biopsy) employ the largest needles: the 14-gauge, 2 3/8-inch or 3 1/8-inch Silverman needle or the 14 gauge, 3-inch Tru-Cut disposable needle is the most studied. The operator maintains sterility as a skin nick is performed to permit the insertion of the relatively large needle. LNCB is performed in ambulatory patients and provides a tissue cylinder for histological evaluation from nodules of about 20 mm or larger. LNCB does not typically require ultrasound (US) guidance.

The LNAB technique was first reported in 1930 and is very similar to that used for FNAB. No incision of skin is performed; skin is anesthetized as for FNAB. The syringe can contain heparin to prevent coagulation of blood around the tissue specimen. The needle is inserted into the nodule through the same skin puncture already made for the previous FNAB and is then rotated within the nodule so that the sharp end severs the tissue fragments, which are then aspirated into the barrel of the syringe (hence the term LNAB). A simpler method to obtain tissue fragments is to perform the same procedure as for FNAB, only more vigorously.

Needles of different sizes can be used according to the dimensions and consistency of the nodule. Needle sizes range from 16–20 gauge for the largest nodules (>35 mm) to 22 gauge for the smallest nodules (about 10 mm). LNAB provides tissue fragments of variable size, depending on the needle size and the nodule pathology. The tissue fragments are usually easily visible. Figure 12.1 shows tissue fragments obtained with LNAB from a palpable thyroid nodule. Both LNAB and LNCB can be performed with or without US guidance. Ultrasound-guided LNAB or LNCB with a biopsy gun (both employing a 18-gauge needle) have been reported.[44] Ultrasound guidance permits examination of small nodules, even smaller than 10 mm. LNAB obtains more tissue than FNAB and is safer than LNCB.[44]

Eight needles of different size (25, 23, 22, 21, 20, 18, 16, and 15 gauge) and a 14-gauge Tru-Cut needle were studied.[44] The internal and external diameters of these needles were measured by an optic microscope (Olympus AX 70, Tokyo, Japan). The corresponding cross-sectional area and volume (for a length of 20 mm) were calculated. For the 14-gauge Tru-Cut needle, the cutting section of the internal cylinder was measured. The internal and external cross section areas and volumes were

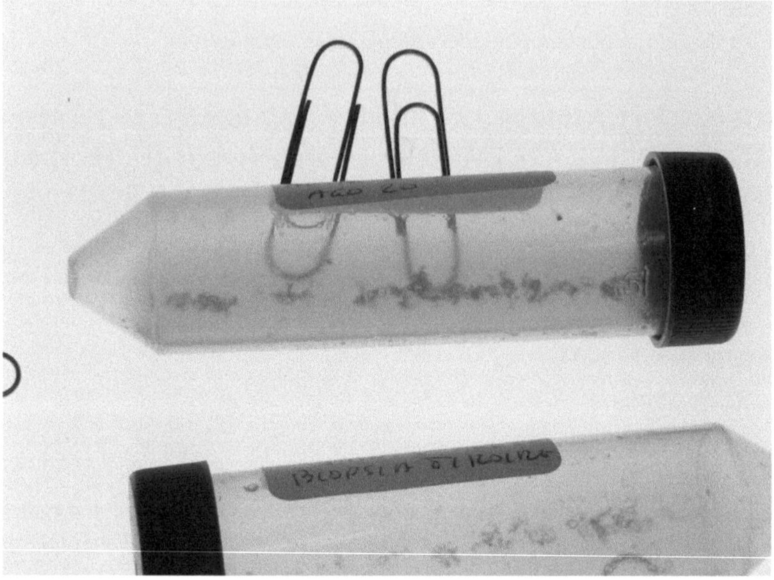

FIGURE 12.1 Tissue fragments obtained with LNAB from a palpable thyroid nodule.

related to the size of the tissue sampled and traumatized, respectively, by the needle. The internal *diameter* of the needles, with progressively increasing size, from 25–15 gauge, was from 0.3–1.65 mm (5-fold increase). In comparison, the external diameter of the same needles only increased from 0.5–1.85 mm (3.7-fold increase). The internal and external *cross section area* of these needles increased to a greater extent than the diameter. In fact, the internal cross section area increased from 0.07–2.14 mm^2 (30.6-fold increase), with the external cross-sectional area from 0.19–2.68 mm^2 (14-fold increase). In these needles, a progressive increase of the ratio between the internal and external diameters (from 0.6–0.9) and the corresponding cross-sectional areas (from 0.36–0.79) occurred. Additionally, the increase in the *volume* of the internal cylinder, corresponding to the amount of tissue sampled, was proportionately greater than the increase of the external cylinder, which corresponds to the tissue traumatized by the needle. The ratio between the internal and external cross-sectional area or volume values was greater in the 20- (0.47) or 18- (0.53) gauge needles than in the 14-gauge Tru-Cut needle (0.45). This analysis demonstrates the more favorable ratio of tissue sampled per tissue traumatized.[44] Furthermore, it provides insight as to why a recent review on the use of LNAB did not report clinically relevant complications.[42] Thus, based on studies representing a subset of the total clinical experience with FNAB and LNAB, there is evidence to support the use of LNAB in situations where improved sample adequacy is imperative.[21,44] Specifically, among patients found to have FNAB samples with inadequate cellularity or an indeterminate follicular cytological finding, LNAB has facilitated a preoperative diagnosis (Table 12.1).

12.4.1.2 Cytological Sample Inadequacy: LNAB vs. FNAB

LNAB may reduce the number of cytologically inadequate findings.[45] The first FNAB procedure yields inadequate results in 28%, decreasing to 6% after four aspirations.[46] A retrospective study performed on 261 nodules with inadequate FNAB cytology showed that LNAB provides a diagnosis in 130 (49.8%).[44] Furthermore, among the patients with initial inadequate cytological findings from FNAB and LNAB, 61 were subjected to repeated FNAB and 36 to repeated LNAB.[44] More than 60% of the nodules on which FNAB was repeated achieved a cytological diagnosis, whereas more than 80% of the nodules on which LNAB was repeated achieved a histological diagnosis.[45]

TABLE 12.1
LNAB Can Reduce the Number of the Inadequate or Indeterminate Follicular FNAB Findings: Literature Report

		Inadequate at FNAB		Indeterminate Follicular at FNAB	
Reference	LNB Technique	Total, N	Adequate at LNB, N	Total, N	Hyperplastic at LNB, N
107	LNAB or LNCB	n.r.	228	n.r.	69
59	LNAB			260	25
2	LNAB			245	115
45	LNAB	261	130		
55	LNCB			78	35
28	LNCB[a]	11	10		

Note: N = number, n.r. = not reported, FNAB = fine-needle aspiration biopsy, LNB = large-needle biopsy, LNAB = large-needle aspiration biopsy, LNCB = large-needle coarse or cutting-needle biopsy.
[a] Ultrasound-guided LNCB.

In a series of 483 excised nodules that were preoperatively examined by FNAB alone or in combination with LNAB, only 28 (5.8%) were found to have inadequate samples for cytology or histology.[2] Four malignant nodules were found among these excised nodules in two reports[2,47] without a correct preoperative FNAB diagnosis. At postoperative histology, the incidence of malignancy in the nondiagnostic FNAB category varied from 2–19%.[45,48–50] Repetition of FNAB cytology has been used to reduce this frequency.[51–53] LNAB has also been used on these nodules for the same purpose.[2] A more recent paper on about 100 thyroid nodules, which were difficult to palpate or with previously unsatisfactory freehand FNAB cytology, has reported that a LNB technique with 20-gauge needles under US guidance reduces nondiagnostic FNAB cytology from 46.8–15.6%.[28]

12.4.1.3 Indeterminate Follicular Cytology: LNAB vs. FNAB

As mentioned above, the task of discriminating preoperatively hyperplastic and neoplastic (follicular adenoma and FTC) lesions is problematic. A recent study of Hirokawa et al.[24] analyzed the factors contributing to observer variation among eight pathologists from different institutions who reviewed the same postoperative studies of 21 encapsulated follicular lesions. The frequency of diagnosis of adenomatous goiter among Japanese pathologists (31%) was considerably higher than that among American pathologists (6%). In contrast, the frequency of diagnosis of PTC among American pathologists (25%) was considerably higher than that among Japanese pathologists (4%). The analysis revealed three main factors affecting observer variation: (1) interpretation of the significance of microfollicles intimately related to capillaries within the tumor capsule, (2) evaluation of nuclear clearing indicative of PTC, and (3) absence of clear morphologic criteria for separation of adenomatous goiter and follicular adenoma.[24] With this premises, it is not surprising that the postoperative incidence of follicular neoplasms (follicular adenoma plus FTC) in various series of nodules with follicular structure at FNAB varies widely, from 6–80%.[54] Moreover, the relative frequencies of the cancer histotypes observed postoperatively in these nodules are very different.[18,19,54]

Overall, hyperplasia, adenoma, and carcinoma each occur in about one-third of cases,[18,19,46] with follicular variant of PTC (FVPTC) being the most represented.[18,19] These data underscore the importance of follicular architecture in the diagnosis of these nodules. Recent data[11,13,27] have led to our proposal that the LNAB finding of a pure microfollicular structure in a nodule with indeterminate follicular structure at FNAB includes a higher probability of postoperative follicular neoplasia (adenoma or FTC).[2,6,7]

Lo Gerfo[55] showed in 132 encapsulated follicular lesions that the preoperative LNAB finding of pure microfollicular or macrofollicular structure was associated with a postoperative diagnosis of carcinoma in 23 or 0% of cases, respectively. However, he did not report the detailed preoperative FNAB diagnosis from these nodules. Kung et al.[56] suggested that FNAB cell blocks might better distinguish nodular goiter from follicular neoplasm by showing architectural details.[56] Carpi et al.[2,57] and Lo Gerfo[55] reviewed preoperative LNAB histology of a large series of palpable thyroid nodules, which were microfollicular by FNAB, and reported that approximately 50% of these histological results showed important macrofollicular components. In addition, "nuclear clearing," which represents an important feature of PTC, was easily appreciated in the preoperative LNAB histologies from formalin-fixed paraffin-embedded cell blocks.[58]

Two hundred sixty nodules that were diagnosed as microfollicular by FNAB were also examined by LNAB: inadequate specimens were obtained in 17% of cases; pure microfollicular structure was confirmed in 35% of the nodules; and a macrofollicular component suggesting a benign hyperplastic lesion was confirmed in the remaining 48% of cases.[59] Seventeen nodules that were found to be microfollicular with FNAB cytological analysis and micromacrofollicular (both small and large follicles present) with LNAB histological analysis were excised, and the postoperative results were benign in all cases.[59] Twenty-five nodules diagnosed as microfollicular with both FNAB and LNAB were excised, and the postoperative diagnoses were benign in 80% of cases (20 of 25 nodules) or malignant in 20% of cases (5 of 25 nodules).[59] These results were confirmed by a later study,[6] in which 114 nodules with preoperative FNAB diagnosis of microfollicular nodule or suspected cancer also had a preoperative LNAB performed. The prevalence of cancer by surgical pathology was 22% among the microfollicular and 4% among the micromacrofollicular nodules by LNAB ($p < 0.25$).[6] A follow-up study found that the preoperative diagnostic combination of microfollicular nodule or suspected cancer by FNAB and microfollicular nodule by LNAB had a sensitivity of 87.5% for diagnosis of follicular adenoma.[7]

More recently, the diagnostic performance of preoperative LNAB histology was clarified in a study involving 182 thyroid nodule patients.[60] One hundred thirty-two patients had a preoperative FNAB diagnosis of microfollicular nodule (atypical cells in 50 and suspected carcinoma in 82) and had surgery. Among the 50 nodules with atypical cells by FNAB, the preoperative LNAB finding was inadequate in 8, benign in 15, microfollicular in 20, microfollicular with atypical cells in 5, and suspected cancer in 2.[60] The postoperative carcinoma incidences in these 50 nodules with atypical cells by FNAB were benign LNAB histology diagnosis, 0%; microfollicular LNAB histology, 10%; microfollicular with atypical cells LNAB histology, 60% ($p = 0.008$); and suspected cancer LNAB histology, 100% ($p = 0.007$).[59] Among the 82 nodules with suspected carcinoma by FNAB, the preoperative LNAB finding was inadequate in 21, benign in 21, microfollicular in 15, microfollicular with atypical cells in 15, and suspected cancer in 10. The postoperative carcinoma incidences in these 82 nodules with suspected carcinoma by FNAB were benign LNAB histology, 14%, and suspected carcinoma by LNAB histology, 80% ($p = 0.0007$).[60] It is well known that some adenomas or differentiated cancers will harbor a macrofollicular structure; however, they are relatively rare and their malignant potential is very low.[61–66] In addition, even though capsular and vascular invasion, which characterizes FTC, do not appear in most LNAB histological analyses,[61] LNAB can still reduce the number of indeterminate FNAB diagnoses and avoid unnecessary surgeries for benign thyroid nodules.

12.4.1.4 Summary

LNAB can reduce the number of surgical operations for benign nodules because:

- It diagnoses as benign about 50% of the nodules with inadequate or indeterminate follicular FNAB findings.
- These effects are explained by the larger needle size (LNAB vs. FNAB), which is more adequate for the investigation of these nodules.

12.5 MOLECULAR TUMOR MARKERS FOR IMPROVING THE PREOPERATIVE DIAGNOSIS OF BENIGN THYROID NODULES

12.5.1 General Aspects

Recent advances in molecular biology have produced different tests for discriminating benign from malignant thyroid tumors, particularly those with follicular structure (hyperplastic nodules, follicular adenomas, FVPTC, FTC, and follicular tumors with uncertain malignant potential (FT-UMP)).[67–69] These molecular tests would complement the use of FNAB cytology and LNAB histology. The burgeoning interest in molecular medicine for thyroid cancer reflects an increased knowledge base regarding pathogenesis, biological treatments, and prognosis.[66] More specifically, the reported utility of these molecular markers in preoperative FNAB-derived biological material has generated remarkable enthusiasm for more accurate diagnoses. This could optimize medical and surgical decision making in the comprehensive multimodality care of the thyroid cancer patient.

Fifty potential markers of thyroid tumors have been analyzed, and five of these (thyroid peroxidase (TPO), telomerase, gal-3, RET-PTC, and P_{53}) were investigated due to a relatively good accuracy for detection of thyroid cancer in nodules with indeterminate FNAB findings.[68] The expression of many individual genes[70,71] and seven microRNA expression profiles[72] have been analyzed on FNAB specimens.

In addition, new applications of microarray technology have been reported: first, the combination of chromatin immunoprecipitation (ChIP) with hybridization of microarrays (ChIP-on-chip) to explore sites of DNA-protein interaction across the whole genome, and second, the analysis of the methylation status of CpG islands in promoter regions. On the basis of these high-resolution techniques and emerging biotechologies on the horizon, novel molecular mechanisms of thyroid cancer are likely to be discovered and impact routine care.[69]

12.5.2 Limits of Molecular and Genetic Testing

Notwithstanding the evidence surrounding molecular markers of thyroid cancer, expert opinion still asserts that "until a specific marker or panel of markers is devised which can effectively distinguish between benign and malignant follicular lesions of the thyroid in FNAB specimens, morphology remains the gold standard"[18] or "additional diagnostic markers of malignancy are greatly needed."[73] At present, clinical practice has not yet been significantly modified by translation of molecular and genetic studies to routine pathology practice.[67,69] So, what is the reason for this delay?

The antibodies available for many molecular markers demonstrate considerable variability in sensitivity and specificity.[74,75] No doubt, the scarcity of many FNAB specimens will tax any assay.[19] Some markers, like gal-3, require an optimal substrate, such as formalin-fixed and paraffin-embedded cell block preparations, rather than conventional FNAB smears.[74] However, many thyroid aspirates have low cellularity and are not suitable for cell block immunochemistry.[76] The amount and quality of RNA available from FNAB are also limiting. Therefore, an assay based on a limited number of differentiating genes, identified by sophisticated algorithms in comparative studies, may be applicable to FNAB specimens.[77,78] Nevertheless, in contrast to the analysis of tumor-specific mutations, the approach of quantitatively measuring RNA markers is more susceptible to potential limitations of FNAB, such as limited and variable numbers of follicular cells obtained in each biopsy and the potential contamination by other cell types, such as macrophages[79] or activated lymphocytes in patients with lymphocytic thyroiditis. This may be addressed with a correction for mRNA yield (e.g., by measuring a housekeeping gene like β-actin) and thyroid specificity of mRNA extracted from an FNA sample (e.g., by measuring a thyroid-specific gene like thyroglobulin (TG)).[69]

More quantitative molecular methods lack a very important element: the morphology of the specimen. This lack exacerbates problems of sensitivity and specificity since only a small fraction of the FNAB sample consists of thyroid epithelial cells.[19] Chen et al.[80] reported that for microRNA analysis, paraffin sections were a better substrate than FNAB samples. Siddiqui et al.[81] reported that FNAB slides older than 3 years were inadequate for evaluation of telomerase reverse transcriptase gene expression. Samija et al.[82] reported that 13%, in the case of one puncture, or 21%, in the case of two punctures, of the FNAB samples were inadequate for the reverse transcriptase polymerase chain reaction (RT-PCR) expression analysis of glyceraldehyde-3-phosphate deydrogenase and TG.

Some tests are not yet ready for widespread clinical use due to the limited availability of the specific antibody to the molecular marker[68] or its elevated cost and complexity, as in the case of the gene expression profiling techniques.[69] Results from these studies performed on FNAB specimens are limited due to paucity of cytological material, which limits morphological diagnoses and biological marker determinations.[81–83]

12.5.3 Summary

- Many molecular and genetic tests have been recently developed and proposed for improving diagnostic accuracy of preoperative FNAB cytology.
- A major limiting factor is insufficient FNAB substrate.

12.5.4 Focus on Galectin-3

12.5.4.1 General Aspects

The development of preoperative molecular analyses for the selection of patients for thyroid nodule surgery will require technical standardization, a tissue substrate larger than FNAB, and large prospective clinical trials. To this end, the best performing test and substrate should be identified. Due to the limitation outlined in the section above, molecular markers are preferred to gene expression analysis. Among these markers, TPO, telomerase, and gal-3 have been selected.[68] Very recently, TPO and gal-3 immunostaining in differentiated thyroid carcinoma has been correlated to the biological aggressiveness of the carcinoma.[84]

Gal-3 has been among the few molecular markers subjected to wide multicenter studies.[74,85] Gal-3 is a member of the β-galactoside-binding family of lectins and has been implicated in various biologic processes, such as cell adhesion, cell recognition, proliferation, differentiation, immunomodulation, organization of the extracellular matrix, and metastasis.[86,87] Cell surface gal-3 mediates cell-cell and cell-matrix interaction, nuclear gal-3 mediates pre-mRNA splicing, and cytoplasmatic gal-3 mediates antiapoptosis and neoplasia.[87,88] Alteration of gal-3 expression has been reported in a variety of tumors, including colon, breast, stomach, brain, and thyroid.[89–92]

In thyroid malignancy, tissue gal-3 is overexpressed, while it is not usually expressed in normal thyroid tissue. Galectin-3 is therefore considered a marker of early malignant transformation, nearly all PTC, and most to all (82–100%) FTC.[68,75] Some nonthyroid cells (squamous cells, fibroblast, and inflammatory cells) can also express gal-3 and account for the small false positive rate.

12.5.4.2 Studies on Preoperative Percutaneous Biopsy

There are seven exemplary studies on preoperative detection of gal-3 in thyroid tissue:[58,74,79,85,93,94] six from a single institution and one[74] multicenter study (Table 12.2). Different techniques were used in these studies, with the most common being FNAB formalin-fixed paraffin-embedded cell blocks and immunochemistry. The size of the series with indeterminate follicular nodules at FNAB varied from 34–465 cases. The specificity for gal-3 to diagnose thyroid nodules with indeterminate follicular cytology by FNAB varied in the studies. The highest value (97.2%) was from a single institution study using LNAB substrate and biotin-free immunoperoxidase staining.[58,93] The lowest value (74.8%) used the RT-PCR method, which was associated with false positive results due to macrophages and Hürthle cells.[79]

TABLE 12.2
Selection of Seven Studies on Preoperative Detection of Galectin-3 in Thyroid Nodule Tissue

		Galectin-3 Determination		Accuracy Data				
Study: Reference	FNAB Diagnosis, N	Substrate	Technique	Sens %	Spec %	PPV %	NPV %	Postoperative Findings, N
93, r, s.i.	Follicular indeterminate, 125	FNA cell blocks (alcool fixed)	Immunochemistry	92	94			FTC, 33 PTC, 42 Adenoma, 50
85, r, s.i.	Benign and malignant, 165	Cell blocks formalin fixed paraffin embedded	Immunochemistry	90	94	94	91	Hyperplasia, 32 Chronic thyroiditis, 4 Adenoma, 46 Ca, 83
94, p, s.i.	Benign, 63 Malignant,17 Follicular indeterminate, 34	FNA slide						PTC, 9 FTC, 3 Adenoma, 32
79, r, s.i.	Benign, 113	FNAB cellular RNA	Total RT-PCR		74.8			Nodular goiter, 79 Adenoma, 17 Chronic thyroiditis, 27
58, r, s.i.	Follicular indeterminate, 85	LNAB paraffin-embedded blocks, formalin fixed	Immunochemistry biotin-free immunoperoxidase staining	91.6	97.2			Hyperplasia, 54 Differentiated cancer, 12 Adenoma, 19
74, p, mc	Follicular indeterminate, 465	FNAB cell blocks •	Immunochemistry biotin-free immunoperoxidase staining	78	93	82	91	Malignant, 131 Benign, 302 Borderline, 33
95, p, s.i.	All types, 245	LNAB paraffin-embedded blocks formalin fixed	Immunochemistry biotin-free immunoperoxidase staining	100	80			Malignant, 10 Hyperplasia, 19 Adenoma, 11

Note: Rt-PCR = reverse transcriptase polymerase chain reaction, N = number, p = prospective, r = retrospective, s.i. = single institution, mc = multicenter, Sens = sensitivity, Spec = specificity, PPV = positive predictive value, NPV = negative predictive value, Ca = carcinoma, PTC = papillary thyroid carcinoma, FTC = follicular thyroid carcinoma; • = one of the 21 study centers used LNAB-derived blocks.

Another important issue is gal-3 expression in preoperative percutaneous biopsy specimens from encapsulated follicular tumors harboring minimal morphologic features of malignancy, such as follicular adenoma, minimally invasive FTC, and FT-UMP. Follicular adenomas are benign encapsulated tumors that exhibit a variety of morphologic patterns and, in some cases, cannot be easily differentiated from minimally invasive FTC and the FVPTC.[61] In Table 12.2, the proportion of the diagnoses of follicular adenoma among all the postoperative diagnoses varied from 15 to 40%. In one study, gal-3 expression was observed in 3 (50%) of 6 adenomas studied.[58] In a large, prospective, multicenter Italian study on 465 nodules, 19 (11%) gal-3 positive follicular adenomas, among 176 tested, were found.[74] Matesa et al.[79] reported gal-3 expression using RT-PCR in preoperative FNAB from 5 of 17 (29%) follicular adenomas. Saggiorato et al.[93] reported a lower proportion (6%) of adenomas that were positive for gal-3 on preoperative cytology. However, they defined positive gal-3 immunostaining when it occurred in at least 10% of neoplastic cells.[93] This contrasts with the above studies, in which a smaller amount of gal-3 in follicular thyroid cell was considered a positive finding. According to the hypothesis that gal-3 expression is a marker of the thyroid cell transformation toward malignancy,[85,96] the adenomas with focally positive gal-3 may in fact be tumors undergoing malignant transformation or in other premalignant stages.[85,96]

Minimally invasive FTC is a grossly encapsulated tumor. The pattern of growth usually resembles that of an adenoma of embryonal, fetal, or atypical type. It has been suggested that some of these cases represent malignant transformation of an adenoma.[61] Since blood vessel invasion is almost never evident grossly,[61] diagnosis relies on minimal, but entire thickness, infiltration of the capsule.[86] In a retrospective series, 16 of 17 preoperative FNAB cell blocks from this histologically verified tumor type expressed gal-3.[86]

Follicular tumors of uncertain malignant potential are capsulated tumors with questionable capsular invasion and without PTC type nuclear changes.[61,97] In two series, all,[92] or almost all,[85] of the FT-UMPs examined expressed gal-3 in preoperative FNAB cell blocks[92] or postoperative tissue sections.[85] Thus, the data from these three particular follicular tumor types (adenoma, minimally invasive FTC, and FT-UMP) suggest that gal-3 increases its expression proportionally to the morphologic probability or evidence of malignancy.

12.5.4.3 Galectin-3 Immunodetection on LNAB: Facts and Perspectives

LNAB substrate is more abundant and adequate for gal-3 immunodetection than FNAB substrate.[74,98] The residual biological material available in 150 FNAB-derived cell blocks and 200 LNAB-derived histological blocks after gal-3 determination was compared. Only one or two sections were still available for further study in only 10% of the FNAB cell blocks, whereas > 5 sections could be obtained from 97% of the LNAB blocks.[44] As a result of the increased tissue obtained with LNAB for gal-3 immunodetection, there was more likelihood of having adequate formalin-fixed cell blocks rather than isolated cells, less nonspecific gal-3 immunoreactivity, and hence a lower false positive rate.[74] Furthermore, LNAB-derived cell block preparations allow a comparative immunocytochemical assessment, on the same cytological slides, of different antigens associated with thyroid cancer.[74]

The most recent clinical trials agree on the high accuracy and specificity rates of preoperative gal-3 determinations despite different techniques[74,79] or different substrates[58,74,79] used. Nevertheless, since almost all of the studies selected cases with adequate preoperative substrate, the impact of the cell paucity in FNAB on the routine clinical use of gal-3 has not yet been sufficiently evaluated.[76]

At present, neither LNAB nor gal-3 has been included in official CPG of scientific societies. Tissue gal-3 determination has been briefly discussed by the American Thyroid Association, but without any involvement in a clinical algorithm.[99] Recent guidelines considered gal-3 for patients with indeterminate cytology on FNAB.[100] In 1999, we proposed a protocol for diagnoses and management of palpable thyroid nodules in which LNAB was suggested when there was an FNAB finding of indeterminate follicular (pure microfollicular) nodule or an inadequate finding.[59] Saggiorato et al.[93] also proposed a clinical algorithm in which an indeterminate follicular FNAB finding was followed by gal-3 immunodetection on FNAB material. If the gal-3 finding was positive, then surgical excision was recommended. If the gal-3 finding was negative, then detection of cytokeratin CK 19 or HBME-1 on FNAB was recommended according to the oncocytic or nononcocytic FNAB cytology picture, respectively. This decision tree was based on a retrospective analysis of 125 FNAB cell blocks from cytologically indeterminate follicular neoplasms evaluated for gal-3, HBME-1

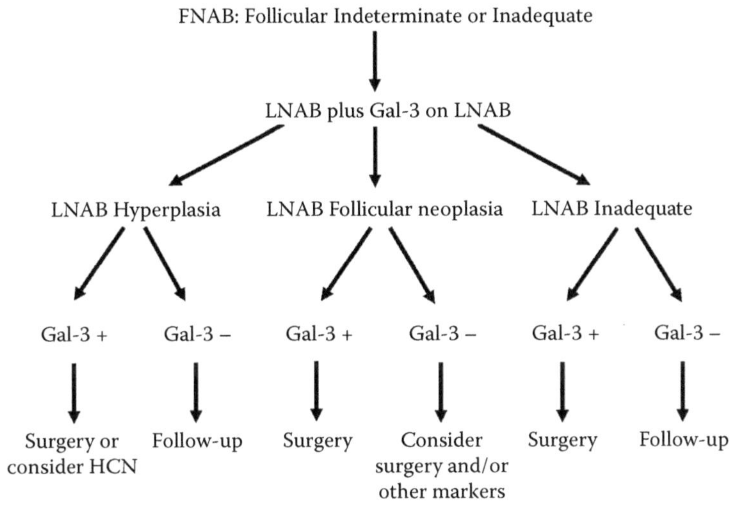

FIGURE 12.2 Decision tree to evaluate inadequate or indeterminate follicular FNAB cytology findings. LNAB Inadequate indicates that in the LNAB specimen there are not enough follicles for histological diagnosis, while there are sufficient follicular cells for galectin-3 immunodetection. HCN = Hürthle cell nodule.

thyroperoxidase, cytokeratin-19. and keratin-sulfate. Importantly, there has yet to be a clinical study that prospectively manages thyroid nodule patients with a protocol including preoperative molecular tumor marker evaluation as decisional elements. Since one or more molecular markers can be considered for decision trees, and since FNAB can be an insufficient substrate, we recommended LNAB as a better source of substrate.[44] Furthermore, if we consider that LNAB can be obtained from many nodules with inadequate FNAB, there will be more nodular tissue to provide a morphologic diagnosis and allow gal-3 testing to establish a diagnosis of benign nodule. Therefore, we believe a rational decision tree and clinical algorithm (Figure 12.2) recommends LNAB for histology and gal-3 immunodetection in cases of inadequate or indeterminate follicular FNAB cytology. This approach can increase the number of the benign nodules diagnosed preoperatively and permits tissue storage for further analysis if results remain indeterminate.

The probability that a microfollicular or indeterminate follicular cytology at FNAB and hyperplastic histology at LNAB results as a malignancy postoperatively varies from 0 to 11%.[2,6,7,58,60] However, when a nodule with this FNAB + LNAB diagnosis is also negative by gal-3 immunodetection on LNAB, the probability of postoperative malignancy becomes zero.[58] On the other hand, the diagnostic combination of indeterminate follicular at FNAB + follicular neoplasm at LNAB was followed by a postoperative finding of malignancy in 15–20% of cases.[6,58–60] However, when a nodule with this FNAB + LNAB diagnosis is positive by gal-3 immunodetection on LNAB, the probability of postoperative malignancy becomes 100%.[58]

If the operator is experienced enough, the ambulatory procedures of FNAB and LNAB can be performed simultaneously at the initial evaluation of a patient with a thyroid nodule, and the decision of performing or not performing gal-3 immunodetection can be made after having examined the cytological and histological specimens. This protocol can reduce the number of patients recalled to the clinic.[2,58] Prospective studies on this protocol, as well as other improvements in the above decision tree, including newer molecular methods and US guidance for LNAB in addition to FNAB, are in progress.

12.5.4.4 Summary

- Among the tissue molecular markers, gal-3 is the most widely investigated for improving accuracy of preoperative diagnosis of thyroid nodules.

- LNAB provides a much more abundant and adequate substrate for gal-3 immunodetection than FNAB.
- Decision trees and clinical algorithms on the evaluation of thyroid nodules with inadequate or indeterminate follicular FNAB findings can include LNAB and gal-3 immunodetection on LNAB.

12.6 DISCUSSION

Ideally, the objective of preoperative selection of thyroid nodule patients is to excise as few nodules as possible without missing malignant nodules among the nonoperated patients. At present, only two large studies[2,101] have reported a proportion of operated thyroid nodule patients lower than 10%, with a 3% proportion of cancers among all the patients examined. One study,[101] which is relatively dated, included 1479 consecutive patients with a palpable thyroid nodule and recorded 68 operations, 4.5% of all the examined patients. In contrast to other series, the authors reported that at FNAB, benign diagnoses were 97% and malignant plus suspect diagnoses were 3%; inadequate findings were not considered at all. Another study,[2] more recent, included 5403 patients with palpable thyroid nodules examined with FNAB cytology, of which 1668 were examined with FNAB plus LNAB histology. This study reported that among all the FNAB findings, 11% were inadequate, 75% benign, 4.3% malignant or suspicious, and 7.7% indeterminate. There were 483 patients (8.9%) undergoing surgery with a cancer rate of 3.1%. This finding highlights the assertion that in a large controlled study, where LNAB is added to the preoperative evaluation with FNAB, a relatively low (<10%) cancer rate is demonstrable.

Ultrasonography is also considered to provide valuable information that can be added to the information gleaned from an FNAB-derived cytological examination. A large study[49] on 4986 thyroid nodule patients reported that FNAB selected 10.7% of the patients for surgery and compared these results with those obtained in another, similar series of 4697 patients examined with UG-FNAB. The proportion of adequate FNAB findings was higher 91.5% vs. 85.9%, and diagnostic accuracy was slightly higher, 75.9% vs. 72.6%, when US guidance was employed for FNAB. However, the proportion of all the patients who were operated on was similar (about 11.5%) with or without US guidance. This is due to the fact that diagnostic specificity was also similar between the two series, 70.9% and 68.8%.

The role of US in the nodules with inadequate or indeterminate follicular FNAB findings has also been investigated. The proportion of inadequate FNAB findings was 21.5 or 3.4% in a study[102] that investigated 292 FNAB specimens obtained with clinical guidance or 600 specimens from UG-FNAB, respectively. However, in the same study, the proportion of indeterminate FNAB results was similar, 12 and 11.9%, in the two series.[106] A retrospective study on 329 thyroid nodules confirmed that for 64 postoperatively confirmed follicular neoplasms (follicular adenoma and FTC), US diagnosis was unreliable.[103]

The occurrence of more than one particular US finding in some follicular neoplasms has been reported to have predictive value in recent studies.[104,105] However, a recent expert opinion has reported that "in follicular lesions, conventional echographic patterns were found to be of minor relevance for predicting carcinoma."[106] Therefore, unlike the use of LNAB and gal-3 immunodetection, the use of US has not had significant impact on the reduction of unnecessary thyroid surgeries for nodules with indeterminate FNAB results.

12.7 CONCLUSION

Molecular tumor markers represent an emerging tool for improved preoperative selection and characterization of thyroid nodules. LNAB is a safe technique that can increase the specificity of FNAB cytological findings by providing an optimal substrate for histological analysis and molecular tumor marker determinations. It is realistic to envision the incorporation of LNAB and various molecular tumor marker determinations into official professional medical society-sponsored CPG within the next 5 years.

REFERENCES

1. Gharib H, Goellner JR. 1993. Fine-needle aspiration biopsy of the thyroid: An appraisal. *Ann Intern Med* 118:282–89.
2. Carpi A, Ferrari E, Toni MG, Sagripanti A, Nicolini A, Di Coscio G. 1996. Needle aspiration techniques in preoperative selection of patients with thyroid nodules: A long term study. *J Clin Oncol* 14:1704–12.
3. Yang J, Schnadig V, Logrono R, Wasserman PG. 2007. Fine-needle aspiration of thyroid nodules: A study of 4703 patients with histologic and clinical correlations. *Cancer* 111:306–15.
4. Baloch ZW, Fleisher S, LiVolsi VA, Gupta PK. 2002. Diagnosis of "follicular neoplasm": A gray zone in thyroid fine-needle aspiration cytology. *Diagn Cytopathol* 26:41–44.
5. Miller B, Burkey S, Lindberg G, Snyder WH III, Nwariaku FE. 2004. Prevalence of malignancy within cytologically indeterminate thyroid nodules. *Am J Surg* 188:459–62.
6. Carpi A, Nicolini A, Sagripanti A, Righi C, Fabris FM, Di Coscio G. 2000. Large needle aspiration biopsy for preoperative selection of palpable thyroid nodules diagnosed by fine-needle aspiration as a microfollicular nodule or suspected cancer. *Am J Clin Pathol* 113:872–77.
7. Carpi A, Nicolini A, Sagripanti A, Menchini Fabris F, Righi C, Romani R, et al. 2002. Large-needle aspiration biopsy for the preoperative selection of follicular adenoma diagnosed by fine-needle aspiration as a microfollicular nodule or suspected cancer. *Am J Clin Oncol (CCT)* 2:209–12.
8. Gharib H, Goellner JR, Zinsmeister AR, Grant CS, Van Heerden JA. 1984. Fine-needle aspiration biopsy of the thyroid. The problem of suspicious cytologic findings. *Ann Intern Med* 101:25–28.
9. Goldstein RE, Netterville JL, Burkey B, Johnson JE. 2002. Implications of follicular neoplasms, atypia, and lesions suspicious for malignancy diagnosed by fine-needle aspiration of thyroid nodules. *Ann Surg* 235:656–62.
10. Hall TL, Layfield LJ, Philippe A, et al. 1989. Sources of diagnostic error in fine needle aspiration of the thyroid. *Cancer* 63:718–25.
11. DeMay RM. 1995. *The art & science of cytopathology aspiration*, 713–78. Chicago: American Society of Clinical Pathologists.
12. Carling T, Udelsman R. 2005. Follicular neoplasms of the thyroid: What to recommend. *Thyroid* 15:583–87.
13. Gutman PD, Henry M. 1998. Fine needle aspiration cytology of the thyroid. *Clin Lab Med* 18:461–82.
14. Kini SR. 1984. Differential diagnosis in cytopathology of thyroid. In *Clinical and pathological advances in thyroid tumors*, ed. A Carpi, 23–35. Pisa: ETS.
15. Basu D, Jayaram G. 1992. A logistic model for thyroid lesions. *Diagn Cytopathol* 8:23–27.
16. Solomon D. 1993. Fine needle aspiration of the thyroid: An update. *Thyroid Today* XVI:1–9.
17. Greaves TS, Olvera M, Florentine BD, Raza AS, Cobb CJ, Tsao-Wei DD, Groshen S, Singer P, Lopresti J, Martin SE. 2000. Follicular lesions of thyroid: A 5-year fine-needle aspiration experience. *Cancer* 90:335–41.
18. Deveci MS, Deveci G, LiVolsi VA, Baloch ZW. 2006. Fine-needle aspiration of follicular lesions of the thyroid. Diagnosis and follow-up. *Cytojournal* 3:9.
19. Segev DL, Clark DP, Zeiger MA, Umbricht C. 2003. Beyond the suspicious thyroid fine needle aspirate. A review. *Acta Cytol* 47:709–22.
20. Mowschenson PM, Hodin RA, Wang HH, Upton M, Silen W. 1994. Fine-needle aspiration of normal thyroid tissue may result in the misdiagnosis of microfollicular lesions. *Surgery* 116:1006–9.
21. Carpi A, Di Coscio G, Iervasi G, Antonelli A, Mechanick J, Sciacchitano S, Nicolini A. 2008. Thyroid fine needle aspiration: How to improve clinicians' confidence and performance with the technique. *Cancer Lett* 264:163–71.
22. Clary KM, Condel JL, Liu Y, Johnson DR, Grzybicki DM, Raab SS. 2005. Interobserver variability in the fine needle aspiration biopsy diagnosis of follicular lesions of the thyroid gland. *Acta Cytol* 49:378–82.
23. Stelow EB, Bardales RH, Crary GS, Gulbahce HE, Stanley MW, Savik K, Pambuccian SE. 2005. Interobserver variability in thyroid fine-needle aspiration interpretation of lesions showing predominantly colloid and follicular groups. *Am J Clin Pathol* 124:239–44.
24. Hirokawa M, Carney JA, Goellner JR, DeLellis RA, Heffess CS, Katoh R, Tsujimoto M, Kakudo K. 2002. Observer variation of encapsulated follicular lesions of the thyroid gland. *Am J Surg Pathol* 26:1508–14.
25. Cersosimo E, Gharib H, Suman SJ, et al. 1993. "Sospicious" thyroid cytologic findings: Outcome in patients without immediate surgical treatment. *Mayo Clin Proc* 68:343–48.
26. Caraway NP, Sneige N, Samann NA. 1993. Diagnostic pitfalls in thyroid fine-needle aspiration: A review of 394 cases. *Diagn Cytopathol* 9:345–50.

27. Geisinger KR, Raab SS, Stanley MW, Silverman JF, Abati A. 2004. Thyroid gland fine needle aspiration. In *Modern cytopathology*, ed. RR Geisinger, SS Raab, MW Stanley, JF Silverman, A Abati, 731–69. Philadelphia: Churchill Livingstone.
28. Mehrotra P, Hubbard JG, Johnson SJ, Richardson DL, Bliss R, Lennard TW. 2005. Ultrasound scan-guided core sampling for diagnosis versus freehand FNAC of the thyroid gland. *Surgeon* 3:1–5.
29. Orija IB, Hamrahian AH, Reddy SS. 2004. Management of nondiagnostic thyroid fine-needle aspiration biopsy: Survey of endocrinologists. *Endocr Pract* 10:317–23.
30. Cusick EL, MacIntosh CA, Krukowski ZH, Williams VM, Ewen SW, Matheson NA. 1990. Management of isolated thyroid swellings: A prospective six year study of fine needle aspiration cytology in diagnosis. *BMJ* 301:318–21.
31. Burch HB, Burman KD, Reed HL, Buckner L, Raber T, Ownbey JL. 1996. Fine needle aspiration of thyroid nodules. Determinants of insufficiency rate and malignancy yield at thyroidectomy. *Acta Cytol* 40:1176–83.
32. Hölzer S, Reiners C, Mann K, Bamberg M, Rothmund M, Dudeck J, Stewart AK, Hundahl SA. 2000. Patterns of care for patients with primary differentiated carcinoma of the thyroid gland treated in Germany during 1996. U.S. and German Thyroid Cancer Group. *Cancer* 89:192–201.
33. Hundahl SA, Cady B, Cunningham MP, Mazzaferri E, McKee RF, Rosai J, Shah JP, Fremgen AM, Stewart AK, Hölzer S. 2000. Initial results from a prospective cohort study of 5583 cases of thyroid carcinoma treated in the United States during 1996. U.S. and German Thyroid Cancer Study Group. An American College of Surgeons Commission on Cancer Patient Care Evaluation study. *Cancer* 89:202–17.
34. Giard RW, Hermans J. 2000. Use and accuracy of fine-needle aspiration cytology in histologically proven thyroid carcinoma: An audit using a national pathology database. *Cancer* 90:330–34.
35. Cramer H. 2000. Fine-needle aspiration cytology of the thyroid: An appraisal. *Cancer* 90:325–29.
36. Baldet L, Manderscheid JC, Glinoer D, Jaffiol C, Coste-Seignovert B, Percheron C. 1989. The management of differentiated thyroid cancer in Europe in 1988. Results of an international survey. *Acta Endocrinol* 120:547–58.
37. Bennedbaek FN, Perrild H, Hegedüs L. 1999. Diagnosis and treatment of the solitary thyroid nodule. Results of a European survey. *Clin Endocrinol* 50:357–63.
38. Bennedbaek FN, Hegedüs L. 2000. Management of the solitary thyroid nodule: Results of a North American survey. *J Clin Endocrinol Metab* 85:2493–8.
39. Tangpricha V, Hariram SD, Chipkin SR. 1999. Compliance with guidelines for thyroid nodule evaluation. *Endocr Pract* 5:119–23.
40. Bennedbaek FN, Perrild HJ, Hegedüs L. 1999. Investigation and treatment of solitary thyroid gland nodules by Danish endocrinologists. A questionnaire study. *Ugeskr Laeger* 161:1264–69.
41. Führer D, Mügge C, Paschke R. 2005. Questionnaire on management of nodular thyroid disease (Annual Meeting of the Thyroid Section of the German Society of Endocrinology 2003). *Exp Clin Endocrinol Diabetes* 113:152–89.
42. AACE/AME Task Force on Thyroid Nodules. 2006. Thyroid nodule guidelines. *Endocr Pract* 12:64–102.
43. American Thyroid Association Guidelines Taskforce. 2006. Management guidelines for patients with thyroid nodules and differentiated thyroid cancer. *Thyroid* 16:1–33.
44. Carpi A, Nicolini A, Marchetti C, Iervasi G, Antonelli A, Carpi F. 2007. Percutaneous large-needle aspiration biopsy histology of palpable thyroid nodules: Technical and diagnostic performance. *Histopathology* 51:249–57.
45. Carpi A, Sagripanti A, Nicolini A, Santini S, Ferrari E, Romani R, Di Coscio G. 1998. Large needle aspiration biopsy for reducing the rate of inadequate cytology on fine needle aspiration specimens from palpable thyroid nodules. *Biomed Pharmacother* 52:303–7.
46. Hooft L, Hoekstra OS, Boers M, Van Tulder MW, Van Diest P, Lips P. 2004. Practice, efficacy, and costs of thyroid nodule evaluation: A retrospective study in a Dutch university hospital. *Thyroid* 14:287–93.
47. Ylagan LR, Farkas T, Dehner LP. 2004. Fine needle aspiration of the thyroid: A cytohistologic correlation and study of discrepant cases. *Thyroid* 14:35–41.
48. Castro MR, Gharib H. 2003. Thyroid fine-needle aspiration biopsy: Progress, practice, and pitfalls. *Endocr Pract* 9:128–36.
49. Danese D, Sciacchitano S, Farsetti A, Andreoli M, Pontecorvi A. 1998. Diagnostic accuracy of conventional versus sonography-guided fine-needle aspiration biopsy of thyroid nodules. *Thyroid* 8:15–21.
50. Burch HB. 1995. Evaluation and management of the solid thyroid nodule. *Endocrinol Metab Clin North Am* 24:663–710.

51. Castro MR, Gharib H. 2005. Continuing controversies in the management of thyroid nodules. *Ann Intern Med* 142:926–31.
52. McHenry CR, Walfish PG, Rosen IB. 1993. Non-diagnostic fine needle aspiration biopsy: A dilemma in management of nodular thyroid disease. *Am Surg* 59:415–19.
53. Schmidt T, Riggs MW, Speights VO Jr. 1997. Significance of nondiagnostic fine-needle aspiration of the thyroid. *South Med J* 90:1183–86.
54. Carpi A, Nicolini A, Gross MD, Fig LM, Shapiro B, Fanti S, Rampin L, Polico C, Rubello D. 2005. Controversies in diagnostic approaches to the indeterminate follicular thyroid nodule. *Biomed Pharmacother* 59:517–20.
55. Lo Gerfo P. 1994. The value of coarse needle biopsy in evaluating thyroid nodules. *Thyroid Clin Exp* 6:1–4.
56. Kung ITM, Yuen RWS. 1989. Fine needle aspiration of the thyroid. Distinction between colloid nodules and follicular neoplasm using cell blocks and 21-gauge needles. *Acta Cytol* 33:53–59.
57. Carpi A, Toni MG, Nicolini A, et al. 1992. Progress in the management of thyroid nodule patients. In *Progress in clinical oncology*, ed. A Carpi, A Sagripanti, and C Mittermayer, 204–21. Munich: Sympomed.
58. Carpi A, Naccarato AG, Iervasi G, Nicolini A, Bevilacqua G, Viacava P, Collecchi P, Lavra L, Marchetti C, Sciacchitano S, Bartolazzi A. 2006. Large needle aspiration biopsy and galectin-3 determination in selected thyroid nodules with indeterminate FNA-cytology. *Br J Cancer* 95:204–9.
59. Carpi A, Fabris F, Ferrari E, Sagripanti A, Nicolini A, Romani R, Di Coscio G. 1999. Aspiration needle biopsy in preoperative selection of thyroid nodules defined at fine needle aspiration as microfollicular lesion. *Am J Clin Oncol* 22:65–69.
60. Carpi A, Nicolini A, Righi C, Romani R, Di Coscio G. 2004. Large needle aspiration biopsy results of palpable thyroid nodules diagnosed by fine-needle aspiration as a microfollicular nodule with atypical cells or suspected cancer. *Biomed Pharmacol* 58:351–55.
61. Rosai J. 2004. Thyroid gland. In *Rosai J. Rosai and Ackerman's surgical pathology*, 515–68. 9th ed. Edinburgh, UK: Mosby-Elsevier.
62. Rojeski MT, Gharib H. 1985. Nodular thyroid disease. Evaluation and management. *N Engl J Med* 313:428–36.
63. Ross DS. 1991. Evaluation of the thyroid nodule. *J Nucl Med* 32:2181–92.
64. Mazzaferri EL. 1993. Current concepts. Management of a solitary thyroid nodule. *N Engl J Med* 328:553–59.
65. Albores-Saavedra J, Gould E, Vardaman C, et al. 1991. The macrofollicular variant of papillary thyroid carcinoma: A study of 17 cases. *Hum Pathol* 22:1195–205.
66. Mesonero CE, Jungle JE, Wilbur DC, et al. 1998. Fine-needle aspiration of the macrofollicular subtypes of the follicular variant of papillary carcinoma of the thyroid. *Cancer Cytopathol* 84:235–44.
67. Giordano TJ. 2008. Genome-wide studies in thyroid neoplasia. *Endocrinol Metab Clin North Am* 37(2):311–31.
68. Haugen BR, Woodmansee WW, McDermott MT. 2002. Towards improving the utility of fine-needle aspiration biopsy for the diagnosis of thyroid tumours. *Clin Endocrinol* (Oxf) 56:281–90.
69. Eszlinger M, Krohn K, Kukulska A, Jarzab B, Paschke R. 2007. Perspectives and limitations of microarray-based gene expression profiling of thyroid tumors. *Endocr Rev* 28:322–38.
70. Lubitz CC, Ugras SK, Kazam JJ, Zhu B, Scognamiglio T, Chen YT, Fahey TJ 3rd. 2006. Microarray analysis of thyroid nodule fine-needle aspirates accurately classifies benign and malignant lesions. *J Mol Diagn* 8:490–98.
71. Karger S, Engelhardt C, Eszlinger M, Tönjes A, Herrmann F, Müller P, Schmidt T, Weiss CL, Dralle H, Lippitzsch F, Tannapfel A, Führer D. 2006. Cytology and mRNA expression analysis of fine needle aspirates of thyroid nodules in an East German region with borderline iodine deficiency. *Horm Metab Res* 38:662–67.
72. Nikiforova MN, Tseng GC, Steward D, Diorio D, Nikiforov YE. 2008. MicroRNA expression profiling of thyroid tumors: Biological significance and diagnostic utility. *J Clin Endocrinol Metab* 93:1600–8.
73. Zeiger MA. 2008. Distinguishing molecular markers in thyroid tumors: A tribute to Dr. Orlo Clark. *World J Surg* 33:375–377.
74. Bartolazzi A, Orlandi F, Saggiorato E, Volante M, Arecco F, Rossetto R, Palestini N, Ghigo E, Papotti M, Bussolati G, Martegani MP, Pantellini F, Carpi A, Giovagnoli MR, Monti S, Toscano V, Sciacchitano S, Pennelli GM, Mian C, Pelizzo MR, Rugge M, Troncone G, Palombini L, Chiappetta G, Botti G, Vecchione A, Bellocco R. 2008. Italian Thyroid Cancer Study Group (ITCSG). Galectin-3-expression analysis in the surgical selection of follicular thyroid nodules with indeterminate fine-needle aspiration cytology: A prospective multicentre study. *Lancet Oncol* 9:543–49.

75. Serra S, Asa SL. 2008. Controversies in thyroid pathology: The diagnosis of follicular neoplasms. *Endocr Pathol* 19:156–65.
76. Mills LJ, Poller DN, Yiangou C. 2005. Galectin-3 is not useful in thyroid FNA. *Cytopathology* 16:132–38.
77. Weber F, Shen L, Aldred MA, Morrison CD, Frilling A, Saji M, Schuppert F, Broelsch CE, Ringel MD, Eng C. 2005. Genetic classification of benign and malignant thyroid follicular neoplasia based on a 3-gene combination. *J Clin Endocrinol Metab* 90:2512–21.
78. Jarzab B, Wiench M, Fujarewicz K, Simek K, Jarzab M, Oczko-Wojciechowska M, Wloch J, Czarniecka A, Chmielik E, Lange D, Pawlaczek A, Szpak S, Gubala E, Swierniak A. 2005. Gene expression profile of papillary thyroid cancer: Sources of variability and diagnostic implications. *Cancer Res* 65:1587–97.
79. Matesa N, Samija I, Kusić Z. 2007. Galectin-3 and CD44v6 positivity by RT-PCR method in fine needle aspirates of benign thyroid lesions. *Cytopathology* 18(2):112–16.
80. Chen YT, Kitabayashi N, Zhou XK, Fahey TJ 3rd, Scognamiglio T. 2008. MicroRNA analysis as a potential diagnostic tool for papillary thyroid carcinoma. *Mod Pathol* 21:1139–46.
81. Siddiqui MT, Greene KL, Clark DP, Xydas S, Udelsman R, Smallridge RC, Zeiger MA, Saji M. 2001. Human telomerase reverse transcriptase expression in Diff-Quik-stained FNA samples from thyroid nodules. *Diagn Mol Pathol* 10:123–29.
82. Samija I, Matesa N, Lukac J, Kusic Z. 2008. Thyroid fine-needle aspiration samples inadequate for reverse transcriptase-polymerase chain reaction analysis. *Cancer* 114:187–95.
83. Cerutti JM, Latini FR, Nakabashi C, Delcelo R, Andrade VP, Amadei MJ, Maciel RM, Hojaij FC, Hollis D, Shoemaker J, Riggins GJ. 2006. Diagnosis of suspicious thyroid nodules using four protein biomarkers. *Clin Cancer Res* 12:3311–18.
84. Savin S, Cvejic D, Isic T, Paunovic I, Tatic S, Havelka M. 2008. Thyroid peroxidase and galectin-3 immunostaining in differentiated thyroid carcinoma with clinicopathologic correlation. *Hum Pathol* 39:1656–1663.
85. Bartolazzi A, Gasbarri A, Papotti M, Bussolati G, Lucante T, Khan A, Inohara H, Marandino F, Orlandi F, Nardi F, Vecchione A, Tecce R, Larsson O, Thyroid Cancer Study Group. 2001. Application of an immunodiagnostic method for improving preoperative diagnosis of nodular thyroid lesions. *Lancet* 357:1644–50.
86. Saggiorato E, Cappia S, De Giuli P, Mussa A, Pancani G, Caraci P, Angeli A, Orlandi F. 2001. Galectin-3 as a presurgical immunocytodiagnostic marker of minimally invasive follicular thyroid carcinoma. *J Clin Endocrinol Metab* 86:5152–58.
87. Inohara H, Raz A. 1995. Functional evidence that cell surface galectin-3 mediates homotypic cell adhesion. *Cancer Res* 55:3267–71.
88. Akahani S, Nangia-Makker P, Inohara H, Kim HR, Raz A. 1997. Galectin-3: A novel antiapoptotic molecule with a functional BH1 (NWGR) domain of Bcl-2 family. *Cancer Res* 57:5272–76.
89. Cvejic D, Savin S, Paunovic I, Tatic S, Havelka M, Sinadinovic J. 1998. Immunohistochemical localization of galectin-3 in malignant and benign human thyroid tissue. *Anticancer Res* 18:2637–41.
90. Irimura T, Matsushita Y, Sutton RC, Carralero D, Ohannesian DW, Cleary KR, Ota DM, Nicolson GL, Lotan R. 1991. Increased content of an endogenous lactose-binding lectin in human colorectal carcinoma progressed to metastatic stages. *Cancer Res* 51:387–93.
91. Lotan R, Ito H, Yasui W, Yokozaki H, Lotan D, Tahara E. 1994. Expression of a 31-kDa lactoside-binding lectin in normal human gastric mucosa and in primary and metastatic gastric carcinomas. *Int J Cancer* 56:474–80.
92. Kim MJ, Kim HJ, Hong SJ, Shong YK, Gong G. 2006. Diagnostic utility of galectin-3 in aspirates of thyroid follicular lesions. *Acta Cytol* 50:28–34.
93. Saggiorato E, De Pompa R, Volante M, Cappia S, Arecco F, Dei Tos AP, Orlandi F, Papotti M. 2005. Characterization of thyroid "follicular neoplasms" in fine-needle aspiration cytological specimens using a panel of immunohistochemical markers: A proposal for clinical application. *Endocr Relat Cancer* 12:305–17.
94. Collet JF, Hurbain I, Prengel C, Utzmann O, Scetbon F, Bernaudin JF, Fajac A. 2005. Galectin-3 immunodetection in follicular thyroid neoplasms: A prospective study on fine-needle aspiration samples. *Br J Cancer* 93:1175–81.
95. Carpi A, Rossi G, Di Coscio G, Iervasi G, Nicolini A, Carpi F, Mechanick JI, Bartolazzi A. 2009. Galectin-3 detection on large needle aspiration biopsy improves preoperative selection of thyroid nodules: A prospective cohort study. *Ann Med* 92:70–78.
96. Bartolazzi A. 2000. Improving accuracy of cytology for nodular thyroid lesions. *Lancet* 355:1661–62.

97. Papotti M, Rodriguez J, De Pompa R, Bartolazzi A, Rosai J. 2005. Galectin-3 and HBME-1 expression in well-differentiated thyroid tumors with follicular architecture of uncertain malignant potential. *Mod Pathol* 18:541–46.
98. Gasbarri A, Marchetti C, Iervasi G, et al. 2004. From the bench to the bedside. Galectin-3 immunodetection for improving the preoperative diagnosis of the follicular thyroid nodules. *Biomed Pharmacother* 58:356–59.
99. American Thyroid Association. 2003. Thyroid fine needle aspiration (FNA) and cytology. *Thyroid* 13(1):80–86.
100. American Thyroid Association (ATA) Guideliness Taskforce on Thyroid Nodules and Differentiated Thyroid Cancer, Cooper DS, Doherty GM, Haugen BR, Kloos RT, Lee SL, Mandwel SJ, Mazzaferri EL, Mclever B, Pacini F, Schlumberger M, Sherman SI, Steward DL, Tuttle M. 2009. Revised American Thyroid Association management guideliness for patients with thyroid nodules and differentiated thyroid cancer. *Thyroid* 19(11):1167–98.
101. Stavric GD, Karanfilski BT, Kalamaras AK, Serafimov NZ, Georgievska BS, Korubin VH. 1980. Early diagnosis detection of clinically non-suspected thyroid neoplasia by the cytologic method. *Cancer* 45:340–44.
102. Robinson IA, Cozens NJ. 1999. Does a joint ultrasound guided cytology clinic optimize the cytological evaluation of head and neck masses? *Clin Radiol* 54:312–16.
103. Koike E, Noguchi S, Yamashita H, Murakami T, Ohshima A, Kawamoto H, Yamashita H. 2001. Ultrasonographic characteristics of thyroid nodules: Prediction of malignancy. *Arch Surg* 136:334–37.
104. Paramo JC, Mesko T. 2008. Age, tumor size, and in-office ultrasonography are predictive parameters of malignancy in follicular neoplasms of the thyroid. *Endocr Pract* 14:447–51.
105. Gulcelik NE, Gulcelik MA, Kuru B. 2008. Risk of malignancy in patients with follicular neoplasm: Predictive value of clinical and ultrasonographic features. *Arch Otolaryngol Head Neck Surg* 134:1312–15.
106. Rago T, Vitti P. 2008. Role of thyroid ultrasound in the diagnostic evaluation of thyroid nodules. *Best Pract Res Clin Endocrinol Metab* 22:913–28.
107. Miller JM. 1984. Large needle biopsy in a thyroid biopsy program. In *Clinical and pathological advances in thyroid tumors*, ed. A Carpi, 37–48. Pisa: ETS.

13 Image-Guided Mini-Invasive Ablation of Thyroid Tumors and Distant Metastases

Enrico Papini, Antonio Bianchini, Rinaldo Guglielmi, Claudio M. Pacella, Irene Misischi, Laura Papini, and Giancarlo Bizzarri

CONTENTS

ABSTRACT: Postsurgical recurrences or distant metastases of poorly differentiated and medullary thyroid carcinomas are usually managed without the relevant therapeutic aid of RAI treatment. In cases of metastatic thyroid cancer not amenable to a further surgical treatment, a multimodality adjunctive therapeutic approach may be employed to eradicate or control local or distant foci. Several loco-regional procedures have been proposed for inducing a clinically significant debulking of primary tumors or distant metastases. Ultrasound (US)-guided percutaneous ethanol injection, transarterial embolization and chemoembolization, and image-guided procedures based on thermal ablation (percutaneous laser ablation (PLA), radiofrequency ablation (RFA), high-intensity focused US (HIFU), microwaves (MVs), and cryoablation) have been tested on clinical or experimental grounds. On the basis of available clinical evidence, laser ablation may be considered for inducing a rapid and effective cytoreduction of the neoplastic burden in cervical recurrences of thyroid cancer that are not amenable to surgical or RAI treatment. Both PLA and RFA are effective procedures for the palliative treatment of liver and bone metastases of poorly differentiated and medullary thyroid carcinoma with disease progression. At present, the major limitation of their clinical use is the impossibility of an accurate real-time US monitoring of the margins of the area of necrosis induced by thermal ablation. Clinical data concerning treatments with cryoablation, HIFU, MW, and electroporation are still insufficient to recommend their use in current clinical practice. Loco-regional procedures should always be used together with external beam radiotherapy, chemotherapy, or targeted therapy as part of an integrated multidisciplinary approach to the management of advanced thyroid cancer.

13.1 INTRODUCTION

The great majority of differentiated thyroid cancers are successfully cured with surgery and radioiodine (RAI) treatment, but a few cases of thyroid malignancy have a poor prognosis and present relevant therapeutic problems.[1,2] Some patients have resectable thyroid tumors or local recurrences but are not amenable to surgery because they are at surgical risk or refuse repeat surgery after multiple neck dissections or external beam radiotherapy (EBRT). Poorly differentiated thyroid carcinoma, anaplastic thyroid carcinoma (ATC), and certain types of medullary thyroid carcinoma (MTC) manifest aggressive behavior. Their postsurgical recurrences or distant metastases require management without the benefit of RAI treatment because they do not have or lose the ability to trap RAI.[3] Moreover, a few patients with distant metastases may have already approached their maximally safe cumulative dose of RAI.[4–6]

In these cases, a multimodality adjunctive therapeutic approach may be designed to eradicate or control local or distant foci of the thyroid malignancy. This may improve the outcome of the disease and performance status of the patient. Several loco-regional procedures may be considered for inducing a clinically significant debulking of thyroid tumors or distant metastases: US-guided percutaneous ethanol injection (PEI), transarterial embolization (TAE) and chemoembolization (TACE), as well as image-guided procedures that are based on thermal ablation, such as percutaneous laser ablation (PLA), radiofrequency ablation (RFA), high-intensity focused ultrasound (HIFU), microwaves (MWs), and cryoablation. Another promising treatment modality under development is irreversible electroporation (IRE), but it has not yet been tested in thyroid cancer patients.

The present section reviews both the well-established and the most recent mini-invasive procedures for a nonsurgical approach to thyroid malignancies that are refractory to conventional treatments. Loco-regional procedures should be used together with EBRT, chemotherapy, or targeted therapy as part of an integrated multidisciplinary approach to the management of advanced thyroid cancer.

13.2 TECHNIQUES OF TREATMENT

13.2.1 Percutaneous Ethanol Injection

Percutaneous ethanol injection (PEI) was first proposed for the treatment of benign nodules of the thyroid gland.[7] Several initial papers reported satisfactory results, but the procedure showed some technical limitations,[8–10] and PEI is currently indicated only for the treatment of cysts or predominantly cystic lesions.[11] Successively, PEI has been evaluated as a possible treatment of thyroid cancer metastases in cervical lymph nodes of patients who are not candidates for surgical resection or RAI therapy.[12] After local anesthesia with 1.5% buffered lydocaine, a 25-gauge needle is inserted under US guidance into the center of the lymph node, and a small amount of 95% ethanol is injected into the lesion. The needle is then withdrawn and repositioned, performing some additional injections of 0.1 ml of alcohol, until a complete treatment is achieved.[13] PEI is rapid and inexpensive and can be performed on outpatients as an office-based procedure.[14]

13.2.2 Thermal Tissue Ablation

Cellular homeostasis can be maintained only within a narrow range of temperature, and when the temperature level arrives at 42–45°C, cells become more susceptible to damage induced by various agents.[15,16] When temperature is increased between 60 and 100°C, there is a rapid induction of protein coagulation that irreversibly damages cytosolic and mitochondrial enzymes as well as nuclear DNA complexes. Temperatures greater than 105°C result in tissue vaporization and carbonization, processes that may hamper an optimal ablation treatment as a result of their insulating effect on energy diffusion. Hence, a key aim for ablative therapies is achieving and maintaining a temperature range between 50 and 100°C throughout the whole target lesion.

Thermal energy sources such as laser,[15,17–19] radiofrequency,[20,21] microwave,[22] high-intensity focused ultrasound,[23] and cryotherapy[24] are receiving increasing attention for the treatment of benign and malignant lesions.

13.2.2.1 Percutaneous Laser Ablation

Light may be delivered interstitially by implanting a flexible fiber optic directly into various organs.[15] PLA is based on the conversion of light into heat within a particular tissue. In this case, the light source is a neodimium–yttrium aluminum garnet (ND-YAG) laser, with a wavelength of 1064 nm, or a continuous wave infrared (820 nm) diode laser. The delivered photons induce an increase in temperature followed by denaturation of the irradiated cells. By using low-energy output, tissue close to the implanted fiber tip reaches temperatures exceeding 100°C, resulting in vaporization of the core of the lesion. However, sufficient energy is diffused throughout the surrounding tissue to heat the cells within a radius of several millimeters to temperatures exceeding the protein denaturation threshold (60°C). Hence, PLA induces a zone of coagulative necrosis with well-defined margins and fairly predictable size without viable tissue inside the treated area.[25,26]

The procedure is performed by inserting, under US guidance and after a mild local anesthesia, from one to four 21-gauge spinal needles into thyroid lesions. After correct positioning of needle tips has been assessed, a 300 μm diameter plane-cut quartz optical fiber is introduced through the sheath of each needle until the fiber tip is placed in direct contact with thyroid tissue. Illuminations performed with a power output from 2 to 5 watts and a period ranging from 5 to 10 minutes induce histological changes characterized by a central area of cavitation surrounded by a thin layer of carbonization and encircled by a broad area of coagulative necrosis.[16,27]

PLA was first proposed in 2000 for the debulking of benign thyroid lesions that caused local pressure symptoms in patients not eligible for or refusing surgical treatment or RAI.[27] PLA, in contrast to PEI, induces an area of necrosis with predictable size and limits, reducing the risk of damage to surrounding cervical structures.[27] Several uncontrolled series[28–35] and two randomized clinical trials established that PLA may induce a satisfactory clinical response in the majority of patients with a benign thyroid nodule.[36,37] A single PLA session is followed by a median reduction in nodule volume of about 40% and a steady decrease of pressure symptoms. Repeated PLA procedures seem necessary only for large nodules or malignant thyroid lesions that need complete ablation.[38]

Pain is usually mild and subsides a few hours after illumination. Transient dysphonia, lasting from one week to two months, self-resolving neck hematoma, skin infections, and mild burns are rare complications[36,37] and are usually related to the training period of the operator. For benign thyroid nodules, PLA is an outpatient procedure that lasts about 20 minutes, and the patient can be dismissed shortly after the treatment.[39] In patients with unresectable thyroid tumors or local recurrences, the procedure is usually more cumbersome and painful due to the close proximity of the target lesion to the cervical nervous and vascular structures. For these reasons, a more protracted observation period and an effective pain-relieving treatment are appropriate for 24 hours after PLA.[40]

13.2.2.2 Radiofrequency Ablation

Needle-like electrodes (usually from 14 to 18 gauge) are placed directly into the target lesion with the use of either US, computed tomography (CT), or magnetic resonance imaging (MRI) guidance.[41–43] An RFA generator supplies RFA power to the tissue and is connected to both the RFA electrode tip and a large conductive pad in contact with the patient skin. An RFA voltage is produced between the active (intralesional) electrode and the reference electrode, thus establishing lines of electric field within the patient's body. The mechanism of tissue heating for RFA ablation is frictional and is due to an oscillatory movement of ions within the tissue that is proportional to the electric field intensity. The shape of the electric field and the area of thermal necrosis depend on tissue conductivity and may be to some extent unpredictable. Patients are usually treated under conscious sedation. RFA ablation was performed using a 17- and 18-gauge internally cooled electrode in a series of nine patients with toxic and pretoxic thyroid nodules that were not eligible for surgery or RAI therapy.[44] The mean baseline volume was 14.9 ± 25.5 ml, and it decreased to 7.5 ± 19.9 ml 12 months after RFA. After ablation, four nodules became cold or isofunctioning at thyroid scan and five remained as hot nodules. The mean symptom and cosmetic score was reduced significantly, and no major complications were encountered.[45] In a second series of 94 elderly patients with cytologically benign compressive thyroid nodules or goiters, RFA was performed by using a RITA hook-umbrella needle inserted in every single nodule under sonographic guidance. The mean decrease in nodule volume after RFA was from 24.5 ± 2.1 ml to 7.5 ± 1.2 ml at 12 months, with a mean percent decrease of 78.6 ± 2.0%. Two years after RFA, a 79.4 ± 2.5% decrease of nodule size was observed. Compressive symptoms improved in all patients and hyperthyroidism resolved in most. In this series as well, no major complications were observed. A larger study[46] assessed volume reduction induced by RFA on 302 benign thyroid nodules. RFA was carried out using an internally cooled electrode under local anesthesia. The baseline mean volume of index nodules was 6.1 ± 9.5 ml, and after ablation decreased to a mean of 1.1 ± 2.9 ml. The complications encountered were pain, hematoma, and transient voice changes.

13.2.2.3 Microwaves

MWs represent another mean of inducing thermal tissue ablation.[47] Tissue heating is caused by friction through the vibration of water molecules acting as magnetic dipoles. Microwave energy is radiated from the active portion of an antenna into the surrounding tissue. In comparison with RFA, the main advantages are a more predictable volume, shape of tissue ablation, and the absence of grounding pads. As a result, the risk of skin burning is completely eliminated, and the procedure appears more manageable. However, an important drawback of the procedure is due to the large dimension of the antenna that appears more suitable for the insertion into the liver or bone than into the narrow and critical space of the neck.

13.2.2.4 High-Intensity Focused Ultrasound

HIFU can be produced by arrays of piezoelectric elements driven by a high-frequency amplifier.[48] The US beam is focused geometrically (by means of a lens or a curved transducer) or electronically (through a phase-array transducer) in order to obtain an energy concentration that is able to induce an elevated local hyperthermia and tissue coagulation. As the US focus is very small (about one or a few mm^3), multiple shots are necessary to obtain an ablation volume of clinical value. This result is achieved by moving the focal spot through the target tissue. The treatment is controlled via a computer that acquires data from an imaging modality (MRI or US) in order to aim precisely the focused energy to the target and obtain feedback signals regarding the completeness of ablation. The procedure is performed under conscious sedation and requires from one to a few hours for achieving a significant volume ablation in kidney tumors,[49] uterine fibroids,[50] and prostate cancer.[51]

13.2.2.5 Cryotherapy

The cryoprobe is basically a sealed 2.4 mm diameter (13-gauge) needle that generates an "ice ball" measuring up to 3.7 cm in its greatest diameter and up to 5.7 cm along the needle shaft.[52] Percutaneous cryoprobes are based on the delivery of a gas (usually argon) through a segmentally insulated probe and on the rapid expansion of the gas in a sealed probe tip. Gas expansion induces a rapid cooling of the surrounding tissue that reaches a temperature of −100°C within several seconds. Defrosting of the ice ball is achieved by instilling helium gas instead of argon into the cryoprobes. One or more cryoprobes are introduced through a skin nick under CT or US guidance and placed into the lesion in a parallel arrangement approximately 2 cm apart. The limits of the ablation zone may be readily identified with a CT scan.[52] Cryoablation has been used extensively to treat neoplasms in different organs.[46,53,54] Patients need to be treated under general anesthesia.

13.2.3 Transarterial Embolization

TAE was first proposed for the nonsurgical treatment of vascular malformations and hemangiomas and successively for the palliative management of various benign tumors.[55] The treatment is performed during a selective angiography and is based on the injection of polyvinyl alcohol beads or gel foam into a feeding artery to a tumor. A careful preliminary imaging study is needed to rule out the possible presence of a downstream supply to vital structures. In 2002, a bilateral embolization of superior thyroid arteries with or without unilateral embolization of an inferior thyroid artery was proposed for the nonsurgical treatment of Graves' disease.[56] In a series of 22 cases the outcome was satisfactory and the majority of patients achieved long-term remission. The therapeutic effect was obtained by selecting an appropriate size of embolizing granules, based on the diameters of thyroid arteries. Although no complications were reported and the clinical outcome was claimed to be similar to that of a surgical subtotal thyroidectomy, no further contributions on the treatment of benign thyroid lesions are presently available.

Currently, embolization may be used for the treatment of unresectable liver and bone metastases from a variety of tumors.[57,58] TAE and TACE are rather expensive inpatient procedures. A few days of observation are needed after TAE because most patients experience fever and local symptoms of variable severity.[59]

13.2.4 Electroporation

The permeabilization of the cell membrane may be induced with the application of strong but very short (a few hundred microseconds duration) electric fields across the tissues.[60] The permeabilization can be reversible or permanent, as a consequence of the magnitude and duration of the electrical field and the number of generated pulses. Reversible electroporation is used to produce the access of otherwise nonpermeable molecules into the target cells. Irreversible electroporation

(IRE) can permanently permeabilize cell membranes and causes the ablation of significant volumes of tissue without inducing a thermal effect.[61] *In vivo* experiences in the liver of rats demonstrated a complete, well-defined, and predictable ablation zone in the targeted tissue. Histology demonstrated that IRE damages cell membranes only, spares connective tissue, and can reliably ablate tissue along the margin of large blood vessels with a negligible heat production.[62] Using much shorter pulsed electric fields with a duration of a few hundred nanoseconds and electric field strengths of 20 kV/cm, Nuccitelli et al. found a complete pathological response after treatment of basal cell carcinomas and melanomas.[63] Cell death is not due to immediate cell membrane destruction but seems to be the result of an efflux of Ca^{++} from endoplasmic reticulum that eventually induces cell apoptosis.[64]

13.3 CLINICAL USE OF MINI-INVASIVE PROCEDURES FOR DEBULKING OF UNRESECTABLE THYROID CANCER AND NECK RECURRENCES

13.3.1 Percutaneous Ethanol Injection

In a series of 14 patients with 29 metastatic nodes, US follow-up after PEI showed a mean volume decrease from 492 mm^3 at baseline to 76 mm^3 after 12 months, and to 20 mm^3 after 24 months.[12] In a second series[13] of 20 patients with 23 metastatic cervical lymph nodes, six lesions disappeared completely after PEI, while seven lymph nodes required a second treatment. Complete control was reported in 15 patients, with an average injection of only 0.7 ml of ethanol.

In a third trial, six patients with biopsy-proven neck recurrences of well-differentiated thyroid cancer were treated with PEI and had 18.7 months of clinical and US follow-up.[65] Four patients showed a rapid reduction of the volume of their metastatic lymph nodes, while two of them needed repeated treatments. No neck disease persistence was reported, and serum thyroglobulin levels dropped from a mean pretreatment value of 6.1 ng/ml to 2.0 ng/ml.

Although the procedure is reported as quite safe, PEI treatment of neck lesions devoid of a capsule is usually characterized by a transient sharp pain radiating to the jaw and the chest due to ethanol seeping into cervical tissues. A posttreatment local fibrosis is frequent, and in rare cases, damage to the recurrent laryngeal nerve may occur.[14] A relevant limitation of PEI for ablation of thyroid cancer recurrences is the difficulty of achieving a definite area of coagulative necrosis that is, with high certainty, associated with complete ablation of the target tissue.[14] Ethanol diffusion is erratic, and there is no precise correlation between the amount of ethanol injected into the lesion and the size of the zone of coagulative necrosis. Moreover, part of the ethanol effect is damage and thrombosis of small vessels, and hence is unpredictable.[10] Due to these limitations, PEI cannot currently be recommended as a first-line procedure for the treatment of neck recurrences of thyroid cancer.

13.3.2 Percutaneous Laser Ablation

PLA has been tested in undifferentiated thyroid carcinomas and local recurrences that are not amenable or are refractory to traditional surgical or RAI treatments targeting local compressive symptoms or reduction of neoplastic tissue volume prior to EBRT or chemotherapy (Figures 13.1 through 13.6). The first reported case was a 75-year-old woman with a rapidly progressive ATC.[29] After achieving the ablation of a large area of the tumor bulk, EBRT was performed. The volume of the tumor and local symptoms (dysphagia and cervical pain) showed a marked reduction, followed by a near-complete stability during the following four months. A similar improvement was reported in a further inoperable ATC with aggressive course, that showed a favorable, even if transient, clinical response to PLA.[66]

PLA treatment of four cases of local recurrence from poorly differentiated thyroid carcinoma has been described.[40,67] All patients were elderly and had prior total thyroidectomy followed by

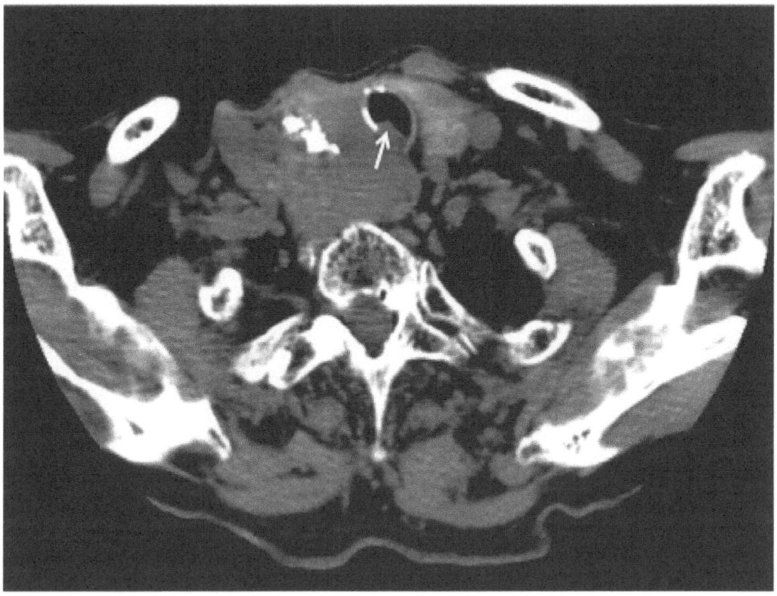

FIGURE 13.1 Unenhanced CT scan of the neck in a patient with medullary thyroid carcinoma. A large tumor mass replaces the right thyroid lobe, infiltrating and displacing the trachea and the surrounding cervical structures (white arrow). Coarse calcifications are detectable within the thyroid lesion.

cervical lymphoadenectomy and EBRT therapy for repeated cervical recurrences. Neck metastases were not RAI avid, and symptoms of local invasion (cervical pain, dysphagia, and dysphonia) were present and progressive. In all cases, from two to five PLA treatments performed during a mean period of 20 months induced a marked tumor debulking and a clinically significant improvement of local symptoms. PLA was combined with a further cycle of EBRT in two patients and with a broncoscopic laser treatment for control of tracheal invasion in one case. PLA procedures were well tolerated, with a mild cervical pain that was controlled by prescription of betametasone and ketoprophene for 24 hours. No major complications were registered.

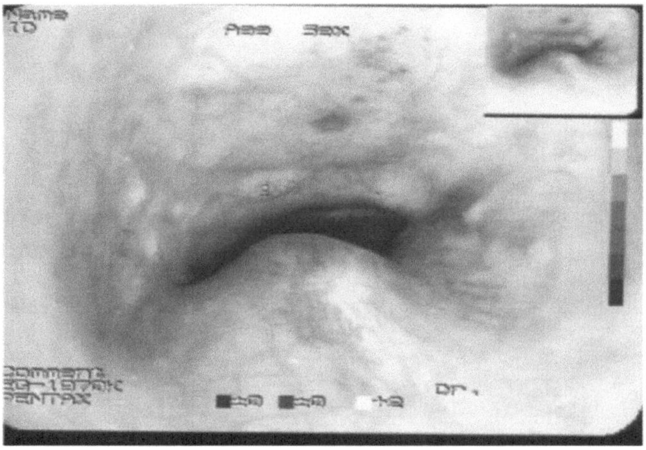

FIGURE 13.2 Fiber optic bronchoscopy in the same patient. A severe stenosis of the airway due to extrinsic infiltration and compression of the trachea is present. Normal aspect of the mucosa.

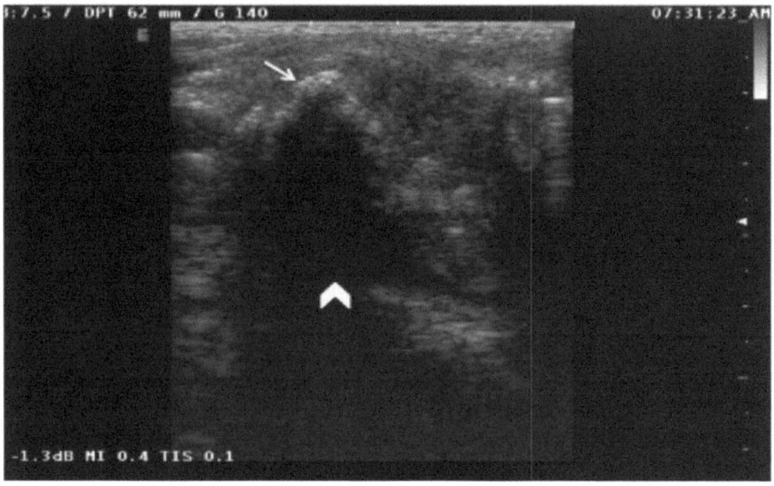

FIGURE 13.3 Thyroid US examination before laser ablation procedure: A solid hypoechoic and inhomogeneous neck mass with irregular and ill-defined margins. A coarse calcification (white arrow) inside the pathological tissue induces a posterior beam shadowing (white arrowhead).

13.3.3 Radiofrequency Ablation

RFA was first applied to a group of eight patients with locally recurrent well-differentiated thyroid carcinoma.[68] The mean size of the lesions was 2.4 cm, and the treatment was performed with US guidance and under intravenous conscious sedation. The RFA electrode was inserted into the site of recurrence and treated with the maximum allowable current for between 2 and 12 minutes. All patients were treated as outpatients. A minor skin burn and one case of vocal cord paralysis occurred. No recurrent disease at the treatment site was detected, with a mean follow-up of 10.3 months. Histological examination showed no evidence of a tumor in the treated lymph nodes in 6 patients. The same center reported a second series (not specified if in part coincident)[65] of 12 patients who underwent RFA treatment of biopsy-proven recurrent thyroid carcinoma in the neck. No recurrent disease was detected at the treatment site in over

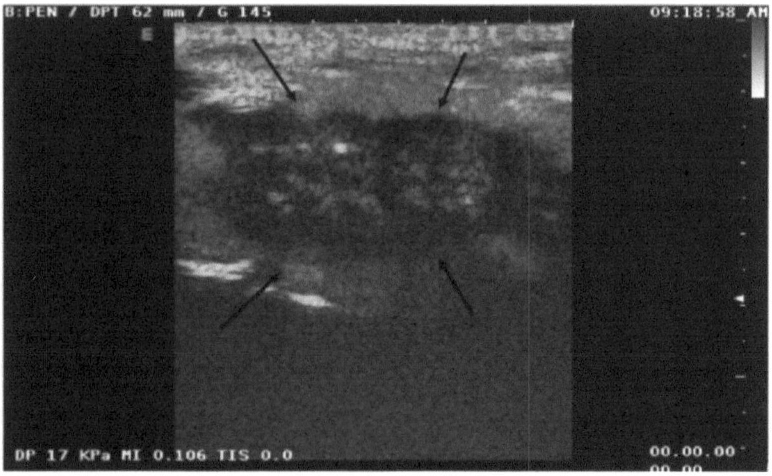

FIGURE 13.4 Contrast-enhanced US examination of the right thyroid lobe performed 60 minutes after laser ablation in the same patient. Black arrows delineate a large, oval-shaped avascular area corresponding to the necrotic zone produced by laser ablation.

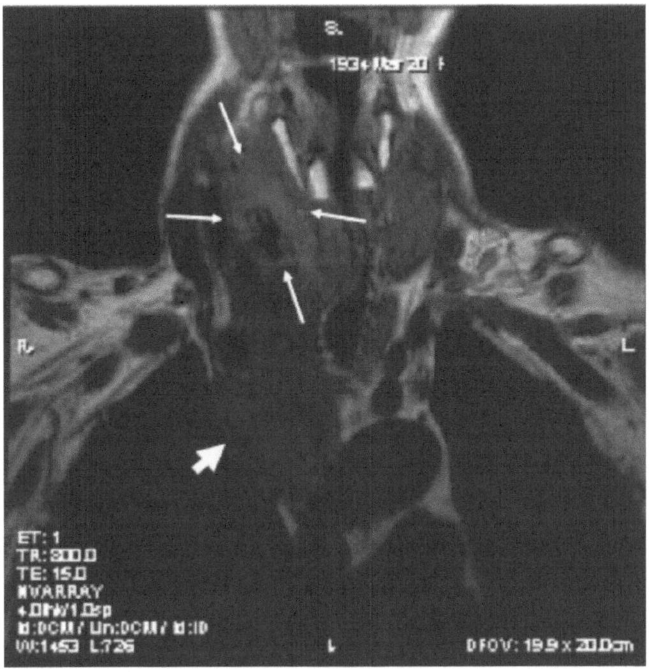

FIGURE 13.5 Neck MRI: Coronal unenhanced T1W image performed 24 hours after laser ablation. A spherical area of signal loss (small white arrows) due to necrosis, cavitation, and gas bubbles collection is visible in the cranial portion of the enlarged right thyroid lobe. Compression, infiltration, and dislocation of trachea are still well appreciable. Tumor extends downward into the upper mediastinum (large white arrow).

80% of the patients at a mean follow-up of 40.7 months. A minor skin burn and one permanent vocal cord paralysis occurred after RFA treatment. RFA ablation shows promise as an alternative to surgical treatment of recurrent differentiated thyroid carcinoma in patients at surgical risk, but controlled long-term studies are necessary to determine the possible role of RFA in the treatment of recurrent thyroid tumors.

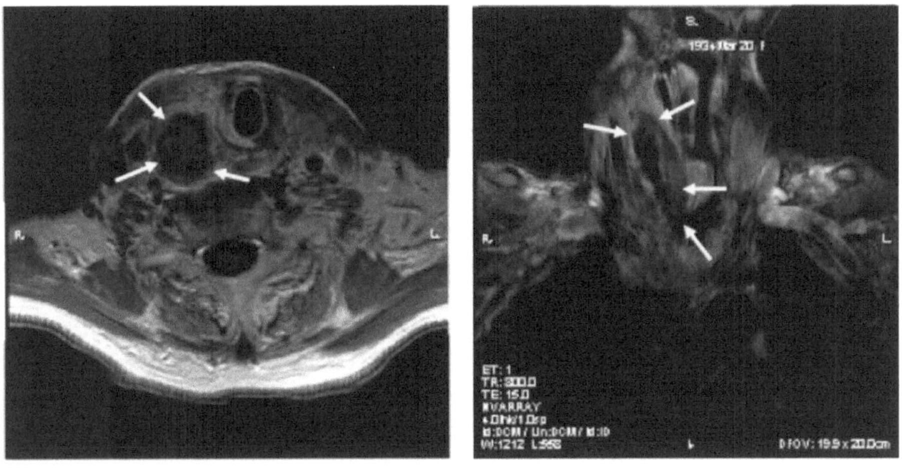

FIGURE 13.6 Neck MRI performed in the same patient one month after laser ablation (two treatments): axial- and coronal-enhanced T1W images. A large area of necrosis with absence of contrast enhancement is detected within the tumor mass (white arrows). The airway compression and dislocation are now less pronounced.

13.3.4 High-Intensity Focused Ultrasound

A feasibility study of using HIFU to obtain a localized ablation of thyroid tissue with no damage to the surrounding structures was performed under general anesthesia on eight ewes.[48] HIFU was generated by a 3 MHz spherical piezocomposite transducer that delivered an average of 24 US pulses per lobe, covering a mean volume of 0.7 cm^3. Ewes were sacrificed 6–13 days after HIFU treatment, and a characteristic histologic lesion of the thyroid was obtained: central coagulative necrosis with ghost vesicular structures and disappearance of the nuclei. At the periphery of the necrotic zone, a cell reaction was observed with fibroblastic granulation tissue, mononuclear cell infiltrate, and regenerating thyroid tissue. Adverse events (death or severe damage to cervical structures) occurred in some of the first treated ewes. This preliminary study confirms the possibility of using HIFU to destroy a well-defined area of thyroid tissue. However, no clinical data on benign or malignant thyroid lesions are at this time available.

13.3.5 Transarterial Embolization

Tumors of the head and neck may be effectively treated with embolization, but at present clinical data on selective TAE of thyroid arteries for presurgical or palliative local embolization in thyroid cancer are scarce. A group of 20 patients with thyroid tumors (7 unresectable ATC and 13 differentiated thyroid carcinoma) underwent selective embolization of the superior or inferior thyroid arteries.[69] After TAE, a selective angiography of thyroid arteries was performed to ensure that the targeted arteries had been occluded. In all cases TAE decreased the blood flow to the thyroid gland, limited bleeding during surgery, and decreased the operating time. TAE had no influence on the overall survival of ATC, but after treatment, patients reported improvement in swallowing, breathing, and pain. If this report will be confirmed, TAE may become an option for palliative treatment of unresectable thyroid tumors with cervical pain, tracheal, or esophageal compression.

13.3.6 Electroporation

IRE was successfully applied for the destruction of prostate[70] and breast cancer[71] cells *in vitro*. No data regarding the IRE treatment of thyroid nodules or malignancies are currently available. However, *in vitro* studies on cultured thyroid cancer cells demonstrated that thyroid cancer cells lacking the iodine transporter protein incorporated significant amounts of RAI after reversible electroporation.[72] The use of electroporation to improve the tumoricidal properties of RAI in poorly differentiated carcinomas has not yet been evaluated *in vivo*.

13.4 CLINICAL USE OF MINI-INVASIVE PROCEDURES FOR TREATMENT OF DISTANT METASTASES OF THYROID TUMORS

Distant metastases from follicular and poorly differentiated thyroid carcinoma are present in up to 10% of patients with thyroid malignancy.[4,5] The presence of bone metastasis decreases the 10-year survival of patients by nearly 50% and severely affects their quality of life, causing pain, pathologic fractures, spinal cord compression, and disability.[4,5] Follicular and poorly differentiated thyroid carcinomas presenting with bone metastases sometimes demonstrate a poor response to RAI treatment. In these cases, treatment options include EBRT, chemotherapy, and surgery, but a few lesions are, or become with time, not amenable to traditional palliative treatments. EBRT is the standard of care for patients with localized bone pain, but this is reported in only a fraction of patients and is frequently transient.[73,74]

13.4.1 Bone Metastases

RFA, PLA, and cryotherapy have been proposed as minimally invasive palliative treatments for bone metastases in advanced thyroid malignancies, providing patients with clinically significant pain relief without significant morbidity.

13.4.1.1 Radiofrequency Ablation

Forty-three patients with painful osteolytic metastases (two with thyroid carcinoma) were treated with RFA under conscious sedation or general anesthesia.[75] A decrease in pain was experienced by 95% of patients, with a mean worst pain score reduction from 7.9 to 1.4 (out of 10) after 24 weeks. Mean opioid requirements peaked at week 1 and subsequently decreased over time. Adverse events occurred in three patients: skin burn, transient bowel and bladder incontinence, and an acetabular fracture. Major limitations were the impossibility to depict the ablation margins with CT scans and the difficulty to penetrate deeply into bone. Two of the three patients treated for bone metastases were free of disease at the treatment site at 44 and 53 months of follow-up, respectively.

In a second series,[76] 22 patients with 28 lesions located in the thoracic and lumbar spine, sacrum, pelvis, acetabulum, femur, and tibia were treated with RFA. Underlying tumors were breast, lung, renal cell, thyroid, cancer of unknown primary, and multiple myeloma. RFA ablation was performed with the patient under moderate sedation with CT fluoroscopy-guided cement injection. Pain relief was evaluated with the visual analog scale (VAS) score and the extent to which analgesics could be reduced. A distribution of cement between both end plates of a vertebral body or at least 75% filling of osteolysis in other bones was achieved in all patients. Pain ratings with the VAS decreased from a mean of 8.5 to a mean of 5.5 after 24 hours, and a further decrease to 3.5 was detected after 3 months. Analgesics were reduced in 15 patients and remained unchanged in 5. No major complications, no clinically obvious fracture of a formerly treated bone, and no treatment-related deaths occurred.

13.4.1.2 Cryoablation

Cryoablation was performed on a series of 13 patients with painful metastatic lesions involving the bone from a variety of primary neoplasms and one case of thyroid carcinoma.[52] A statistically significant decrease of the worst pain score (from 6.7 to 3.8, out of 10) and mean pain interference with the activities of daily living (from 5.5 to 3.2, of 10) was reported. A reduction of the daily dosage of narcotics was also registered after the procedure. All patients were treated under CT guidance after the administration of general anesthetic, and the immediate pain after the procedure was controlled with intravenous fentanyl and midazolam. There were no general complications, and 4–6 weeks after the procedure CT or MRI showed an area of unenhanced low-attenuating tissue, consistent with necrosis and corresponding to the placement of the cryoablation probes. The advantages of cryoablation were the rapid control of pain and a clear depiction of the ablation margins at contrast-enhanced CT evaluation. However, the procedure was time-consuming (over 120 minutes), rather expensive, and required the use of general anesthesia.

13.4.1.3 Laser Ablation

Two patients with bone metastases (vertebral and sacral, respectively) from poorly differentiated thyroid carcinoma have been treated in our center with PLA under CT guidance.[77] Both patients had been unsuccessfully treated with surgery, RAI, and EBRT irradiation, but bone disease accompanied a progression with invasion of surrounding soft tissues. Skeletal pain was continuous and was not relieved by analgesic treatment. PLA was performed under conscious sedation with a total energy delivery of 7200 joules, and the treatment was repeated after one month. The vertebral metastasis had a maximum diameter of 6.5 cm and was ablated by 90%, as defined by the lack of enhancement after contrast medium injection on the follow-up CT. The sacral metastasis had a maximum diameter of 5.0 cm and was ablated by 80% after two sessions. In both cases, treatment led to an improvement of performance status and a marked decrease in daily analgesic dosage. No major complications were registered during and after the procedure.

13.4.1.4 Transarterial Embolization

The effects of selective TAE were investigated in patients with symptomatic bone metastases of differentiated thyroid carcinoma.[78] Forty-one TAE were performed in 16 patients. Success was defined as an improvement in clinical symptoms without tumor progression. The procedure was successful in 24 of 41 occasions (59%). Twenty-six TAE were preceded or followed by additional therapies, consisting of surgery (laminectomy), EBRT, or RAI. Subgroup analysis revealed that these additional therapies did not influence the success rate. However, an effect on success duration was present: for TAE without additional RAI or EBRT, the median success duration was 6.5 months; for TAE combined with additional RAI or EBRT, this was 15 months. The ultimate outcome of the patients was unfavorable: nine patients died and five patients have progressive disease. Selective TAE of bone metastases may be considered a palliative therapy that may induce rapid, but transient, relief of symptoms. The combination of TAE with RAI or EBRT may prolong the duration of success.

13.4.2 Liver Metastasis

13.4.2.1 Radiofrequency Ablation

Three patients with liver metastases of thyroid carcinoma were retrieved from a database of 125 patients who had been treated with RFA for liver tumors.[79] In all three patients, the metastases were a sign of widespread disease, and several other treatment modalities had been performed earlier. Two patients had metastases from MTC and had severe diarrhea. The third patient had a rapidly progressive metastasis of a FTC. The aim of the treatment was cytoreduction with amelioration of symptoms and debulking with increased sensitivity for subsequent RAI treatment. The ablation was performed via laparotomy, laparoscopically and percutaneously. One patient experienced superficial burn wounds after a long-lasting RFA procedure. Severity of symptoms was reduced significantly after RFA for a prolonged period of time. RFA induced only partial tumor necrosis in one patient, but after TAE, the second RFA treatment induced complete tumor necrosis. Local recurrences at the site of the ablated liver metastases were not encountered during follow-up.

13.4.2.2 Percutaneous Laser Ablation

PLA has been successfully employed in the treatment of liver metastases from differentiated thyroid carcinoma or endocrine tumors.[80,81] The combination of PLA and RAI therapy enabled management of a severely ill patient with a hyperfunctioning and surgically untreatable massive liver FTC metastasis.[82] A preliminary treatment with PLA reduced the risk of side effects due to high doses of therapeutic RAI in a thyrotoxic patient and increased the efficacy of successive RAI tumor ablation. After three PLA sessions, the patient showed a marked improvement in her performance status and local pressure symptoms. Two cases of solitary liver metastases from MTC were treated at our center with a complete ablation of the metastatic lesions and reduction of general symptoms (Figure 13.7).

13.4.2.3 Transarterial Embolization of Liver Metastasis

The efficacy and tolerability of bland embolization and chemoembolization have been assessed in several series of patients with liver metastases of advanced neuroendocrine tumors[58] (Figure 13.8). Distant metastases are the main cause of cancer-related death for MTC and are a component of diffuse metastatic disease of poorly differentiated thyroid tumors.[59] In these cases, systemic chemotherapy produces only rare tumor responses and somatostatin analogs, and other available modalities are poorly effective to control the growth of secondary tumors. Twelve MTC patients with

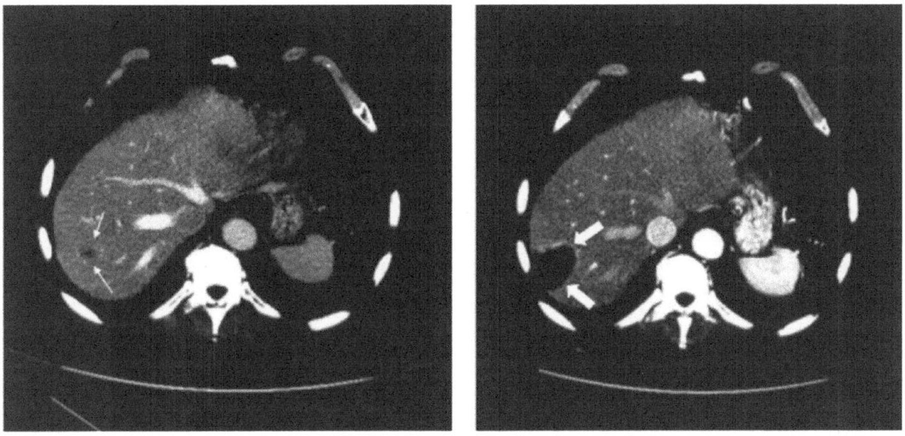

FIGURE 13.7 Contrast-enhanced CT scans before (left) and after (right) laser ablation. On the left CT scan a small liver metastasis (small white arrows) of a medullary thyroid carcinoma is visible. The lesion is completely ablated and replaced by a necrotic area (large white arrows) after a single session of laser treatment, as demonstrated by a CT scan performed 24 hours after the procedure (on the right).

predominant and progressive liver metastases underwent 18 liver TACE courses (mean, 1.5; range, 1–2). Response evaluation criteria in solid tumors (RECIST) were used to evaluate tumor responses. Symptomatic responses were defined by more than a 25% decrease of symptoms' intensity. Partial radiological tumor response was obtained in 5 patients (42%), with a median duration of 17 months; stabilization in 5 (42%), with a median duration of 24 months; and progression in the remaining 2 (16%). The 5 partial tumor responses were observed in the 9 patients with less than 30% liver involvement. Clinical response was observed in 2 of the 5 patients with diarrhea. Significant grade 3–4 toxicity was observed in one patient who had a major tumor necrosis after TACE.

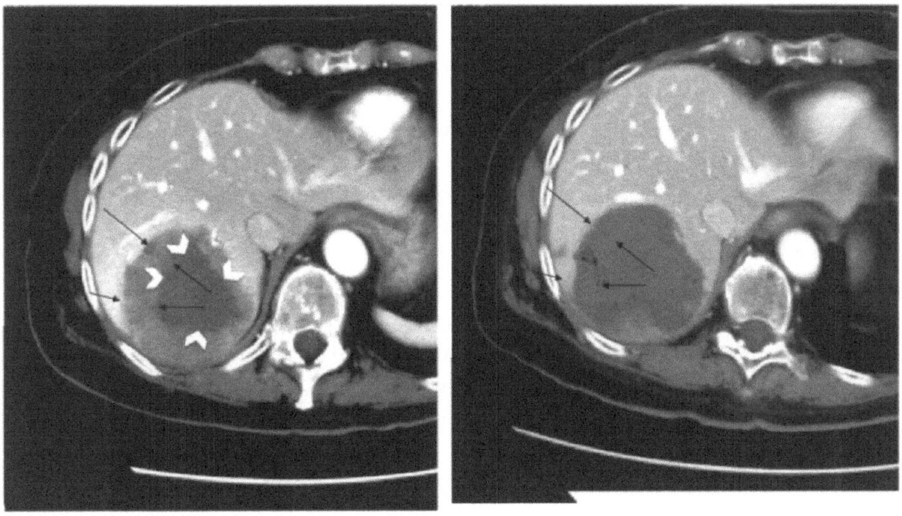

FIGURE 13.8 Combined treatment of a large liver metastasis in a patient with lung neuroendocrine tumor. Contrast-enhanced CT scans before (left) and after (right) precision TAE performed with new-generation embolic particles. A huge metastasis with a central necrotic area (large white arrow heads) and a thick viable rim (black arrows) of neoplastic tissue is still visible in the right hepatic lobe after laser ablation. Twenty-four hours after pTAE (image on the right) the residual neoplastic rim appears completely necrotic, with the presence of some gas bubbles (black arrows).

13.5 FUTURE PERSPECTIVES OF MINI-INVASIVE PROCEDURES WITH RESPECT TO EMERGING US GUIDANCE TECHNOLOGY

At present, the major limitation in the clinical use of PLA and RFA is the current unavailability of an accurate real-time US monitoring system of the area of necrosis induced by thermal ablation. During ablation, the zone under treatment appears as an ill-defined hyperechoic area that correlates poorly with the actual size of thermal necrosis. US and color Doppler studies demonstrate a well-demarcated loss of vascular signals in the treated area only a few hours after the procedure.[27] Unfortunately, the exact definition of the margins of thermal-induced necrosis is critical when performing the treatment of cervical or distant recurrences of thyroid cancer or endocrine neoplasms. These malignant lesions are usually close to, encase, or invade vital structures in the neck, liver, or bones. Thus, the greatest extent of debulking should be achieved with the least risk of major damage to the contiguous nervous or vascular structures. In selected cases, PLA and cryoablation treatments may be performed in real time under MR guidance using fast gradient echo sequences.[43,83] The clinical advantages of the MR monitoring are relevant, but the use of an efficient US guidance system would be much more convenient, more accessible, less expensive, and less cumbersome than that of an MR-guided technique.

Harmonic and pulse inversion imaging are sensitive to the nonlinear effects of US interactions with contrast agents.[67] A US contrast agent is usually composed of gas microbubbles enclosed in a lipidic or proteinous shell with elevated resistance to external pressure changes. Intravenous injection of contrast media induces after a few seconds a sharp increase of the signals reflected from parenchymal vessels. Additional US information about the size and limits of the area of tissue ablation may thus be provided. Contrast-enhanced US (CEUS) was tested by our group as a tool for an early definition of the volume of ablated tissue in 10 patients with benign and malignant thyroid lesions.[67] The area of coagulative necrosis was evaluated after the procedure with a CEUS examination. The results were compared with the volume of ablated tissue as demonstrated by 24-hour contrast-enhanced CT and power Doppler examination. CEUS was more effective than conventional US and power Doppler in the early definition of the size and margins of the ablated area, which appeared devoid of blood supply and well demarcated from the still viable tissue.

13.6 CONCLUSIONS

Poorly differentiated and medullary thyroid carcinomas frequently behave aggressively, and their postsurgical recurrences or distant metastases usually cannot be treated with RAI therapy. Several mini-invasive loco-regional procedures are available for inducing a clinically significant cytoreduction in these settings. In addition, these procedures may be safe and effective and may be combined with EBRT, chemotherapy, or targeted therapies as part of a multimodal therapeutic management of advanced cancer disease.

Based on the available clinical evidence, PLA may be considered for inducing a rapid and effective cytoreduction of the neoplastic burden in cervical recurrences of thyroid malignancy that are not amenable to surgical or RAI treatment. Thermal ablation is followed by amelioration of local compressive symptoms, control of hormonal production, and reduction of the volume of neoplastic tissue prior to traditional palliative treatment. PLA and RFA are an effective and well-tolerated procedure for the palliative treatment of liver and bone metastases of ATC, poorly differentiated thyroid carcinoma, MTC, and endocrine tumors in nonsurgical patients as part of a multimodal approach to the treatment of a progressive disease. Thermal ablation induces a rapid reduction of pain and the improvement of performance status. Clinical data and controlled studies concerning treatments with cryoablation, HIFU, MV, and IRE are still insufficient to recommend their use in current clinical practice.

Mini-invasive treatments should be performed only by well-trained operators in specialized centers due to the risk of relevant damage to the patients. Currently, the major limit in the routine use of thermal ablation techniques is the impossibility of an accurate real-time US or CT monitoring of the area of necrosis induced by the procedure. Long-term follow-up studies and controlled trials are necessary to determine the precise role that PLA and RFA may play in the treatment of local or distant metastases from thyroid carcinoma. However, an aggressive multidisciplinary approach to the management of patients with advanced thyroid tumors and rapidly progressive disease should always be considered at an early stage in order to provide a better quality of life and to improve the poor prognosis of these patients.

REFERENCES

1. De Groot LJ, Kaplan EL, McCormick M, Straus FH. 1990. Natural history treatment and course of papillary thyroid carcinoma. *J Clin Endocrinol Metab* 71:414–24.
2. Vassilopoulou-Sellin R, Schultz PN, Haynie TP. 1996. Clinical outcome of patients with papillary thyroid carcinoma who have recurrence after initial radioactive iodine therapy. *Cancer* 78:493–51.
3. Rosai J, Saxèn EA, Woolner L. 1985. Session III: Undifferentiated and poorly differentiated carcinoma. *Seminar Diag Pathol* 2:123–36.
4. American Thyroid Association Guidelines Taskforce. 2006. Management guidelines for patients with thyroid nodules and differentiated thyroid cancer. *Thyroid* 16:109–142.
5. Pacini F, Schlumberger M, Dralle H, Elisei R, Smit JWA, Wiersinga W, European Thyroid Cancer Taskforce. 2006. European consensus for the management of patients with differentiated thyroid carcinoma of the follicular epithelium. *Eur J Endocrinol* 154:787–803.
6. British Thyroid Association and Royal College of Physicians. 2007. *Guidelines for the management of thyroid cancer in adults*. www.british-thyroid-association.org.
7. Livraghi T, Paracchi A, Ferrari C, Bergonzi M, Garavaglia G, Raineri P, Vettori C.1990. Treatment of autonomous thyroid nodules with percutaneous ethanol injection: Preliminary results—Work in progress. *Radiology* 175:827–29.
8. Bennedbaek FN, Nielsen LK, Hegedus L. 1998. Effect of percutaneous ethanol injection therapy versus suppressive doses of L-thyroxine on benign solitary cold thyroid nodules: A randomized trial. *J Clin Endocrinol Metab* 83:830–35.
9. Valcavi R, Frasoldati A. 2004. Ultrasound-guided percutaneous ethanol injection therapy in thyroid cystic nodules. *Endocr Pract* 10:269–75.
10. Guglielmi R, Pacella CM, Bianchini A, Bizzarri G, Rinaldi R, Graziano FM, Petrucci L, Toscano V, Palma E, Poggi M, Papini E. 2004. Percutaneous ethanol injection treatment in benign thyroid lesions: Role and efficacy. *Thyroid* 14:125–31.
11. Gharib H, Papini E. 2007. Thyroid nodules: Clinical importance, assessment, and treatment. *Endocrinol Metab Clin North Am* 36:707–35.
12. Hay ID, Charboneau JW, Lewis BD, et al. 2002. Successful ultrasound-guided percutaneous ethanol ablation of neck metastases in 20 patients with postoperative TNM stage I papillary thyroid carcinoma resistant to conventional therapy, abstract 176. In *74th Meeting, American Thyroid Association*, Los Angeles, CA.
13. Lewis BD, Hay ID, Charboneau JW, et al. 2002. Percutaneous ethanol injection for the treatment of cervical lymph node metastases in patients with papillary thyroid carcinoma. *Am J Roentgenol* 178:699–704.
14. Papini E, Pacella CM. 2000. Percutaneous ethanol injection of benign thyroid nodules and cysts using ultrasound. In *Thyroid ultrasound and ultrasound-guided FNA biopsy*, ed. HJ Baskin, 169–213. Boston: Kluwer Academic Publishers.
15. Bown SG. 1983. Phototherapy of tumors. *World J Surg* 7:700–9.
16. Goldberg SN, Dupuy D. 2001. Image-guided radiofrequency tumor ablation: Challenges and opportunities—Part I. *J Vasc Interv Radiol* 12:1021–32.
17. Soler-Martínez J, Vozmediano-Chicharro R, Morales-Jiménez P, Hernández-Alcaraz D, Vivas-Vargas E, Santos García-Vaquero I, Baena-González V. 2007. Holmium laser treatment for low grade, low stage, noninvasive bladder cancer with local anesthesia and early instillation of mitomycin C. *J Urol* 178:2337–39.
18. Kumar SM. 2007. Rapid communication: Holmium laser ablation of large prostate glands: An endourologic alternative to open prostatectomy. *J Endourol* 21:659–62.

19. Pacella CM, Bizzarri G, Francica G, Bianchini A, De Nuntis S, Pacella S, Crescenzi A, Taccogna S, Forlini G, Rossi Z, Osborn J, Stasi R. 2005. Percutaneous laser ablation in the treatment of hepatocellular carcinoma with small tumors: Analysis of factors affecting the achievement of tumor necrosis. *J Vasc Interv Radiol* 16:1447–57.
20. Watanabe F, Kawasaki T, Hotaka Y, Ishiyama M, Fuwa S, Nagata M, Saida Y. 2008. Radiofrequency ablation for the treatment of renal cell carcinoma: Initial experience. *Radiat Med* 26:1–5.
21. Tranberg KG. 2004. Percutaneous ablation of liver tumors. *Best Pract Res Clin Gastroenterol* 18:125–45.
22. Abbas G, Schuchert MJ, Pennathur A, Gilbert S, Luketich JD. 2007. Ablative treatments for lung tumors: Radiofrequency ablation, stereotactic radiosurgery, and microwave ablation. *Thorac Surg Clin* 17:261–71.
23. Li YY, Sha WH, Zhou YJ, Nie YQ. 2007. Short and long term efficacy of high intensity focused ultrasound therapy for advanced hepatocellular carcinoma. *J Gastroenterol Hepatol* 22:2148–54.
24. Callstrom MR, Atwell TD, Charboneau JW, Farrel Ma, Goetz MP, Rubin J, Sloan JA, Novotny PJ, Welch TJ, Maus TP, Wong GY, Brown KJ. 2006. Painful metastases involving bone: Percutaneous image-guided cryoablation—Prospective trial interim analysis. *Radiology* 241:572–80.
25. Pacella CM, Rossi Z, Bizzarri G, Papini E, Marinozzi V, Paliotta D, Castaldo D, Ziparo A, Cinti G, Muzzi R. 1993. Ultrasound-guided percutaneous laser ablation of liver tissue in a rabbit model. *Eur Radiol* 3:26–32.
26. Cakir B, Topaloglu O, Gul K, Agac T, Aydin C, Dirikoc A, Gumus M, Yazicioglu K, Ersoy RU, Ugras S. 2006. Effects of percutaneous laser ablation treatment in benign solitary thyroid nodules on nodule volume, thyroglobulin and antithyroglobulin levels, and cytopathology of nodule in 1 yr follow-up. *J Endocrinol Invest* 29:876–84.
27. Pacella CM, Bizzarri G, Guglielmi R, Anelli V, Bianchini A, Crescenzi A, Pacella S, Papini E. 2000. Thyroid tissue: US-guided percutaneous interstitial laser ablation—A feasibility study. *Radiology* 217:673–77.
28. Dossing H, Bennedbaek FN, Karstrup S, Hegedus L. 2002. Benign solitary solid cold thyroid nodules: US-guided interstitial laser photocoagulation—Initial experience. *Radiology* 225:53–57.
29. Pacella CM, Bizzarri G, Spiezia S, Bianchini A, Guglielmi R, Crescenzi A, Pacella S, Toscano V, Papini E. 2004. Thyroid tissue: US-guided percutaneous laser thermal ablation. *Radiology* 232:272–80.
30. Papini E, Guglielmi R, Bizzarri G, Pacella CM. 2004. Ultrasound-guided laser thermal ablation of benign thyroid nodules. *Endocr Pract* 10:276–83.
31. Dossing H, Bennedbaek FN, Hegedus L. 2006. Effect of ultrasound-guided interstitial laser photocoagulation on benign solitary solid cold thyroid nodules: One versus three treatments. *Thyroid* 16:763–68.
32. Gambelunghe G, Fatone C, Ranchelli A, Fanelli C, Lucidi P, Cavaliere A, Avenia N, d'Ajello M, Santeusanio F, De Feo P. 2006. A randomized controlled trial to evaluate the efficacy of ultrasound-guided laser photocoagulation for treatment of benign thyroid nodules. *J Endocrinol Invest* 29:23–26.
33. Filetti S, Durante C, Torlontano M. 2006. Nonsurgical approaches to the management of thyroid nodules. *Nat Clin Pract Endocrinol Metab* 2:384–94.
34. Spiezia S. Vitale G, Di Somma C, Pio Avanti A, Ciccarelli A, Lombardi G, Colao A. 2003. Ultrasound-guided laser thermal ablation in the treatment of autonomously hyperfunctioning thyroid nodules and compressive nontoxic nodular goiter. *Thyroid* 13:941–47.
35. Barbaro D, Orsini P, Lapi P, Pasquini C, Tuco A, Righini A, Lemmi P. 2007. Percutaneous laser ablation in the treatment of toxic and pretoxic nodular goiter. *Endocr Pract* 13(1):30–36.
36. Dossing H, Bennedbaek FN, Hegedus L. 2005. Effect of ultrasound guided interstitial laser photocoagulation on benign solitary cold thyroid nodules—A randomised study. *Eur J Endocrinol* 152:341–45.
37. Papini E, Guglielmi R, Bizzarri G, Graziano F, Bianchini A, Brufani C, Pacella S, Valle D, Pacella CM. 2007. Treatment of benign cold thyroid nodules: A randomized clinical trial of percutaneous laser ablation versus levothyroxine therapy or follow-up. *Thyroid* 17:229–35.
38. Døssing H, Bennedbaek FN, Hegedüs L. 2006. Effect of ultrasound-guided laser photocoagulation on benign solitary thyroid nodules: One versus three treatments. *Thyroid* 16:763–68.
39. Valcavi R, Bertani A, Pesenti M, et al. 2008. Laser and radiofrequency ablation procedures. In *Thyroid ultrasound and ultrasound-guided FNA biopsy*, ed. HJ Baskin, DS Duick, RA Levine. New York: Springer, 191–218.
41. Owen RP, Silver CE, Ravikumar TS, Brook A, Bello J, Breining D. 2004. Techniques for radiofrequency ablation of head and neck tumors. *Arch Otolaryngol Head Neck Surg* 130:52–56.

40. Pacella CM, Bizzarri G, Bianchini A, Guglielmi R, Pacella S, Papini E. 2003. US-guided laser thermal ablation of benign and malignant thyroid lesions, 1474 (abstract). In *RSNA 89th Scientific Assembly and Annual Meeting*, Chicago, November 30–December 5.
42. Vogl TJ, Lehnert T, Eichler K, Proshek D, Floter J, Mack MG. 2007. Adrenal metastases: CT-guided and MR-thermometry-controlled laser-induced interstitial thermotherapy. *Eur Radiol* 17:2020–27.
43. Pech M, Wieners G, Freund T, Dudeck O, Fischbach F, Ricke J, Seemann MD. 2007. MR-guided interstitial laser thermotherapy of colorectal liver metastases: Efficiency, safety and patient survival. *Eur J Med Res* 12:161–68.
44. Spiezia S, Garberoglio R, Di Somma C, Deandrea M, Basso E, Limone PP, Milone F, Ramundo V, Macchia PE, Biondi B, Lombardi G, Colao A, Faggiano A. 2007. Efficacy and safety of radiofrequency thermal ablation in the treatment of thyroid nodules with pressure symptoms in elderly patients. *J Am Geriatr Soc* 55:1478–79.
45. Spiezia S, Garberoglio R, Milone F, Ramundo V, Caiazzo C, Assanti AP, Deandrea M, Limone PP, Macchia PE, Lombardi G, Colao A, Faggiano A. 2009. Thyroid nodules and related symptoms are stably controlled two years after radiofrequency thermal ablation. *Thyroid* 19:219–25.
46. Jeong WK, Baek JH, Rhim H, Kim YS, Kwak MS, Jeong HJ, Lee D. 2008. Radiofrequency ablation of benign thyroid nodules: Safety and imaging follow-up in 236 patients. *Eur Radiol* 18:1244–50.
47. Simon CJ, Dupuy DE, Mayo-Smith WW. 2005. Microwave ablation: Principles and applications. *Radiographics* 25(Suppl 1):S69–83.
48. Esnault O, Franc B, Monteil JP, Chapelon JY. 2004. High-intensity focused ultrasound for localized thyroid-tissue ablation: Preliminary experimental animal study. *Thyroid* 14:1072–76.
49. Klingler HC, Susani M, Seip R, Mauermann J, Sanghvi N, Marberger MJ. 2008. A novel approach to energy ablative therapy of small renal tumours: Laparoscopic high-intensity focused ultrasound. *Eur Urol* 53:810–16; discussion, 817–18.
50. Zhang L, Chen WZ, Liu YJ, Hu X, Zhou K, Chen L, Peng S, Zhu H, Zou HL, Bai J, Wang ZB. 2008. Feasibility of magnetic resonance imaging-guided high intensity focused ultrasound therapy for ablating uterine fibroids in patients with bowel lies anterior to uterus. *Eur J Radiol*, December 22 (e-pub ahead of print).
51. Hou AH, Sullivan KF, Crawford ED. 2009. Targeted focal therapy for prostate cancer: A review. *Curr Opin Urol* 19:283–89.
52. Callstrom MR, Atwell TD, Charboneau JW, et al. 2006. Painful bone metastasis involving bone: Percutaneous image-guided cryoablation—Prospective trial interim analysis. *Radiology* 241:572–80.
53. Shingleton WB, Sewell PE Jr. 2001. Percutaneous renal tumor cryoablation with magnetic resonance imaging guidance. *J Urol* 163:773–76.
54. Zhou XD, Tang ZY. 1998. Cryotherapy for primary liver cancer. *Seminar Surg Oncol* 14:171–74.
55. Wartofsky L. 2006. Adjunctive local approaches to metastatic thyroid cancer. In *Thyroid cancer. A comprehensive guide to clinical management*, ed. L Wartofsky, D Van Nostrand, 509–13. 2nd ed. Totowa, NJ: Humana Press.
56. Xiao H, Zhuang W, Wang S, et al. 2002. Arterial embolization: A novel approach to thyroid ablative therapy for Graves' disease. *J. Clin Endocrinol Metab* 87:3583–89.
57. Ho AS, Picus J, Darcy MD, Tan B, Gould JE, Pilgram TK, Brown DB. 2007. Long term outcome after chemoembolization and embolization of hepatic metastatic lesions from neuroendocrine tumors. *Am J Roentgenol* 188:1201–7.
58. Ruutiainen AT, Soulen MC, Tuite CM, Clark TW, Mondschein JI, Stavropoulos SW, Trerotola SO. 2007. Chemoembolization and bland embolization of neuroendocrine tumor metastases to the liver. *J Vasc Interv Radiol* 18:847–55.
59. Fromiguè J, De Baere T, Baudin E, Dromain C, Lebouillex S, Schlumberger M. 2006. Chemoembolization for liver metastases from medullary thyroid carcinoma. *J Clin Endocrinol Metab* 91:2496–99.
60. Orlowski S, Mir LM. 1993. Cell electropermeabilization: A new tool for biochemical and pharmacological studies. *Biochim Biophys Acta* 1154:51–63.
61. Davalos RV, Mir LM, Rubinski B. 2005. Tissue ablation with irreversible electroporation. *Ann Biomed Eng* 33:223–31.
62. Edd JF, Horowitz L, Davalos RV, Mir LM, Rubinski B. 2006. *In vivo* results of a new focal tissue ablation technique: Irreversible electroporation. *IEEE Trans Biomed Eng* 53:1409–15.
63. Nuccitelli R, Chen X, Pakhomov AG, Baldwin WH, Sheikh S, Pomicter JL, Ren W, Osgood C, Swanson RJ, Kolb JF, Beebe SJ, Schoenbach KH. 2009. A new pulsed electric field therapy disrupts tumor's supply and caused complete remission without recurrence. *Int J Cancer* 125:438–45.
64. Garon E, Sawcer D, Vernier PT, Tang T, Sun Y, Marcu L, Gundersen MA, Koeffler HP. 2007. *In vitro* and *in vivo* evaluation and a case report of intense nanosecond pulsed electric field as a local therapy for human malignancies. *Int J Cancer* 121:675–82.

65. Monchick JM, Donatini G, Iannocilli J, Dupuy E. 2006. Radiofrequency ablation and percutaneous ethanol injection treatment for recurrent local and distant well-differentiated thyroid carcinoma. *Am J Surg* 244:296–304.
66. Cakir B, Topaloglu O, Gul K, Agac T, Aydin C, Dirikoc A, Ersoy RU, Gumus M, Yazicioglu K, Yalcin B, Demirkazik A, Icli F, Ceyhan K. 2007. Ultrasound-guided percutaneous laser ablation treatment in inoperable aggressive course anaplastic thyroid carcinoma: The introduction of a novel alternative palliative therapy—Second experience in the literature. *J Endocrinol Invest* 30:624–25.
67. Papini E, Bizzarri G, Bianchini A, et al. 2008. Contrast-enhanced ultrasound in the management of thyroid nodules. In *Thyroid ultrasound and ultrasound-guided FNA biopsy*, ed. HJ Baskin, Duick DS, Levine RA. New York: Springer, 151–71.
68. Dupuy DE, Monchick JM, Decrea C, Pisharodi L. 2001. Radiofrequency ablation of regional recurrence from well-differentiated thyroid malignancy. *Surgery* 130:901–7.
69. Dedecjus M, Tazbir J, Kaurzel B, et al. 2007. Selective embolization of thyroid arteries as a preresective and palliative treatment of thyroid cancer. *Endocr Relat Cancer* 14:847–52.
70. Rubinski J, Onik G, Mikus P, Rubinski B. 2008. Optimal parameters for the destruction of prostate cancer using irreversible electroporation. *J Urol* 180:2668–74.
71. Neal RE 2nd, Davalos RV. 2009. The feasibility of irreversible electroporation for the treatment of breast cancer and other heterogeneous systems. *Ann Biomed Eng* 37:2615–25.
72. Gopal R, Narkar AA, Mishra KP, Samuel AM, Nair N. 2003. Electroporation: A novel approach to enhance the radioiodine uptake in a human thyroid cancer cell line. *Appl Radiat Isot* 59:305–10.
73. Tong D, Gillick L, Hendrickson FR. 1982. The palliation of symptomatic osseous metastases: Final results of the study by the Radiation Therapy Oncology Group. *Cancer* 50:893–99.
74. Steenland E, Leer JW, van Hauvelingen H, et al. 1999. The effect of a single fraction compared to multiple fractions on painful bone metastases: A global analysis of the Dutch Bone Metastasis Study. *Radiother Oncol* 52:101–9.
75. Goetz MP, Callstrom MR, Charboneau JW, et al. 2004. Percutaneous image-guided radiofrequency ablation of painful metastases involving bone: A multicenter study. *J Clin Oncol* 22:300–6.
76. Hoffmann RT, Jakobs TF, Trumm C, Weber C, Helmberger TK, Reiser MF. 2008. Radiofrequency ablation in combination with osteoplasty in the treatment of painful metastatic bone disease. *J Vasc Interv Radiol* 19:419–25.
77. Papini E, Bizzarri G, Bianchini A, Anelli V, Guglielmi R, Graziano F, Pacella CM. 2008. Percutaneous laser ablation of unresectable bone metastases from poorly differentiated thyroid carcinoma. *Hormones* 7(Suppl 1):101.
78. Eustatia-Rutten CFA, Romijn JA, Guijt MJ, et al. 2003. Outcome of palliative embolization of bone metastases in differentiated thyroid carcinoma. *J Clin Endocrinol Metab* 88:3184–89.
79. Wertenbroek MW, Links TP, Prins TR, Plukker JT, van der Jagt EJ, de Jong KP. 2008. Radiofrequency ablation of hepatic metastases from thyroid carcinoma. *Thyroid* 18:1105–10.
80. Pacella CM, Stasi R, Bizzarri G, Pacella S, Graziano FM, Guglielmi R, Papini E. 2008. Percutaneous laser ablation of unresectable primary and metastatic adrenocortical carcinoma. *Eur J Radiol.*
81. Grimaldi F, Bianchini A, Bizzarri G, Nasoni S, Noacco C, Pacella CM, Misischi I, Rossi Z, Papini E. 2009. Percutaneous laser ablation for palliative treatment of neuroendocrine liver metastases. *J Endocrinol Invest* 32(Suppl):78.
82. Guglielmi R, Pacella CM, Dottorini ME, Bizzarri GC, Todino V, Crescenzi A, Rinaldi R, Panunzi C, Rossi Z, Colombo L, Papini E. 1999. Severe thyrotoxicosis due to hyperfunctioning liver metastasis from follicular carcinoma: Treatment with ^{131}I and interstitial laser ablation. *Thyroid* 9:173–77.
83. Eyrich GK, Bruder E, Hilfiker P, Dubno B, Quick HH, Patak MA, Gratz KW, Sailer HF. 2000. Temperature mapping of magnetic resonance-guided laser interstitial thermal therapy (LITT) in lymphangiomas of the head and neck. *Lasers Surg Med* 26:467–76.

14 Thyroidectomy and Neck Exploration

New Surgical Approaches

Jean-Christophe Lifante and William B. Inabnet III

CONTENTS

ABSTRACT: Well-differentiated thyroid carcinoma (WDTC) has an excellent prognosis, especially for low-risk patients. Due to the indolent natural history, low recurrence, and mortality rates of WDTC, the surgical strategy is not determined with a high level of evidence. Total or near-total thyroidectomy seems to be the primary purpose of the initial surgery. There is also a debate concerning the role for routine neck dissections, central with or without lateral compartments, since many authors do not agree about the impact of the lymph node metastases on survival. Ultrasonography has insufficient specificity to define the metastatic nature of an enlarged lymph node, necessitating innovation and implementation of new technologies. Sentinel lymph node techniques have been developed to assist the surgeon with these decisions, but at present, the false negative rates are too high. Clinical decision making for an optimal surgical approach, as well as laryngeal nerve monitoring, is also discussed. Minimally invasive thyroidectomy and other new surgical techniques, initially used for benign thyroid disease, have proven their efficacy and ability to achieve excellent cosmetic results.

14.1 INTRODUCTION

Well-differentiated thyroid carcinoma (WDTC) has an excellent prognosis. For low-risk thyroid carcinoma, total thyroidectomy with selective use of radioactive iodine (RAI) ablation is recommended, taking into consideration risk stratification to determine the extent of treatment. For high-risk cancers, the prognosis of WDTC, in terms of disease-specific survival and loco-regional recurrence, is worse, and an aggressive surgical and postoperative management is warranted. This chapter will first address the issues concerning the extent of the surgery and necessity of systematic lymph node dissection. Then, new technologies, such as intraoperative nerve monitoring and minimally invasive surgery, will be discussed.

14.2 EXTENT OF PRIMARY SURGERY

14.2.1 Total Thyroidectomy vs. Lobectomy

The extent of the initial surgical resection is still an important subject of debate in the field of thyroid carcinoma. Most surgical teams advocate a bilateral procedure (total or near-total thyroidectomy) for all patients with WDTC.[1,2] That being said, some authors continue to prefer thyroid lobectomy for low-risk patients.[3,4] The best surgical strategy will probably not be determined with strong scientific evidence due to the indolent natural history, low recurrence, and mortality rates of WDTC. Wong et al.[5] wrote that a trial assessing the efficacy of RAI treatment should include 4000 patients in each cohort to have sufficient power to show a reduction of 10% in mortality with a 25-year follow-up.[6] Udelsman et al.[6] showed in 1996 that a randomization of 12,000 patients should be necessary to carry out a randomized trial comparing total thyroidectomy and lobectomy in terms of survival outcomes. The same authors, using the disease-specific mortality rates reported by Mazzaferri and Jhiang,[7] showed the necessity of randomizing 3100 patients for comparing total thyroidectomy and lobectomy in terms of mortality.

Moreover, comparing numerous retrospective series from single-surgeon and single-institution series does not provide sufficient evidence. Surgical techniques are not standardized, and the definition of a unilateral procedure (lobectomy, lobectomy and isthmectomy, or resection of a nodule) and of bilateral procedures (total thyroidectomy, subtotal thyroidectomy, near-total bilateral thyroidectomy, or near-total unilateral thyroidectomy) varies from study to study.[6,8–11] In addition, the extent of lymph node dissection is also difficult to assess. The effect of the postoperative RAI treatment and L-thyroxine suppression must be taken into account when assessing the results of the different surgical strategies. Of note, the utilization of postoperative radioiodine ablation and TSH suppression therapy is not equivalent according to the group or to the time of the study for the same authors.[6,8–11]

In general, the proponents of total or near-total thyroidectomy argue the following:

- Bilateral cancer rate is common with WDTC (30 to 85%).
- Recurrence in the contralateral lobe is not rare (4.7 to 85%).
- Postoperative RAI administration is possible for the diagnosis and treatment of recurrence.

- Surveillance and early detection with serum thyroglobulin (TG) levels facilitate the follow-up of thyroid carcinoma and the early detection of recurrences.

The proponents of unilateral surgery support the following:

- Multifocal microcarcinomas are often asymptomatic without clinical significance.
- Half of the recurrences may be treated by surgical resection.
- Total thyroidectomy without lymph node resection does not remove lymph node micrometastases.
- The complication rate is greater after total thyroidectomy than after thyroid lobectomy.
- In low-risk patients, there is no difference in survival between patients treated with total or less than total thyroidectomy.

For the reason previously cited, conducting a randomized clinical trial is not possible. In patients with WDTC, about 80% are cured irrespective of the extent of the thyroidectomy.

In 1987, Hay et al.[8] retrospectively studied 860 patients operated on for thyroid carcinoma between 1946 and 1970. Postoperative RAI ablation was used in 3% of patients during this period, with a mean follow-up of 18 years (maximum 40 years).[8] Fifty-six patients (6.4%) died from their cancer.[8] In this study, the patients were divided into two groups according to the AGES classification: a high-risk group and a low-risk group. In each group, the survival was studied according to the initial surgical procedure. The specific mortality was 4.8% at 10 years and 6% at 20 years for the two groups together.[8] The low-risk patients' group had a disease-specific mortality of 0.2% at 5 years and 0.5% at 15 years.[8] Six hundred three patients (81%) in this group underwent a bilateral surgery, and 131 underwent a unilateral resection (18%).[8] The disease-specific mortality rate at 25 years was 1% after lobectomy and 2% after bilateral thyroidectomy ($p = 0.15$).[8] The high-risk patients' group (14% of the patients) had a disease-specific mortality rate of 14% at 5 years and 23% at 15 years.[8] In this group, 25% of patients underwent a lobectomy and 75% underwent a total thyroidectomy.[8] The disease-specific mortality rate for thyroid carcinoma at 25 years was 65% after lobectomy and 36% after bilateral thyroidectomy ($p = 0.07$).[8] The difference approached but did not reach statistical significance. In conclusion, the authors of this study advocate lobectomy for low-risk patients and bilateral surgery for high-risk patients.[8]

Following this initial report, the same group again studied the same cohort of patients with longer follow-up.[10] In this study, high-risk and low-risk designations were based on the AMES classification. There were 1606 low-risk patients in the study with a mean follow-up of 16 years (maximum 54 years). The patients were again divided into two groups according to their surgical treatment: bilateral or unilateral surgery. The disease-specific mortality at 30 years was 2.6% after unilateral surgery and 2.4% after bilateral surgery ($p = 0.27$).[10] The recurrence rate at 20 years was 22% in the unilateral surgery group versus 8.3% in the bilateral surgery group ($p = 0.0001$).[10] The local recurrence rate at 20 years was 2% after bilateral surgery versus 14% after unilateral surgery.[10] The same results were found when the data were adjusted for the administration of RAI treatment, the initial presence of lymph node metastases, and a finding of local invasion by the tumor. The authors concluded that the unilateral surgery resulted in a 7-fold increase in recurrence and advocated for total thyroidectomy.[10]

In 1994, Mazzaferri and Jhiang[7] studied the impact of the surgical procedure on a cohort of 1355 patients. They found a recurrence rate at 30 years of 26% in the bilateral surgery group versus 40% in the unilateral surgery group ($p = 0.002$).[7] The disease-specific mortality rates were 6 and 9%, respectively ($p = 0.02$).[7] Local invasion, the presence of lymph node metastases, the size of the tumor, and the administration of RAI treatment were independent prognostic factors for recurrence in multivariate analysis. The time before treatment, local invasion, the presence of lymph node metastasis, female sex, the size of the tumor, and the extent of the initial surgical procedure were each independent prognostic factors for mortality.[7]

De Groot et al.[12] advocate an initial extensive surgery in association with systematic RAI ablation irrespective of the stage of the thyroid cancer. In this study, the number of deaths as well as the number of recurrences were lower for patients who underwent bilateral surgery.[12]

In 2001, the team in Houston published an analysis using a Markov model of the optimal surgical treatment of WDTC.[13] High-risk and low-risk patients were classified using the AMES classification. The authors concluded that total thyroidectomy was the most appropriate initial surgical procedure for high-risk patients. Lobectomy remained acceptable only for low-risk patients who wanted to avoid thyroid hormonal dependency and the risk of RAI-induced malignancy.[13]

In 2007 Bilimoria et al.[14] analyzed the National Cancer Database (NCDB) to study survival in patients undergoing surgery for papillary thyroid cancer (PTC) between 1985 and 1998. They identified 52,173 patients and studied the relative survival rates according to the size of the tumor and the extent of the resection: total thyroidectomy (including total, near total, or subtotal) vs. lobectomy (including lobectomy with or without isthmectomy).[14] Of the 52,173 patients operated on with a diagnosis of PTC, 43,227 (82.9%) underwent a total thyroidectomy and 8946 (17.1%) underwent a lobectomy.[14] Twenty-three percent of the patients had a tumor <1 cm.[14] In the evaluation of all patients, lobectomy was associated with a 57% higher risk of recurrence and a 21% higher risk of death than total thyroidectomy ($p = 0.001$ and 0.027, respectively).[14] For patients with tumors ≥1 cm, lobectomy was associated with a 15% higher risk of recurrence and a 31% higher risk of death ($p =$ 0.04 and 0.04, respectively).[14] This study was the first to demonstrate improvement of disease-specific survival with total thyroidectomy. Critics of this study point out that the NCDB is an administrative database that lacks information on extrathyroidal extension, thyroid hormone suppression therapy, RAI administration, or the need for completion thyroidectomy in a non-NCDB institution.[14] There are several current clinical practice guidelines (CPG) on thyroid cancer sponsored by professional societies of endocrinologists and endocrine surgeons. The American Thyroid Association (ATA) recommends total thyroidectomy or near-total thyroidectomy for most patients with thyroid cancer. However, the ATA states that lobectomy alone may be sufficient treatment for low-risk patients with a small, intrathyroidal PTC in the absence of cervical node metastases. (Grade A recommendation: "The recommendation is based on good evidence that the service of intervention can improve important health outcomes. Evidence includes consistent results from well designed, well conducted studies in representative population that directly assess effects on health outcomes."[15]) The National Comprehensive Cancer Network (NCCN) CPG recommend total thyroidectomy for PTC in patients ages <15 to >45 years, with a radiation history, known distant metastases, extrathyroidal extension, tumor of >4 cm, cervical lymph node metastases, or aggressive variant. Apart from these situations, the NCCN CPG do not discriminate between the two procedures. (Category 2A of evidence and consensus: "The recommendation is based on lower level evidence and there is uniform NCCN consensus."[16]) The British Association of Endocrine and Thyroid Surgeons and the European Thyroid Association recommend a total thyroidectomy for differentiated thyroid cancers greater than 1 cm, and lobectomy only for microcarcinoma without lymph node metastasis and without history of previous neck irradiation in childhood.[17,18] The Societe Francaise d'Endocrinologie recommends total thyroidectomy for all cases of WDTC.[19]

In conclusion, most authors and CPG advocate for total thyroidectomy for WDTC, though lobectomy is acceptable under certain circumstances.

14.2.2 Lymphadenectomy

14.2.2.1 General Statements

There still exists a debate concerning the extent of lymphadenectomy for WDTC. The extent of lymph node dissection should depend on whether lymph node metastases influence recurrences and survival, the pattern of lymph node metastases, and whether lymphadenectomy can improve outcome. However, there is controversy about the prognostic consequences in patients with PTC and nodal metastases. Notwithstanding whether there exists consensus for lymph node dissection with *macroscopic* lymph node metastasis, the indication of prophylactic lymph node dissection with potential *microscopic* lymph node metastases is still questionable. This is particularly important since this latter scenario involves 90% of patients with PTC in some series.[20] In any case, the possible benefits of systematic lymph node dissection should be weighed against its potential risks.

Lymph node metastases are common in patients with PTC, but are relatively uncommon in those with follicular thyroid cancer (FTC). Clinically apparent cervical lymphadenopathy has been found at initial presentation in 23–56% of cases of PTC and in 5–13% of cases of FTC.[21] Using prophylactic central compartment (level VI) dissection, Noguchi et al.[22] found lymph node metastases in more than 80% of patients with PTC. Nevertheless, the evolution of these occult metastases as well as the efficacy of the RAI ablation and of the use of TSH suppression remains controversial.

Several authors showed the influence of lymph node metastases on recurrences and survival. Sellers et al.[23] carried out a retrospective study of 212 patients with a median follow-up of 15 years (range, 4 months to 36 years). Out of the 33 patients with lymph node metastases at the time of initial diagnosis, 5 patients died during the follow-up (15%).[23] No death occurred in patients without initial lymph node metastases ($p < 0.02$).[23] Despite the low number of patients, the authors concluded that cervical lymph node metastases were prognostic, especially in older patients.[23] In a study by Mazzaferri and Jhiang,[7] 42% of patients initially presented with lymph node metastases. The recurrence rates at 30 years (32% vs. 28%, $p = 0.01$) and the disease-specific mortality rates ($p = 0.02$) were greater in cases of lymph node involvement.[7]

14.2.2.2 Central Compartment (Level VI ± VII) Dissection

In 1994, Scheumann et al.[24] analyzed the evolution of 342 patients with WDTC. A central compartment dissection was performed systematically in association with a posterior lateral dissection or with a mediastinal dissection according to the operative findings. The median follow-up was 11.8 years (from 1 to 23 years). Out of the 46 patients who died from thyroid cancer, 29 (63%) presented with lymph node metastases at the time of the primary surgery. Concerning T2 and T3 tumors, the recurrence rate and the mortality rate increased significantly in cases of lymph node metastases regardless of the age at surgery. In this work, the authors compared the prospective follow-up of 60 patients who underwent a systematic central compartment dissection during the primary surgery with that of 135 patients who underwent a selective central compartment dissection because of the presence of lymph nodes palpable either pre- or intraoperatively. Lymph node metastases were present in 73.3% of patients in the systematic dissection group and in 81.5% of patients in the selective dissection group. The disease-specific survival and the survival without recurrence were greater in the systematic dissection group. The authors concluded that prophylactic dissection of the central compartment was beneficial. The main limitation of this work was the absence of patient without dissection in the selective dissection group for assessing the disease-specific survival and the survival without recurrence.[24] More recently, Koo et al.[25] published a prospective study of 111 patients with PTC without lymph node metastases who underwent total thyroidectomy with bilateral central compartment dissection. They found that 54.4% of patients had occult central neck lymph node metastasis. Out of these patients, 50% had bilateral central neck lymph node metastases, 43.3% had unilateral ispsilateral central neck lymph nodes metastases, and 6.7% had unilateral contralateral central neck lymph node metastases. In multivariate analysis, the tumor size was an independent risk factor for the presence of unilateral ipsilateral lymph node metastases, and ipsilateral lymph nodes metastases was the only independent predictor for the presence of contralateral central neck lymph node metastases. In conclusion, the authors advocated performing a bilateral central neck dissection for all PTC greater than 1 cm.[25]

In 1992, Proye et al.[26] advocated a selective central compartment dissection. They retrospectively studied 753 patients with thyroid cancer with a follow-up greater than 7 years.[26] This study showed that the absence of a systematic lymph node dissection did not increase the disease-specific mortality. They argued that a prophylactic central compartment dissection does not prevent lymph node recurrence, which may reach 22% in the literature.[26] It should be noted that even though first-echelon nodal metastases are typically in the central compartment, this may not always be the case. For instance, lateral lymph node metastases may exist without involvement of the first-echelon central compartment (skipped metastases), and in these cases, the lymph node recurrence rate without systematic lymph node dissection can still be very low (from 5 to 7% with a follow-up of 20 years).[26]

In 2003 Shah et al.[27] studied the location and incidence of lymph node recurrences according to the initial type of lymph node resection. The study involved 103 patients with lymph node metastases and without extrathyroidal extension at the time of primary surgery, operated on by the three same surgeons in one institution.[27] All patients underwent a total thyroidectomy, with central neck dissection (compartments VI and VII (superior mediastinal)) alone (N = 6), posterior lateral neck dissection alone (N = 47), or both procedures according to the presence of macroscopic lymph node metastases (N = 21).[27] The remaining 29 patients did not undergo lymph node dissection due to the absence of macroscopic lymph node metastases at the time of the primary surgery.[27] However, all of these 29 patients were found to have microscopic metastasis in the perithyroid tissue removed with the total thyroidectomy specimen.[27] With a follow-up median duration of 104 months (from 4 to 456 months), patients who underwent only a posterior lateral neck dissection without treatment of the central compartment did not have a significantly different rate of recurrence in the central compartment than patients who underwent a central neck dissection.[27] Furthermore, of the 29 patients with microscopic perithyroidal lymph node metastases who did not undergo any lymph node resection, the recurrence rate was not significantly different than for patients who underwent a central neck dissection.[27] The authors concluded that lymph node dissection is advocated only in cases of macroscopic lymph node metastases.[27]

Other authors advocate a systematic lymph node dissection for WDTC. In 1996, Tisell et al.[28] studied 342 patients operated on for PTC. Out of the 342 patients, 60 had a systematic central neck dissection and a dissection of the lateral compartment only in cases of macroscopic lymph node metastasis.[28] With a mean follow-up of 11.8 years, a systematic compartment dissection was associated with a decrease of recurrence ($p < 0.001$) and with an increase of disease-specific survival ($p < 0.005$). In their 2007 technical review of the literature, White et al.[29] supported systematic central neck dissection and concluded that this may decrease recurrence and increase disease-free survival (based on low levels of evidence and a low recommendation grade).

Permanent hypoparathyroidism remains a major concern in patients undergoing central lymph node dissection. Henry et al.[30] compared the complication rate between patients who underwent a thyroidectomy for nontoxic multinodular goiter and patients who underwent total thyroidectomy with central neck dissection for PTC. Transient and permanent hypoparathyroididsm rates were 8% and 0%, respectively, in the total thyroidectomy alone group, and 14% and 4%, respectively, in the group with central neck dissection.[30]

The National Comprehensive Cancer Network (NCCN) CPG in oncology recommend performing central neck dissection (level VI) only if nodes are palpable or biopsy positive during total thyroidectomy for PTC, FTC, or Hürthle cell thyroid carcinoma (HTC). (Recommendation 2A: There is uniform NCCN consensus, based on lower-level evidence including clinical experience, that the recommendation is appropriate.[16]) Recommendation 27 of the ATA CPG advocates routine central compartment neck dissection for patients with PTC and suspected HTC. However, near-total and total thyroidectomy without central neck dissection may be appropriate for FTC and, when followed by RAI therapy, may represent an alternative approach for PTC and HTC. (Recommendation B: The recommendation is based on fair evidence that the service or intervention can improve important health outcomes. The evidence is sufficient to determine effects on health outcomes, but the strength of the evidence is limited by the number, quality or consistency of the individual studies; generalizability to routine practice; or indirect nature on the health outcomes.[15])

The British Association of Endocrine and Thyroid Surgeons recommend performing level VI dissection in cases of palpable metastatic nodes and in cases of clinically uninvolved nodes in the setting of high-risk PTC (male sex, age >45 years, tumors greater than 4 cm in diameter, extracapsular or extrathyroidal disease). (Grade C of Recommendation: Requires evidence from expert committee reports or opinions and/or clinical experience of respected authorities. Indicates absence of directly applicable studies of good quality.[18]) The European Association of Endocrinologists and the Société Française d'Endocrinologie recommend performing a central lymph node dissection only in cases of macroscopic lymph node disease.[17,19] At Columbia University in New York City, selective central node dissection in cases of WDTC is advocated.

14.2.2.3 Lateral Compartment (Levels II to IV) Dissection

There also exists a debate concerning the necessity of a lateral lymph node dissection. The lateral compartment includes fat tissue and lymph nodes around the internal jugular vein reaching from the carotid artery to the trapezoid muscle and from the subclavian artery to the hypoglossic nerve.[31] Some authors advocate a modified lateral neck dissection guided by the intraoperative finding or by the preoperative US.[32] This strategy consists in removing only lymph nodes visualized during the surgical exploration from the preoperative US. The US characteristics of metatstatic nodes are well described in the literature.[33] So, a diameter of 1 cm or more, a clear hypoechoic pattern or inhomogeneous pattern with alternating hypoechoic and hyperechoic areas, an irregular cystic appearance, the presence of internal calcification, and a rounded or building shape with increased antero-posterior diameter are characteristics of metastatic nodes by US.[27,33] Unfortunately, with a positive predictive value of US for lateral node metastases reaching 98%, the negative predictive value is only 42.6% and sensitivity is only 27.2%.[34] In Europe and the United States, this strategy is prevalent even though the clinical significance of nodal metastases may not be an important prognostic factor influencing survival. In other words, most lymph node recurrences can be cured by reoperation, and final outcome is therefore unaffected. This strategy is not employed in Japan, where the use of RAI is restricted by law. Furthermore, the effectiveness of RAI ablation of the cervical lymph node is limited. For these reasons, Japanese endocrine surgeons advocate prophylactic lateral lymph node dissection.[31]

Polarization of these strategies has prompted identification of factors to guide decision making regarding lateral neck compartment dissection. Machens et al.[35] showed that the more lymph node metastases that are found in the central neck, the greater the chance of lateral neck involvement. The rate of ipsilateral lateral neck compartment involvement increased from between 45% (0 node involved in the central compartment) and 69% (1 to 5 nodes involved in the central neck compartment) to 100% (>5 nodes involved in the central compartment) ($p < 0.001$).[35] Furthermore, in the absence of positive central compartment nodes, 45% of patients presented with positive lymph nodes in the lateral neck compartment on the side of the largest primary tumor.[35] In conclusion, these authors advocated performing central neck dissection associated with lateral neck dissection ipsilateral to the primary PTC when more than five lymph nodes metatstases are found in the central neck compartment or when lateral lymph nodes metastases are identified during or before the operation. The conclusion of these authors was not able to advocate routine lateral lymph node metastases in patients who fail to demonstrate any disease in the lateral neck on preoperative US. These findings are limited by the retrospective study design since only patients who had preoperative lateral lymph node metastases by US were included.[35]

Marchesi et al.[36] found that patients with a high-risk AMES score should undergo a prophylactic modified lateral neck dissection. Noguchi et al.[37] recommended a prophylactic lateral neck dissection for patients with extracapsular invasion and for women over 60 years. Ito et al.[38] advocate this prophylactic strategy in patients who present two or more of the following factors: male, age 55 years or older, tumor size greater than 3 cm, and extrathyroidal extension.

Due to the large number of false positive US findings, the sentinel lymph node technique was developed.[39] Two techniques may be used to detect the sentinel lymph node. In the first technique, the patient undergoes an injection of a radiotracer one day before surgery, and then lymphoscintigraphy is performed on the day of surgery. During the operation the surgeon uses a gamma probe to detect the sentinel lymph node, which corresponds to the hottest node positive by preoperative lymphoscintigraphy. In the second technique, blue dye is injected around the tumor at the time of surgery. Within seconds or minutes, blue dye is seen to pass through lymphatic channels toward the sentinel lymph node. Once identified, all blue-stained nodes are removed and then submitted for frozen analysis. In the literature, the rate of sentinel lymph node localization varies from 65% to 100%, and the rate of false negative results varies from 0% to 25%, according to the method used. Lee et al.[40] combined the two methods and found a sensitivity of 91%, specificity of 100%, and false negative rate of 9%. According to the literature, the goals of the sentinel lymph node technique are a target lymph node biopsy, identification of undetected lymph node metastases by US, and identification of micrometastases by histopathology. The major drawback is the false negative rate that may reach 11.7%.

The National Comprehensive Cancer Network (NCCN) CPG in oncology recommend performing lateral neck dissection (levels II to IV) only if nodes are palpable or biopsy positive during total thyroidectomy for PTC, FTC, or Hürthle cell thyroid carcinoma (HTC). (Recommendation 2A: "There is uniform NCCN consensus, based on lower level evidence including clinical experience, that the recommendation is appropriate."[16])

ATA CPG advocate lateral compartment lymph node dissection for patients with biopsy-proven metatstatic cervical lymphadenopathy detected clinically or by imaging, especially when they are likely to fail RAI treatment based on lymph node size, number, or other factors, such as aggressive histology of the primary tumor. (Recommendation B: "The recommendation is based on fair evidence that the service or intervention can improve important health outcomes. The evidence is sufficient to determine effects on health outcomes, but the strength of the evidence is limited by the number, quality, or consistency of the individual studies; generalizability to routine practice; or indirect nature on the health outcomes."[15])

The British Association of Endocrine and Thyroid Surgeons recommends performing selective lateral neck dissection (levels II-a to V-b) with the level VI dissection in cases of palpable metastatic nodes and in cases of clinically uninvolved nodes in the setting of high-risk PTC (male sex, age > 45 years, tumors greater than 4 cm in diameter, or extracapsular or extrathyroidal disease). (Grade C of recommendation: "Requires evidence from expert committee reports or opinions and/or clinical experience of respected authorities. Indicates absence of directly applicable studies of good quality.[18])

14.3 MONITORING OF THE RECURRENT LARYNGEAL NERVE (RLN) AND OF THE EXTERNAL BRANCH OF THE SUPERIOR LARYNGEAL NERVE (EBSLN)

14.3.1 Technical Elements

The three most important complications occurring during thyroidectomy are recurrent laryngeal nerve palsy (RLNP), hypoparathyroidism, and hematoma. The first complication, RLNP, can jeopardize a patient's quality of life due to dysphonia and swallowing trouble. This leads to psychological and work-related problems. It is therefore not surprising that RLNP is a common reason for litigation. In the literature, factors influencing the occurrence of RLNP are the type of thyroid disease, the extent of resection, surgical training, and surgeon experience.[41] Clearly, a major goal of thyroid surgery is avoiding RLNP. The gold standard technique consists in visualizing and dissecting the recurrent laryngeal nerve systematically, which allows a significant reduction of the RLNP rate. Extensive dissection rather than simple visualization is associated with improved outcome.[42] Despite the adoption of this recommendation by the majority of thyroid surgeons, the problem of RLNP is not resolved. Some vocal cord paralysis still occurs even with routine identification of the RLN. The classical causes of RLN injury result from transection, clamping, stretching, electrothermal injury, ligature entrapment, or ischemia.[43]

In the past decade, intraoperative nerve monitoring (IONM) has been used with increasing frequency to help surgeons localize or identify the RLN and assess nerve function during surgery.[43]

Numerous techniques have been developed to facilitate the utilization of nerve monitoring. IONM of the RLN can be accomplished by visual or electromyographic (EMG) techniques. The technique without EMG consists of observing by visualization or palpation the contraction of the posterior cricothyroid muscle, monitoring glottic pressure response, or checking vocal cord mobility during the operation. For EMG, electrodes may be either placed in contact with the vocal cord, inserted endoscopically into the vocal muscle through the cricoid ligament, or situated using the endotracheal surface.[44]

14.3.2 Evidence Base

In 2004, Dralle et al.[44] published a multi-institutional study assessing complications in thyroid surgery occurring in 63 German hospitals. The study involved 16,448 consecutive operations (29,998 RLN at risk).[44] The highest rate of permanent RLNP was observed in patients with recurrent benign and malignant goiter (3.6% and 5.1%, respectively).[44] In multivariable analysis, these two diagnoses were found to be major risk factors for postoperative permanent RLNP.[44] The study did not disclose any significant benefit using RLN monitoring compared with systematic nerve visualization despite a trend toward reduction of paralysis.[44] Furthermore, in a subgroup of low-volume surgeons, RLNP monitoring helped to reduce the permanent RLNP rate.[44]

In 2006, Chan et al.[45] prospectively compared IONM with visualization and dissection of the nerve only. There were 501 RLNs at risk in the without IONM group and 499 in the IONM group.[45] The only patient characteristic that significantly differed between the two groups was the weight of the specimen (higher in the IONM group).[45] Postoperative vocal cord paralysis was identified in 4.7% of the nerves at risk, with a complete recovery in 84% of these cases (median period of 6 months).[45] The adoption of IONM was not associated with a significant reduction in the occurrence of postoperative RLNP ($p = 0.21$).[45] The RLNP rates were 4.2% in the IONM group versus 5.2% when IONM was not used.[45] In short, this study may not advocate the routine use of IONM, but in a subgroup of high-risk thyroidectomy cases, IONM may be associated with an outcome advantage.[45]

In 2008, Dralle et al.[46] published an important systematic review assessing the results of IONM in the literature. They reported six studies comparing IONM with visual identification of the RLN only. The RLNP rate was lower, with IONM in five of them, but the difference never reached statistical significance.[46] According to the authors, this review could not prove superiority of IONM over RLN visualization for avoiding RLNP due to inadequate powering of the studies. However, there was still a trend toward improved RLNP protection with the use of the IONM technique.[46]

Despite technological advances, avoiding alteration of the voice remains a challenge in thyroid surgery. In 2002, Stojadinovic et al.[47] published a prospective functional voice assessment in patients undergoing thyroid surgery. Fifty patients without RLNP were studied and underwent voice testing using a standardized voice grading scale test and videostroboscopic test before, one week after, and 3 months after thyroidectomy. Thirty percent of these patients reported early subjective voice change, and 14% reported late voice change. Therefore, voice change is common after thyroidectomy and is caused by factors other than RLN injury.[47] These factors include trauma following endotracheal intubation, surgical trauma such as laryngofixation of the strap muscles with impairment of vertical movement, lesions of perithyroidal plexus, or an EBSLN lesion.[48]

The superior laryngeal nerve is one of the first branches of cranial nerves. The surgical importance of the external branch of the superior laryngeal nerve relates to its proximity to the superior thyroid vessels. In 68% of cases, the external branch is intimately associated with the superior pole vessels and is at risk during thyroidectomy.[49] The integrity of the EBSLN is important for the quality of the voice, and its injury can manifest as ipsilateral paralysis of the cricothyroid muscle. Clinical symptoms include a hoarse, breathy voice, increases in throat clearing, vocal fatigue, or diminished vocal frequency range, especially in regard to raising pitch.[48] The manner to protect patients from an injury of the EBSLN is still a subject of debate. For some authors, the skeletonization and individual ligation of the superior pole vessels close to the capsule of gland is sufficient to prevent EBSLN injury.[50]

For others, the systematic identification of the EBSLN is necessary.[51] Recently, we published a new technique describing monitoring of EBSLN during minimally invasive thyroidectomy under local anesthesia. After opening the skin and the flap muscles, a bipolar electrode is placed directly in the cricothyroid muscle through the skin. The EBSLN is stimulated by applying the stimulation probe directly to the EBSLN or in close proximity within the dissection area in order to verify its absence. A positive signal was determined by observing contractions of the cricoid muscle or hearing an auditory message. Therefore, using this neuromonitoring procedure, we could exclude the presence of the EBSLN in each anatomical field, step by step, before dissection and ligation.[52]

14.4 MINIMALLY INVASIVE SURGERY AND THYROID CARCINOMA

Minimally invasive thyroidectomy may be divided into four types, minimally invasive video-assisted thyroidectomy (MIVAT), complete endoscopic techniques, mini-incision open thyroidectomy, and robotic-assisted endoscopic surgery.

Minimally invasive surgical techniques have been applied since 1996.[53] In 2003, Brunaud et al.[54] suggested the term *minimally invasive thyroidectomy* for incision lengths less than 3 cm, arguing that the majority of thyroid procedures are performed with an incision size between 3.5 and 5.5 cm. Other characteristics can be used to define minimally invasive surgery, such as pain, duration of surgery, type of anesthesia (general or local anesthesia), cost, cosmetic result, and cure of the disease.[54] Surgeons must not compromise the aims and principles of thyroid surgery for an improved cosmetic result of minimally invasive surgery. The thyroid surgery technique mandates meticulous surgical dissection, absolute hemostasis, *en bloc* tumor resection, and adequate visualization of the laryngeal nerves and parathyroid glands for avoiding complications.[55] Minimally invasive surgery must also guarantee the same oncologic security as the conventional approach. The completeness of the resection must not be compromised by the limited access.[55,56]

14.4.1 Minimally Invasive Video-Assisted Thyroidectomy (MIVAT)

14.4.1.1 Technique

The MIVAT procedure is performed through a 1.5 cm incision in the middle area of the neck, approximately 2 cm above the sternal notch, as described by Miccoli et al.[57] The middle line is divided and the strap muscles are separated. From this point, the procedure is performed with a videoendoscope with an external retraction. The surgeon introduces the videoendoscope through the small skin incision. Due to the optical magnification, the surgeon obtains excellent visualization of the operative field, including the RLN and parathyroid glands. The vessels are ligated using clips or ultrasonic/bipolar energy. Once the thyroid lobe has been completely freed, it is extracted through the skin incision. The isthmus is then dissected from the trachea and divided. Usually this operation is carried out under general anesthesia.[57,58]

14.4.1.2 Evidence Base

Two Italian teams (Rome and Pisa) have developed extensive experience with this technique. In 2001, Miccoli et al.[57] reported the results of their prospective randomized study. They compared 25 patients operated on by MIVAT with 24 patients operated on by conventional thyroidectomy. There was a significant difference concerning cosmetic results ($p = 0.003$) in favor of the video-assisted technique, while there was a significantly shorter operative duration in the conventional group ($p = 0.01$).[57] The authors concluded that MIVAT was safe and offered a cosmetic advantage and decreased postoperative pain.[57] In 2002, Bellantone et al.[58] compared the video-assisted approach to conventional thyroid lobectomy. They studied 62 patients randomly assigned to MIVAT or conventional lobectomy. A conversion to conventional surgery was necessary in four cases in the MIVAT group due to difficulties identifying the RLN.[58] Despite a longer mean operating time in the MIVAT group, the authors advocated minimally invasive surgery. Indeed, at 12 weeks after lobectomy, 87% of patients who underwent MIVAT thought the outcome of their operation was excellent vs. only 45% in the conventional surgery group ($p < 0.01$).[58] Moreover, patients in the MIVAT group were more satisfied with the appearance of their scars (94% vs. 39%; $p < 0.01$).[58] The first and the second day after surgery, postoperative pain was lower in the MIVAT group ($p < 0.001$), as demonstrated by the postoperative anesthesia requirements.[58] No complication occurred in any group, but the length of postoperative hospitalization was shorter for patients operated on by minimally invasive surgery.[58] The authors concluded that MIVAT was a valid alternative to conventional surgery for treating small solitary nodules (<2 cm).[58]

Miccoli et al.[57] compared patients with thyroid cancer undergoing conventional near-total thyroidectomy group vs. MIVAT and used postoperative RAI uptake in the neck and serum thyroglobulin levels to assess the completeness of the MIVAT procedure. Thirty-three patients with nonmetastatic PTC smaller than 3.5 cm, confirmed by fine-needle aspiration, were selected for this study. All patients were hypothyroid at the time of the RAI administration and thyroglobulin assay. Although three cases of transient RLNP were immediately observed after MIVAT, no permanent deficits were documented in either group at 6 months, but one case of permanent hypoparathyroidism occurred in the MIVAT group. The authors found that MIVAT had the same completeness of resection as conventional near-total thyroidectomy in patients with low-risk PTC.[56]

14.4.2 Complete Endoscopic Techniques

14.4.2.1 Technique

Complete endoscopic thyroidectomy, as described by Gagner and Inabnet[59] in 2001, is performed with the patient under general anesthesia, using 3–5 mm instruments and a 5 mm endoscope. A monitor is placed at each shoulder of the patient. The operation begins with a 1.5 cm incision along a natural skin crease in the superior lateral area of the neck. A lateral approach is used to open the deep cervical fascia, which permits identification and exposure of the carotid sheath. Once the correct tissue plane is entered, a gauze sponge is placed in the wound to bluntly develop the space around the lateral and posterior aspect of the thyroid gland. A 10 mm trocar is inserted through the 1 cm incision; carbon dioxide insufflation is initiated to a maximum pressure of 10–12 mmHg. The endoscope is used to perform the initial dissection. Once a sufficient working space has been created, three additional small trocars are inserted under direct vision. The important anatomic structures (upper and lower parathyroid gland and recurrent laryngeal nerve) are dissected under direct endoscopic visualization. The ultrasonic scalpel is used for ligation of the thyroid vessels. The thyroid lobe is then retracted in an anteromedial direction. With the RLN and parathyroid glands in full view, inferior pole vessels, ligament of Berry, and isthmus are divided with the ultrasonic scalpel.[59,60]

14.4.2.2 Results

In 2003, Inabnet et al.[61] published the results of complete endoscopic thyroidectomy by cervical approach in 38 cases, of which 35 (92%) were completed endoscopically, with a mean surgery duration of 190 minutes. The three converted cases were completed through a small incision because of insufficient working space. There was one permanent RLNP despite the excellent visualization and dissection of the nerve. The authors concluded that this technique was safe and led to an improved cosmetic result and quicker recovery for selected patients with a small solitary nodule despite the longer duration of surgery and absence of cost benefit.[61] The same conclusions were found by the team of Marseille, France,[60] in their report of 38 patients operated on with the complete endoscopic thyroidectomy. The authors agreed with the indications by Inabnet et al.[61] and confirmed that complete endoscopic thyroidectomy should coexist with, rather than replace, conventional thyroidectomy.

14.4.3 Mini-Incision Open Thyroidectomy

14.4.3.1 Technique

Mini-incision open thyroidectomy is carried out through an incision of 3 cm or less that is placed high in the neck near the cricoid cartilage in a natural skin crease. Due to the higher incision location, large myocutaneous flaps are not required. The first step of this procedure is the ligation of the superior thyroid vessels, which is facilitated by the higher incision. The lobe is then delivered through the small incision in a similar fashion. Visualization of the recurrent laryngeal nerve and parathyroid glands is possible after retraction of the thyroid lobe. This minimally invasive surgical procedure may be performed under local anesthesia using a combined deep and superficial cervical block.[55]

14.4.4 Robotic-Assisted Thyroidectomy

Recently, a team from Seoul published a series of 100 first patients with thyroid carcinoma operated on with the Da Vinci S surgical robotic system (Intuitive Surgical, Sunnyvale, California).[62] These robotics use a transaxillary endoscopic technique that allows a three-dimensional magnifield of view, precise and multiarticulated handlike motions, a hand tremor filtering system, and an ergonomically designed operative space.[63]

14.4.4.1 Technique

The patients are placed in the supine position under general anesthesia. The neck is slightly extended and the lesion-side arm is raised and fixed for the shortest distance from the axilla to the anterior neck. A 5–6 cm vertical skin incision is made in the axilla, and the subplatysma skin flap from the axilla to the anterior neck area is dissected over the anterior surface of the pectoralis major muscle and clavicle using an electrical cautery under direct vision. The dissection is continued beneath the strap muscle until the contralateral lobe of the thyroid is exposed. After having inserted an external retractor to maintain a good working space, a second mini-incision is performed on the medial side of the anterior chest wall for insertion of the fourth robot arm. Four robot arms are used during the operation, one for the endoscope (central arm) and three for the manipulation of the surgical instruments. The dissection and ligations are usually performed using the Harmonic curved shears. The operation, consisting in total or near-total thyroidectomy with central neck dissection, is conducted in the same way as an open thyroidectomy.[64]

14.4.4.2 Results

To our knowledge, only one article concerning this robotic-assisted technique in thyroid cancer has been published.[62] A really total thyroidectomy has been performed only in 16 patients; the other patients underwent unilateral total thyroidectomy and contralateral partial lobectomy. All patients underwent a prophylactic ipsilateral compartment central node dissection. There was no conversion in conventional thyroidectomy. T1 stage was noted in 52 patients, T3 stage in 47 patients, and one patient presented with a T4 stage.[62] Sixty-nine patients were classified N0, and 31 N1.[62] The mean operative time was 136.5 ± 36.6 min.[62] The postoperative complications included one case of transient hypocalcemia and two cases of transient hoarseness. This study showed the feasibility and safety of the robotic-assisted thyroidectomy in thyroid cancer. However, the interest of the technique is principally the cosmetic result. The robotic-assisted thyroidectomy is time-consuming and is more invasive than the conventional technique because of the wide dissection from the axilla to the anterior neck. Furthermore, it is very important to take into consideration the costs associated with this technique, such as initial purchase price, disposable supplies, and maintenance. In conclusion, this technique needs further evaluation to be recommended in the surgical treatment of thyroid cancer.[62]

14.5 THYROIDECTOMY UNDER HYPNOSEDATION OR LOCAL ANESTHESIA

The definition of minimally invasive surgery should take into account not only the surgical technique, but also the anesthetic conditions. Specifically, this includes the possibility of performing a thyroidectomy without general anesthesia.

14.5.1 Hypnosedation

In cervical surgery, hypnosedation was developed by a Belgian team initially for parathyroid surgery (i.e., bilateral neck exploration for hyperparathyroidism).[65] Once this approach was shown to be feasible and safe, thyroidectomy under hypnosedation was investigated. The anesthesiologist creates a condition that allows the patient to reach a specific hypnotic state, induced by eye fixation and progressive muscle relaxation. The patient is invited to focus his or her attention on a single freely chosen memory with positive connotations, which help to detach the patient from ambient reality and to concentrate on the inner self. The anesthesiologist continually gives permissive and induced suggestion of well-being to maintain this hypnotic process. A monotonous voice is used, with intentional use of repetitive metaphoric language. A moderate degree of sensory isolation is necessary to accomplish the hypnotic state, provided in part by reducing the activity level in the operating room, eliminating unnecessary conversation, and reducing the volume level of equipment. In the hypnotic state, the patient appears immobile and relaxed, with slow roving eye movement intermingled with ocular saccades; a decrease in respiratory and heart rate is frequently observed. The patient remains conscious. At the end of the procedure, the anesthesiologist invites the patient to reestablish contact with the outside world. This serves to restore a fully conscious state within several seconds.[65]

In 1999, the same team carried out a randomized study comparing thyroidectomies using general anesthesia vs. hypnosedation.[66] The two groups were compared in terms of inflammatory, endocrine (serum cortisol and catecholamine levels), hemodynamic parameters, and postoperative variables (stress, fatigue, pain, and muscle strength). Operative time, bleeding, and surgical comfort were comparable. However, there were significant differences in inflammatory and hemodynamic parameters that were in favor of hypnosedation. Moreover, patients receiving hypnosedation had less postoperative pain, faster recovery, and less fatigue. Even though this type of anesthesia is advocated by these research teams, it has not gained in popularity.[66]

14.5.2 Local/Regional Anesthesia

Thyroidectomy under local/regional anesthesia has gained in popularity over the last 10 years. The Columbia University protocol for administering local/regional anesthesia has been well described in several publications.[67–69] Using a 1:1 mixture of 0.5% lidocaine and 0.25% bupivicaine, 40–60 ml of local/regional anesthesia is administrated under light sedation of propofol. For patients undergoing thyroid lobectomy, an ipsilateral deep/superficial field block and a contralateral superficial field block are performed, whereas patients undergoing total thyroidectomy receive a bilateral deep/superficial field block. For the deep block, 10 ml of the local anesthesia solution is administered at the posterior border of the sternocleidomastoid muscle at the level of C2 and again at C4 (respectively, two- and four-finger breadths inferior to the mastoid process). In each case, the surgeon administered the local anesthesia immediately before prepping the surgical field. The level of sedation was lessened to permit communication between the patient and surgical team.

In 2006, Spanknebel et al.[70] compared thyroidectomy under local/regional anesthesia with thyroidectomy performed under general anesthesia by a single surgeon, Paul LoGerfo, between 1996 and 2003. Out of 1194 patients, 79% were operated on under local/regional anesthesia. General anesthesia was more commonly used in patients with increased comorbidity, but there was no difference in the body mass index (BMI) between the two groups. Patients with preoperatively identified substernal or retrotracheal esophageal goiter, or patients with thyroid cancer requiring cervical lymphadenectomy, were typically not offered local/regional anesthesia. There was no difference in terms of operative time, but general anesthesia was associated with increased mean times of operating room utilization. Significantly, more patients who were operated on under local/regional anesthesia were able to benefit from a same-day discharge (82% vs. 34%). After cost analysis of the two anesthesia procedures, the authors concluded that due to an early discharge and decreased operating room utilization, thyroidectomy under local anesthesia appears to have a lower cost.[70]

In 2008, Inabnet et al.[69] published a comparison between local and general anesthesia for thyroidectomy in 224 consecutive patients operated on using minimally invasive open surgery. Eighty-four percent of patients were operated on under local/regional anesthesia during the study period. Four patients undergoing thyroidectomy under local anesthesia required conversion to general anesthesia due to the need for a deeper state of sedation. In cases of malignant pathology, there was no significant difference in mean tumor size or MACIS score between the two groups. The mean specimen weight was larger in the general anesthesia group. No difference was found in the complications rate. Patients undergoing thyroidectomy under local/regional anesthesia were more likely to be discharged within six hours.[69]

The different minimally invasive thyroidectomy techniques have proved to be feasible and safe. However, utilization of these techniques is still limited to nodules <4 cm, goiters with a volume <30 cc, thyroid cancers <2 cm, low-risk WDTC, follicular neoplasms <4 cm, and experienced thyroid surgeons. There is also a clear contraindication for using these techniques in aggressive high-risk and poorly differentiated thyroid cancers.[55] Thus, minimally invasive thyroid surgery techniques are useful procedures that should coexist with, rather than replace, conventional thyroidectomy.

REFERENCES

1. Sosa JA, Udelsman R. 2006. Total thyroidectomy for differentiated thyroid cancer. *J Surg Oncol* 94(8):701–15.
2. Caron NR, Clark OH. 2006. Papillary thyroid cancer. *Curr Treat Options Oncol* 7(4):309–19.
3. Cross S, Wei JP, Kim S, Brams DM. 2006. Selective surgery and adjuvant therapy based on risk classification of well-differentiated thyroid cancer. *J Surg Oncol* 94(8):678–82.
4. Shaha A. 2006. Treatment of thyroid cancer based on risk groups. *J Surg Oncol* 94(8):678–82.
5. Wong JB, Kaplan MM, Meyer KB, Pauker SG. 1990. Ablative radioactive iodine therapy for apparently localized thyroid carcinoma. A decision analytic perspective. *Endocrinol Metab Clin North Am* 19(3):742–60.
6. Udelsman R, Lakatos E, Ladenson P. 1996. Optimal surgery for papillary thyroid carcinoma. *World J Surg* 20(1):88–93.
7. Mazzaferri EL, Jhiang SM. 1994. Long term impact of initial surgical and medical therapy on papillary and follicular thyroid cancer. *Am J Med* 97(5):418–28.
8. Hay ID, Grant CS, Taylor WF, McConahey WM. 1987. Ipsilateral lobectomy versus bilateral lobar resection in papillary thyroid carcinoma: A retrospective analysis of surgical outcome using a novel prognostic scoring system. *Surgery* 102(6):1088–95.
9. Hay ID, Bergstralh EJ, Goelner JR, Ebersold JR, Grant CS. 1993. Predicting outcome in papillary thyroid carcinoma: Development of a reliable prognostic scoring system in a cohort of 1779 patients surgically treated at one institution during 1940 through 1989. *Surgery* 114(6):1050–57.
10. Hay ID, Grant CS, Bergstralh EJ, Thompson GB, van Heerden JA, Goelner JR. 1998. Unilateral total lobectomy; is it sufficient surgical treatment for patients with AMES low-papillary thyroid carcinoma? *Surgery* 124(6):958–64.
11. Mazzaferri EL, Kloos RT. 2001. Clinical review 128: Current approaches to primary therapy for papillary and follicular thyroid cancer. *J Clin Endocrinol Metab* 86(4):1447–63.
12. De Groot LJ, Kaplan EL, Straus FH, Shukla MS. 1994. Does the method of management of papillary thyroid carcinoma make a difference in outcome? *World J Surg* 18(1):123–30.
13. Esnaola NF, Cantor SB, Sherman SI, Lee JE, Evans DB. 2001. Optimal treatment strategy in patients with papillary thyroid cancer: A decision analysis. *Surgery* 130(6):921–30.
14. Bilimoria KY, Bentrem DJ, Clifford YK, Stewart AK, Winchester DP, Talamonti MS, Sturgeon C. 2007. Extent of surgery affects survival for papillary thyroid cancer. *Ann Surg* 246(3):375–81.

15. Cooper SD, Doherty GM, Haugen BR, Kloos RT, Lee SL, Mandel SJ, Mazzaferri EL, McIver B, Sherman SI, Tuttle R. 2006. Management guidelines for patients with thyroid nodules and differentiated thyroid cancer. *Thyroid* 16(2):1–33.
16. NCCN clinical practice guideline in oncology. Thyroid carcinoma V.I.2008 (October 18, 2008).
17. Pacini F, Schlumberger M, Dralle H, Elisei R, Smit JW, Wiersinga W, Taskforce TEtc. 2006. European consensus for the management of patients with differentiated thyroid carcinoma of the follicular epithelium. *Eur J Endocrinol* 154(6):787–803.
18. BAETS. 2007. Guidelines for the management of thyroid cancer. In Report of the thyroid cancer guidelines update group, ed. P Perros. 2nd ed. London: Royal College of Physicians.
19. Guidelines for the management of differentiated thyroid cancers. 2007. *Ann Endocrinol* (Paris) 68(Suppl 2):S57–72.
20. Noguchi S, Noguchi A, Murakami N. 1970. Papillary carcinoma of the thyroid I. Developing pattern of metastasis. *Cancer* 26(5):1053–60.
21. Wang TS, Dubner S, Sznyter LA, Heller KS. 2004. Incidence of metastatic well-differentiated thyroid cancer in cervical lymph nodes. *Arch Otolaryngol Head Neck Surg* 130:110–13.
22. Noguchi M, Kumaki T, Taniya T, Miyazaki I. 1990. Bilateral cervical lymph node metastases in well-differentiated thyroid cancer. *Arch Surg* 125(6):804–6.
23. Sellers M, Beenken S, Blankenship A, Soong SJ, Turbat-Herrera E, Urist M, Maddox W. 1992. Prognostic significance of cervical lymph node metastases in differentiated thyroid cancer. *Am J Surg* 164(6):578–81.
24. Scheumann GF, Gimm O, Wegener G, Hundeshagen H, Dralle H. 1994. Prognostic significance and surgical management of locoregional lymph node metastases in papillary thyroid cancer. *World J Surg* 18(4):559–67; discussion, 567–58.
25. Koo BS, Choi CC, Yoon YH, Kim DH, Kim EH, Lim YC. 2009. Predictive factors for ipsilateral or contralateral central lymph node metastasis in unilateral papillary thyroid carcinoma. *Ann Surg* 249:840–44.
26. Proye C, Carnaille B, Vix M, Gontier A, Franz C, Goropoulos A. 1992. [Recurrence of cervical lymph node involvement in surgically treated thyroid cancer. Uselessness of routine cervical lymph node excision (medullary carcinoma excluded)]. *Chirurgie* 118(8):448–52; discussion, 453.
27. Shah MD, Hall FT, Eski SJ, Witterick IJ, Walfish PG, Freeman JL. 2003. Clinical course of thyroid carcinoma after neck dissection. *Laryngoscope* 113(12):2102–7.
28. Tisell LE, Nilsson B, Molne J, Hansson G, Fjalling M, Jansson S, Wingren U. 1996. Improved survival of patients with papillary thyroid cancer after surgical microdissection. *World J Surg* 20(7):854–59.
29. White ML, Gauger PG, Doherty GM. 2007. Central lymph node dissection in differentiated thyroid cancer. *World J Surg* 31:895–904.
30. Henry JF, Gramatica L, Denizot A, Kvachenyuk A, Pucini M, Defechereux T. 1998. Morbidity of prophylactic lymph node dissection in the central neck area in patients with papillary thyroid carcinoma. *Langenbeck's Arch Surg* 383:167–69.
31. Ito Y, Miyachi A. 2007. Lateral and mediastinal lymph node dissection in differentiated thyroid carcinoma: Indications, benefits and risks. *World J Surg* 31:905–15.
32. Cady B. 2007. Regional lymph node metastases, a singular manifestation of the process of clinical metastases in cancer: Contemporary animal research and clinical reports suggest unifying concepts. *Ann Surg Oncol* 14:1790–800.
33. Antonelli A, Miccoli P, Ferdeghini M, Di Coscio G, Alberti B, Iacconi P, Baldi V, Fallahi P, Bashieri L. 1995. Role of neck ultrasonography in the follow up of patients operated on for thyroid cancer. *Thyroid* 5:25–28.
34. Sugitani I, Fujimoto Y, Yamada K, Yanamoto N. 2008. Prospective outcomes of selective lymph node dissection for papillary thyroid carcinoma based on preoperative ultrasonography. *World J Surg* 32:2494–502.
35. Machens A, Hauptmann S, Dralle H. 2009. Lymph node dissection in the lateral neck for completion in central node-positive papillary thyroid cancer. *Surgery* 145:176–81.
36. Marchesi M, Biffoni M, Biancari F, Berni A, Campana FP. 2003. Predictors of outcome for patients with differentiated and aggressive thyroid. *Eur J Surg* 588(Suppl):546–50.
37. Noguchi S, Murakami N, Yamashita H, Toda M, Kawamoto H. 1998. Papillary thyroid carcinoma: Modified radical neck dissection improves prognosis. *Arch Surg* 133:276–80.
38. Ito Y, Higashimaya T, Takamura Y, Miya A, Kobayashi K, Matsuzuka F, Kuma K, Miyauchi A. 2007. Risk factors for recurrence to the lymph node in papillary thyroid carcinoma patients without preoperatively detectable lateral node metastasis: Validity of prophylactic modified radical neck dissection. *World J Surg* 31:2085–91.

39. Carcoforo P, Feggi L, Trasforini S, Lanzara S, Sortini D, Zulian V, Pansini GC, Degli Uberti E, Liboni A. 2007. Use of preoperative lymphoscintigraphy and intra operative gamma-probe detection for identification of the sentinel lymph node in patients with papillary thyroid carcinoma. *EJSO* 33:1075–80.
40. Lee SK, Choi JH, Lim HI, KIim WW, Choe JH, Lee JE, Shin JH, Choi JY, Kim JH, Kim JS, Nam SJ, Yang JH. 2009. Sentinel lymph node biopsy in papillary cancer: Comparison study of blue dye method and combined radioisotope and blue dye method in papillary thyroid cancer. *EJSO* 35:974–79.
41. Thomusch O, Machens A, Sekulla C, Ukkat J, Lippert H, Gastinger I, Dralle H. 2000. Multivariate analysis of risk factors for postoperative complications in benign goiter surgery: Prospective multicenter study in Germany. *World J Surg* 24(11):1335–41.
42. Hermann M, Alk G, Roka R, Glaser K, Freissmuth M. 2002. Laryngeal recurrent nerve injury in surgery for benign thyroid diseases: Effect of nerve dissection and impact of individual surgeon in more than 27,000 nerves at risk. *Ann Surg* 235(2):261–68.
43. Chiang FY, Lu IC, Kuo WR, Lee KW, Chang NC, Wu CW. 2008. The mechanism of recurrent laryngeal nerve injury during thyroid surgery—The application of intraoperative neuromonitoring. *Surgery* 143(6):743–49.
44. Dralle H, Sekulla C, Haerting J, Timmermann W, Neumann HJ, Kruse E, Grond S, Muhlig HP, Richter C, Voss J, Thomusch O, Lippert H, Gastinger I, Brauckhoff M, Gimm O. 2004. Risk factors of paralysis and functional outcome after recurrent laryngeal nerve monitoring in thyroid surgery. *Surgery* 136(6):1310–22.
45. Chan W, Lang BH, Lo C. 2006. The role of intraoperative neuromonitoring of recurrent laryngeal nerve during thyroidectomy: A comparative study on 1000 nerves at risk. *Surgery* 140:866–73.
46. Dralle H, Sekulla C, Lorenz K, Brauckhoff M, Machens A. 2008. Intraoperative monitoring of the recurrent laryngeal nerve in thyroid surgery. *World J Surg* 32(7):1358–66.
47. Stojadinovic A, Shaha AR, Orlikoff RF, Nissan A, Kornak MF, Singh B, Boyle JO, Shah JP, Brennan MF, Kraus DH. 2002. Prospective functional voice assessment in patients undergoing thyroid surgery. *Ann Surg* 236(6):823–32.
48. Friedman M, LoSavio P, Ibrahim H. 2002. Superior laryngeal nerve identification and preservation in thyroidectomy. *Arch Otolaryngol Head Neck Surg* 128:296–303.
49. Cernea CR, Ferraz AR, Nishio S, Dutras AJ, Hojaij FC, Dos Santos LR. 1992. Surgical anatomy of the external branch of the superior laryngeal nerve. *Head Neck* 14:380–83.
50. Bellantone R, Boscherini M, Lombardi CP, Bossola M, Rubino F, De Crea C, Alesina P, Traini E, Cozza T, D'Alatri L. 2001. Is the identification of the external branch of the superior laryngeal nerve mandatory in thyroid operation? Results of a prospective randomized study. *Surgery* 130(6):1055–59.
51. Loch-Wilkinson TJ, Stalberg PL, Sidhu SB, Sywak MS, Wilkinson JF, Delbridge LW. 2007. Nerve stimultion in thyroid surgery: Is it really useful? *ANZ J Surg* 77:377–80.
52. Inabnet WB, Dhiman SV, Murry T, Aviv J, Lifante J. 2008. Neuromonitoring of the external branch of the superior laryngeal nerve (EBSLN) during minimally invasive surgery under local anesthesia: A prospective study of 10 patients. *Laryngoscope* 119(3):597–601.
53. Miccoli P, Pinchera A, Cecchini G, Conte M, Bendinelli C, Vignali E, Picone A, Marcocci C. 1997. Minimally invasive, video-assisted parathyroid surgery for primary hyperparathyroidism. *J Endocrinol Invest* 20(7):429–30.
54. Brunaud L, Zarnegar R, Wada N, Ituarte P, Clark OH, Duh QY. 2003. Incision length for standard thyroidectomy and parathyroidectomy: When is it minimally invasive? *Arch Surg* 138(10):1140–43.
55. Dhiman SV, Inabnet WB. 2008. Minimally invasive surgery for thyroid diseases and thyroid cancer. *J Surg Oncol* 97(8):665–68.
56. Miccoli P, Elisei R, Materazzi G, Capezzone M, Galleri D, Pacini F, Berti P, Pinchera A. 2002. Minimally invasive video-assisted thyroidectomy for papillary carcinoma: A prospective study of its completeness. *Surgery* 132(6):1070–73; discussion, 1073–74.
57. Miccoli P, Berti P, Raffaelli M, Materazzi G, Baldacci S, Rossi G. 2001. Comparison between minimally invasive video-assisted thyroidectomy and conventional thyroidectomy: A prospective randomized study. *Surgery* 130(6):1039–43.
58. Bellantone R, Lombardi CP, Bossola M, Boscherini M, De Crea C, Alesina PF, Traini E. 2002. Video-assisted vs conventional thyroid lobectomy: A randomized trial. *Arch Surg* 137(3):301–4; discussion, 305.
59. Gagner M, Inabnet WB 3rd. 2001. Endoscopic thyroidectomy for solitary thyroid nodules. *Thyroid* 11(2):161–63.
60. Palazzo FF, Sebag F, Henry JF. 2006. Endocrine surgical technique: Endoscopic thyroidectomy via the lateral approach. *Surg Endosc* 20(2):339–42.

61. Inabnet WB 3rd, Jacob BP, Gagner M. 2003. Minimally invasive endoscopic thyroidectomy by a cervical approach. *Surg Endosc* 17(11):1808–11.
62. Kang SW, Jeong JJ, Yun JS, Sung TY, Lee SC, Lee SL, Nam KH, Chamg HS, Chung WY, Park CS. 2009. Robot-assisted endoscopic surgery for thyroid cancer: Experience with the first 100 patients. *Surg Endosc* 5: e-pub ahead of print.
63. Miyano G, Lobe TE, Wright SK. 2008. Bilateral transaxillary endoscopic total thyroidectomy. *J Pediatr Surg* 43:299–303.
64. Kang SW, Lee SC, Lee SH, Lee KY, Jeong JJ, Lee YS, Nam KH, Chang HS, Chung WY, Park CS. 2009. Robotic thyroid surgery using a gasless, transaxillary approach and the da Vinci S system: The operative outcomes of 338 consecutive patients. *Surgery*, epub ahead of print.
65. Meurisse M, Hamoir E, Defechereux T, Gollogly L, Derry O, Postal A, Joris J, Faymonville ME. 1999. Bilateral neck exploration under hypnosedation: A new standard of care in primary hyperparathyroidism? *Ann Surg* 229(3):401–8.
66. Defechereux T, Degauque C, Fumal I, Faymonville ME, Joris J, Hamoir E, Meurisse M. 2000. Hypnosedation, a new method of anesthesia for cervical endocrine surgery. Prospective randomized study. *Ann Chir* 125(6):539–46.
67. Lo Gerfo P. 1998. Local/regional anesthesia for thyroidectomy: Evaluation as an outpatient procedure. *Surgery* 124(6):975–78; discussion, 978–79.
68. Lo Gerfo P, Ditkoff BA, Chabot J, Feind C. 1994. Thyroid surgery using monitored anesthesia care: An alternative to general anesthesia. *Thyroid* 4(4):437–39.
69. Inabnet WB, Shifrin A, Ahmed L, Sinha P. 2008. Safety of same day discharge in patients undergoing sutureless thyroidectomy: A comparison of local and general anesthesia. *Thyroid* 18(1):57–61.
70. Spanknebel K, Chabot JA, DiGiorgi M, Cheung K, Curty J, Allendorf J, LoGerfo P. 2006. Thyroidectomy using monitored local or conventional general anesthesia: An analysis of outpatient surgery, outcome and cost in 1194 consecutive cases. *World J Surg* 30(5):813–24.

15 Thyroid Surgery Performed under Local Anesthesia and Same-Day Surgery

Joseph DiNorcia and John D. Allendorf

CONTENTS

ABSTRACT: The surgical management of benign and malignant nodular thyroid disease can now be performed under locoregional anesthesia and with same-day discharge. Newer anesthetic agents and locoregional block types (superficial cervical plexus, deep cervical plexus, and anterior neck field) are reviewed. Concerns that have traditionally precluded same-day surgery are now better addressed with new hemostasis technologies, intraoperative parathyroid hormone assays, and improved methods for patient selection. These procedures demonstrate benefit for the hospital, surgeon, and patient, and should find increased use, especially with future innovations and optimization.

15.1 INTRODUCTION

The number of thyroid operations performed for both benign and malignant disease has increased steadily over the past several decades. Patient preference and movement toward more efficient, economic healthcare delivery have driven a shift from performing thyroid surgery under general anesthesia with planned hospital admission to performing it under local anesthesia as a same-day procedure. Nevertheless, the central tenets of thyroid surgery remain: extirpate the primary disease, preserve the parathyroid glands, avoid injury to the laryngeal nerves, and achieve excellent hemostasis and cosmesis without morbidity or mortality.[1] This chapter explores the art of operating on the thyroid gland under locoregional anesthesia and safely discharging the patient within hours of the operation's end. Innovations such as these optimize surgical options and will certainly shape clinical algorithms for the management of thyroid nodules and thyroid cancer. The first section of this chapter specifically addresses the use of locoregional anesthesia in select patients; the second section addresses thyroid surgery as a same-day procedure.

15.2 LOCOREGIONAL ANESTHESIA

15.2.1 Historical Perspective

Traditionally, surgeons used locoregional anesthesia to operate on the thyroid gland in patients with thyrotoxicosis.[2] Early in the twentieth century, the morbidity and mortality associated with general anesthesia, particularly for patients with this disease, were considerable.[3] Sir Thomas Peel Dunhill first reported the use of locoregional anesthesia for thyroidectomy on patients with thyrotoxicosis.[4] By the end of 1907, he had performed a total of seven thyroid operations under locoregional anesthesia.[5] Dunhill's contemporary, George W. Crile, had performed more than 20,000 thyroidectomies under locoregional anesthesia by 1932.[6] As the medical management of thyrotoxicosis advanced and as general anesthesia improved, thyroid operations were performed almost exclusively under general anesthesia.[7] Today, thyroidectomy for both benign and malignant thyroid disease is performed most commonly with the patient under general anesthesia.[8]

Within the past 30 years, there has been renewed interest in operating on the thyroid gland with patients under locoregional anesthesia. The reasons for this renaissance are many, including patient preference for avoiding the risks of general anesthesia and an international movement toward more efficient and economical surgery.[5,9] Initial experiences involved patients at two ends of a spectrum: healthy patients with unilateral benign disease on one end and poor operative risk patients with contraindications to general anesthesia on the other.[8] More recent experiences, however, have shown that locoregional anesthesia can be used for a variety of thyroid gland pathology in diverse patient populations with minimal morbidity and mortality.[10,11] It is relatively simple, safe, and effective, with the added benefits of fewer side effects, shorter recovery periods, and decreased costs when compared to general anesthesia. Before exploring the indications, techniques, and benefits of locoregional anesthesia, we briefly comment on the use of general anesthesia in thyroid surgery.

15.2.2 General Anesthesia

General anesthesia remains the primary anesthetic method in many modern endocrine surgery practices.[8] General anesthesia has benefits for the surgeon, anesthesiologist, and patient. General anesthesia allows the surgeon to operate in a controlled, motionless field, provides the anesthesiologist with a secure airway, and keeps the patient unaware of the surgical procedure throughout its duration.[10,12] However, there are well-known side effects and risks. Major side effects include postoperative nausea, vomiting, sore throat, and disorientation. Major risks include myocardial infarction, cerebral vascular accident, and death.[10] Despite these drawbacks, certain patient and disease characteristics are better served by general anesthesia. For patients with severe claustrophobia, anxiety, or history of panic attacks, or for patients who prefer complete unconsciousness, general anesthesia should be used over locoregional anesthesia. Similarly, for large multinodular goiters that may require significant manipulation, substernal disease that may require sternotomy, and widely invasive, locally metastatic carcinomas that may require extensive dissection, general anesthesia may be the wiser choice.[12,13] Moreover, when performing a thyroid operation under locoregional anesthesia, general anesthesia always should be available as an alternative should the patient not tolerate the procedure, become uncooperative or difficult to sedate, or manifest signs or symptoms of instability.[10] Inadequate locoregional nerve block and unexpected intraoperative findings that necessitate a more involved procedure also may require conversion to general anesthesia. It is important to note, however, that complex neck operations such as laryngectomy with bilateral neck dissection can be performed under locoregional anesthesia, and a high-quality bilateral cervical plexus block allows most thyroid operations to be accomplished in an oncologically sound manner, regardless of intraoperative surprises.[7] Finally, even patients having thyroid surgery under general anesthesia benefit from the use of a preoperative locoregional nerve block, which can significantly reduce postoperative incision pain.[14]

15.2.3 Patient and Disease Selection for Locoregional Anesthesia

The nature of locoregional anesthesia requires an excellent doctor-patient relationship as well as active patient participation in the operating room and throughout the perioperative period. Although there are no absolute indications for thyroidectomy under locoregional anesthesia, some patients may derive considerable benefit.[15] Patients must meet basic requirements and surgeons must select them carefully to ensure a safe and effective thyroid operation with this technique. The ideal candidates are young, healthy, emotionally stable patients who are well informed and motivated.[3,6] A history of tolerating local anesthesia for a dental procedure or of a positive experience with other locoregional anesthesia is a good predictor of success.[15] Patients who have a morbid fear of general anesthesia or those who prefer locoregional anesthesia for other reasons are also good candidates, as are patients with compromised cardiopulmonary function or other contraindications to general anesthesia who otherwise meet the criteria for locoregional anesthesia.[15,16]

Candidates for locoregional anesthesia must be able to communicate openly with the surgeon. In the preoperative period, communication enables proper patient education about what to expect before, during, and after the operation. In the operating room, communication is equally important. The surgeon must be able to provide reassurance, assess the adequacy of the locoregional block, and monitor the patient's voice, particularly when identifying and preserving the laryngeal nerves. The surgeon also needs patient cooperation with positioning and a motionless operative field, which may be particularly challenging when the patient is in a sedated, dissociative state. Thus, the presence of communication barriers such as foreign language or deafness may preclude the use of locoregional anesthesia unless medically trained interpreter services are readily available. Likewise, inability to interact with the operating team in a cooperative manner, as may be the case with children, the mentally impaired, or otherwise immature individuals is a relative contraindication for using this technique.[15,17]

Patient allergy to local anesthetic agents is an obvious contraindication to locoregional anesthesia. Other contraindications include patients with physical characteristics that may present difficulties with positioning and exposure. Patients who cannot lie supine or extend their neck for extended periods of time because of arthritis, deformity, or other reasons are not good candidates.[15] Similarly, patients with morbid obesity, large neck circumference, or sleep apnea may pose problems for both the surgeon and anesthesiologist.[6,10] The patient who is a known difficult intubation deserves careful consideration, as rapid intubation for airway protection acts as a safety net during thyroid surgery under locoregional anesthesia. Strong patient preference for general anesthesia can be considered a relative contraindication, provided the patient meets the requirements for general anesthesia.

Preoperative knowledge of the nature of the thyroid disease as well as the type and extent of operation needed should be taken into consideration. Several reports in the literature suggest that locoregional anesthesia can be used to perform thyroidectomy across a wide range of pathologies, from hemithyroidectomy for benign nodule to total thyroidectomy with cervical lymphadenectomy for metastatic carcinoma.[8,13] One unique application of locoregional anesthesia involves patients with amiodarone-induced hyperthyroidism. These patients are at greater risk of complications from general anesthesia because of their thyrotoxicosis and significant medical comorbidities. Williams and Lo Gerfo[18] reported a series of seven patients with amiodarone-induced hyperthyroidism, all of whom had successful thyroidectomy under locoregional anesthesia.

The use of locoregional anesthesia to manage certain thyroid disease states is contraindicated. Examples include those with the potential need for sternotomy (e.g., retrosternal multinodular goiter), extensive dissection (e.g., invasive, metastatic carcinoma), or other simultaneous procedures that will require general anesthesia.[6,13] The length of the operation and the potential need for large volumes of local anesthetic agent to provide adequate analgesia should figure into the surgeon's decision as well.[15] A patient cannot be expected to tolerate any procedure under locoregional anesthesia, let alone one involving neck structures, much longer than 2 hours. The surgeon also must consider the preoperative status of the laryngeal and phrenic nerves. Saxe et al.[16] reported a patient with known preoperative vocal cord paralysis secondary to involvement of the recurrent laryngeal nerve (RLN) by invasive thyroid carcinoma. The patient experienced respiratory distress after cervical plexus block and required intubation, suggesting that preoperative RLN or phrenic nerve dysfunction is a relative contraindication to locoregional anesthesia. By avoiding these scenarios, the surgeon can maximize patient cooperation and comfort and minimize complications that might otherwise result from performing a lengthy operation under locoregional anesthesia on an unsuitable patient.[9]

15.2.4 Patient Preparation for Locoregional Anesthesia

Thorough preoperative patient preparation is essential for successful thyroid surgery under locoregional anesthesia and requires a detailed, step-by-step explanation of the process from the preoperative period, through the operating room, to the postoperative period.[15] Consultation with the anesthesiologist, placement of an intravenous catheter, and premedication with an anxiolytic agent are important preoperative events to discuss. Positioning, expected sensations, and length of procedure are important intraoperative issues to address as well. Lo Gerfo[9] suggests laying the patient on an exam table during a preoperative office visit and applying gentle pressure to the trachea to mimic the sensation of intraoperative manipulation and retraction. To allay common fears, the surgeon should emphasize that the locoregional block in combination with intravenous sedation will be titrated to comfort, and that general anesthesia can be administered at any time should the procedure become intolerable under locoregional anesthesia.[15] A general description of operating room activity further encourages the patient's mental preparation. Finally, the duration of in-hospital observation, potential postoperative complications, and normal recovery process should be addressed. By so preparing the patient, the surgeon optimizes the conditions for successful thyroid surgery under locoregional anesthesia.

15.2.5 Monitored Anesthesia Care

Locoregional anesthesia alone may be sufficient to perform limited thyroid resections, but nearly always is combined with monitored anesthesia care to ensure favorable operative conditions and patient comfort. Monitored anesthesia care requires diligence and attention from the anesthesiologist and patience and willingness from the surgeon to operate under changeable conditions.[10]

Indeed, the success of performing thyroid surgery under locoregional anesthesia depends on both the coordination of intravenous sedation by the anesthesiologist and the efficacy of the anesthetic block by the surgeon. While complete details of monitored anesthesia care are beyond the scope of this chapter, we briefly describe the most relevant aspects here.

Once in the operating room, the patient is premedicated with anxiolytic, analgesic, and antiemetic agents. Appropriate monitoring devices are placed, and supplemental oxygen is administered via nasal prongs. The anesthesiologist then induces a state of light, conscious sedation using a combination of intravenous medications, the most common including midazolam, fentanyl, and propofol.[10] Propofol is a particularly useful agent for induction and maintenance of sedation during thyroid surgery because its rapid onset and short duration of action allow for minute-to-minute titration to meet the patient's comfort needs and the surgeon's technical needs.[12] Additionally, propofol is not associated with the nausea and vomiting commonly induced by other anesthetic agents, an important characteristic for both patient safety and comfort throughout the perioperative period.[19,20]

Newer agents such as dexmedetomidine, an alpha-2 receptor agonist with similar sedative and analgesic properties, may be useful adjuncts for thyroid surgery under locoregional anesthesia. Plunkett et al.[21] report a case of a patient with significant medical comorbidities, including congestive heart failure, chronic obstructive pulmonary disease, obstructive sleep apnea, and unilateral vocal cord dysfunction secondary to disease involvement of the right RLN. This patient successfully underwent total thyroidectomy for papillary thyroid carcinoma using local anesthesia supplemented with dexmedetomidine. The surgeon preferred to avoid intubation and operate with the patient under minimal sedation in order to monitor the patient's voice during difficult dissection around the right RLN. The surgeon used an anterior neck field block alone because of concern about the potential for phrenic or RLN paralysis associated with cervical plexus block in this patient with little respiratory reserve and preexisting vocal cord dysfunction. The surgeon completed the operation without complications, and the patient recovered well with high satisfaction scores. This case report suggests that advances in monitored anesthesia care make thyroidectomy under minimal locoregional anesthesia feasible and safe even for the most challenging patients.

15.2.6 Patient Positioning for Locoregional Anesthesia

While the anesthesiologist prepares and delivers monitored anesthesia care, the surgeon positions the patient and delivers the locoregional block. The patient lies supine on the operating room table with his or her head in a neutral position, arms resting at the side. A rolled towel is commonly placed underneath the patient at the level of the shoulder blades to help hyperextend the neck for optimal exposure of the thyroid gland. The surgeon then delivers the locoregional block under sterile technique, as discussed below. It is important for the surgeon to administer the block prior to the operative scrub to allow sufficient time for the anesthetic to take effect.[9] Moreover, it has been shown that delivering local anesthetic agent prior to incision provides more effective postoperative analgesia.[22] The surgeon then scrubs, preps the patient with antiseptic solution, and places sterile drapes in the usual fashion, taking care to keep the drapes away from the patient's face. An ether screen can help elevate the drapes and limit feelings of claustrophobia in the patient.

15.2.7 Types of Local Anesthesia

There are several different types and concentrations of local anesthetic agent that can be used alone or in combination with similar effects. The choice depends mostly on surgeon preference, with consideration of the anticipated length of operation.[10] Epinephrine can be added to prolong the duration of local anesthetic action, but should be used with great care, particularly in patients with severe arteriosclerotic cardiovascular disease.[15] The volume of local anesthetic agent used determines the success of the locoregional block. Large volumes of dilute local anesthetic agent work most effectively, provided they are below toxic levels. The authors, for example, use a total volume of 40–60 ml of a 1:1 mixture of 0.5% lidocaine and 0.25% bupivicaine without epinephrine when delivering an anterior neck field block combined with superficial and deep cervical plexus block.

Patients with a personal or family history of malignant hyperthermia usually are good candidates for the locoregional technique, but a word of caution is deserved. An ester-based local anesthetic agent such as procaine or tetracaine should be used in these patients over amide-based agents such as lidocaine or bupivicaine. Via an effect on the sarcoplasmic reticulum, the amide-based agents raise the concentration of calcium in muscle cells, which ultimately can precipitate malignant hyperthermia.[15]

15.2.8 Relevant Anatomy

The surgeon must be intimately familiar with the anatomy of the neck and its pertinent neurovascular structures to deliver an effective locoregional nerve block while avoiding potential complications. The cervical plexus, which provides sensory and motor innervation to the neck and posterior scalp, arises from the ventral nerve roots of the second, third, and fourth cervical nerves. The cervical nerves exit the spinal canal through the intervertebral foramina and travel between the anterior and posterior tubercles of the transverse processes of the cervical vertebrae. The anterior and middle scalene muscles attach to the anterior and posterior tubercles, respectively, forming a fascial compartment through which the cervical nerves pass. Within this compartment, the cervical nerves give off roots that form the cervical plexus.[23] Roots from the third, fourth, and fifth cervical nerves also give rise to the phenic nerve, which travels caudally along the anterior scalene muscle to innervate the ipsilateral hemidiaphragm. The brachial plexus emerges between the anterior and middle scalene muscles immediately caudal to the cervical plexus and innervates the ipsilateral arm. Finally, the carotid sheath, which houses the carotid artery, internal jugular vein, and vagus nerve, lies anterior to the anterior scalene muscle and deep to the sternocleidomastoid muscle. Knowledge of these structures in relation to the cervical plexus is critically important for avoiding the common complications of cervical plexus block.

The cervical plexus is unique because its sensory and motor fibers separate into superficial and deep branches early in their course between the scalene muscles.[17] The superficial branches of the cervical plexus provide sensory innervation to the anterior neck; the deep branches provide motor innervation to muscles inserting on the first four cervical vertebrae.[23] Block of the superficial branches thus provides anesthesia to the relevant structures in the anterior neck, whereas block of the deep branches provides some muscle relaxation to the neck. It is rare to block the deep branches in isolation because they travel in the same fascial compartment as the superficial branches. Block of the deep branches most often provides some anesthesia to the superficial branches as well.[17]

The superficial branches of the cervical plexus emerge from the mid-posterior border of the sternocleidomastoid muscle and fan out to form the lesser occipital (C2), greater auricular (C2, C3), transverse cervical (C2, C3), and supraclavicular (C3, C4) nerves.[16] Branches of the transverse cervical nerves supply the main sensory innervation to the platysma, strap muscles, thyroid gland, and surrounding tissues. In theory, a unilateral block of the transverse cervical nerves should provide adequate analgesia for a hemithyroidectomy, but in reality many of the transverse cervical nerve endings cross to the opposite side. A bilateral block thus provides more complete analgesia.[10]

15.2.9 Locoregional Block Types

The three types of locoregional blocks used in thyroid surgery include the superficial cervical plexus block, the deep cervical plexus block, and the anterior neck field block. There is literature in support of each of these blocks, individually or in concert, depending on the planned thyroid resection. The three blocks can be administered unilaterally or bilaterally as well. There is no consensus, and the

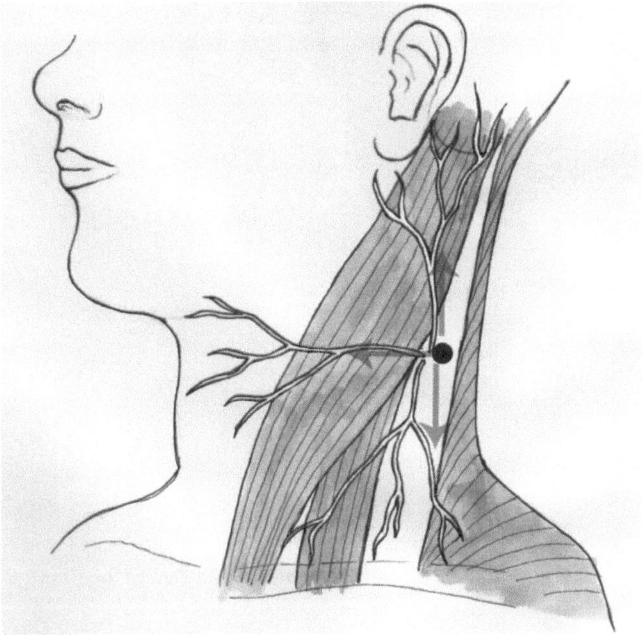

FIGURE 15.1 The superficial branches of the cervical plexus emerge from the posterior border of the sternocleidomastoid muscle. The superficial cervical plexus block is delivered in three directions to anesthetize the greater auricular, transverse cervical, and supraclavicular nerves.

thyroid disease, patient anatomy, and surgeon preference dictate the block. The authors routinely use an anterior neck field block in combination with bilateral superficial and deep cervical plexus block, depending on the type of thyroidectomy.

The superficial cervical plexus block is a relatively simple, safe, and effective method for providing analgesia during thyroid surgery.[24] The patient's head is turned to the opposite side, and the sternocleidomastoid muscle is identified by asking the patient to lift his or her head.[23] The point where the external jugular vein crosses the posterior border of the sternocleidomastoid muscle is a useful landmark for the emergence of the superficial branches of the cervical plexus.[3] After a sterile prep, a superficial skin wheal is placed to anesthetize the point of needle entry. Local anesthetic agent then is injected in the subcutaneous plane above the sternocleidomastoid muscle in three directions at its posterior border to block all of the superficial branches of the cervical plexus: superiorly the lesser occipital and greater auricular nerves, medially the transverse cervical nerves, and inferiorly the supraclavicular nerves (Figure 15.1).[10,24,25] Adequate block of the transverse cervical nerves is essential. Local anesthetic agent next is injected in the subfascial plane below the sternocleidomastoid muscle at its posterior border to block deeper sensory branches of the cervical plexus.[16] The needle must pass through the superficial layer of the cervical fascia to reach the proper plane below the muscle to anesthetize these branches, with care taken not to penetrate too deeply as to infiltrate the structures of the carotid sheath or phrenic nerve.[25]

The deep cervical plexus block provides a more complete regional block, but is technically more difficult to perform, and thus carries greater potential for complications.[10] For this reason, some authors advocate the superficial over the deep cervical plexus block.[16,17,24,26] Various techniques are described, including multiple separate injections at the transverse processes of C2, C3, and C4, or a single injection at C3.[17,23] Because the nerves travel in a common fascial compartment between the scalene muscles, a single injection at the C3 level usually suffices because the local anesthetic agent spreads within the compartment to affect the branches of C2, C3, and C4.[23] The patient's head should be in the midline position while the mastoid process superiorly and the transverse process of

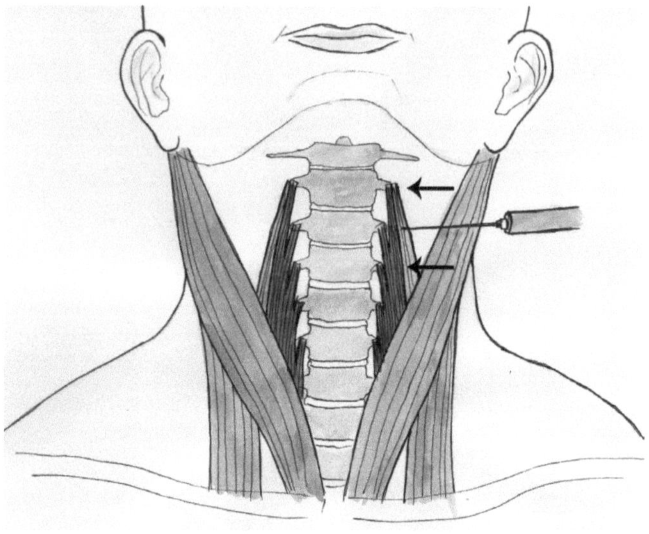

FIGURE 15.2 The deep cervical plexus block is administered lateral to the transverse processes of C2, C3, and C4 in the interscalene groove.

C6 inferiorly are identified (Figure 15.2). The transverse process of C6, also known as Chassaignac's tubercle, is the most prominent and easily palpated cervical transverse process.[7,16] The transverse processes of C2, C3, and C4 can be palpated posterior to a line drawn from the mastoid process to the transverse process of C6. If the transverse processes are difficult to palpate, centimeter approximations can be used. C2 is located about 1.5 cm below the mastoid process, C3 is located about 1.5 cm below C2, and so on.[16] After a sterile prep and skin wheal, the needle is inserted perpendicularly over the transverse process of interest and advanced until it contacts bone. As the needle is walked caudally, it will slip off the bone if it is on the transverse process, as opposed to continuing to contact bone if it is on the vertebral body.[23] As the needle is walked posteriorly, the most superficial point of contact with bone will be the terminal aspect of the transverse process, where the cervical nerves exit the vertebral column. Once this terminal aspect is contacted, the needle is withdrawn about 1 cm to place its tip within the interscalene groove, and the local anesthetic agent is injected.[23] As with the superficial block, an adequate deep cervical plexus block provides anesthesia to the C2, C3, and C4 dermatomes.[16]

The final locoregional block is the anterior neck field block. It is a useful adjunct to the cervical plexus block, but can be used alone in certain cases. The anterior neck in the region of the incision, preferably in a natural skin crease, is anesthetized using a small-gauge needle in multiple directions to create a broad field block. Care should be taken to infiltrate the subcutaneous tissues and not the dermis to avoid a painful burning sensation.[9] The anterior neck field block, used alone, is primarily indicated for central thyroid resections (i.e., isthmusectomy). Total thyroidectomy has been performed under an anterior neck field block without a cervical plexus block, but the anesthesia provided by a cervical plexus block facilitates deeper dissection and reduces the need for repeated injections of local anesthetic agent during the operation.[10,12]

Three technical points deserve emphasis. First, prior to injecting any local anesthetic agent, it is crucial to aspirate and check for cerebrospinal fluid or blood to avoid intrathecal or intravascular injection. Second, adequacy of the locoregional block should be tested by pinprick or other method before making incision. Finally, additional local anesthetic agent always should be at the ready to relieve any unanticipated patient discomfort caused by the dissection.[9] The superior poles of the thyroid gland often are not adequately anesthetized by the locoregional block.[27] The authors routinely inject local anesthetic agent in the region of the superior poles prior to mobilization to avoid the patient discomfort often associated with dissection in this area.

15.2.10 Complications and Disadvantages of Locoregional Anesthesia

Complications of locoregional anesthesia generally are associated with the administration of the locoregional block. Inadvertent penetration through the intervertebral foramina can lead to intrathecal injection and total spinal anesthesia, or can lead to vertebral artery injection and seizure.[9,23] Misdirection of the needle into the carotid sheath can lead to carotid artery or internal jugular vein injection with subsequent seizure.[9] Carotid sheath injection also can affect the vagus nerve and RLN, causing possible hemodynamic instability and vocal cord dysfunction, respectively. This latter complication may lead to temporary hoarseness and difficulty swallowing if unilateral, or complete airway obstruction if bilateral, necessitating emergent conversion to general anesthesia.[10]

Infiltration of the phrenic nerve with local anesthesia can temporarily paralyze the diaphragm, resulting in respiratory compromise if unilateral, or respiratory failure if bilateral. Certain authors advise against bilateral deep cervical plexus block in order to avoid bilateral phrenic nerve paralysis.[23,24] Infiltration of the brachial plexus can temporarily paralyze the ipsilateral arm and mimic a cerebral vascular accident. By staying at or above the level of the C4 transverse process, the surgeon can avoid anesthetizing the phrenic nerve and brachial plexus.[9,16] Other complications include injection into the thyroid or external jugular veins leading to local anesthetic toxicity and injection into the substance of the thyroid gland causing a localized hematoma. Knowledge of the relevant anatomy, meticulous attention to needle depth and direction, and routine aspiration of the syringe prior to injection are fundamental technical points to ensure safe and effective administration of locoregional anesthesia.

Many disadvantages of locoregional anesthesia manifest in the operating room for the surgeon, anesthesiologist, and patient. The surgeon may contend with occasional motion of the operative field due to swallowing, coughing, or other patient movement.[13] The anesthesiologist must monitor the patency of the airway and adequacy of the sedation more closely than in the case of general anesthesia, making frequent adjustments to optimize oxygenation, operative conditions, and patient comfort.[12] In addition, should rapid conversion to general anesthesia be necessary, intubation can be difficult without contaminating the sterile field.[28] Finally, the patient is aware of being in the operating room and may sense pulling and pressure in the surgical wound.

Most of the large series in the literature, however, demonstrate very low rates of adverse events for thyroid surgery performed under locoregional anesthesia, with morbidity and mortality rates comparable to those described for thyroid surgery performed under general anesthesia (Table 15.1). In a review of the literature from 1975 to 2006, Arora et al.[10] reported a 4.7% overall complication rate, a 1.6% conversion to general anesthesia rate, and a 0.04% mortality rate for patients who had thyroid surgery under locoregional anesthesia. Furthermore, Snyder et al.[12] performed a prospective randomized study comparing locoregional anesthesia with monitored anesthesia care versus general anesthesia in 58 patients undergoing thyroidectomy. They performed all of the thyroid operations using an anterior neck field block without cervical plexus block and found no significant differences in postoperative morbidity and mortality between the two groups. Together, these data suggest that, despite specific challenges, thyroid surgery under locoregional anesthesia is feasible and safe.

15.2.11 Advantages of Locoregional Anesthesia

There are several advantages to using locoregional anesthesia. Foremost, use of locoregional anesthesia circumvents most of the risks and side effects associated with general anesthesia.[12,25,28] It is useful for pregnant and other high general anesthetic risk patients. No intubation spares patients

TABLE 15.1
Low Morbidity and Mortality with the Use of Locoregional Anesthesia in Thyroid Surgery

Reference	Year	No. Patients	No. Complications (%)	No. Conversions to General (%)	Mortality (%)
Inabnet et al. (29)	2008	184	2 (1.2)	4 (1.8)	0 (0)
Snyder et al. (12)	2006	29	5 (17)	0 (0)	0 (0)
Spanknebel et al. (8)	2005	1,025	39 (3.8)	34 (3.3)	0 (0)
Hisham & Aino (5)	2002	65	2 (3)	0 (0)	0 (0)
Specht et al. (6)	2001	58	0 (0)	2 (3.4)	0 (0)
Samson et al. (30)	1997	809	37 (4.6)	0 (0)	1 (0.1)
Hochman & Fee (15)	1991	21	6 (28)	1 (4.8)	0 (0)
Saxe et al. (16)	1988	17	0 (0)	2 (12)	0 (0)
Fernandez (31)	1984	433	27 (6.2)	0 (0)	0 (0)
Cunningham & Lee (32)	1975	43	2 (4.6)	1 (2.3)	0 (0)

from complete unconsciousness and potential vocal cord trauma, making it particularly useful for patients who fear the vulnerability associated with general anesthesia and for patients whose careers depend on voice (e.g., singers, public speakers).[8,13] In addition, locoregional anesthesia allows immediate intraoperative assessment of the laryngeal nerves because the patient is awake and can phonate if necessary. Critics of the locoregional technique express concern that a patient who is awake and can move has increased risk of inadvertent RLN injury.[8,13] Large series in the literature, however, have demonstrated no correlation between rate of RLN injury and type of anesthesia used.[8,13]

Locoregional anesthesia is associated with shorter total operating room times when compared to general anesthesia.[10,13,28] This effect is partly attributed to the extra time required to prepare, administer, and awaken the patient from general anesthesia.[14] Locoregional anesthesia also is associated with shorter recovery room times for several reasons. First, patients regain alertness and function more quickly. Second, patients experience less postoperative nausea and vomiting. Third, the analgesia provided by the locoregional block continues into the postoperative period, tempering the need for nauseating oral narcotics and modifying the body's stress response to surgery.[15,17,24] These effects translate into more rapid recovery, shorter hospital stays, and reduced costs.[2,10–12,30] Lo Gerfo[14] noted an additional benefit of the prolonged effect of the locoregional block; reoperation for postoperative bleeding, when necessary, usually can be performed without additional anesthetic if done early.

Finally, locoregional anesthesia offers patients a sense of autonomy. Several authors report higher overall patient satisfaction with the locoregional anesthesia experience.[3,14,30] Snyder et al.[12] reported significantly shorter recovery times in the locoregional group compared to the general anesthesia group. Patients in each group reported equivalent satisfaction with the type of anesthesia and overall surgical experience as well. Thyroid surgery under locoregional anesthesia thus has definite advantages, including avoidance of general anesthesia, more rapid recovery, and greater potential for same-day discharge.

15.3 SAME-DAY THYROID SURGERY

15.3.1 Basic Rationale

The length of postoperative hospital stays continues to decrease as operations ranging from the relatively simple (e.g., hemorrhoidectomy and herniorrhaphy) to the more complex (e.g.,

TABLE 15.2
Feasibility of Same-Day Discharge after Thyroid Surgery[a]

Reference	Year	No. Patients	Time to Discharge		
			Same Day (%)	<24 h (%)	>24 h (%)
Teoh et al.[35]	2008	50	49 (98)	1 (2)	0 (0)
Materazzi et al.[36]	2007	1,571	0 (0)	1,543 (98.2)	28 (1.7)
Terris et al.[33]	2007	91	52 (57.1)	26 (28.6)	13 (14.3)
Snyder et al.[12]	2006	58	51 (88)	ND	ND
Spanknebel et al.[13]	2006	1,194	778 (65)	232 (19.4)	ND
Sahai et al.[37]	2005	104	ND	104 (100)	0 (0)
Specht et al.[6]	2001	175	ND	143 (81.7)	ND
Lo Gerfo[14]	1998	203	134 (66)	ND	ND
McHenry[38]	1997	80	ND	71 (88.8)	ND
Samson et al.[30]	1997	809	ND	809 (100)	0 (0)
Mowschenson and Hodin[39]	1995	100	61 (61)	ND	ND
Marohn and LaCivita[40]	1995	150	ND	145 (96.7)	ND
Lo Gerfo et al.[41]	1991	134	76 (56.7)	53 (39.5)	5 (3.7)
Steckler[42]	1986	48	41 (85.4)	3 (6.2)	4 (8.3)

[a] Literature review demonstrates the feasibility of same-day discharge after thyroid surgery, with the majority of patients leaving the hospital within 24 hours of their operation. ND, not documented.

appendectomy and cholecystectomy) are performed on a same-day basis.[33,34] Driven in part by the insurance industry and in part by patient advocacy, approximately 70% of all elective surgical procedures in the United States are performed on an outpatient basis.[34] Same-day surgery benefits the hospital, the surgeon, and the patient. For the hospital, it represents a more efficient use of resources; for the surgeon, a more effective use of time. For the patient, same-day surgery means reduced exposure to nosocomial pathogens and iatrogenesis, while also providing the psychosocial benefit and convenience of convalescence in the patient's own home.[33–35] The key issue for same-day surgery, however, remains balancing the benefits of early discharge with optimal patient recovery and safety.

Maintaining this balance is the crux of performing thyroid surgery as a same-day procedure. Traditionally, patients who underwent thyroid surgery were observed in the hospital for 72 hours or longer to monitor for potentially lethal complications such as cervical hematoma, hypocalcemia, and RLN injury.[33,36,37] In recent years, several authors have published large series demonstrating that same-day thyroid surgery is feasible with minimal morbidity, readmission, and mortality (Tables 15.2 and 15.3). They maintain that same-day thyroid surgery is practical, efficacious, and safe with the benefit of reduced costs. Other authors adamantly oppose the notion, arguing that postoperative complications render same-day thyroid surgery unnecessary and dangerous.[36,43] However, with improvements in anesthesia, the advent of dedicated same-day surgery units, and the creation of thorough postoperative discharge criteria, thyroid operations are being performed successfully on an outpatient basis with increasing frequency.

15.3.2 Historical Perspective

The first report of same-day thyroid surgery was published by Steckler in 1986.[42] The author assessed the feasibility of same-day thyroid surgery in a review of 48 highly selected patients. Forty-one of the patients underwent thyroidectomy under general anesthesia for mostly benign disease and were discharged the day of surgery without morbidity or mortality. The remaining seven

TABLE 15.3
Low Postoperative Morbidity and Mortality after Same-Day Discharge[a]

Reference	Year	No. Patients	Bleed	Hypocalcemia	RLN	Readmission	Death
Teoh et al.[35]	2008	50	1	0	1P	0	0
Materazzi et al.[36]	2007	1,571	10	112T, 3P	13 T, 4P	4	0
Terris et al.[33]	2007	91	0	0	4T	4	0
Snyder et al.[12]	2006	58	2	6T, 1P	2T	7	0
Spanknebel et al.[13]	2006	1,194	8	1P	20 T, 9P	1	0
Sahai et al.[37]	2005	104	0	8T	4T	4	0
Specht et al.[6]	2001	175	1	1T	2T	1	0
Lo Gerfo[14]	1998	203	2	3T	1P	0	0
McHenry[38]	1997	80	1	8T	2T	1	0
Samson et al.[30]	1997	1,178	1	16T	19T	1	1
Mowschenson and Hodin[39]	1995	100	0	2T	0	0	0
Marohn and LaCivita[40]	1995	150	1	3T, 1P	1P	1	0
Lo Gerfo et al.[41]	1991	134	0	8T	1P	0	0
Steckler[42]	1986	48	0	0	0	0	0

[a] RLN, recurrent laryngeal nerve injury; P, permanent; T, transient.

patients who were admitted either needed more extensive surgery to resect cancer or elected to stay overnight despite being deemed fit for discharge by the surgeon. In 1991, Lo Gerfo[41] published his experience with 134 patients who underwent thyroidectomy under general anesthesia for both benign and malignant disease in an ambulatory surgery setting with low morbidity and no mortality. Seventy-six of his patients went home the day of and 53 the day following surgery, with an average length of postoperative hospital stay of 0.49 day.[41] In 1995, Mowschenson and Hodin[39] reported their experience with 61 patients who underwent thyroid and parathyroid operations under general anesthesia, all of whom went home the day of surgery. The authors also demonstrated a 30% savings in hospital costs when comparing same-day with admission thyroid surgery.[39] In subsequent years, studies from the Philippines, Britain, Italy, Singapore, and China reported similar results, indicating that same-day thyroid surgery was a practical and safe option.[30,35–37,44]

15.3.3 Same-Day Thyroid Surgery: The Controversy

It is not surprising that same-day thyroid surgery is gaining popularity.[33] From an anatomic standpoint, the thyroid gland is readily accessible. From a physiologic standpoint, removing the thyroid gland, in part or in whole, does not cause physiologic debilitation.[30] Furthermore, when performed by an experienced surgeon, thyroid surgery boasts short operative times, minimal morbidity, and almost no mortality. Even under conventional guidelines, most patients go home the day after their thyroid operation with little or no discomfort. These characteristics give momentum to the same-day thyroid surgery movement.

Despite advances, the well-known complications of thyroid surgery remain as obstacles. The vast majority of complications occur within 24 hours of surgery. In fact, 75% occur within the first 6 hours.[43] These statistics make an overnight hospital admission seem excessive. Potentially lethal complications, however, can occur after 24 hours and, though rare, pose definite risks to the patient. The three main postoperative complications that undermine the absolute safety of same-day thyroid surgery are bleeding, hypocalcemia, and RLN injury.[43] Other problems, such as pain, nausea and vomiting, and difficulty voiding, also can act as barriers to same-day discharge.[34]

TABLE 15.4
Postoperative Cervical Hematoma Following Thyroid Surgery[a]

Reference	Year	No. Patients	Total Hematomas (%)	Time Interval to Hematoma (%)		
				6 hours	7–24 hours	>24 hours
Rosenbaum et al.[45]	2008	1,050	6 (0.6)	4 (66.7)	1 (16.7)	1 (16.7)
Materazzi et al.[36]	2007	1,571	10 (0.6)	3 (30)	7 (70)	0 (0)
Spanknebel et al.[13]	2006	1,194	8 (0.7)	6 (75)	1 (12.5)	1 (12.5)
Hurtado-Lopez et al.[46]	2002	1,131	9 (0.8)	9 (100)	0 (0)	0 (0)
Abbas et al.[47]	2001	1,268	10 (0.8)	1 (10)	5 (50)	4 (40)
Burkey et al.[48]	2001	13,817	42 (0.3)	18 (43)	16 (38)	8 (19)
Bergamaschi et al.[49]	1998	1,192	10 (0.8)	7 (70)	2 (20)	1 (10)
Shaha and Jaffe[50]	1994	600	10 (1.3)	7 (87.5)	1 (12.5)	0 (0)
Lo Gerfo et al.[41]	1991	134	0 (0)	0 (0)	0 (0)	0 (0)

[a] Literature review highlights the low incidence of clinically significant postoperative cervical hematoma following thyroid surgery. The majority occur within 24 hours, but delayed bleeding is a risk.

15.3.3.1 Bleeding

Bleeding is the main concern of most surgeons who perform thyroid surgery as a same-day procedure. It is a life-threatening, albeit uncommon, complication, with a reported incidence of less than 2% (Table 15.4).[45] Causes include slipping of ligatures, increased blood pressure from pain, coughing or retching that reopens previously sealed vessels, and bleeding from residual thyroid tissue.[51,52] Risk factors for postoperative bleeding include substernal goiter, blood dyscrasia, patients taking antiplatelet or anticoagulation medications, and patients who undergo extensive neck dissection.[38,52] The increased vascularity of the thyroid glands in patients with Graves' disease also may be a risk factor for postoperative bleeding.[38] The surgeon should carefully consider these factors when evaluating a patient for same-day thyroid surgery.

Bleeding into the confined neck space can rapidly lead to tracheal obstruction, respiratory failure, and death. The hematoma unlikely compresses the rigid, cartilaginous trachea in most cases, but rather impedes venous and lymphatic drainage of the neck with subsequent laryngeal and pharyngeal edema.[45,50,52] It is this edema that likely obstructs the airway. The signs and symptoms of cervical hematoma include progressive neck swelling, bleeding at the incision, stridor, dysphagia, dyspnea, and pain or pressure sensation in the patient's neck. More subtle signs of early hypoxia and impending airway obstruction include tachycardia, sweating, irritability, and confusion.[52] Surgeons, house staff, and recovery room nurses must be adequately taught to recognize and manage these signs and symptoms in the postoperative period because early detection, close observation, proper airway management, and appropriate surgical intervention are vital to prevent complete airway obstruction and death.[45]

The majority of life-threatening postoperative cervical hematomas occur within the first few hours of surgery (Table 15.4). In 1998, LoGerfo[43] performed a retrospective review and identified 21 cases of postoperative bleeding in patients who underwent thyroidectomy over a 20-year period at Columbia University Medical Center. All 21 patients exhibited signs and symptoms of bleeding and airway compromise within 4 hours.[43] Other authors report similar experiences.[45] Accordingly, advocates of same-day thyroid surgery suggest that there is a critical period during which most postoperative bleeding occurs; most patients can be discharged safely after 6 hours of careful postoperative observation.

Life-threatening bleeding, however, can occur beyond the typical postoperative observation period, with reports of delayed cervical hematomas occurring 7 hours to as late as 7 days after thyroid surgery.[13,45,47] In 2001, Burkey et al.[48] identified 42 out of 13,817 patients in a 24-year retrospective review who required reexploration for bleeding after thyroid or parathyroid surgery at the Mayo Clinic. Fifty-seven percent of the patients bled more than 6 hours after the operation.[48] In 2008, Rosenbaum et al.[45] performed a similar retrospective review of 1,050 thyroidectomy or parathyroidectomy patients and identified 6 life-threatening cervical hematomas. Four of these hematomas developed within 4 hours and one within 24 hours. The final hematoma developed in a patient on anticoagulation therapy 7 days after the operation.[45] All patients were successfully re-explored and survived.

Although rare, delayed cervical hematomas following thyroid surgery are a concern. There are reports of hospitalized patients who develop delayed cervical hematomas and live because they are treated immediately.[39] There also are reports of discharged patients who develop delayed cervical hematoma and die because they cannot receive treatment.[43] Critics of same-day discharge thus suggest that patients should be observed in the hospital for longer periods of time to minimize risks and maximize survival. Advocates counter these arguments by noting that, following traditional hospital admission protocols, most patients will still be home within 24 to 72 hours.[41] Any bleeding that occurs after 24 hours either is slow enough to allow adequate time for the patient to recognize the problem and seek treatment or is not significant enough to cause clinical symptoms and will resorb without treatment.[38] Advocates conclude that the low incidence and early presentation of life-threatening cervical hematomas do not justify routine hospital admission after thyroid surgery.[8,13,28,43,45]

Certain intraoperative maneuvers can help reduce the risk of life-threatening postoperative cervical hematoma and ensure safe same-day discharge. Meticulous attention to hemostasis in the operating room is paramount. It has been suggested that knowledge of the possibility of same-day discharge may foster even closer attention to hemostasis in the operating room.[13] The conventional clamp-and-tie technique largely has been replaced by hemostatic clips, electrocautery, and more recently, ultrasonic cutting and coagulating systems, as well as electrothermal bipolar vessel sealing systems, with encouraging results.[1,29] For patients under general anesthesia, it is advisable to test for potential bleeding at the end of the operation by placing the patient in a 30° head-down position while the anesthesiologist hyperinflates the lungs.[53] Similarly, patients under locoregional anesthesia can be asked to cough or perform a Valsalva maneuver.[13] These maneuvers produce venous hypertension in the neck, test the integrity of hemostasis, and expose potential bleeding problems.[13,53,54] Another intraoperative technique is to leave the inferior strap muscles open.[41] This safeguard against airway obstruction works by two means. First, it allows decompression of bleeding into the superficial soft tissue planes. Second, decompression into the subcutaneous tissue allows easier and earlier recognition of swelling, thus facilitating prompt correction of the problem.[13] Finally, by leaving the wound undressed, caregivers can inspect the wound easily and recognize bleeding early.[12,29,41] Taken together, these techniques provide a means for thyroid surgeons to decrease the incidence of postoperative bleeding and safely discharge patients after an appropriate observation period.

15.3.3.2 Hypocalcemia

Hypocalcemia is a common complication after thyroid surgery, particularly bilateral thyroid surgery, and is a potential barrier to same-day discharge.[55] The incidence in the literature varies from as low as 0% in patients who undergo hemithyroidectomy to as high as 50% in patients who undergo total thyroidectomy.[35,56] Most series report an incidence around 20 to 30%.[57] The exact cause of hypocalcemia after thyroid surgery is variable and multifactorial. Most often, it can be attributed to hypoparathyroidism as a result of injury, devascularization, or removal of the parathyroid glands.[58] Less common etiologies include preexisting vitamin D deficiency and hemodilution secondary to intraoperative intravenous fluid administration.[59] Postoperative hypocalcemia usually is only a problem when there is known parathyroid gland impairment (i.e., devascularization), when the operation jeopardizes all of the parathyroid glands (i.e., total thyroidectomy), or when previous thyroid surgery leaves the status of the parathyroid glands in question (i.e., reoperative thyroidectomy).[35,43] Patients with enlarged thyroid glands secondary to Graves' disease have an additional risk of developing marked postoperative hypocalcemia due to thyrotoxic osteodystrophy.[38] These factors must figure into the surgeon's consideration of a patient for same-day thyroid surgery.

Hypocalcemia following thyroid surgery can be transient or permanent. The majority of cases of hypocalcemia are self-limiting and do not require treatment, and in experienced thyroid surgery centers, the incidence of permanent hypocalcemia is less than 2%.[55,60] Hypocalcemia also can be asymptomatic or symptomatic. Asymptomatic hypocalcemia is of little clinical significance, discovered incidentally by postoperative serum calcium measurements. Symptomatic hypocalcemia, on the other hand, is more clinically relevant and usually manifests later in the postoperative period. Whereas bleeding and RLN injury typically present early, symptomatic hypocalcemia generally presents 24 to 48 hours after the operation.[34,43,55] Hence, patients undergoing thyroid surgery traditionally were admitted to the hospital for serial serum calcium level monitoring.[60] Hospital admission was advocated to diagnose and treat hypocalcemia quickly in order to avoid the possible sequelae of tetany, seizure, and laryngospasm with potential airway obstruction and death.[2,38] Ampules of calcium replacement were kept at thyroid patients' bedsides to facilitate rapid intravenous infusion should these potentially lethal symptoms develop.[38]

The move toward same-day thyroid surgery has prompted significant changes in the recognition and management of postoperative hypocalcemia. The traditional model of inpatient serial serum calcium level monitoring is being used less frequently.[60] Several studies have attempted to validate the use of perioperative serum calcium and intact parathyroid hormone (PTH) levels to identify patients at risk for developing hypocalcemia and facilitate safe same-day discharge after thyroid surgery.[1,34,36,55,61,62] Some studies suggest that a rapid decline in postoperative serum calcium levels can predict impending postoperative hypocalcemia.[40,63] Other studies propose that a certain percentage of decrease in pre- and postoperative intact PTH levels can identify at-risk patients.[64,65] Still other studies advocate measuring both serum calcium and intact PTH levels in the perioperative period and interpreting different combinations of the results to forecast postoperative hypocalcemia.[57,58] Various protocols have been described in the literature, but they remain largely institution specific, and there is no standard of practice.[34,58]

More recently, several authors advocate the use of intraoperative measurement of intact PTH to identify patients at risk for developing hypocalcemia after thyroid surgery.[55,58,61,62,66,67] The intraoperative PTH assay is widely used in the surgical management of hyperparathyroidism to help determine the removal of all hyperfunctioning parathyroid tissue.[60] When applied to the surgical management of thyroid disease, the assay helps monitor parathyroid gland function intraoperatively and can predict impending postoperative hypocalcemia.[60] At-risk patients then can be treated prospectively in the postoperative period with supplemental calcium or vitamin D, allowing for a safe and early discharge.[55,61,62] An additional benefit is that low intraoperative PTH levels may indicate vascular injury to the parathyroid glands and thus promote earlier, more aggressive assessment of gland viability and consideration of autotransplantation.[60] Intraoperative PTH monitoring holds promise, yet further research is needed to resolve contradictory data regarding the accuracy, cost-effectiveness, and optimal perioperative timing of the assay.[59,60,66,68]

While these innovations become validated, many thyroid surgeons continue to prescribe routine prophylactic oral calcium or vitamin D supplementation to patients following thyroidectomy.[28,33,58,69,70] Routine calcium/vitamin D supplementation is easy, inexpensive, and effective. It prevents symptomatic hypocalcemia, eliminates excessive venipuncture, use of hospital resources, and time spent in the hospital, and ultimately permits safe same-day thyroid surgery.[33,69,70] Moore[71] first proposed giving postoperative thyroid patients prophylactic calcium in 1994. LoGerfo[43] began a similar protocol shortly thereafter, emphasizing the need for thorough patient education about the symptoms of hypocalcemia and close doctor-patient communication in the postoperative period. LoGerfo believed that symptoms are easily recognizable by an educated, responsible patient, and that there is little threat of calcium overdose. He therefore empowered his patients to treat themselves should symptoms develop. It is important to note that self-treatment is not for every patient. Prophylactic calcium/vitamin D supplementation may be particularly problematic for the elderly, mentally impaired, or noncompliant patient, as well as for patients with extensive medication lists.[34,43] Overtreatment, possible inhibition of parathyroid function, and interference with the accurate detection of postoperative hypocalcemia are other criticisms of routine calcium/vitamin D supplementation.[59–62] The surgeon thus must carefully consider the patient's circumstances when formulating a treatment regimen for potential hypocalcemia after thyroidectomy. A high-dose, tapering regimen of oral calcium/vitamin D supplementation is a reasonable option to help prevent symptomatic hypocalcemia and ensure safe same-day thyroid surgery.

15.3.3.3 RLN Injury

Vocal cord dysfunction and subsequent airway obstruction secondary to RLN injury is another potentially life-threatening complication of thyroid surgery. Vocal cord dysfunction can be a paresis or a paralysis involving one or both vocal cords and can be transient or permanent, depending on the degree of nerve injury. The incidence of transient unilateral vocal cord dysfunction ranges from 0 to 10%, whereas the incidence of permanent unilateral dysfunction is less than 1%.[41] Bilateral vocal cord dysfunction resulting in airway obstruction is an extremely rare event, occurring in approximately 1 out of 30,000 thyroid operations.[53] Unilateral paresis or paralysis will manifest immediately as a change in voice or dyspnea. If transient, the dysfunction will resolve with time; if permanent, the dysfunction may require laryngoscopic intervention at a later date.[35] Unilateral vocal cord dysfunction should not prevent safe same-day discharge unless it causes significant respiratory compromise, which will be recognized and managed immediately and expeditiously. Similarly, bilateral vocal cord paresis or paralysis will manifest early as complete airway obstruction. Again, immediate presentation of this complication early in the postoperative period means that patients will be managed quickly, and thus does not represent a significant problem for patients who fulfill discharge criteria after 6 to 8 hours of close observation.[41] There is considerable anatomic variation in the RLN, and meticulous intraoperative nerve identification is necessary to minimize the complications related to nerve injury.[54]

15.3.3.4 Pain

In general, pain after thyroid surgery is not severe, but must be understood and managed effectively to ensure successful same-day surgery.[43,54] A patient cannot be discharged from the hospital without adequate postoperative analgesia. There are multiple causes of postoperative pain in thyroid surgery patients, each of which needs to be addressed. Expectedly, patients experience pain at the site of incision. Use of the anterior neck field block in conjunction with regional or general anesthesia has been shown to help alleviate postoperative incision pain.[43] Patients also may experience laryngeal discomfort caused by endotracheal intubation and surgical manipulation.[54] The use of locoregional rather than general anesthesia avoids the former, and minimal, careful dissection lessens the latter. Intraoperative neck hyperextension for exposure can result in postoperative cervical pain and occipital headache. Gentle neck extension and attentive patient positioning are necessary. Additionally, Han et al.[72] showed that preoperative block of the greater occipital nerve with local anesthetic effectively reduced both posterior neck pain and occipital headache after thyroid surgery. Oral analgesics likely will be necessary once the locoregional anesthesia wears off for patients to leave the hospital comfortably. Nonsteroidal anti-inflammatory drugs (e.g., ibuprofen and ketorolac) can be prescribed in addition to opioid-based drugs (e.g., oxycodone and hydrocodone) and have been shown to reduce the use of other opioids, and thus minimize their well-known adverse effects. Acetaminophen offers another class of oral analgesic. In combination with locoregional anesthesia, these medications provide multimodality analgesia with greater efficacy and fewer side effects, helping facilitate a patient's recovery at home.[73]

15.3.3.5 Nausea and Vomiting

Patients who have thyroid surgery experience a high rate of postoperative nausea and vomiting. Two likely causes include intraoperative stimulation of the vagus nerve during manipulation of the neck and postoperative effects of general anesthesia.[20] Nausea and vomiting are significant sources of patient discomfort that may lead to delayed discharge and decreased satisfaction with the overall surgical experience.[19] More importantly, nausea and vomiting put undue stress on the operative site and increase the chance of postoperative bleeding. The increased thoracic and abdominal pressures from retching can cause arterial and venous hypertension, with subsequent bleeding at the operative site and development of a cervical hematoma.[20]

Several developments have helped decrease the incidence of nausea and vomiting after thyroid surgery to facilitate same-day discharge. First, patients often are given preemptive anti-emetic medication prior to the start of the operation, as well as doses throughout the postoperative period. Second, more surgeons are performing thyroid surgery under locoregional anesthesia. Locoregional anesthesia prevents the nausea and vomiting associated with certain general anesthetic agents and also provides sustained analgesia, which reduces the patient's oral narcotic requirement in the immediate postoperative period.[54,74] Third, the intravenous sedative propofol is used more frequently for the induction and maintenance of anesthesia in thyroid surgery. Propofol is believed to exert an anti-emetic effect independent of its other actions. Sonner et al.[20] demonstrated in a prospective, randomized trial that propofol for the maintenance of anesthesia significantly reduced postoperative nausea and vomiting in women undergoing thyroid or parathyroid surgery. Gauger et al.[19] corroborated these results in a prospective, randomized trial, demonstrating that propofol decreases nausea and vomiting in the early postoperative period following thyroid and parathyroid surgery. These advances have curtailed the impact that postoperative nausea and vomiting have on same-day discharge after thyroid surgery.

15.3.3.6 Urinary Retention

With the increasing use of locoregional anesthesia, postoperative urinary retention is an uncommon complication of same-day thyroid surgery. Unlike general anesthesia, locoregional anesthesia rarely affects a patient's ability to void or to be discharged the same day as the procedure.

15.3.4 Patient Selection for Same-Day Thyroid Surgery

Attempts to understand, minimize, and manage the common postoperative complications of thyroid surgery have spurred the creation of specific patient selection criteria, comprehensive patient education, and stringent postoperative discharge protocols.[41,44,54] All patients must be agreeable, motivated, and compliant. Poor candidates for same-day thyroid surgery include those with known bleeding disorders, significant medical comorbidities, the need for simultaneous neck and mediastinal procedures that require hospitalization, or psychosocial circumstances not conducive to same-day discharge (e.g., mental impairment, lack of external support, or distant home).[33,34] Patient education is vital for establishing effective doctor-patient communication, reasonable expectations, and knowledge of the potential postoperative complications.[41] Finally, same-day surgery units must have mechanisms in place to ensure the utmost vigilance in postoperative patient observation, discharge, and follow-up.[54] These mechanisms provide the necessary means by which postoperative concerns and complications are identified and addressed in the immediate and distant postoperative periods, safeguarding patient safety and augmenting patient satisfaction.[12,30,34,36,39]

15.4 CONCLUSION

Thyroid surgery has returned to its roots with the resurgence of locoregional anesthesia. As in the early twentieth century, surgeons once again are performing thyroid surgery using this method.

The pathologic indications, however, have expanded from thyrotoxicosis to include nearly all thyroid disease. Thoughtful patient selection and preparation as well as meticulous attention to relevant anatomy and proper technique help ensure the successful use of locoregional anesthesia. Randomized trials that evaluate the safety and efficacy of the three types of locoregional blocks as they pertain to thyroid surgery are needed to inform surgeon preference. Although challenges exist, locoregional anesthesia has proven to be a safe, feasible, and advantageous option for select patients with thyroid disease.

Thyroid surgery has evolved to meet the demands of the modern world with its maturation as a same-day procedure. Increasingly, surgeons are discharging patients within hours of the operation's end. Improvements in perioperative protocols have helped identify and neutralize most of the common postoperative complications of thyroid surgery that can act as barriers to discharge. Although risks are inherent, same-day thyroid surgery has proven to be safe, feasible, and advantageous. The combination of locoregional anesthesia and same-day thyroid surgery is intuitive and synergistic, leading to greater patient satisfaction, shorter hospital stays, and reduced costs. Economic incentive, however, should never eclipse the importance of surgeon expertise, operation efficacy, or patient safety. General anesthesia and hospital admission still have places in thyroid surgery, but locoregional anesthesia and same-day discharge can be judiciously employed to the mutual benefit of the hospital, surgeon, and patient.

ACKNOWLEDGMENTS

The authors thank Patrick L. Reavey, MD, Resident, General Surgery, Department of Surgery, Columbia University College of Physicians and Surgeons, for the illustrations.

REFERENCES

1. Dionigi G, Rovera F, Carrafiello G, Bacuzzi A, Boni L, Dionigi R. 2009. New technologies in ambulatory thyroid surgery. *Int J Surg* xxx:1–4. doi:10.1016/j.ijsu.2008.12.029/
2. Shukla VK, Narayan S, Chauhan VS, Singh DK. 2005. Thyroid surgery under local anaesthesia: An alternative to general anaesthesia. *Indian J Surg* 67(6):316–19.
3. Stephen E, Nayak S, Salins SR. 2008. Thyroidectomy under local anaesthesia in India. *Tropical Doctor* 38:20–21.
4. Taylor S. 1997. Sir Thomas Peel Dunhill (1876–1957). *World J Surg* 21:660–62.
5. Hisham AN, Aina EN. 2002. A reappraisal of thyroid surgery under local anaesthesia: Back to the future? *ANZ J Surg* 72:287–89.
6. Specht MC, Romero M, Barden CB, Esposito C, Fahey TJ. 2001. Characteristics of patients having thyroid surgery under regional anesthesia. *J Am Coll Surg* 193:367–72.
7. Prasad KC, Shanmugam VU. 1998. Major neck surgeries under regional anesthesia. *Am J Otolaryngol* 19(3):163–69.
8. Spanknebel K, Chabot JA, DiGiorgi M, Cheung K, Lee S, Allendorf J, Lo Gerfo P. 2005. Thyroidectomy using local anesthesia: A report of 1,025 cases over 16 years. *J Am Coll Surg* 201:375–85.
9. Lo Gerfo P, Ditkoff BA, Chabot J, Feind C. 1994. Thyroid surgery using monitored anesthesia care: An alternative to general anesthesia. *Thyroid* 4(4):437–39.
10. Arora N, Dhar P, Fahey TJ. 2006. Seminars: Local and regional anesthesia for thyroid surgery. *J Surg Oncol* 94:708–13.
11. Martins Mamede RC, Raful H. 2008. Comparison between general anesthesia and superficial cervical plexus block in partial thyroidectomy. *Rev Bras Otorrinolaringol* 74(1):99–105.
12. Snyder SK, Roberson CR, Cummings CC, Rajab MH. 2006. Local anesthesia with monitored anesthesia care vs general anesthesia in thyroidectomy: A randomized study. *Arch Surg* 141:167–73.
13. Spanknebel K, Chabot JA, DiGiorgi M, Cheung K, Curty J, Allendorf J, Lo Gerfo P. 2006. Thyroidectomy using monitored local or conventional general anesthesia: An analysis of outpatient surgery, outcome and cost in 1,194 consecutive cases. *World J Surg* 30:813–24.
14. Lo Gerfo P. 1998. Local/regional anesthesia for thyroidectomy: Evaluation as an outpatient procedure. *Surgery* 124:975–79.

15. Hochman M, Fee WE Jr. 1991. Thyroidectomy under local anesthesia. *Arch Otolaryngol Head Neck Surg* 117:405–7.
16. Saxe AW, Brown E, Hamburger SW. 1988. Thyroid and parathyroid surgery performed with patient under regional anesthesia. *Surgery* 103(4):415–20.
17. Materazzi G, Monchik J. 1999. Local anesthesia in traditional and minimally invasive approach to the parathyroid. *Acta Chir Austriaca* 31(4):232–34.
18. Williams M, Lo Gerfo P. 2002. Thyroidectomy using local anesthesia in critically ill patients with amiodarone-induced thyrotoxicosis: A review and description of the technique. *Thyroid* 12(6):523–25.
19. Gauger PG, Shanks A, Morris M, Greenfield MLVH, Burney RE, O'Reilly M. 2008. Propofol decreases early postoperative nausea and vomiting in patients undergoing thyroid and parathyroid operations. *World J Surg* 32:1525–34.
20. Sonner JM, Hynson JM, Clark O, Katz JA. 1997. Nausea and vomiting following thyroid and parathyroid surgery. *J Clin Anesth* 9:398–402.
21. Plunkett AR, Shields C, Stojadinovic A, Buckenmaier CC. 2009. Awake thyroidectomy under local anesthesia and dexmedetomidine infusion. *Military Med* 174(1):100–2.
22. Bagul A, Taha R, Metcalfe MA, Brook NR, Nicholson ML. 2005. Pre-incision infiltration of local anesthetic reduces postoperative pain with no effects on bruising and wound cosmesis after thyroid surgery. *Thyroid* 15(11):1245–48.
23. Tobias JD. 1999. Cervical plexus block in adolescents. *J Clin Anesth* 11:606–8.
24. Andrieu G, Amrouni H, Robin E, Carnaille B, Wattier JM, Pattou F, Vallet B, Lebuffe G. 2007. Analgesic efficacy of bilateral superficial cervical plexus block administered before thyroid surgery under general anesthesia. *Br J Anaesth* 99:561–66.
25. Ross DE. 1950. Thyroidectomy using local anesthesia. *Am J Surg* 20(2):211–15.
26. Eti Z, Irmak P, Gulluoglu BM, Manukyan MN, Gogus FY. 2006. Does bilateral superficial cervical plexus block decrease analgesic requirement after thyroid surgery? *Anesth Analg* 102:1174–76.
27. Lombardi CP, Raffaelli M, Modesti C, Boscherini M, Bellantone R. 2004. Video-assisted thyroidectomy under local anesthesia. *Am J Surg* 187:515–18.
28. Allendorf J, DiGiorgi M, Spanknebel K, Inabnet W, Chabot J, Lo Gerfo P. 2007. 1112 consecutive bilateral neck explorations for primary hyperparathyroidism. *World J Surg* 31:2075–80.
29. Inabnet WB, Shifrin A, Ahmed L, Sinha P. 2008. Safety of same day discharge in patients undergoing sutureless thyroidectomy: A comparison of local and general anesthesia. *Thyroid* 18(1):57–61.
30. Samson PS, Reyes FR, Saludares WN, Angeles RP, Francisco RA, Tagorda ER Jr. 1997. Outpatient thyroidectomy. *Am J Surg* 173:499–503.
31. Fernandez FH. 1984. Cervical block anesthesia in thyroidectomy. *Int Surg* 69:309–11.
32. Cunningham IG, Lee YK. 1975. The management of solitary thyroid nodules under local anesthesia. *ANZ J Surg* 45: 285–89.
33. Terris DJ, Moister B, Seybt MW, Gourin CG, Chin E. 2007. Outpatient thyroid surgery is safe and desirable. *Otolaryngol Head Neck Surg* 136:556–59.
34. Mirnezami R, Sahai A, Symes A, Jeddy T. 2007. Day-case and short-stay surgery: The future for thyroidectomy? *Int J Clin Pract* 61(7):1216–22.
35. Teoh AYB, Tang YC, Leong HT. 2008. Feasibility study of day case thyroidectomy. *ANZ J Surg* 78:864–66.
36. Materazzi G, Dionigi G, Berti P, Rago R, Frustaci G, Docimo G, Puccini M, Miccoli P. 2007. One-day thyroid surgery: Retrospective analysis of safety and patient satisfaction on a consecutive series of 1,571 cases over a three-year period. *Eur Surg Res* 39:182–88.
37. Sahai A, Symes A, Jeddy T. 2005. Short-stay thyroid surgery. *Br J Surg* 92:58–59.
38. McHenry CR. 1997. "Same-day" thyroid surgery: An analysis of safety, cost savings, and outcome. *Am Surg* 63:586–89.
39. Mowschenson PM, Hodin RA. 1995. Outpatient thyroid and parathyroid surgery: A prospective study of feasibility, safety, and costs. *Surgery* 118:1051–54.
40. Marohn MR, LaCivita KA. 1995. Evaluation of total/near total thyroidectomy in a short-stay hospitalization: Safe and cost-effective. *Surgery* 118:943–48.
41. Lo Gerfo P, Gates R, Gazetas P. 1991. Outpatient and short-stay thyroid surgery. *Head Neck* 13:97–101.
42. Steckler RM. 1986. Outpatient thyroidectomy: A feasibility study. *Am J Surg* 152:417–19.
43. Schwartz AE, Clark OH, Ituarte P, Lo Gerfo P. 1998. Therapeutic controversy: Thyroid surgery—The choice. *J Clin Endocrinol Metab* 83(4):1097–105.
44. Chin CWD, Loh KS, Tan KSL. 2007. Ambulatory thyroid surgery: An audit of safety and outcomes. *Singapore Med J* 48(8):720–24.

45. Rosenbaum MA, Haridas M, McHenry CR. 2008. Life-threatening hematoma complicating thyroid and parathyroid surgery. *Am J Surg* 195:339–43.
46. Hurtado-Lopez LM, Zaldivar-Ramirez FR, Basurto KE, et al. 2002. Causes for early reintervention after thyroidectomy. *Med Sci Monit* 8:CR247–50.
47. Abbas G, Dubner S, Heller KS. 2001. Re-operation for bleeding after thyroidectomy and parathyroidectomy. *Head Neck* 23:544–46.
48. Burkey SH, van Heerden JA, Thompson GB, Grant CS, Schleck CD, Farley DR. 2001. Reexploration for symptomatic hematomas after cervical exploration. *Surgery* 130:914–20.
49. Bergamaschi R, Becourarn G, Ronceray J, Arnaud JP. 1998. Morbidity of thyroid surgery. *Am J Surg* 176:71–75.
50. Shaha AR, Jaffe BM. 1994. Practical management of post-thyroidectomy hematoma. *J Surg Oncol* 57:235–38.
51. Savargaonkar AP. 2004. Post-thyroidectomy haematoma causing total airway obstruction—A case report. *Indian J Anaesth* 48(6):483–85.
52. Harding J, Sebag F, Sierra M, Palazzo FF, Henry JF. 2006. Thyroid surgery: Postoperative hematoma—Prevention and treatment. *Langenbecks Arch Surg* 391:169–73.
53. Farrar WB. 1983. Complications of thyroidectomy. *Surg Clin North Am* 63:1353–61.
54. Dionigi G, Rovera F, Carrafiello G, Boni L, Dionigi R. 2009. Ambulatory thyroid surgery: Need for stricter patient selection criteria. *Int J Surg* xxx:1–3. doi:10.1016/j.ijsu.2008.12.029.
55. Lombardi CP, Raffaelli M, Princi P, Santini S, Boscherini M, De Crea C, Traini E, D'Amore AM, Carrozza C, Zuppi C, Bellantone R. 2004. Early prediction of postthyroidectomy hypocalcemia by one single iPTH measurement. *Surgery* 136(6):1236–41.
56. Pattou F, Combemale F, Fabre S, Carnaille B, Decoulx M, Wemeau JL, Racadot A, Proye C. 1998. Hypocalcemia following thyroid surgery: Incidence and prediction of outcome. *World J Surg* 22:718–24.
57. Grodski S, Serpell J. 2008. Evidence for the role of perioperative PTH measurement after total thyroidectomy as a predictor of hypocalcemia. *World J Surg* 32:1367–73.
58. Payne RJ, Hier MP, Tamilia M, MacNamara E, Young, J, Black MJ. 2005. Same-day discharge after total thyroidectomy: The value of 6-hour serum parathyroid hormone and calcium levels. *Head Neck* 27:1–7.
59. Di Fabio F, Casella C, Bugari G, Iacobello C, Salerni B. 2006. Identification of patients at low risk for thyroidectomy-related hypocalcemia by intraoperative quick PTH. *World J Surg* 30:1428–33.
60. Roh JL, Park CI. 2006. Intraoperative parathyroid hormone assay for management of patients undergoing total thyroidectomy. *Head Neck* 28:990–97.
61. Lo CY, Luk JM, Tam SC. 2002. Applicability of intraoperative parathyroid hormone assay during thyroidectomy. *Ann Surg* 236(5):564–69.
62. Lo CY. 2003. Postthyroidectomy hypocalcemia. *J Am Coll Surg* 196:497–98.
63. Adams J, Andersen P, Everts E, Cohen J. 1998. Early postoperative calcium levels as predictors of hypocalcemia. *Laryngoscope* 108:1829–31.
64. Warren FM, Andersen PE, Wax MK, Cohen JI. 2004. Perioperative parathyroid hormone levels in thyroid surgery: Preliminary report. *Laryngoscope* 114:698–93.
65. Alia P, Moren P, Rigo R, Francos JM, Navarro MA. 2007. Postresection parathyroid hormone and parathyroid hormone decline accurately predict hypocalcemia after thyroidectomy. *Am J Clin Pathol* 127:592–97
66. Higgins KM, Mandell DL, Govindaraj S, Geden EM, Mechanick JI, Bergman DA, Diamond EJ, Urken ML. 2004. The role of intraoperative rapid parathyroid hormone monitoring for predicting thyroidectomy-related hypocalcemia. *Arch Otolaryngol Head Neck Surg* 130:63–67.
67. Quiros RM, Pesce CE, Wilhelm SM, Djuricin G, Prinz RA. 2005. Intraoperative parathyroid hormone levels in thyroid surgery are predictive of postoperative hypoparathyroidism and need for vitamin D supplementation. *Am J Surg* 189:306–9.
68. Del Rio P, Arcuri MF, Ferreri G, Sommaruga L, Sianesi M. 2005. The utility of serum PTH assessment 24 hours after total thyroidectomy. *Otolaryngol Head Neck Surg* 132:584–86.
69. Roh J, Park C. 2006. Routine oral calcium and vitamin D supplements for prevention of hypocalcemia after total thyroidectomy. *Am J Surg* 192(5):675–78.
70. Bellantone R, Lombardi CP, Raffaelli M, Boscherini M, Alesina PF, De Crea C, Traini E, Princi P. 2002. Is routine supplementation therapy (calcium and vitamin D) useful after total thyroidectomy? *Surgery* 132:1109–12.
71. Moore FD Jr. 1994. Oral calcium supplements to enhance early hospital discharge after bilateral surgical treatment of the thyroid gland or exploration of the parathyroid glands. *J Am Coll Surg* 178:11–16.

72. Han DW, Koo BN, Chung WY, Park CS, Kim SY, Palmer PP, Kim KJ. 2006. Preoperative greater occipital nerve block in total thyroidectomy patients can reduce postoperative occipital headache and posterior neck pain. *Thyroid* 16(6):599–603.
73. Joshi GP. 2005. Multimodal analgesia techniques and postoperative rehabilitation. *Anesthesiol Clin North Am* 23(1):185–202.
74. Black MJ, Ruscher AE, Lederman J, Chen H. 2007. Local/cervical block anesthesia versus general anesthesia for minimally invasive parathyroidectomy: What are the advantages? *Ann Surg Oncol* 14(2):744–49.

16 The Use of Therapeutic Radiotracers in Thyroid Cancer

Josef Machac

CONTENTS

ABSTRACT: The use of radiotracers in the therapy of thyroid cancer begins with thyroid remnant ablation therapy and the treatment of differentiated thyroid cancer with radioiodine (RAI). It includes commonly used strategies for patient preparation as well as accepted alternatives. The role of targeted dosimetry, particularly important in higher-risk situations, is discussed. There is extensive discussion of the efficacy and limitations of RAI therapy and continuing challenges, including the use of RAI therapy in patients without detectable uptake of RAI in thyroid cancer metastases. Side effects and toxicity to the gastrointestinal tract, the salivary glands, and their prophylaxis are discussed as well. The potential for carcinogenesis, teratogenicity, and effects on fertility are addressed. This chapter also reviews ongoing work in alternative approaches, attempting to compensate for limitations in RAI, in the form radiosensitization therapy, and the use of non-RAI treatment strategies.

16.1 INTRODUCTION

Radioiodine (RAI) therapy with I-131 has been a mainstay of thyroid cancer management since its first application was described by Seidlin et al. in 1946[1] and has spearheaded the development of nuclear medicine as a medical specialty. However, because of the variety in thyroid cancer cell types, various degrees of differentiation, and variable clinical expression, there is no single approach to the radioisotope treatment of thyroid cancer.

Initial therapy of thyroid cancer consists of surgical removal of the thyroid gland and the primary tumor, along with identification and removal of involved lymph nodes. Those patients with well-differentiated thyroid cancer who are not considered cured by surgery frequently undergo RAI therapy. In the appropriate setting, RAI therapy reduces the rate of cancer recurrence, extends the disease-free interval, and helps to control inoperable disease. Most patients with metastatic disease can be managed successfully for many years with RAI therapy.

16.2 RISK ASSESSMENT

Various systems of clinical risk assessment for well-differentiated carcinoma exist.[2–4] Because of its utility in predicting disease mortality and its requirements for cancer registries, the AJCC/UICC staging is currently recommended for all patients with differentiated thyroid cancer.[5] The most benign setting occurs in a young woman with a small, well-differentiated papillary thyroid carcinoma (PTC) that is well encapsulated, without lymphovascular invasion or abnormal lymph nodes, and with no history of radiation exposure. Such patients exhibit a benign course with risk of death of less than 2% in 20 years after initial treatment.[6] Increasing age, male gender, increasing size of the primary lesion, penetration of the thyroid capsule, invasion of surrounding tissues, invasion of blood vessels, the presence of metastases in local lymph nodes, and the presence of distant metastases progressively increase the likelihood of PTC recurrence and disease-related mortality.[7–9] Those with all of these risk factors have a 20-year mortality of 25 to 40%. Similar risk stratification holds for follicular thyroid carcinoma (FTC). Low-risk patients have about 5% risk of mortality in 10 years, and 10 to 12% at 20 years. Patients at high risk have a 70% mortality at 10 years and about 90% at 20 years.[10] According to the prognostic scheme of Mazzaferri and Jhiang,[12] patients with stage I PTC or FTC with primary tumor size less than 1.5 cm have a low risk of recurrence. Those with either stage II and III PTC or FTC with primary tumor size of 1.5 cm or more, with or without cervical lymph node metastases or localized tumor invasion, have a risk of recurrence of about 25% at 10 years, 30% at 20 years, and 40% at 35 years. These patients were treated with surgery and

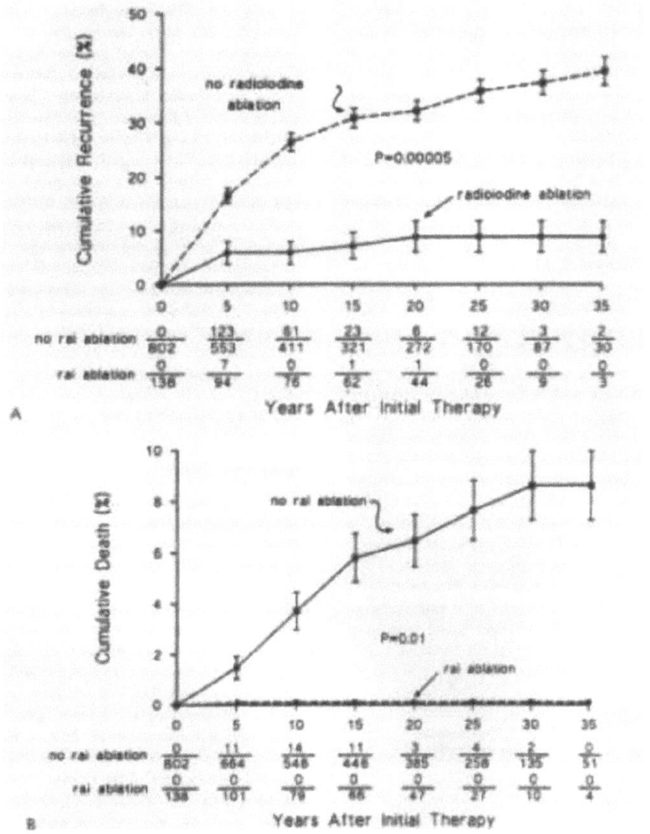

FIGURE 16.1 The benefit of I-131 ablation on tumor recurrence (A) and mortality (B) in 940 patients presenting with either PTC or FTC larger than 1.5 cm without distant metastases. (Reproduced from Mazzaferri EL, Jhiang SM. 1994. Long-term impact of initial surgical and medical therapy on papillary and follicular thyroid cancer. *Am J Med* 97:418–28. With permission from Elsevier.)

thyroid hormone suppression alone (no RAI ablation) and had a cumulative risk of death of 6% at 20 years and 8% at 35 years. In this setting, Mazzaferri and Young[11] and Mazzaferri and Jhiang[12] showed that RAI therapy had a beneficial effect on survival, cutting the death rate down to no reported deaths out of 138 patients, and the rate of recurrence to about 7 to 8% in 20–35 years in 802 patients (Figure 16.1). These results have served as a rationale for routine RAI therapy in stage II or stage III patients.

In addition to reducing the risk of disease of recurrence in stage II or stage III patients, and reducing cumulative mortality, RAI ablation therapy removes residual normal thyroid tissue. This tissue is almost always seen even after "complete" thyroidectomy,[13] interferes with the interpretation of RAI scans, and may result in slightly elevated serum thyroglobulin (TG) levels. Thus, the goals of postoperative RAI ablation therapy are to destroy residual thyroid tissue in an effort to decrease the risk of recurrent loco-regional disease and to facilitate long-term surveillance with whole body RAI imaging and TG measurements.[14] This has been a rationale for ablation in patients with stage I disease. In larger remnants that secrete sufficient amounts of thyroid hormone to prevent therapy of known residual or new metastatic disease, RAI ablation removes this possible cause of ineffective therapy. In addition to complete surgical removal of primary tumor and local extension or local lymph node metastases, RAI treatment is used, together with TSH suppression therapy and local external beam radiation therapy (EBRT), to control disease.[15,16] A number of large, retrospective studies on RAI treatment show a significant reduction in the rates of disease recurrence,[17–20] as well as cause-specific mortality.[21,22] On the other hand, other studies show that the majority of patients with PTC, those at the lowest risk for mortality, did not find any clear benefit of RAI ablation therapy.[23–28] In those studies that show benefit, the advantage appeared to be restricted to patients with larger tumors (>1.5 cm), or with residual disease after surgery, while lower-risk patients did not show evidence for benefit. Therefore, the current consensus recommendation is that RAI ablation is recommended for patients with stage III and IV disease, all patients with stage II disease younger than age 45, most patients with stage II disease 45 years or older, and selected patients with stage I disease, especially those with multifocal disease, nodal metastases, extrathyroidal or vascular invasion, or more aggressive histologies.

16.3 USE OF RHTSH FOR ABLATION THERAPY

Recombinant human TSH (rhTSH) stimulation was FDA approved for diagnostic I-131 or I-123 imaging or a stimulated TG measurement. Patients who are unable to tolerate thyroid hormone withdrawal or who are unable to generate an elevated TSH have been shown to undergo successful remnant ablation with rhTSH.[29,30] In another study, remnant ablation was equally successful with 30 mCi of I-131 in patients after conventional thyroid hormone withdrawal, compared to rhTSH stimulation.[31] Pacini et al.[32] conducted a multicenter randomized study comparing conventional thyroid hormone withdrawal with rhTSH stimulation with respect to efficacy of remnant ablation after 100 mCi (3.76 GBq) of I-131, judged 8 months after treatment. They found 100% success in both groups by RAI imaging; they also found 96% success in the rhTSH group and 86% success in the withdrawal group, using a TG level of less than 2 ng/ml. Moreover, Borget et al.[33] demonstrated that by increasing iodine excretion, the use of rhTSH, in comparison to thyroid hormone withdrawal, shortened the length of stay in patients hospitalized for therapy, thus helping to offset the cost of the rhTSH. Robbins et al.[34] found no significant difference in outcome between 42 patients with thyroid hormone withdrawal and 45 patients who received rhTSH in preparation for ablation therapy, as complete ablation was achieved in 81% with withdrawal and 84% with rhTSH. Hanscheid et al.[35] studied the biokinetics and dosimetry in RAI therapy of thyroid cancer after either rhTSH or hormone withdrawal in 63 randomized patients. They were treated with 3.7 GBq (100 mCi) of I-131 with blood activity and whole-body probe measurements, followed by whole-body scans in serial fashion and dosimetry calculations. They found that the effective half-time in the remnant thyroid tissue was significantly longer after rhTSH (68 hrs) than after hormone withdrawal (48 hrs), while the observed differences of the mean 48-hour uptake of 0.5% with rhTSH vs. 0.9% after hormone withdrawal, and the residence times of 0.9 versus 1.4 hrs, respectively, were not statistically significant. The absorbed dose to the blood was significantly lower after rhTSH, indicating that higher activities of RAI might be safely administered after stimulation with rhTSH.

16.4 PRECAUTIONS REGARDING PREGNANCY

Care must be taken to determine if a woman is pregnant prior to the administration of RAI, as it has been determined that the I-131 concentration is 6–7 times higher in the fetal thyroid than in the maternal thyroid. Dose estimates to the fetal thyroid range from 888 rad/mCi for the 3-month fetus to as high as 4,070 rad/mCi for the 6-month fetus.[36] Because of this fact, it is absolutely necessary to establish that a woman of childbearing age is not pregnant, by the use of standard pregnancy tests, unless the woman has previously undergone a bilateral oophorectomy, hysterectomy, or tubal ligation.

16.5 THE DECISION TO PERFORM THYROID REMNANT ABLATION WITH RAI

The decision to ablate a thyroid remnant is based on the risk of recurrence and eventual death. Additionally, ablation therapy can be justified as simplifying the follow-up of patients who have some risk of recurrence. Ablation eliminates the presence of thyroid remnants that confound interpretation of RAI imaging and TG results: both tests can be positive in patients with a remnant regardless of the presence or absence of disease after thyroidectomy. Performing an ablation allows monitoring with RAI imaging and TG levels with increased sensitivity and specificity.

16.6 RAI DOSAGE REGIMENS AND EFFECTIVENESS

There is a wide range of therapeutic options available for the treatment of thyroid cancer—from simple observation to aggressive treatment with emerging biotechnologies—so patients should routinely undergo risk stratification analysis to determine optimal management. RAI activities ranging from 30 to 100 mCi generally show similar rates of successful remnant ablation,[37–40] as defined by the absence of visible RAI uptake on a subsequent diagnostic RAI scan, although there is a trend toward higher success rates with higher activity (Figure 16.2). The traditional dose of I-131 administered for ablation therapy has been 50–75 mCi.[41] The traditional limit in the past for outpatient therapy was 30 mCi, or 5 milliroentgens per hour (mR/h) measured at 1 m as set by the U.S. Nuclear Regulatory Commission (NRC). At that rate, a person in casual proximity to the subject would still receive, over several days, less than 100 millirads (or 100 millirem), the upper limit for exposure to the general public. Patients requiring higher amounts were admitted to the hospital or received an outpatient ablation dose of 29.9 mCi. This lower dose was found to be effective in most patients, although it was not an entirely reliable method, sometimes requiring retreatment, with a range of success ranging from 20% to 83%.[42–44] Maxon et al.[45] also studied the calculated dosimetry to the residual thyroid tissue and noted an ablation success rate of 79–83% when a calculated dosimetry of 30,000 Gy was achieved. The activity required to achieve this target dosimetry ranged from below 30 mCi to several hundred mCi. By this approach, some 47% of the population studied by Maxon et al.[45] could be treated with a dose as low as 30 mCi. These results suggest that the use of remnant dosimetry is to be encouraged. The additional cost of this procedure is balanced by reduction in the expenses of hospitalization, higher doses, and inconvenience of additional radiation safety precautions. The goal is to use the minimum activity necessary to achieve successful thyroid remnant ablation, particularly for low-risk patients.

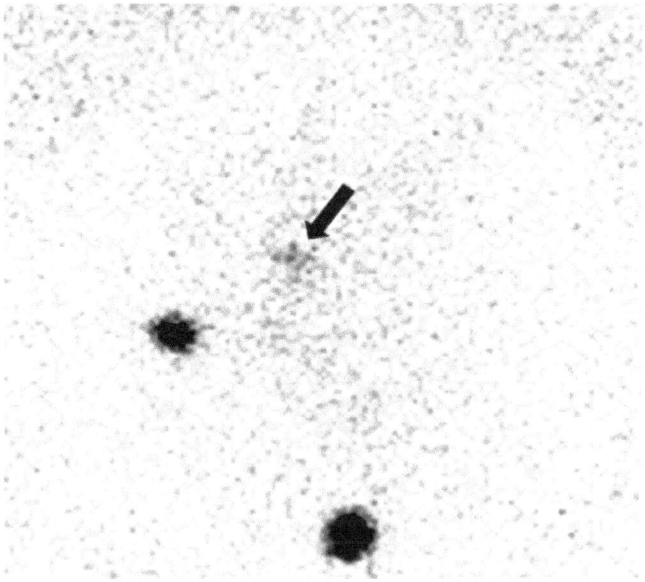

FIGURE 16.2 Pinhole diagnostic I-131 image of patient (Figure 10.6 in Chapter 10) showing slight residual thyroid tissue one year after therapy with 150 mCi of I-131.

Rosario et al.[46] demonstrated the usefulness of diagnostic I-131 imaging in predicting the success of therapeutic thyroid remnant ablation. They found that after 100 mCi of an I-131 ablation dose, thyroid remnants were eradicated in 46 of the 48 patients who had an uptake of less than 1%, 30 of 32 patients with an uptake between 1 and 2%, 25 of 30 patients with an uptake of 2 to 5%, 14 of 20 patients with an uptake between 5 and 10%, and 6 of 12 patients with an uptake greater than 10%. Hence, uptake and ablation efficacy were inversely related. Karam et al.[47] studied the success of ablation in 109 patients, as measured by a negative whole body scan on follow-up and TG levels. Success was associated with stage and female gender. Ablation was successful in 62 out of 109 patients, and it failed by both RAI follow-up scan and TG level in 10/109, TG level measurement alone in 21/109, and whole body scanning alone in 16/109. Thus, success was also dependent on the method of measurement (RAI scanning or TG levels).

16.7 TREATMENT OF RESIDUAL OR RECURRENT CANCER

This setting usually occurs at some time after the initial thyroid remnant ablation, as a result of detection of elevated TG levels, detection of regional neck lymph node enlargement on physical exam or by US or CT (Figure 16.3), or distant metastases by follow-up imaging (typically RAI). Sometimes, evidence of metastases is present at the time of initial ablation therapy. Generally, higher activities of 100 to 200 mCi are used for ablation of thyroid remnants when residual microscopic disease is suspected or documented, or if there is aggressive tumor histology, such as tall-cell, insular, or

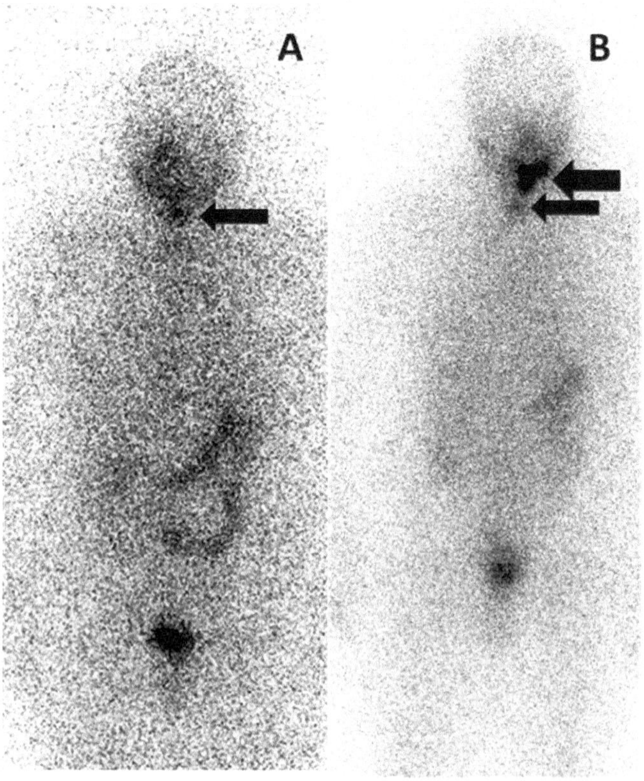

FIGURE 16.3 Pre- (A) and five-day post- (B) 200 mCi anterior whole body I-131 scans one year after initial RAI ablation therapy with 150 mCi in a 94-year-old woman. She now presents with a slightly elevated serum TG level and a suspicious lymph node by US in the perilaryngeal region. The posttherapy scan (B) shows several foci of uptake (arrows) in the laryngeal region that were poorly demonstrated by pretreatment, diagnostic RAI scanning.

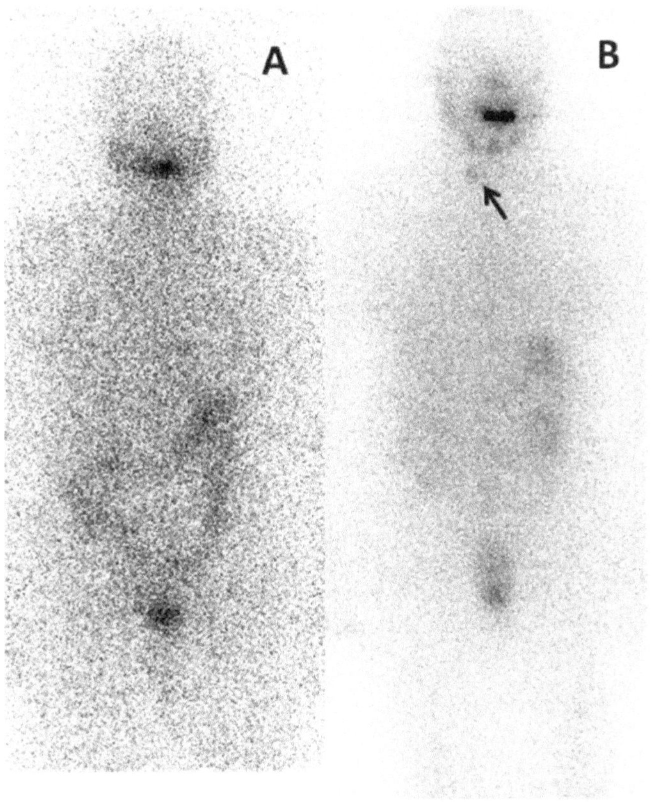

FIGURE 16.4 The same patient as in Figure 16.3 one year later. The lymph nodes in the neck had regressed. (A) Anterior whole body diagnostic I-131 scan shows resolution of local neck uptake. Because of the appearance of pulmonary nodules on a chest CT scan, which did not appear to take up RAI on the diagnostic scan, the patient received another 190 mCi of I-131. (B) The posttherapy anterior whole body I-131 scan showed a faint focus (arrow) in the neck, consistent with lymph node uptake, but no uptake in the chest.

columnar-cell PTC variant. Metastases that appear on follow-up scans may be difficult to eradicate completely with RAI therapy alone and should therefore be considered for surgical resection or adjuvant EBRT. Many patients may benefit from additional RAI treatment with reduction of tumor burden that may improve survival or provide palliative benefit[48–51] (Figures 16.3 and 16.4).

There are several approaches to treatment. These include the use of fixed empirical doses based on the disease, doses based on maximal tolerated dose, and doses based on dosimetry to the tumor. These approaches presume that higher RAI activity absorbed by the tumor leads to better outcome as suggested by some data,[52] while not confirmed by other data.[53,54]

16.7.1 Empirical Fixed Doses

This approach uses standard fixed doses from 100 to 200 mCi, based on the site of tumor location. The recommended treatment of recurrent tumor in the thyroid bed is 100 to 150 mCi, while treatment of neck lymph nodes calls for 150 to 180 mCi. Treatment of tumor outside of the neck usually involves doses of 200 mCi.[55] The main argument for the use of simple empirical doses is that this approach is simple and the least expensive, while little evidence compels that more patient- or tumor-specific approaches are more effective.[56]

Rosario et al. studied 274 patients who received ablation therapy without previous diagnostic imaging, with the dose being based on surgical staging: 100 mCi for stage I, and 150 mCi for stage

II and III. Among stage I patients, the efficacy of ablation treatment was 79%. A 47% failure rate was associated with metastases, which are less likely to respond to treatment. Of the 20 patients without metastases that did not have successful remnant ablation, 70% had thyroid bed uptake of >5%.[57] It is useful to point out that a large remnant with a high uptake could have been detected with a pretherapy diagnostic scan, and perhaps treated with a two-stage treatment approach.

16.7.2 Maximum Safe Dose Method

This approach is generally applied to patients with peripheral metastases on the grounds that such metastases often have weak RAI uptake, and doses in excess of the 200 mCi dose recommended in the empirical approach are desired. One would also like to avoid delivering a dose to the marrow in excess of the "safe" limit, even if the activity given is 200 mCi or less. This is particularly relevant for patients with a small body mass, and the elderly, with decreased renal clearance. The basis of the safe dose is the observation that bone marrow suppression is a problem only when the delivered dose to the bone marrow (calculated as discussed above) exceeds 200 rads (200 Cgy).[58–60] Generally, a rapid turnover in I-131 results in a higher-tolerated dose, while slow turnover results in a low-tolerated dose. As a result, 80% of patients can tolerate activities greater than 200 mCi of I-131, sometimes reaching up to 400–500 mCi. In 20% of patients, this maximal activity is less than 200 mCi. The use of rhTSH stimulation results in a more rapid turnover of I-131, resulting in a higher maximal tolerated dose.

16.7.3 Individual Lesion Dosimetry

Differences in the radiosensitivity of the primary thyroid lesion, nodal metastases, pulmonary metastases, and bone metastases were described by Maxon et al.,[45] who studied treatment success with I-131 as a function of radiation dose delivered. They demonstrated that cervical lymph node metastases were successfully ablated with doses of 10,000 to 12,000 cGy. The observation that dosimetry lower than that which is effective for native thyroid tissue in malignant lymph nodes may be due to the tumor occupying a smaller volume than the total volume of the lymph node used for the calculation. Thus, an ablation dose for normal remnant ablation is also effective in the treatment of loco-regional lymph nodes, even if they are too small to be seen on I-131 imaging. Small size favors a successful response, as the radiation absorbed dose increases for a given amount of radioactivity delivered as the tumor mass decreases.[61]

The use of dosimetry for individual lesions, including the use of iodine-124 positron emission tomography (PET) imaging, has not been widely applied, with the exception of only a few sites (due to the large amount of work required for measuring not only the plasma kinetics, but also individual tumor uptake and kinetics to arrive at the actual tumor dosimetry).[62] It is not known if this approach changes prognostic outcome. In addition, even if one sees a single or few lesions on the diagnostic scan, there are often additional lesions on the posttherapy scan. Thus, it may still be more appropriate to deliver the maximal tolerated dose for peripheral metastases.

16.8 THE USE OF RHTSH FOR RECURRENT OR METASTATIC DISEASE

While no randomized studies exist comparing the use of rhTSH stimulation with thyroid hormone withdrawal for the treatment of metastatic disease, there have been a number of studies regarding its usefulness.[63–65] Early on, rhTSH stimulation was targeted in patients with brain or spinal cord metastases to avoid prolonged tumor stimulation that could cause neurological signs or symptoms from tumor swelling. In our experience, this turned out to be true, with fewer symptoms of neurological deterioration with rhTSH stimulation followed by tumor shrinkage by magnetic resonance imaging (MRI). Nevertheless, the use of rhTSH does not eliminate the danger of tumor swelling causing neurological sequalae.[66–68] More recently, many centers have used rhTSH stimulation in patients with other medical comorbidities where thyroid hormone withdrawal might cause a significant deterioration in health status.

There are several theoretical considerations regarding the comparative effectiveness of thyroid hormone stimulation versus rhTSH stimulation. The prolonged TSH stimulation during thyroid hormone withdrawal favors increased uptake in tumor metastases, compared with the brief stimulation by serial rhTSH injections. On the other hand, the peak levels of TSH resulting from rhTSH injections are higher,[69] thus favoring use of rhTSH. Since the clearance of RAI is faster in the euthyroid state, compared to the hypothyroid state during thyroid hormone withdrawal, nontarget dosimetry is lower with rhTSH use. Thus, for a fixed dose of I-131, whole body and organ and salivary gland dosimetry is expected to be lower, thus minimizing side effects. When attempting to deliver a maximal tolerable dose limited by bone marrow dosimetry, the total administered activity, and therefore delivered dose, to the target cancer is expected to be higher. It is important to note that one must conduct a diagnostic dosimetry study with the same conditions, i.e., the same type of TSH stimulation as for the therapy, since the dosimetry and blood clearance will be different for rhTSH stimulation from withdrawal. Ultimately, the respective merits of the two approaches of stimulation can be assessed, first by using data from the diagnostic use of the two methods in terms of uptake in thyroid remnants and in metastases with diagnostic doses of RAI.[69] Finally, long-term comparative therapeutic studies using the two methods will be necessary, but difficult to carry out.

16.9 TREATMENT OF METASTATIC DISEASE OUTSIDE OF THE NECK

Peripheral metastases may be evident at the time of diagnosis or may become evident during subsequent follow-up. Morbidity and mortality are increased in patients with distant metastases, although other factors, such as distribution (e.g., lung, brain, and bone), the number of metastatic lesions, the amount of tumor burden, and age, are important. Responsiveness to surgery or RAI is associated with improved survival.[70–77] Even if improved survival might not be demonstrated, significant palliative objectives and decreased morbidities can be achieved with therapy.[78–79] If isolated, pulmonary and bone metastases can be treated surgically to decrease tumor bulk, and in the case of bone metastases, to provide stabilization. EBRT is also used for this purpose to stabilize critical weight-bearing bones. This approach is often combined with RAI therapy, since other disseminated disease is likely, even if not visible on the I-131 scan. The ablation of tumor outside of local cervical lymph nodes may require larger absorbed radiation doses.

16.9.1 Treatment of Pulmonary Metastases

Pulmonary metastases can be detected by chest radiography and CT if macronodular, and by CT and RAI if micronodular. While the efficacy of I-131 ablation therapy postthyroidectomy has been well established, the efficacy of treatment of metastatic disease is less substantiated partly due to the varying approaches and methods of different medical centers. Although postoperative ablation of thyroid remnants undoubtedly destroys some micrometastases, the common site of recurrence is in the cervical lymph nodes. However, good results have been reported in the treatment of metastatic disease with I-131.[80–82]

Pulmonary micrometastases are the most responsive to RAI treatment among peripheral metastases and can be treated every 6–12 months as long as responsiveness is observed.[83] The amount of I-131 activity used empirically ranges from 100 to 300 mCi, subject to limitation by whole body retention of 80 mCi at 48 hours and 200 Gy dosimetry to the bone marrow. Macronodular pulmonary metastases can also be treated and retreated with RAI if they are RAI avid.

Generally, survival is related to early detection of localized or low-volume residual or metastatic disease and high tumor uptake of I-131. Survival is better if the tumor is detected by I-131 imaging rather than by chest x-ray or CT alone. As a corollary, micronodular disease has a better prognosis than macronodular disease when treated with I-131. In one study, the percentage of patients surviving 5 years increased from 8% for macronodular disease to 72% for micronodular disease.[83] In this study, 50% of patients with lung metastases treated with I-131 survived for 8 years, while none survived without treatment. Hindie et al.[84] demonstrated 11 out of 20 patients with RAI-avid pulmonary metastases, but negative chest x-ray findings, had good outcomes with I-131 therapy. The 10-year survival was 84%. Even in the presence of advanced disease, long-term survival is possible with continued I-131 treatments.

Macronodular pulmonary metastases that do not demonstrate RAI uptake on diagnostic scans often demonstrate FDG uptake on PET imaging and have a relatively poor prognosis.[85,86] RAI negative but FDG positive pulmonary lesions have been shown to be poorly responsive to RAI treatment.[87] In one study, the presence of non-RAI-avid pulmonary metastases treated with 200–300 mCi of I-131 during thyroid hormone withdrawal reported a high mortality within 4 years.[88] At the same time, no chemotherapy agents have been found to be effective with more than 25% partial response rates.[89]

16.9.2 Treatment of Bone Metastases

I-131 therapy of RAI-avid bone metastases has been found to be associated with improved survival with doses ranging from 150–300 mCi, or estimated by dosimetry.[90] Usually, high-dose RAI treatment is given, since these lesions are relatively resistant to RAI therapy compared to micronodular pulmonary metastases. Bone lesions are frequently also treated with EBRT, especially when pathologic fracture threatens, as well as other palliative measures. The presence of bone metastases generally predicts a poorer prognosis than soft tissue metastases and a lower response to I-131 therapy.[91] For this reason, surgical therapy or EBRT is used for control of the tumor (Figure 16.5). I-131 therapy is used because there is usually additional disseminated tumor, even if not seen on I-131

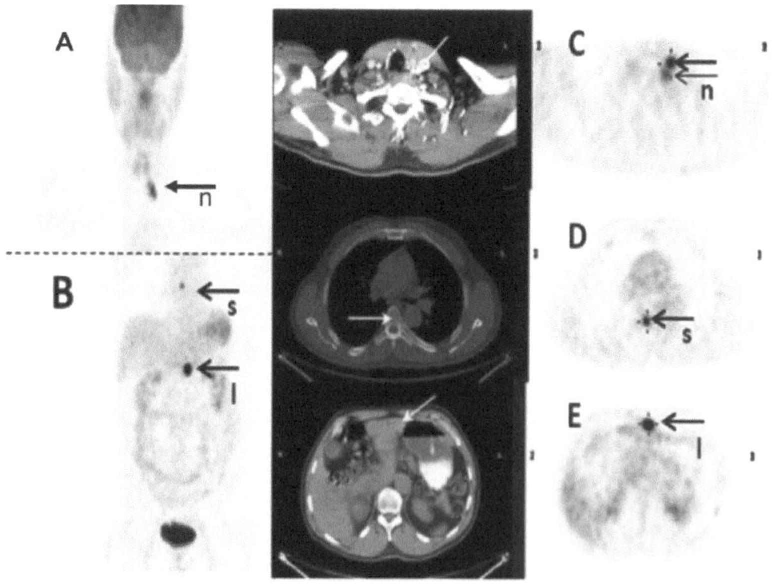

FIGURE 16.5 FDG PET anterior three-dimensional views of the neck (A) and torso (B) and PET-CT single-slice images of the neck (C), liver (D), and vertebra (E) in a 67-year-old male status post total thyroidectomy for a 5.5 cm poorly differentiated FTC found to be widely locally invasive and treated with 150 mCi of I-131 ablation therapy. Six months later, the patient's TG level was 644. A diagnostic I-131 scan on thyroid hormone withdrawal was negative. The FDG PET-CT scan revealed two FDG-avid tumor foci in the left thyroid bed (n), an FDG-avid focus in the spine (T6 thoracic vertebra) (s), and an FDG-avid metastasis in the left hepatic lobe (l).

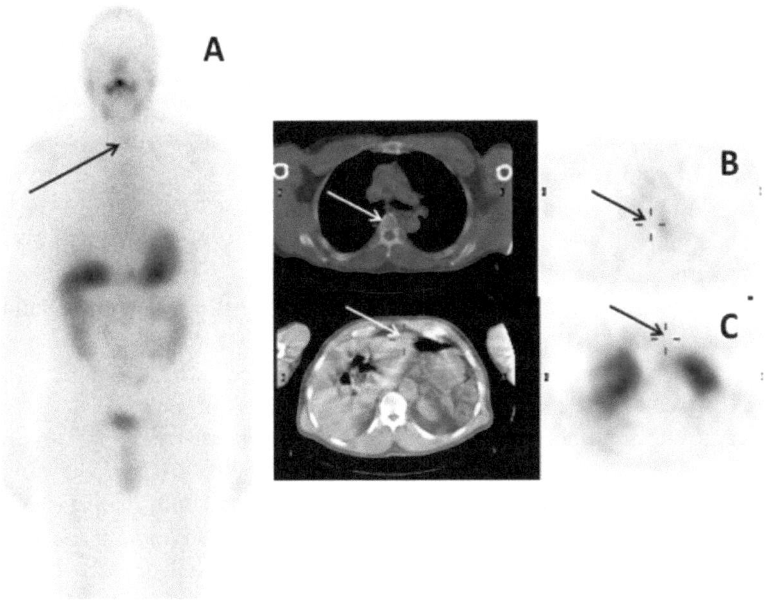

FIGURE 16.6 Anterior (A) whole body image and SPECT-CT I-131 images of the T6 vertebra (B) and liver (C) of the same patient as in Figure 16.5, imaged after receiving 200 mCi of I-131, showing only faint uptake in the left thyroid bed (A, arrow) corresponding to the hypermetabolic foci on FDG PET imaging. There is no uptake on the SPECT-CT images (E) corresponding to the previously seen liver and mid-thoracic spine lesions (B and C, arrows). The patient underwent surgical resection of the neck and liver lesions and EBRT of the T6 lesion.

imaging (Figure 16.6). The response to treatment of bone metastases is still related to the degree of I-131 uptake.[91] In general, this calls for as aggressive approach as possible. The observation that there was a 30% increased survival of patients diagnosed with distal metastases after 1976 compared to those diagnosed and treated before 1976, and a 140% increase in survival in patients diagnosed after 1960 compared to patients diagnosed before 1960, supports aggressive I-131 therapy.[76]

16.10 TREATMENT FOR ELEVATED TG EVEN THOUGH THE RAI SCAN IS NEGATIVE

A difficult situation comes up when a patient with PTC or FTC has an elevated TG level or a positive FDG PET scan or other evidence of tumor, when there is no RAI uptake on a RAI diagnostic scan (Figures 16.3 and 16.6). One can make an argument that it is not worth treating such disease with I-131 due to the low likelihood that one can deliver sufficient radiation dose to elicit some benefit, preferring to follow the patient and wait until more tumor appears that can be seen on RAI scan and more activity can be delivered with I-131 therapy.

If the TG is elevated, particularly after thyroid hormone withdrawal or rhTSH stimulation in a patient that has not yet had thyroid ablation therapy, it is often revealed that this is due to residual thyroid tissue in the thyroid bed. In such a case, ablation therapy is indicated. However, if a rise in TG or development of anti-TG antibodies occurs after thyroid remnant ablation, particularly rising TG levels on serial testing, then the presence of thyroid cancer is concerning.[92]

A counterargument can be made that after a therapeutic RAI dose is given, one can see the tumor on the posttherapy scan (Figure 16.4). Pineda et al.[93] and Pacini et al.[94] have found that treating these patients with 100–150 mCi of I-131 results in visualization of the thyroid cancer metastases in one-third of such patients. In another third of such patients, the TG levels decrease significantly after therapy,[95] although serum TG levels may at times decline without specific therapy. Kabasakal et al.[96] studied 27 patients in this category, who received high-dose RAI therapy, with a mean follow-up of 6.3 years. Posttreatment whole body scans revealed localized accumulation in 19 of the 24 patients (79%). Serum TG levels decreased in 8 of 16 (50%). Among patients with micrometastases, 5 out of 7 (71%) showed decreased TG levels, while among patients with macrometastases, 3 out of 9 (33%) showed decreased TG levels, with 3 (33%) dying due to metastatic thyroid cancer. Ma et al.[97] reviewed the world literature, and in 314 patients in 10 observed series and 3 nonrandomized controlled trials, 62% displayed pathological uptake in the thyroid bed, lung, bone, mediastinum, or lymph nodes. In studies with TG level measurements, 63% of 271 patients achieved decreases in TG levels. Pacini et al.[98] studied outcomes in 42 treated patients, with elevated TG levels and negative diagnostic scans, and 28 untreated patients. In the treated patients, the first posttherapy diagnostic I-131 whole body scan was negative in 12 patients and positive in 30 patients. These 30 patients in the treated group received further treatment. Among the treated patients, a complete remission was observed in 10 patients (33%). The posttherapy diagnostic scan became negative in 9 patients, and serum TG was reduced but still detectable. In 11 patients, serum TG was detectable and the posttherapy scan was positive. The resolution of I-131 uptake was evident in 8 of 9 patients with lung metastases, and in 11 of 18 patients with cervical node metastases. Among the 28 untreated patients, serum TG off thyroid hormone therapy became undetectable in 19 patients and was significantly reduced in 6 patients. Thus, patients with detectable TG levels but negative diagnostic I-131 scans may benefit from high-dose I-131 treatment, and to some extent, that benefit can occur in patients with cervical node metastases. However, untreated patients with elevated TG levels may show normalization of TG as part of the natural history of the disease levels.

In short, empiric I-131 therapy with 100–200 mCi should be considered to aid in localization of the responsible tumor, possibly leading to partial reduction in tumor volume, or possible surgical resection. Such an approach has been reported to aid in localization of persistent disease in up to 50% of patients.[99,100] Unfortunately, there is no reported evidence of improved survival in the setting of empiric treatment of elevated TG levels in the setting of a negative RAI scan.[101] If the posttherapy scan after an empiric dose of RAI does not localize disease, in spite of an elevated TG level, F-18 FDG PET imaging should be considered.

16.11 EFFECT OF AGE

Age is an important covariant that affects the likelihood of response to I-131 therapy. While more than 95% of younger patients with limited disease and good I-131 uptake responded to I-131 therapy, only 56% of patients over 40 years of age had good uptake, and only 19% of those responded to I-131 therapy. The higher prevalence of poor uptake in the older population suggests alteration of tumor function in this age group.[102]

16.12 TREATMENT OF FDG POSITIVE METASTASES

In general, patients with increasing uptake and volume of F-18 FDG-avid lesions on PET imaging are less likely to respond to RAI therapy, with a higher mortality in the next three years.[103] In another study, RAI therapy of lesions positive on FDG PET imaging was of no benefit.[104] Still, another study had not found FDG PET imaging helpful in predicting response to therapy.[105] In such patients, thyroid hormone suppression therapy, along with surgical resection, or treatment with EBRT, experimental chemotherapy or biological therapy, and other palliative treatments, should be considered. The overall poor outcome in patients with clinically or radiographically evident metastases that do not respond to RAI and are not accessible to surgery or EBRT suggests a referral to tertiary centers with expertise and investigational protocols in the area of novel treatments.

16.13 USE OF EBRT IN COMBINATION WITH RAI THERAPY

Unresectable disease in regional lymph nodes or in bone (Figures 16.5 and 16.6) is frequently unresponsive or only partially responsive to RAI therapy alone. Local sites of metastasis are frequently amenable to EBRT with successful control[106,107] and improved relapse-free survival.[108] Painful bone lesions that cannot be resected can be treated with combinations of EBRT, RAI, as well as other palliative approaches, including bone-seeking strontium-89 or samarium-153 if the lesions show uptake of bone-seeking radiotracers such as Tc-99m phosphonates. Likewise, brain metastases that occur in more advanced disease are treated with surgical resection and EBRT, along with RAI treatment of other lesions.[109,110]

16.14 SIDE EFFECTS AND COMPLICATIONS OF RAI THERAPY

It is important to ensure that the benefits of RAI treatment outweigh the possible risks. The side effects stemming from thyroid hormone withdrawal and rhTSH stimulation are discussed in Section 16.3. The side effects and risks from the RAI treatment itself pose a number of considerations.

16.14.1 Bone Marrow Toxicity

Bone marrow toxicity is a prime concern when considering I-131 therapy. For this reason, younger patients who are being prepared for treatment with 200 mCi or more of I-131, and older patients with doses of 180 mCi or more, should undergo blood dosimetry calculation. Patients with blood dosimetry at or below 200 cGy (200 rads) rarely show serious bone marrow toxicity, such as aplastic anemia or prolonged bone marrow suppression, manifested as depressed white cell counts or platelet counts.[111] Transient mild anemia, leukopenia, or thrombocytopenia for up to 3 months in duration may be seen. The lymphocyte count is reported to show more prolonged suppression. A persistent mild decrease in leukocyte or platelet counts is common in patients after multiple RAI therapies. In such cases, a baseline blood count measurement before each additional treatment is advised. In patients requiring large-dose repeated treatments, bone marrow growth factors (Neutrogen, Epogen) may be used to mitigate the more chronic effects of multiple high doses of RAI therapy. Immonodeficiency syndromes have not been reported, even after multiple large-dose treatments over many years.[6]

16.14.2 Gastric Toxicity

The most acute effect of RAI therapy is acute queasiness, usually after more than 100 mCi, and more prevalent after more than 200 mCi, starting several hours after therapy, and lasting until several days after the therapy. Oral, IM, or rectal antinausea medications are helpful. In general, oral medications are useful with light symptoms, while IM or rectal medications are more useful for more severe symptoms. Vomiting may occur, but is uncommon, in the evening on the day of RAI therapy. For that reason, patients are instructed to avoid eating for several hours before and after treatment, to ensure rapid absorption of the treatment dose and avoid losing a significant part of the treatment dose with acute vomiting. From a radiation safety point of view, the dose has by several hours later been absorbed, although the stomach does contain some radioactivity, as iodide is excreted from the gastric mucosa. Patients are instructed to vomit in the toilet bowl in order to facilitate cleanup. Patients are instructed to observe good hydration and to eat only lightly the first afternoon and evening.

16.14.3 Salivary Gland Toxicity

Patients who undergo RAI treatment are expected to experience a cumulative dose-related risk of early and late onset of damage to the salivary glands and the nasolacrimal ducts,[112] due to concentration of I-131 in these glands. Of note, uptake in one or both salivary glands may be increased when there is retention due to preexisting parotitis, which leads to increased accumulation of I-131 due to retention of RAI, as shown in individual case reports.[113]

Loss of taste and parotid and submandibular gland swelling and tenderness may be seen in some 50% of patients starting on the second day of therapy, and lasting for several days. The likelihood increases with greater administered doses of I-131, and is less likely with doses of less than 100 mCi. The symptoms usually resolve after several days. Mild analgesics are all that is usually needed. Some patients complain of loss of taste, dry mouth, with attendant increased tooth decay, and chronic or recurring parotitis. Grewal et al.[114] reviewed the clinical records of 262 patients following RAI therapy with a median dose of 141 mCi. They found 39% of patients reporting salivary side effects. Persistent side effects were noted after a median of 7 years in 5% or less of the entire cohort. In those who develop side effects in the first year, the persistence of symptoms ranged from 5 to 13%. A dose-response was seen between administered activity and salivary gland swelling, but not with dry mouth, altered taste, or salivary gland pain. Side effects occurred in 14% of patients in those receiving activities as low as 30 mCi and in 40% of patients with 75 mCi or more. Patients being treated on thyroid hormone withdrawal had a lower rate of salivary gland swelling than those treated with rhTSH, without differences in the other symptoms. Other authors report salivary gland toxicity after therapy ranging from 7% to 33%, some of them reporting a dose-response relationship, with 20% reporting chronic symptoms at 12 months, and 15% at 36 months.[115–118] The symptoms of loss of taste or unusual taste often resolve in 4–6 weeks, but are more likely to be permanent with large doses or repeated treatment.

Caglar et al.[119] demonstrated objective dose-dependent dysfunction of salivary glands in 31 of 45 patients who had received 100–200 mCi of RAI, using imaging of uptake and excretion of Tc-99m pertechnetate on salivary gland scintigraphy. The impairment was worse in the parotid glands than in the submandibular glands.[119] Thus, it is possible to investigate symptoms of salivary gland dysfunction by objective functional testing.

16.14.3.1 Prophylaxis

Good hydration and avoidance of dehydration optimizes renal excretion of I-131, thus minimizing radiation exposure. Patients have also been traditionally advised to use lemon-flavored lozenges, from the day of therapy for several days afterward, to promote salivary flow, thus reducing exposure to the glands. A recent study demonstrated that sour candy may actually increase salivary gland damage when given within 1 hour of RAI therapy, leading the investigators to recommend starting this treatment at 24 hours after therapy.[120] The incidences of sialoadenitis, taste loss, dry mouth, and permanent xerostomia, for early onset of sour candy use vs. delayed onset of use, were 64% vs. 37%, 39% vs. 26%, 24% vs. 11%, and 14% vs. 6%, respectively. In both groups, bilateral involvement of the parotid glands was seen the most frequently, followed by bilateral involvement of the submandibular gland. Alternatively, some physicians use cholinergic agonists, which pharmacologically enhance salivary flow. An example is pilocarpine hydrochloride (Salagen), 5 mg, taken orally three or four times daily for several days. The resulting persistent salivation, however, can be annoying to some patients. One may also decrease the side effects by splitting a high-dose treatment into multiple smaller doses (fractionation). It is unknown if this approach also limits the effectiveness of treatment. Another remedy reported to have some success is the use of intravenous amifostine, an organic thiophosphate, which is converted into WR-1065, a scavenger of oxygen free radicals. A clinical study investigating this agent administered amifostine in 25 patients and a placebo in 25 patients, and for all patients, ascorbic acid and high-dose dexamethasone. One year after RAI therapy, salivary gland function was reduced 40% in the placebo group and unchanged in the amifostine group.[121] This approach has not been adopted, possibly due to concern for antagonism of amifostine to the therapeutic effect of RAI in thyroid cancer.

16.14.3.2 Treatment of Salivary Gland Symptoms

The development of dry mouth may be treated with artificial saliva. The use of cholinergic agents may be used to increase salivary flow. Good dental hygiene and frequent dental follow-up and dental preventive strategies should be followed, as dental caries is increased. Nasolacrimal obstruction, manifested by frequent tearing and infections, may be treated with surgical correction. Blockage of the salivary glands has been treated with massage, and inflammation with anti-inflammatory agents and antibiotics, good hydration, and even hyperbaric oxygen. Some physicians have used pre- and posttreatment steroids in patients with preexisting sialoadenitis from previous RAI treatments.

16.14.4 Pulmonary Toxicity

In patients with extensive, particularly diffuse pulmonary metastases, there is concern for development of radiation pneumonitis. The precautions relating to prevention of pulmonary toxicity have already been discussed as part of dosimetric considerations. In those patients at risk for pulmonary fibrosis, careful follow-up with pulmonary function tests and expert follow-up is recommended. Pulmonary fibrosis secondary to I-131 treatment poses a limit to further RAI treatment.

16.14.5 Lower GI Tract Toxicity

Swallowed activity in saliva and activity excreted in the GI tract, namely, in the stomach, is not completely absorbed as documented by the common detection of radioactivity in the colon on diagnostic and post-therapy RAI scans. To avoid excessive exposure to the colon, where the activity may stay for several days, or even more in constipated patients, avoidance of constipation through the use of high-fiber diet and laxatives is recommended.

16.14.6 Hyperthyroidism or Functional Thyroid Cancer

Rarely, in patients with large tumor volumes, administration of I-131 may cause hyperthyroidism. Sisson and Carey[122] reported two patients with bulky metastases that sequestered iodine and secreted thyroxine, which predicted higher than usual radiation exposure to normal tissues when I-131 therapy was administered. The I-131 was incorporated into thyroxine that circulated for long periods of time. Even with a relatively small dose of 50 and 60 mCi (1850 and 2200 MBq), the patient developed transient leukopenia and neutropenia. One patient also had thrombocytopenia and anemia. This underlies the usefulness of performing dosimetry to prevent serious toxicity to normal tissues. When I-131 is incorporated into thyroxine, however, the standard dosimetry measured for 4 days may be inadequate, and the diagnostic sampling period should be extended up to several weeks. In another case report, Lawrence et al.[123] described a patient with tall-cell variant of PTC that had metastasized to bone, with evidence that the tumor was producing large amounts of thyroid hormone. After surgical excision, thyroxine levels decreased and TSH levels increased 10-fold. In an even more dramatic case report[124] a patient who had undergone total thyroidectomy developed accelerated thyrotoxicosis 32 years after injection of iodinated contrast media to evaluate a left hemipelvic mass, which on Tl-201 and I-131 imaging was proven to be a functioning metastasis of a differentiated follicular thyroid carcinoma. This was treated first with propylthiouracil and beta blockers, then with palliative EBRT, followed by RAI therapy. Granulocyte colony-stimulating factor was used to mitigate bone marrow toxicity, undoubtedly due to prolonged circulation of I-131 incorporated into circulating thyroxine produced by the functional metastases.

16.14.7 Carcinogenesis

Concerns for carcinogenesis due to I-131 therapy by and large have not been borne out by the literature.[76,80,130] However, there are reports of increased incidence of myelogenous leukemia in patients treated with high-dose radiation to the bone marrow. In nearly 700 cases, the incidence of leukemia was found to be 1–2%.[125] Other long-term studies demonstrate a small additional risk of secondary malignancies, such as bone and soft tissue malignancies, colorectal cancer, salivary tumors, and leukemia, in patients who had undergone RAI therapy, in proportion to the cumulative dose. In a European study of 6841 patients with thyroid cancer diagnosed between 1934 and 1995, there was a small increased risk of both solid tumors and leukemia (14.4 solid tumors and 0.8 leukemia per GBq (27 mCi) of I-131 per 10^5 person-years, or 53 excess cases out of 576 of solid malignancies and 3 cases of leukemia per 10 years among 10,000 patients treated with 100 mCi of I-131).[126] The mean time between RAI therapy and the secondary malignancy was 15 years. The relative risk ranged from 2.2 to 4.0. In another study, there was an increased risk of breast cancer in patients with thyroid cancer, but it is unclear if this was due to RAI therapy or a referral bias.[127]

16.14.8 Effects on Fertility in Women

Another potential concern is the effect of I-131 therapy on fertility. Gonadal tissue is subject to radiation from RAI in the blood, adjacent urine, and feces, and RAI uptake in metastases close to the ovaries. Hypothyroidism from thyroxine withdrawal increases ovary exposure. Mathematical models of exposure to the ovaries in individual patients have yielded exposure up to three times higher than official published standard estimates of 1.4 mGy/37 MBq (1 mCi). For example, a cumulative dose of 237 mCi in one patient resulted in exposure to the ovaries of 1 Gy (100 rads). Nonetheless, a follow-up of 2673 pregnancies after RAI therapy for thyroid cancer resulted in no greater incidence of miscarriage, stillbirths, preterm births, low birth weight, congenital malformations, and death during the first year of life, before or after RAI therapy. The incidences of thyroid and nonthyroid cancers were similar in children born before and after the mother's exposure to RAI.[128] Temporary amenorrhea or oligomenorrhea lasting 4–10 months has been reported in 17–20% of menstruating women after I-131 therapy for thyroid cancer. However, long-term rates of fertility, miscarriage, and fetal malformation do not appear to be elevated in women after RAI therapy.[129] Dottorini et al.[130] reported that women treated with 150 mCi of I-131 did not demonstrate fertility rates different from women with thyroid carcinoma treated with surgery alone, observed over a period of 1–15 years. In one study, out of 40 patients treated prior to the age of 21 years, there was no change in the fertility rate, spontaneous abortions, or genetic birth defects, when pregnancy was achieved 6–12 months or later after therapy.[131] Another study, however, showed an increase in miscarriage rate, suggesting that pregnancy should be postponed for 1 year after therapy.[132] Although one study reported menopause occurring 1 year earlier than in the general population, this result was not associated with cumulative dose administered or the age at therapy.[133]

Although the predominance of evidence does not show problems with fertility or teratogenicity in future children, many experts suggest that women should postpone becoming pregnant by 6–12 months from the last therapy.

16.14.9 Effects on Fertility in Men

Oligospermia and elevated follicle-stimulating hormone (FSH) levels have been observed in men treated with I-131.[134–136] The effect is dose dependent and is an example of dose-related toxicity from radiation. Usually, the oligospermia is transient with low and moderate doses of I-131, but more prolonged with high doses. Nonetheless, fertility problems are not significant after doses ranging from low to high. High cumulative doses (500 to 800 mCi) in men are associated with an increased risk of persistent elevation of serum FSH levels, but fertility and risks of miscarriage or congenital abnormalities in subsequent pregnancies are not changed with moderate RAI doses of approximately 200 mCi.[131,137] Nonetheless, it is recommended by many practitioners for men to avoid fathering children for 3–6 months after receiving therapeutic doses of RAI.

Permanent infertility is unlikely with a single dose, but there is the possibility of cumulative damage with multiple treatments. In such instances, sperm banking while the patient is still euthyroid may be advised for cumulative doses greater than 400 mCi. Gonadal radiation exposure in both men and women is reduced with good hydration, frequent urinary voiding, and avoidance of constipation.[138]

16.15 THERAPY IN POSTPARTUM WOMEN

Since uptake in the breasts is increased in lactating or postpartum women, RAI treatment should be avoided, as significant radiation exposure to the breasts would result, until after breast feeding has been stopped for at least 6–8 weeks.[139] If the mother continues to breast feed, the concentration of RAI in milk is such that as little as 25 mcCi of I-131 may cause adverse effects to the infant's thyroid gland.[140]

16.16 REGULATIONS REGARDING HOSPITALIZATION OF PATIENTS TREATED WITH RAI

The U.S. NRC has recently revised its regulations for release of patients receiving RAI doses. As a result, patients with activities larger than 30 mCi can be released if it can be demonstrated that family members and caregivers do not receive more than 5 mSv (500 mrem), and other members of the public do not receive more than 1 mSv (100 mrem), under regulation 10 CFR 35.75.[141] Recent studies confirmed that exposure of family members of patients treated with 75 to 150 mCi (2.8 to 5.6 GBq) is minimal.[142] As opposed to the liberalized regulations in the United States, the regulatory limits in European countries are more restrictive.

Various approaches have been used. Many centers perform outpatient therapy if the administered activity of I-131 is less than 200 mCi (7400 MBq). Others have developed models for estimating the exposure to others, using variables of I-131 occupancy factors, time after dose administration, fractional uptake of I-131 by residual thyroid tissue or metastases, and duration of constrained activity. The "occupancy factor" is the estimated time that the patient will be near the person with whom the patient spends the most time in the post-treatment period. Using the effective half-life of I-131, effective half-life of I-131 in the thyroid remnant, and extrathyroidal component during the equilibrium period, Coover et al.[143] developed charts for use in calculating the maximum I-131 dose for thyroid remnant and metastatic disease treatment. With this technique, most patients could be treated on an outpatient basis with activities even greater than 200 mCi (7400 MBq). In the absence of patient-specific calculation, one may use group-defined exposures for I-131 dose ranging from 30 to 200 mCi for outpatient management, and hospitalize patients with planned doses greater than 200 mCi for inpatient management.

16.17 INSTRUCTIONS TO PATIENTS AND THEIR CARETAKERS AND FAMILY MEMBERS

Many centers recommend a variety of precautions given to patients and their family members and caretakers. This requires an evaluation of the patient's home and work conditions to determine if these conditions are appropriate for patient release, while following an advised protocol of average distance and duration of proximity to other persons. Some centers conduct a specific patient-based dose calculation determined from pretherapy diagnostic dosimetry calculations to ensure that the dose of radiation exposure to any exposed individual doses not exceed the 500 mrem dose for adults and nonpregnant family members and caretakers, and 100 mrem or less to everyone else. The patient must receive clear and simple written instructions about appropriate behavior and precautions, which usually need a consultation in advance of the treatment, plus a review at the time of the treatment.

For a patient who receives 200 mCi of I-131, the measured exposure is about 30 mrem/h at 1 m and 100 mrem/h at 0.3 m (for I-131 gamma rays, 1 mrem/h is the same as 1 millirad/h). Even minimal precautions result in achievement of the desired goals. A nonsleeping partner or a nonheld child receives 173 mrem, on average. In order to avoid giving an exposure of 100 mrem or less to other persons, patients should avoid any proximity to pregnant women and children for 1 day, not go to work for 2 days, not hold small children for 14 days, and not sleep with a pregnant spouse for 18 days. Leslie et al.[144] have performed phantom measurements of exposures of contacts, adult and children, of patients treated with RAI, at distances of 1 and 2 m. They found doses to adults and children and infants to be lower than previously expected, although still higher for cradled infants. The authors' main restriction is for cradled infants, who are limited to fifteen 35-minute periods of close contact for a total of 8.75 hours for every 24-hour period. Most caregivers would be able to restrict their exposure below this amount. Grigsby et al. monitored 30 patients who received outpatient therapy with 80–200 mCi (mean 120 mCi) and placed radiation monitors on 65 household members and 17 household pets for 10 days. Their instructions were quite minimal, consisting of 2 days of sleeping alone, liberal drinking of fluids, avoiding personal contact, and normal activity after that. The effective dose to maximally exposed persons ranged from 163 to 483 (mean 312) mrem; other household members, 10 to 109 (mean 24) mrem; and exposure to household pets, 20 to 111 (mean 37) mrem. Thus, even a very liberal regimen of 2 days of restrictions results in adherence to the NRC regulations. Since most precaution protocols tend to be more restrictive, the exposure to family members and the public is expected to be even lower.

16.18 GENE THERAPY SENSITIZATION METHODS

The effectiveness of RAI therapy in thyroid cancer depends on the uptake and retention of iodide. Patients with poorly differentiated carcinoma and those in the older age group tend to have poor uptake of I-131, and even patients with initially good uptake of RAI develop loss of iodide uptake with metastatic disease, leading to poorer response to I-131 therapy and worse prognosis. In these cases, there appears to be loss of the human sodium (Na)-iodide (I) symporter (NIS). In this group of patients, it has been suggested to attempt to restore NIS activity through the use of NIS gene transfer into NIS-defective tumors. Problems with this approach include one of delivery of the gene, usually with direct local injection, and temporally limited efficacy while the carrier virus is effective. Moreover, increased uptake of RAI by the restored NIS does not necessarily lead to prolonged retention, if there is also poor organification capability. Smit et al.[145] transfected NIS into NIS-deficient FTC cell line xenografted tumors. They found higher peak iodide accumulation: 17% compared with 4.6% in controls with a half-life of 3.8 hours. Mice on a low-iodide diet showed uptake of 8.1% only, but thyroid accumulation increased. The half-life was increased to 26 hours in thyroid-ablated mice on a low-iodide diet. A lethal dose of RAI could not be delivered, although tumor progression was delayed in thyroid-ablated, low-iodide diet mice treated with RAI therapy.

Another approach to the restoration of sensitivity to RAI therapy utilizes the recent discovery that *trans*-activating protein factors in RAI-resistant tumors represses endogenous NIS transcription, accounting for the loss of RAI uptake. Inhibition of these protein factors increases endogenous NIS expression and presents a novel target to restore NIS expression in dedifferentiated thyroid carcinoma. This is illustrated by a case report[146] of a patient with extensively metastatic non-iodine-concentrating differentiated thyroid carcinoma. The patient was treated with cisplatin and doxorubicin. Repeat I-131 imaging after three cycles of chemotherapy showed significant uptake in previously non-RAI-avid lesions. This suggests the usefulness of combined RAI and targeted chemotherapy.[147]

Given the importance of restoration of sensitivity to iodine in thyroid cancer, this line of research is of obvious importance.

16.19 NON-RAI RADIOTRACER THERAPY

16.19.1 Poorly Differentiated Thyroid Cancer

Patients with ATC and poorly differentiated PTC or FTC, as well as those in the older age group, tend to have poor tumor uptake of I-131, poorer response to I-131 therapy, and worse prognosis. This group of patients is a candidate for more innovative therapy. One example is therapeutic somatostatin (SSTR) analog ligands, in those patients demonstrating somatostatin uptake on In-111 octreotide imaging, with or without experimental chemotherapy or molecular targeted therapy. For example, doxorubicin has been reported to have some radiosensitization action in some tumors of thyroid origin.[148,149]

SST analogs have been labeled with high specific activity gamma-emitting isotopes (indium-111), or beta-emitting isotopes (I-131, Y-90, Re-188). Krenning et al.[150] reported that In-111-DTPA-D-Phel-octreotide in high doses could be used as a therapeutic agent because of the radiation effects of Auger electrons and conversion electrons associated with gamma decay. Because the use of high-activity In-111 leads to problems of radioprotection for personnel and the intracellular action of Auger and conversion electrons limits efficacy when SST receptors (SSTRs) are distributed unevenly, beta-emitting isotopes are preferable. This is due to easier radioprotection and penetration of beta radiation over several millimeters, thus obviating the problem of uneven distribution of SSTR and making radiation exposure of personnel minimal. Such compounds as Y-90-DOTA-Tyr3-octreotide (DOTATOC)[151] and Y-90-DOTA-lanreotide[152] are being investigated in clinical therapeutic trials.

Innovative work currently concerns multimodality therapy with molecular targeted therapies, radiolabeled agents, and minimally invasive surgery.[153] Molecular testing for sensitivity to drugs and radioactive agonists, such as the NIS symporter, are hoped to produce improved therapeutic results. The use of gene therapy, monoclonal antibodies, growth factor blockers, cytokines, and interventions that target thyroid stem cells are being investigated.

16.19.2 Therapy of MTC

The 10-year survival of patients with MTC has improved as a result of total thyroidectomy and central lymph node dissection. However, the prognosis of patients with distal metastases to the lungs, liver, and bone is poor: about 30% 5-year survival.[154–157] Chemotherapy alone has not been very effective, and EBRT is not effective in the setting of multiple peripheral metastases, except for local palliation. MTC does not concentrate RAI, unlike differentiated thyroid epithelial cancers. If one can demonstrate affinity of SST analogs such as Indium-111 octreotide, or with I-123 and I-131 metaiodobenzylguanidine (MIBG), one may treat with these agents.

Attempts have been made to deliver radioisotope in sufficient quantity to enable therapy by the use of radiolabeled antibodies and receptor ligands. Therapy with radiolabeled monoclonal antibodies against CEA antigen, which is expressed in 70–90% of MTC, especially in metastases, has shown promise.[158] Anti-CEA radiolabeled murine antibodies were first used against MTC[159,160]; 46–77 mCi of I-131 labeled intact antibody, or 90–195 mCi of I-131 murine Fab fragments, resulted in moderate antitumor effects in 7 out of 14 patients with MTC. The limiting factor was myelosuppression. Studies with I-131 murine FAB2 anti-CEA antibody fragments, in a dose-escalation manner, resulted in 7 out of 12 patients showing a notable response, and 11 out of 12 showed stable disease.[161] The same group used a dose-escalated regimen of the same radiolabeled anti-CEA antibody fragments, in combination with autologous bone marrow rescue. Of the 12 patients, 1 patient had a partial remission for 1 year, one had a minor response for 3 months, and 10 had stabilization of disease for 1 to 16 months.[162] Humanized murine antibodies labeled with Y90, which is a beta emitter only, were tried, making it easier to handle and allow outpatient therapy. The investigators showed higher tumor-to-normal organ ratios and greater antitumor effects.[163,164] The combined treatment with Y90-radiolabeled antibody plus doxorubicin was shown to be more effective.[165] Another trial used escalating doses of Y90 hMN-anti-CEA antibody and a fixed dose of doxorubicin, along with stem cell rescue. The doses used were based on clearance measurements with In-111 antibody performed before the treatment. Of the 13 patients treated, one had a partial response, 3 had a partial response for 3 months, 2 had a minor response for 22 and 11 months, respectively, and 6 had stable disease for 2–19 months, while 4 had disease progression.[166,167] Hence, the initial results of this avenue of treatment demonstrate the feasibility of concurrent high-dose radioimmunotherapy in combination with bone marrow stem cell rescue.

Innovative work is being investigated using the induction of NIS symporter,[168] as discussed above, and the use of other isotopes, such as Re-188, and alpha particle emitters, such as astatine-211, which can produce higher-density radiation energy delivery, as well as combined chemotherapy, molecular intervention therapy, radioactive ligand therapy, and targeted surgical therapy.[169,170]

16.20 CONCLUSIONS

RAI therapy with iodine-131 continues to be the mainstay of treatment of thyroid cancer following surgical thyroidectomy. There is consensus that all patients except for those at the lowest risk of recurrence should undergo thyroid remnant ablation. For those at low risk of recurrence, ablation therapy is optional, with advantages based on easier follow-up. For residual or metastatic disease that shows evidence of RAI uptake, iodine-131 therapy is the treatment of choice, with good results overall. There are differences of opinion on the I-131 activity that should be used for ablation therapy, disease in the neck, and peripheral metastatic disease, as well as the role of lesion dosimetry. Lung and blood (bone marrow) dosimetry is recommended in all patients in whom high-dose therapy for peripheral metastases is contemplated, as well as in the elderly and those with compromised renal function, in whom pulmonary and bone marrow toxicity is a possibility.

RAI activity selection, on the one hand being aimed to deliver the highest possible dose to the target, is tempered with an effort to minimize gastrointestinal and salivary gland toxicity, and minimize risk of carcinogenesis. The need for thorough patient counseling on preparation, the diagnostic and therapy procedures and logistics, and side effects and their amelioration, and preparation of patients and their families on radiation safety procedures after treatment, cannot be overemphasized. If conducted well, counseling can turn a very stressful and anxiety-provoking process into a more relaxed, reassuring experience.

Important unresolved questions remain, including the proper role of rhTSH in thyroid cancer therapy, presently an off-label indication. The treatment of residual or metastatic disease in the absence of RAI uptake remains an open question, given the low achievable dosimetry but empirical evidence of some partial benefit. The role of sensitizing agents and novel molecular targeted therapies that restore Na-iodine symporter capability to thyroid cancer refractory to RAI treatment is yet to be worked out. Work on the use of non-RAI radioactive agents, and the use of RAI and other radiolabeled compounds in combination with chemotherapy is still in its infancy. Finally, the therapy of ATC and MTC, with lack of any effective therapy to date, still awaits the development of effective first-line therapy. For this subset of thyroid cancer, this bleak picture is, on the other hand, a strong motivation for energetic work on effective therapy.

REFERENCES

1. Seidlin SM, Marinelli LD, Osher E, et al. 1946. Radioactive iodine therapy. Effect on functioning metastases of adenocarcinoma of the thyroid. *JAMA* 232:838–47.
2. Byar DP, Gree SB, Dor P, et al. 1979. A prognostic index for thyroid carcinoma: A study of the E.O.R.T.C. Thyroid Cancer Cooperative Group. *Eur J Cancer* 15:1033–41.

3. Fourquet A, Asselain B, Joly J. 1983. Cancer de la thyroide: Analyse multidimensionelle des facteurs prognostiques. *Ann Endocrinol* (Paris) 44:121–29.
4. McConahey WM, Day ID, Woolner LB, et al. 1986. Papillary thyroid cancer treated at the Mayo Clinic, 1946 through 1970: Initial manifestations, pathologic findings, therapy, and outcome. *Mayo Clin Proc* 61:978–88.
5. Cooper DS, Doherty GM, Haugen BR, et al. 2006. Management guidelines for patients with thyroid nodules and differentiated thyroid cancer. *Thyroid* 16:1–3.
6. Goldsmith SJ. 2001. Thyroid carcinoma. In *Nuclear oncology: Diagnosis and therapy*, ed. I Khalkali, J Maublant, SJ Goldsmith. Philadelphia: Lippincott Williams & Wilkins.
7. Hay D. 1990. Papillary thyroid carcinoma. *Endocrinol Metab Clin North Am* 19:545–76.
8. Bacourt F, Aselain B, Savoie JC, et al. 1986. Multifactorial study of prognostic factors in differentiated thyroid carcinoma and a re-evaluation of the importance of age. *Br J Surg* 73:274–80.
9. Schelfhout LJ, Creutzberg CL, Hamming JF, et al. 1988. Multivariate analysis of survival in differentiated thyroid cancer: The prognostic significance of the age factor. *Eur J Cancer Clin Oncol* 24:331–41.
10. Brennan MD, Bergstrahl EJ, Van Herrden JA, et al. 1991. Follicular thyroid cancer treated at the Mayo Clinic, 1946 through 1970: Initial manifestations, pathologic findings, therapy, and outcome. *Mayo Clin Proc* 66:11–22.
11. Mazzaferri EL, Young RL. 1981. Papillary thyroid carcinoma: A 10-year follow-up report of the impact of therapy in 576 patients. *Am J Med* 70:511–18.
12. Mazzaferri EL, Jhiang SM. 1994. Long-term impact of initial surgical and medical therapy on papillary and follicular thyroid cancer. *Am J Med* 97:418–28.
13. Mazzaferri EL. 1999. An overview of the management of papillary and follicular thyroid carcinoma. *Thyroid* 9:421–27.
14. Shah MD, Hall FT, Eski SJ, et al. 1993. Clinical course of thyroid carcinoma after neck dissection. *Laryngoscope* 113:2102–7.
15. Mazzaferri EL. 2000. Long-term outcome of patients with differentiated thyroid carcinoma: Effect of therapy. *Endocr Pract* 6:469–76.
16. Kim TH, Yang DS, Jung KY, et al. 2003. Value of external irradiation for locally advanced papillary thyroid cancer. *Int J Radiat Oncol Biol Phys* 55:1006–12.
17. Mazzaferri EL, Jhiang SM.1994. Long-term impact of initial surgical and medical therapy on papillary and follicular thyroid cancer. *Am J Med* 97:418–28.
18. DeGroot LJ, Kaplan EL, McCormick M, et al. 1990. Natural history, treatment, and course of papillary thyroid carcinoma. *J Clin Endocrinol Metab* 71:414–24.
19. Samaan NA, Schulz PN, Hickey RC, et al. 1992. The results of various modalities of treatment of well-differentiated thyroid carcinomas: A retrospective review of 1599 patients. *J Clin Endocrinol Metab* 75:714–20.
20. Mazzaferri EL, Jhiang SM. 1994. Differentiated thyroid cancer long-term impact of initial therapy. *Trans Am Clin Climatol Assoc* 106:151–68.
21. Taylor T, Specker B, Robbins J, et al. 1998. Outcome after treatment of high-risk papillary and non-Hurthle-cell follicular thyroid carcinoma. *Ann Intern Med* 129:622–27.
22. Sawka AM, Thephamangkhol K, Brouwers M, et al. 2004. Clinical review 170: A systematic review and metaanalysis of the effectiveness of radioactive iodine remnant ablation for well-differentiated thyroid cancer. *J Clin Endocrinol Metab* 89:3668–76.
23. Hay ID, Thompson GB, Grant CS, et al. 2002. Papillary thyroid carcinoma managed at the Mayo Clinic during six decades (1940–1999): Temporal trends in initial therapy and long-term outcome in 2444 consecutively treated patients. *World J Surg* 26:879–85.
24. Sanders LE, Cady B. 1998. Differentiated thyroid cancer: Re-examination of risk groups and outcome of treatment. *Arch Surg* 133:419–25.
25. Kim S, Wei JP, Braverman JM, et al. 2004. Predicting outcome and directing therapy for papillary thyroid carcinoma. *Arch Surg* 139:390–94.
26. Sugitani I, Fujimoto Y. 1999. Symptomatic versus asymptomatic papillary thyroid microcarcinoma: A retrospective analysis of surgical outcome and prognostic factors. *Endocr J* 46:209–16.
27. Hundahl SA, Fleming ID, Fremgen AM, et al. 1998. A National Cancer Data Base report on 53,856 cases of thyroid carcinoma treated in the U.S., 1985–1995. *Cancer* 83:2638–48.
28. Mazzaferri EL. 1997. Thyroid remnant I-131 ablation for papillary and follicular thyroid carcinoma. *Thyroid* 7:265–71.
29. Robbins RJ, Larson SM, Sinha N, et al. 2002. A retrospective review of the effectiveness of recombinant human TSH as a preparation for radioiodine thyroid remnant ablation. *J Nucl Med* 43:1482–88.

30. Pacini F, Molinaro E, Castagna MG, et al. 2002. Ablation of thyroid residues with 30 mCi I-131: A comparison in thyroid cancer patients prepared with recombinant human TSH or thyroid hormone withdrawal. *J Clin Endocrinol Metab* 87:4063–68.
31. Barbaro D, Boni G, Meucci G, et al. 2003. Radioiodine treatment with 30 mCi after recombinant human thyrotropin stimulation in thyroid cancer: Effectiveness for postsurgical remnants ablation and possible role of iodine content in L-thyroxine in the outcome of ablation. *J Clin Endocrinol Metab* 88:4110–15.
32. Pacini F, Ladenson PW, Schlumberger M, et al. 2006. Radioiodine ablation of thyroid remnants after preparation with recombinant thyrotropin in differentiated thyroid carcinoma: Results of an international, randomized, controlled study. *J Clin Endocrinol Metab* 91:926–32.
33. Borget I, Remy H, Chevalier J, et al. 2008. Length and cost of hospital stay of radioiodine ablation in thyroid cancer patients: Comparison between preparation with thyroid hormone withdrawal and Thyrogen. *Eur J Nucl Med Mol Imaging* 35:1457–63.
34. Robbins RJ, Larson SM, Sinha N, et al. 2002. A retrospective review of the effectiveness of recombinant human TSH as a preparation for radioiodine thyroid remnant ablation. *J Nucl Med* 43:1482–88.
35. Hanscheid H, Lassmann M, Luster M, et al. 2006. Iodine biokinetics and dosimetry in radioiodine therapy of thyroid cancer: Procedures and results of a prospective international controlled study of ablation after rhTSH or hormone withdrawal. *J Nucl Med* 47:648–54.
36. Hodges RE, Evans TC, Bradbury JT, et al. 1955. The accumulation of radioactive iodine by human fetal thyroids. *J Clin Endocrinol Metab* 6:661.
37. Johansen K, Woodhouse NJ, Odugbesan O. 1991. Comparison of 1073 MBq and 3700 MBq iodine-131 in post-operative ablation of residual thyroid tissue in patients with differentiated thyroid cancer. *J Nucl Med* 32:252–54.
38. Rosario PW, Reis JS, Barroso AL, et al. 2004. Efficacy of low and high I-131 doses for thyroid remnant ablation in patients with differentiated thyroid carcinoma based on post-operative cervical uptake. *Nucl Med Commun* 25:1077–81.
39. Bal C, Padhy AK, Jana S, et al. 1996. Prospective randomized clinical trial to evaluate the optimal dose of I-131 for remnant ablation in patient with differentiated thyroid carcinoma. *Cancer* 77:2574–80.
40. Creutzig H. 1997. High or low dose radioiodine ablation of thyroid remnants? *Eur J Nucl Med* 12:500–2.
41. Beierwaltes VM, Rabbani R, Dmuchowski C, et al. 1984. An analysis of "ablation of thyroid remnants" with I-131 in 511 patients from 1947–1984: Experience at University of Michigan. *J Nucl Med* 25:1287–93.
42. Sisson JC. 1983. Applying the radioactive eraser: I-131 to ablate normal thyroid tissue in patients from whom thyroid cancer has been resected. *J Nucl Med* 24:743–45.
43. Goolden AW. 1985. The indications for ablating normal thyroid tissue with I-131 in differentiated thyroid cancer. *Clin Endocrinol* 23:81–86.
44. Hurley JR, Becker DV. 1983. The use of radioiodine in the management of thyroid cancer. In *Nuclear medicine annual*, ed. LM Freeman, Weissman HS, 329–84. New York: Raven Press.
45. Maxon HR III, Englaro EE, Thomas SR, et al. 1992. Radioiodine-131 therapy for well-differentiated thyroid cancer—A quantitative radiation dosimetric approach: Outcome and validation in 85 patients. *J Nucl Med* 33:1132–36.
46. Rosario PWS, Ribeiro Maia FF, Cardoso LD, et al. 2004. Correlation between cervical uptake and results of post-surgical radioiodine ablation in patients with thyroid carcinoma. *Clin Nucl Med* 29:358–61.
47. Karam M, Feustel PJ, postal ES, et al. 2005. Successful thyroid tissue ablation as defined by a negative whole-body scan or an undetectable thyroglobulin: A comparative study. *Nucl Med Commun* 26:331–36.
48. Mazzaferri EL, Kloos RT. 2002. Current approaches to primary therapy for papillary and follicular thyroid cancer. *J Clin Endocrinol Metab* 86:1447–63.
49. Leeper RD. 1973. The effect of I-131 therapy on survival of patients with metastatic papillary or follicular thyroid carcinoma. *J Clin Endocrinol Metab* 36:1143–52.
50. Beierwaltes WH, Nishiyama RH, Thompson NW, et al. 1982. Survival time and "cure" in papillary and follicular thyroid carcinoma with distant metastases: Statistics following University of Michigan therapy. *J Nucl Med* 23:561–68.
51. Bernier MO, Leenhardt L, Hoang C, et al. 2001. Survival and therapeutic modalities in patients with bone metastases of differentiated thyroid carcinomas. *J Clin Endocrinol Metab* 86:1568–73.
52. Liel Y. 2002. Preparation for radioactive iodine administration in differentiated thyroid cancer patients. *Clin Endocrinol* (Oxf) 57:523–27.

53. Robbins RJ, Schlumberger MJ. 2005. The evolving role of I-131 for the treatment of differentiated thyroid carcinoma. *J Nucl Med* 46:28S–37S.
54. Samuel AM, Rajashekharrao B, Shah DH, et al. 1998. Pulmonary metastases in children and adolescents with well-differentiated thyroid cancer. *J Nucl Med* 39:1531–36.
55. Beierwaltes WH. 1978. The treatment of thyroid carcinoma with radioactive iodine. *Semin Nucl Med* 8:79–94.
56. Harbert JC. 1996. Radioiodine therapy of differentiated thyroid carcinoma. In *Nuclear medicine: Diagnosis and therapy*, ed. JC Harbert, WC Eckleman, RD Neumann, 975–1020. Stuttgart: Thieme Medical Publishers.
57. Rosario PWS, Barroso AL, Rezende LL, et al. 2005. Ablative treatment of thyroid cancer with high doses of I-131 without pre-therapy scanning. *Nucl Med Commun* 26:129–32.
58. Benua RS, Cicale NR, Sonenberg M, et al. 1962. The relation of radioiodine dosimetry to results and complications in the treatment of metastatic thyroid cancer. *AJR* 87:171.
59. Leeper R. 1982. Controversies in the treatment of thyroid cancer. The New York Memorial Hospital approach. *Thyroid Today* 5:104.
60. Leeper RD, Shimaoka K. 1980. Treatment of metastatic thyroid cancer. *Clin Endocrinol Metab* 9:383–91.
61. Schlumberger M, Challeton C, Vathaire FD, et al. 1996. Radioactive iodine treatment and external radiotherapy for lung and bone metastases from thyroid carcinoma. *J Nucl Med* 37:598–605.
62. Furhang EE, Larson SM, Buranapong P, et al. 1999. Thyroid cancer dosimetry using clearance fitting. *J Nucl Med* 40:131–36.
63. Lippi F, Capezzone M, Angelini F, et al. 2001. Radioiodine treatment of metastatic differentiated thyroid cancer in patients on L-thyroxine, using recombinant human TSH. *Eur J Endocrinol* 144:5–11.
64. Pellegriti G, Scollo C, Giuffrida D, et al. 2001. Usefulness of recombinant human thyrotropin in the radiometabolic treatment of selected patients with thyroid cancer. *Thyroid* 11:1025–30.
65. Adler ML, Macapinlac HA, Robbins RJ. 1998. Radioiodine treatment of thyroid cancer with the aid of recombinant human thyrotropin. *Endocr Pract* 4:282–86.
66. Vargas GE, Uy H, Bazan C, et al. 1998. Hemipegia after thyrotropin alfa in a hypothyroid patient with thyroid carcinoma metastatic to the brain. *J Clin Endocrinol Metab* 84:3867–71.
67. Robbins RJ, Voelker E, Wang W, et al. 2000. Compassionate use of recombinant human thyrotropin to facilitate radioiodine therapy: Case report and review of the literature. *Endocr Pract* 6:460–64.
68. Braga M, Ringel MD, Cooper DS. 2001. Sudden enlargement of local recurrent thyroid tumor after recombinant human TSH administration. *J Clin Endocrinol Metab* 86:5148–51.
69. Haugen RB, Pacini F, Reiners C, et al. 1999. A comparison of recombinant human thyrotropin and thyroid hormone withdrawal for the detection of thyroid remnant or cancer. *J Clin Endocrinol Metab* 84:3877–85.
70. Chiu AC, Delpassand ES, Sherman SI. 1997. Prognosis and treatment of brain metastases in thyroid carcinoma. *J Clin Endocrinol Metab* 82:3637–42.
71. Ronga G, Filesi M, Montesano T, et al. 2004. Lung metastases from differentiated thyroid carcinoma. A 40 years experience. *Q J Mucl Med Mol Imaging* 48:12–19.
72. Lin JD, Chao TC, Chou SC, et al. 2004. Papillary thyroid carcinomas with lung metastases. *Thyroid* 14:1091–96.
73. Shoup M, Stojadinovic A, Nissan A, et al. 2003. Prognostic indicators of outcomes in patients with distant metastases from differentiated thyroid carcinoma. *J Am Coll Surg* 197:191–97.
74. Zettinig G, Fueger BJ, Passler C, et al.2002. Long-term follow-up of patients with bone metastases from differentiated thyroid carcinoma—Surgery or conventional therapy? *Clin Endocrinol* 56:377–82.
75. Pittas AG, Adler M, Fazzari M, et al. 2000. Bone metastases from thyroid carcinoma: Clinical characteristic and prognostic variables in one hundred forty-six patients. *Thyroid* 10:262–68.
76. Schlumberger M, Challeton C, De Vathaire F, et al. 1966. Radioactive iodine treatment and external radiotherapy for lung and bone metastases from thyroid carcinoma. *J Nucl Med* 37:598–605.
77. Dinneen SF, Valimaki MJ, Bergstralh EJ, et al. 1995. Distant metastases in papillary thyroid carcinoma: 100 cases observed at one institution during 5 decades. *J Clin Endocrinol Metab* 80:2041–45.
78. Foote RL, Brown PD, Garces YI, et al. 2003. Is there a role for radiation therapy in the management of Hurthle cell carcinoma? *Int Radiat Oncol Biol Phys* 56:1067–72.
79. Pak H, Gourgiotis L, Chang WI, et al. 2003. Role of metastasectomy in the management of thyroid carcinoma: The NIH experience. *J Surg Oncol* 82:10–18.
80. Hurley JR, Becker DV. 1995. Treatment of thyroid cancer with radioactive I-131. In *Diagnostic nuclear medicine*, ed. MP Sandler, RE Coleman, FJ Wackers, et al., 959–90. 3rd ed. Baltimore: Williams & Wilkins.

81. Samaan NA, Schultz PN, Hickey RC, et al. 1992. The results of various modalities of treatment of well-differentiated thyroid carcinoma: A retrospective review of 1599 patients. *J Clin Endocrinol Metab* 75:714–20.
82. Massin JP, Savoie JC, Gamier H, et al. 1984. Pulmonary metastases in differentiated thyroid carcinoma. *Cancer* 53:982–92.
83. Ilgan S, Karacalioglu AO, Pabuscu Y, et al. 2004. Iodine-131 treatment and high-resolution CT: Results in patients with lung metastases from differentiated thyroid carcinoma. *Eur J Nucl Mol Imaging* 3:825–30.
84. Hindie E, Melliere D, Lange F, et al. 2003. Funcitoning pulmonary metastases of thyroid cancer: Does radioiodine influence the prognosis? *Eur J Nucl Med Mol Imaging* 30:974–81.
85. Fatourechi V, Hay ID, Javedan H, et al. 2002. Lack of impact of radioiodine therapy in thyroglobulin-positive, diagnostic whole-body scan-negative patient with follicular cell-derived thyroid cancer. *J Clin Endocrinol Metab* 87:1521–26.
86. Wang W, Larson SM, Fazzari M, et al. 2000. Prognostic value of F-18 fluorodeoxyglucose positron emission tomographic scanning in patients with thyroid cancer. *J Clin Endocrinol Metab* 85:1107–13.
87. Wang W, Larson SM, Tuttle RM, et al. 2001. Resistance of F-18 fluorodeoxyglucose-avid metastatic thyroid cancer lesions to treatment with high-dose radioactive iodine. *Thyroid* 11:1169–75.
88. Fatourechi V, Hay ID, Javedan H, et al. 2002. Lack of impact of radioiodine therapy in thyroglobulin-positive, diagnostic whole-body scan-negative patients with follicular cell-derived thyroid cancer. *J Clin Endocrinol Metab* 87:1521–26.
89. Sarlis NJ. 2001. Metastatic thyroid cancer unresponsive to conventional therapies: Novel management approaches through translational clinical research. *Curr Drug Targets Immune Endocr Metabol Disord* 1:103–15.
90. Maxon HR, Thomas SR, Hertzberg VS, et al. 1983. Relation between effective radiation dose and outcome of radioiodine therapy for thyroid cancer. *N Engl J Med* 309:937–41.
91. Bernier MO, Leenhardt L, Hoang C, et al. 2001. Survival and therapeutic modalities in patients with bone metastases of differentiated thyroid carcinomas. *J Clin Endocrinol Metab* 86:1568–1573.
92. Schaap J, Eustatia-Rutten CF, Stokkel M, et al. 2002. Does radioiodine therapy have disadvantageous effects in non-iodine accumulating differentiated thyroid carcinoma? *Clin Endocrinol* 57:117–24.
93. Pineda JD, Lee T, Ain K, et al. 1995. Iodine-131 therapy for thyroid cancer patients with elevated thyroglobulin and negative diagnostic scans. *J Clin Endocrinol Metab* 80:1488–92.
94. Pacini F, Lippi F, Formica N, et al. 1987. Therapeutic doses of iodine-131 reveal undiagnosed metastases in thyroid cancer patients with detectable serum thyroglobulin levels. *J Nucl Med* 28:1888–91.
95. Pineda JD, Lee T, Ain K, et al. 1995. Iodine-131 therapy for thyroid cancer patients with elevated thyroglobulin and negative diagnostic scan. *J Clin Endocrinol Metab* 80:1488–92.
96. Kabasakal L, Selcuk NA, Shafipour H, et al. 2004. Treatment of iodine-negative thyrogolulin-positive thyroid cancer: Differences in outcome in patients with macrometastases. *Eur J Nucl Med Mol Imaging* 31:1500–4.
97. Ma C, Xie J, Kunag A. 2005. Is empiric I-131 therapy justified for patients with positive thyroglobulin and negative I-131 whole-body scanning results? *J Nucl Med* 46:1164–70.
98. Pacini F, Agate L, Elisei R, et al. 2001. Outcome of differentiated thyroid cancer with detectable serum TG and negative diagnostic I-131 whole body scan: Comparison of patients treated with high I-131 activities versus untreated patients. *J Clin Endocrinol Metab* 86:4092–97.
99. Schlumberger M, Mancusi F, Baudin E, et al. 1997. I-131 therapy for elevated thyroglobulin levels. *Thyroid* 7:273–76.
100. Van Tol KM, Jager PL, de Vries EG, et al. 2003. Outcome in patients with differentiated thyroid cancer with negative diagnostic whole-body scanning and detectable stimulated thyroglobulin. *Eur J Endocrinol* 148:589–96.
101. Kabasakal L, Selcuk NA, Shafipour H, et al. 2004. Treatment of iodine-negative thyroglobulin positive thyroid cancer: Differences in outcome in patients with macrometastases and patients with micrometastases. *Eur J Nucl Med Mol Imaging* 31:1500–4.
102. Fatourechi V, Hay ID. 2000. Treating the patient with differentiated thyroid cancer with thyroglobulin-positive iodine-131 diagnostic scan-negative metastases: Including comments on the role of serum thyroglobulin monitoring in tumor surveillance. *Semin Nucl Med* 30:107–14.
103. Wang W, Larson SM, Fazzari M, et al. 2000. Prognostic value of F-18 fluorodeoxyglucose positron emission tomographic scanning in patients with thyroid cancer. *J Clin Endocrinol Metab* 85:1107–13.
104. Wang W, Larson SM, Tuttle RM, et al. 2001. Resistance of F-18 flurodeoxyglucose-avid metastatic thyroid cancer lesions to treatment with high-dose radioactive iodine. *Thyroid* 11:1169–75.

105. Hooft L, Hoekstra OS, Deville W, et al.2001. Diagnostic accuracy of F-18 fluorodeoxyglucose positron emission tomography in the follow-up of papillary or follicular thyroid cancer. *J Clin Endocrinol Metab* 86:3779–86.
106. Ford D, Giridharan S, McConkey C, et al. 2003. External beam radiotherapy in the management of differentiated thyroid cancer. *Clin Oncol (R Coll Radiol)* 15:337–41.
107. Mazzarotto R, Cesaro MG, Lora O, et al. 2000. The role of external beam radiotherapy in the management of differentiated thyroid cancer. *Biomed Pharmacother* 54:345–49.
108. Tsang RW, Brierley JD, Simpson WJ, et al. 1998. The effects of surgery, radioiodine, and external radiation therapy on the clinical outcome of patients with differentiated thyroid carcinoma. *Cancer* 82:375–88.
109. Chiu AC, Delpassand ES, Sherman SI, et al. 1997. Prognosis and treatment of brain metastases in thyroid carcinoma. *J Clin Endocrinol Metab* 82:3637–42.
110. McWilliams RR, Giannini C, Hay ID, et al. 2003. Management of brain metastases form thyroid carcinoma: A study of 16 pathologically confirmed cases over 25 years. *Cancer* 98:356–62.
111. Benua RS, Cicale NR, Sonenberg M, et al. 1962. The relation of radioiodine dosimetry to results and complications in the treatment of metastatic thyroid cancer. *AJR* 87:171.
112. Mandel SJ, Mandel L. 2003. Radioactive iodine and the salivary glands. *Thyroid* 13:265–71.
113. Kim S, Park CH, Yoon SN, et al. 2001. A false-positive I-131 whole-body scan in chronic parotitis: A case report. *Clin Nucl Med* 26:536–37.
114. Grewal RK, Larson SM, Pentlow CE, et al. 2009. Salivary gland side effects commonly develop several weeks after initial radioactive iodine ablation. *J Nucl Med* 50:1605–10.
115. Solans R, Bosch JA, Galofre P, et al. 2001. Salivary and lacrimal gland dysfunction (sicca syndrome) after radioiodine therapy. *J Nucl Med* 42:738–43.
116. Hyer S, Kong A, Pratt B, et al. 2007. Salivary gland toxicity after radioiodine therapy for thyroid cancer. *Clin Oncol (R Coll Radiol)* 19:83–86.
117. Edmonds CJ, Smith T. 1986. The long-term hazards of the treatment of thyroid cancer with radioiodine. *Br J Radiol* 59:45–51.
118. Hoeltzer S, Steiner D, Bauer R, et al. 2000. Current practice of radioiodine treatment in the management of differentiated thyroid cancer in Germany. *Eur J Nucl Med* 27:1465–72.
119. Caglar M, Tuncel M, Alpar R. 2002. Scintigraphic evaluation of salivary gland dysfunction in patients with thyroid cancer after radioiodine treatment. *Clin Nucl Med* 27:767–71.
120. Nakada K, Ishibashi T, Takei T, et al. 2005. Does lemon candy decrease salivary gland damage after radioiodine therapy for thyroid cancer? *J Nucl Med* 46:261–66.
121. Bohuslavitzky KH, Klutman S, Brenner W, et al. 1999. Radioprotection of salivary glands by amifostine in high-dose radioiodine treatment. *Strahlenther Onkol* 175:6–12.
122. Sisson JC, Carey JE. 2001. Thyroid carcinoma with high levels of function: Treatment with I-131. *J Nucl Med* 42:975–83.
123. Lawrence E, Lord ST, Leon Y, et al. 2001. Tall cell papillary thyroid carcinoma metastatic to femur: Evidence for thyroid hormone synthesis within the femur. *Am J Med Sci* 322:103–8.
124. Lorberboym M, Mechanick JI. 1996. Accelerated thyrotoxicosis induced by iodinated contrast media in metastatic differentiated thyroid carcinoma. *J Nucl Med* 37:1532–35.
125. Brinkner H, Hansen HS, Andersen AP. 1973. Induction of leukemia by I-131 treatment of thyroid carcinoma. *Br J Cancer* 28:232.
126. Rubino C, de Vathaire F, Dottorini ME, et al. 2003. Second primary malignancies in thyroid cancer patients. *Br J Cancer* 89:1638–44.
127. Chen AY, Ley L, Goepfeert H, et al. 2001. The development of breast carcinoma in women with thyroid carcinoma. *Cancer* 92:225–31.
128. Garsi J-P, Schlumberger M, Rubino C, et al. 2008. Therapeutic administration of I-131 for differentiated thyroid cancer: Radiation dose to ovaries and outcome of pregnancies. *J Nucl Med* 49:845–52.
129. Vini L, Hyer S, Al-Saadi A, et al. 2002. Prognosis for fertility and ovarian function after treatment with radioiodine for thyroid cancer. *Postgrad Med J* 78:92–93.
130. Dottorini ME, Lomuscio G, Mazzucchelli L, et al. 1995. Assessment of female fertility and carcinogenesis after I-131 therapy for differentiated thyroid carcinoma. *J Nucl Med* 36:21–27.
131. Sarkar SD, Beierwaltes WH, Gill SP, et al. 1976. Subsequent fertility and birth histories of patients treated with I-131 for thyroid cancer before 1969 at 20 years or less. *J Nucl Med* 17:460–64.
132. Schlumberger M, De Vathaire F, Ceccarelli C, et al. 1996. Exposure to radioactive iodine-131 for scintigraphy does not preclude pregnancy in thyroid cancer patients. *J Nucl Med* 37:606–12.

133. Ceccarelli C, Bencivelli W, Morciano D, Pinchera A, Pacini F. 2001. 131I therapy for differentiated thyroid cancer leads to an earlier onset of menopause: Results of a retrospective study. *J Clin Endocrinol Metab* 86:3512–15.
134. Pacini F, Gasperi M, Fugazzola L, et al. 1994. Testicular function in patients with thyroid carcinoma treated with radioiodine. *J Nucl Med* 35:1418–22.
135. Wichers M, Benz E, Palmedo H, et al. 2000. Testicular function after radioiodine therapy for thyroid carcinoma. *Eur J Nucl Med* 503–7.
136. Hyer S, Vini L, O'Connell M, et al. 2002. Testicular dose and fertility in men following I-131 therapy for thyroid cancer. *Clin Endocrinol* 56:755–58.
137. Lushbaugh CC, Casarett GW. 1976. The effects of gonadal irradiation in clinical radiation therapy: A review. *Cancer* 37:1111–25.
138. Mazzaferri E. 2002. Gonadal damage from I-131 therapy for thyroid cancer. *Clin Endocrinol* 57:313–14.
139. Honor AJ, Myant NB, Rowlands EN. 1952. Secretion of radioiodine in digestive juices and milk in man. *Clin Sci* 11:447.
140. Miller H, Weetch RS, Glasq MB. 1955. The excretion of radioactive iodine in human milk. *Lancet* 2:1013.
141. Code of Federal Regulations. 1997. Title 10, energy: Part 35, medical use of byproduct material: Section 75, release of individuals containing radiopharmaceutical or permanent implants. *Fed Regist* 62:4133.
142. Grigsby PW, Siegel BA, Baker S, et al. 2000. Radiation exposures from outpatient radioactive I-131 therapy for thyroid carcinoma. *JAMA* 283:2272–74.
143. Coover LR, Silberstein EB, Kuhn PJ, et al. 2000. Therapeutic I-131 in outpatients: A simplified method conforming to the Code of Federal Regulations, Title 10, Part 35.75. *J Nucl Med* 41:1868–75.
144. Leslie WD, Hvelock J, Palser R, et al. 2002. Large-body radiation doses following radioiodine therapy. *Nucl Med Commun* 23:1091–97.
145. Smit JWA, Schroder-van der Elst JP, Karperien M, et al. 2002. Iodide kinetics and experimental I-131 therapy in a xenotransplanted human sodium-iodide symporter-transfected human follicular thyroid carcinoma cell line. *J Clin Endocrinol Metab* 87:1247–53.
146. Morris JC, Kim CK, Padilla ML, et al. 1997. Conversion of no-iodine-concentrating differentiated thyroid carcinoma metastases into iodine-concentrating foci after anticancer chemotherapy. *Thyroid* 7:63–66.
147. Li W, Gopalakrishnan M, Venkataraman, et al. 2007. Protein synthesis inhibitors, in synergy with 5-azacytidine, restore sodium/iodide symporter gene expression in human thyroid adenoma cell line, KAK-1, suggesting trans-active transcriptional repressor. *J Clin Endocrinol Metab* 92:1080–87.
148. Kim JH, Leeper RD. 1983. Combination adriamycin and radiation therapy for locally advanced carcinoma of the thyroid gland. *Int J Radiat Oncol Biol Phys* 9:565–67.
149. Kim JH, Leeper RD. 1987. Treatment of locally advanced thyroid carcinoma with combination doxorubicin and radiation therapy. *Cancer* 60:2372–75.
150. Krenning EP, de Jong M, Kooij PPM, et al. 1999. Radiolabeled somatostatin analogue(s) for peptide receptor scintigraphy and radionuclide therapy. *Ann Oncol* 10(Suppl 2):s23–29.
151. Kozak RW, Raubitschek A, Mirzadeh S, et al. 1989. Nature of the bifunctional chelating agent used for radioimmunotherapy with yttrium-90 monoclonal antibodies: Critical factors in determining *in vivo* survival and organ toxicity. *Cancer Res* 49:2639–2644.
152. Virgolini I, Szilvasi I, Kurtaran A, et al. 1998. Indium-111-DOTA-lanreotide: Biodistribution, safety and tumor dose in patients evaluated for somatostatin receptor-mediated radiotherapy. *J Nucl Med* 39:1928–36.
153. Kundra P, Burman KD. 2007. Thyroid cancer molecular signaling pathways and use of targeted therapy. *Endocrinol Metab Clin North Am* 36:839–53.
154. Rossi RL, Cady B, Meissner WA, et al. 1980. Nonfamilial medullary thyroid carcinoma. *Am J Surg* 139:554–60.
155. Samaan NA, Schyultz PN, Hickey RC. 1988. Medullary thyroid carcinoma: Prognosis of familial versus sporadic disease and the role of radiotherapy. *J Clin Endocrinol Metabol* 67:801–8.
156. Shroder S, Bocker W, Baisch H, et al. 1988. Prognostic factors in medullary thyroid carcinomas. Survival in relation to age, sex, stage, histology, immunocytochemistry, and DNA content. *Cancer* 61:806–16.
157. Cance WG, Wells SA Jr. 1985. Multiple endocrine neoplasia type IIa. *Curr Probl Surg* 22:1.
158. Rougier PH, Calmettes C, Laplanche A, et al. 1983. The value of calcitonin and carcinoembryonic antigen in the treatment and management of non-familial medullary thyroid carcinoma. *Cancer* 51:855–62.
159. Juweid M, Sharkey RM, Behr T, et al. 1995. Targeting and initial radioimmunoherapy of medullary thyroid carcinoma with I-131 labeled monoclonal antibodies to carcinoembryonic antigen. *Cancer Res* 55:5946–51.

160. Juweid M, Sharkey RM, Behr T, et al. 1996. Radioimmunotherapy of medullary thyroid carcinoma with I-131-labeled anti-CEA antibodies. *J Nucl Med* 37:905–11.
161. Juweid M, Hajjar G, Swayne LC, et al. 1999. Phase I/II trial of I-131 MN-14 F(ab)2 anti-carcinoembryonic antigen monoclonal antibody in the treatment of patients with metastatic medullary thyroid carcinoma. *Cancer* 85:1828–42.
162. Juweid M, Hajjar G, Stein, et al. 2000. Initial experience with high-dose radioimmunotherapy of metastatic meduallary thyroid carcinoma using I-131-MN-14 F(ab)2 anticarcinoembryonic entigen (CEA) monoclonal antibody and autologous hematopoietic stem cell rescue (AHSCR). *J Nucl Med* 41:93–103.
163. Stein R, Juweid M, Mattes MJ, et al. 1999. Carcinoembryonic antigen as a target for radioimmunotherapy of human medullary thyroid cancer. Antibody processing, targeting, and experimental therapy with I-131- and Y-90-laeled mAbs. *Cancer Biother Radiopharm* 14:37–47.
164. Stein R, Juweid M, Goldenberg DM. 1999. Effects of unlabeled anti-CEA on the growth of medullary thyroid cancer xenografts. *Proc Am Assoc Cancer Res* 40:18.
165. Stein R, Juweid M, Zhang C-H, et al. 1999. Assessment of combined radioimmunoherapy and chemotherapy for treatment of medullary thyroid cancer. *Clin Cancer Res* 5:3199S–206S.
166. Juweid M, Rubin A, Hajjar G, et al. 1999. Preclinical and clinical findings support concurrent RAIT and chemotherapy in patients with medullary thyroid cancer (MTC) (abstract). In *Proceedings of the American Society of Clinical Oncology*, Vol. 18, p. 407A.
167. Juweid M, Rubin A, Hajjar G, et al. 1999. First clinical results of combined RAIT and chemotherapy in patients with medullary thyroid cancer (MTC) (abstract). In *Program and Abstracts of the Endocrine Society 81st Annual Meeting*, pp. 2–726.
168. Riesco-Eizaguirre G, Santisteban P. 2006. A perspective view of sodium iodide symporter research and its clinical implications. *Eur J Endocrinol* 155:495–512.
169. Dionigi G, Tanda ML, Piantanida E. 2008. Medullary thyroid carcinoma: Surgical treatment advances. *Curr Opin Otolaryngol Head Neck Surg* 16:158–62.
170. Mechanick JI, Carpi A. 2008. Thyroid cancer: The impact of emerging technologies on clinical practice guidelines. *Biomed Pharmacother* 62:554–58.

17 External Beam Radiation Therapy in Thyroid Cancer

Walter Choi and Louis Harrison

CONTENTS

ABSTRACT: A multidisciplinary approach to the management of thyroid cancers is crucial to optimal therapy. Clearly, close collaboration among surgeons, endocrinologists, medical oncologists, and radiation oncologists is imperative in this complex disease. In this context, the role of external beam radiation therapy (EBRT) is to provide local therapy, either at the primary site of disease or for metastatic foci. This chapter will focus on the use of EBRT in the locoregional management of thyroid cancer, and the emergent technologies, such as image-guided radiation therapy (IGRT), in the field as they pertain to thyroid cancer management.

17.1 EXTERNAL BEAM RADIATION THERAPY FOR DIFFERENTIATED THYROID CANCER

17.1.1 Clinical Review

Surgery is typically the recommended initial therapy in the management of differentiated thyroid cancer (DTC). In the majority of patients, resection, in conjunction with radioactive iodine (RAI) therapy and thyroid-stimulating hormone (TSH) suppression, is sufficient to offer a high likelihood of cure. However, in patients with high-risk features, external beam radiation therapy (EBRT) can be used to provide improvements in locoregional control of disease. No prospective, randomized trial has examined the role of EBRT in the postoperative management of DTC. As such, practical indications for its use vary. Identification of those features that put a patient at high risk of locoregional recurrence is imperative in determining the cohort of patients who might benefit from EBRT. Typically, these factors include extrathyroidal extension (i.e., T4 disease by the TMN staging system; Table 17.1), older age, and larger size.

TABLE 17.1
TNM Staging of Thyroid Cancer

	Primary Tumor (T)
Differentiated thyroid cancer:	
TX	Primary tumor cannot be assessed
T0	No evidence of primary tumor
T1	Tumor ≤2 cm in greatest dimension limited to the thyroid
T2	Tumor <2 cm but not <4 cm in greatest dimension limited to the thyroid
T3	Tumor <4 cm in greatest dimension limited to the thyroid or any tumor with minimal extrathyroid extension or any tumor with minimal extrathyroid extension (e.g., extension to sternothyroid muscle or perithyroid soft tissues)
T4a	Tumor of any size extending beyond the thyroid capsule to invade subcutaneous soft tissues, larynx, trachea, esophagus, or recurrent laryngeal nerve
T4b	Tumor invades prevertebral fascia or encases carotid artery or mediastinal vessels
Anaplastic thyroid cancer:	
All anaplastic carcinomas are considered T4 tumors	
T4a	Intrathyroidal anaplastic carcinoma—surgically resectable
T4b	Extrathyroidal anaplastic carcinoma—surgically unresectable
	Regional Lymph Nodes (N)
Regional lymph nodes are the central compartment, lateral cervical, and upper mediastinal lymph nodes	
NX	Regional lymph nodes cannot be assessed
N0	No regional lymph node metastasis
N1	Regional lymph node metastasis
N1a	Metastasis to level VI (pretracheal, paratracheal, and prelaryngeal/Delphian lymph nodes)
N1b	Metastasis to unilateral, bilateral, or contralateral cervical or superior mediastinal lymph nodes
	Distant Metastasis (M)
MX	Presence of distant metastasis cannot be assessed
M0	No distant metastasis
M1	Distant metastasis

Source: Greene, F. L., American Joint Committee on Cancer, and American Cancer Society, *AJCC Cancer Staging Manual*, 6th ed., Springer-Verlag, New York, 2002, xiv.

One of the earlier experiences examining the use of EBRT in DTC (Table 17.2) involved 539 patients treated in France.[1] Among these patients, 97 were treated with EBRT after incomplete resection with gross residual disease.[1] At 15 years, the local relapse rate for these patients was 15%, compared with 32% for those not receiving postoperative EBRT ($p < .05$).[1]

Chow et al.[2] retrospectively reviewed 1297 patients with papillary thyroid carcinoma (PTC) who were treated at the Queen Elizabeth Hospital in Hong Kong from 1960 to 2000. They reported an improvement in 10-year locoregional failure-free survival (FFS) with the use of EBRT for those patients with gross residual disease (63.4% vs. 24%, $p < 0.0001$).[2] The 10-year cause-specific survival (CSS) was improved from 49.7% to 74.1% ($p = 0.01$).[2] EBRT was also associated with improved locoregional FFS with positive margins and stage T4 disease (58% vs. 22.8%, $p < 0.001$). There was also an improvement in nodal FFS in patients with N1b (cervical or superior mediastinal nodal) involvement (79.5% vs. 58.1%, $p = 0.02$) or nodes >2 cm (nodal relapse 19% vs. 44%, $p = 0.008$).[2]

Farahati et al.[3] reported on 238 patients with stage T4 DTC who were treated between 1979 and 1992: 169 patients were disease-free at final staging after standard therapy (resection, RAI therapy, and TSH suppression), of whom 99 received EBRT. Overall, EBRT significantly reduced the rate of recurrences (n = 7) compared with the group of patients who did not receive radiation (n = 21,

TABLE 17.2
Summary of Outcomes for Patients with DTC Treated with EBRT

Author	No. Patients	LRFFS w/ EBRT	LRFFS w/o EBRT	LR w/ EBRT	LR w/o EBRT	*p*-Value
Tubiana	539	—	—	15%	32%	<0.05
Chow	1,297	63.4%	24%	—	—	<0.0001
Farahati	169	—	—	N = 7	N = 21	0.0003
Tsang	729	94.2%	83.9%			0.04
Meadows	42	88%	—	—	—	N/A

Note: DTC = differentiated thyroid cancer, EBRT = external beam radiation therapy, LRFFS = locoregional failure free survival, LR = local recurrence.

$p = 0.0003$).[3] Subgroup analysis revealed a significant benefit for patients over 40 years of age with PTC and lymph node involvement. However, there was no significant improvement in patients with follicular thyroid carcinoma (FTC).

Meadows et al.[4] recently reported the experience at the University of Florida, where 42 patients with locally advanced or recurrent DTC received high-dose EBRT. The locoregional control rate for the overall population at 5 years was 88%; the locoregional control rate for those patients without evidence of distant metastases was 89%.[4] Among patients with gross residual disease at the time of EBRT, the locoregional control rate was 70%, compared with 100% for those with microscopic residual disease ($p = 0.0177$).[4] The CSS for the overall population at 5 years was 80%; the CSS for those without distant metastases before EBRT was 86%.[4] Again, patients with gross residual disease had inferior CSS (69%) compared with those with microscopic residual disease only (90%, $p = 0.313$).

The Princess Margaret Hospital analyzed its cohort of 729 patients with DTC and found that using multivariate analysis, tumor size >4 cm, age >60 years, presence of neck metastasis, less extensive surgery, and the lack of use of RAI were risk factors for locoregional failure.[5] In contrast to their prior report in 1998,[6] extrathyroidal extension was not a prognostic factor for locoregional failure.[5] In a subgroup analysis of 154 patients with PTC histology and microscopic residual disease without distant metastases, the use of EBRT improved both 10-year CSS (100% vs. 95.3%, $p = 0.01$) and the local recurrence-free rate (94.2% vs. 83.9%, $p = 0.02$).[5] Additionally, for the 70 patients who were over 60 years of age with stage T4 disease and no gross residual disease, there was a higher 10-year CSS (81% vs. 64.6%, $p = 0.04$) and locoregional FFS rate (86.4% vs. 65.7%, $p = 0.01$) in those patients receiving RT.[5]

17.1.2 Treatment Recommendations

Ultimately, the role of EBRT would best be clarified by a prospective, randomized controlled study. However, no such studies are available. A recent multicenter German trial[7] was attempted, but was prematurely closed due to poor accrual. As such, the available literature on this subject remains relegated to retrospective single-institution reports.[1–6] Results of these differ, due in large part to variations in staging, initial management, and indications for EBRT. Despite disparities in retrospective findings, there are certain patients in whom we recommend EBRT. This includes older patients with stage T4 disease and those who have gross or microscopic residual disease after resection. Patients with extracapsular extension of nodal metastases are also candidates for EBRT. Similarly, the American Thyroid Association's clinical practice guidelines (CPG) for DTC state that "the use of external beam irradiation should be considered in patients over age 45 with grossly visible extrathyroidal extension at the time of surgery and a high likelihood of microscopic residual disease, and for those patients with gross residual tumor in whom further surgery or radioactive iodine would likely be ineffective."[8]

In patients who also receive RAI therapy, typically EBRT is scheduled shortly afterward. Although it has not been studied well, there is the concern that EBRT may result in thyroid stunning, thereby diminishing the capacity of thyroid cells to concentrate iodine, as is the case with prior RAI therapy. The timing of surgery, RAI therapy, and EBRT is important, as delays in initiating EBRT allow for accelerated repopulation of tumor clonogens. This has been well studied in squamous cell carcinomas of the head and neck and other sites,[9] and it follows that the same would apply in thyroid malignancies. Ideally, EBRT should be initiated within 8 weeks of surgery in order to minimize this. Therefore, our preference is for patients to undergo RAI therapy 6 weeks after surgery, and to initiate EBRT 2 weeks later. This schedule allows reasonable time postoperatively for preparation and administration of RAI.

17.2 EBRT FOR ANAPLASTIC THYROID CANCER

17.2.1 Clinical Review

Anaplastic thyroid cancer (ATC) is typically locoregionally advanced at presentation, rendering complete resection an uncommon therapeutic option. EBRT is therefore the usual treatment of choice in this disease. Because of the rapid proliferation of ATC, a combined approach of EBRT with radiosensitizing chemotherapy has been reported since the 1970s (Table 17.3). Promising results have been reported by Kim and Leeper[10,11] using a hyperfractionated EBRT scheme delivering 1.6 Gy (Gray; unit of absorbed radiation dose) per fraction, twice daily, 3 days a week, to a final dose of 57.6 Gy, with weekly administration of doxorubicin (10 mg/m^2). The rationale for hyperfractionation is to counteract the rapid repopulation of tumor cells; the efficacy of this approach has also been successfully described in head and neck cancers and lung cancers.[12–14] Kim and Leeper treated 19 patients with this regimen and reported a 91% complete response within the irradiated field, with 68% (13 of 19) of patients achieving long-term tumor control and a median survival of 1 year (vs. a median survival of 4 months in historical controls).[10,11] Although all patients developed moderate pharyngo-esophagitis as an acute toxicity of treatment, it subsided within 3 to 4 weeks of completing treatment, and severe long-term toxicity was not reported.

Tennvall et al.[15–18] have reported several therapeutic regimens for ATC. Initial use of concurrent methotrexate and 30–40 Gy of daily EBRT provided good response rates, but all patients experienced severe toxicities, including hematemesis, melena, and mucositis, leading the investigators to replace methotrexate with bleomycin, cyclophosphamide, and 5FU (BCF). This regimen was better tolerated; of 9 patients treated, 7 had a complete or partial response.[17] The authors subsequently made additional refinements, adding surgery and hyperfractionating the radiation.[17] Based on the aforementioned publication by Kim and Leeper,[10,11] the chemotherapy was changed to doxorubicin. Patients were treated using a hyperfractionated radiation schedule to 46 Gy, initially splitting the

TABLE 17.3
Summary of Outcomes for Patients with Anaplastic Thyroid Cancer Treated with Chemoradiation ± Surgery

Author	No. Patients	Long-Term Local Control	Complete Response	Median Survival	Death from Local Failure
Kim	19	68%	91%	1 year	0
Tennvall	55	60%	60%	2–4.5 months	24%
De Crevoisier	24	79% (19/24)	29% (7/24)	10 months	5%

course of RT to deliver 30 Gy preoperatively and 16 Gy postoperatively.[17] In a later protocol, all EBRT was delivered preoperatively.[17] Of the 55 patients treated on these regimens, 60% remained locally controlled, while 24% died due to local failure.[17] The median survival remained poor, ranging from 2 to 4.5 months, but with 16% of patients surviving for one year.

De Crevoisier et al.,[19] from the Institut Gustav-Roussy in France, reported their prospective experience treating 30 patients with ATC from 1990 to 2000. In their series, patients underwent maximal resection, followed by chemotherapy consisting of doxorubicin (60 mg/m^2) and cisplatin (120 mg/m^2) every 4 weeks for two cycles.[19] Fifteen days after the second cycle, patients initiated EBRT, 40 Gy at 1.25 Gy B.I.D., with patients after 1998 receiving an additional boost to the tumor bed, to a final dose of 50–55 Gy. Subsequently, additional chemotherapy was given for a total of 6 cycles. This regimen was associated with acute toxicities, including 70% with grade 4 neutropenia and 43% with grade 3 or 4 pharyngo-esophagitis. Nonetheless, of 24 patients without distant metastases at diagnosis, with a median follow-up of 35 months, 8 patients were in complete remission, including 7 of the 9 patients in whom complete resection of tumor was initially achieved. Among 7 patients without distant metastases at diagnosis but who were unable to receive an initial operation, only 3 sustained a partial response after chemotherapy and radiation, allowing a gross total resection. Of these three patients, only one had a complete response at 12 months.[19] The authors concede that the superior results achieved in patients who underwent complete resection may simply have been due to a less aggressive behavior of tumors that are amenable to surgery. Regardless, it is clear that in ATC, a multimodality approach to treatment should be attempted when feasible; this approach is supported by other studies as well.[20–22]

Although there are no randomized trials available, the aforementioned studies show an association between the use of surgery in the multimodality management of ATC and long-term survival.[17,19–22] Therefore, in patients who receive chemoradiation as their initial therapy, reconsideration of surgery for residual disease may be considered, as is indicated in the American Association of Clinical Endocrinologists/American Association of Endocrine Surgeons CPG for management of thyroid carcinoma.[23] However, with this approach, the radiation treatment plan is considered definitive rather than preoperative. To the radiation oncologist, this classification is important. In the former, a patient is treated with full doses of radiation with curative intent. In contrast, with a preoperative course of EBRT, the dose of radiation prescribed is lower than that necessary to eradicate all disease. The goal of treatment in this case is to reduce the bulk of gross disease and exterminate microscopic deposits of cancer cells beyond the gross detectable disease, while relying on surgery as the definitive treatment. In patients with ATC, the majority of patients will not be candidates for surgery, due to either the extent of involvement of residual disease in the primary site or the progression of metastatic disease systemically. Therefore, from a radiation oncologist's perspective, patients must be treated with a complete course of chemoradiation. Those patients who have a good response and are deemed resectable after chemoradiation are restaged and evaluated by their surgeons for resection.

17.2.2 Treatment Recommendations

When possible, gross total resection of disease should be performed as primary treatment. However, a "debulking" procedure that leaves behind gross residual disease has not proven to be helpful in achieving locoregional control and may merely delay chemotherapy and radiation. It is therefore critical to determine which patients are suitable for an operation.

Concomitant chemotherapy or molecular targeted therapy is the recommended treatment for most patients with ATC. However, given its poor prognosis, the goal of treatment is often palliation. An accelerated EBRT schedule may be offered, similar to those described above. If patients do undergo surgery, it is imperative to initiate treatment as quickly as possible due to the aggressiveness of ATC. Otherwise, patients may be reconsidered for resection after chemoradiation is complete, as incorporation of surgery is associated with long-term survival. In patients whose prognosis or performance status is too poor to tolerate aggressive chemoradiation, a palliative course of EBRT can be offered and completed within five to ten treatments.

17.3 EBRT TECHNIQUES

After consultation, a course of EBRT is preceded by a simulation, the first step in planning a patient's radiation treatment course. The patient is brought to a simulator, which is a specialized x-ray unit (or, more often in the modern era, a CT scanner) that is designed to precisely map out the patient's treatment. By the time of the simulation, the radiation oncologist has reviewed all relevant materials, including imaging studies, operative reports, and pathology reports. The patient is placed in the treatment position, and various immobilization devices are used to ensure accuracy of repositioning on a daily basis. In the case of thyroid cancer and other malignancies of the head and neck, a customized immobilization mask is made in order to fix the position of the patient's head in a reproducible fashion (Figure 17.1). A series of x-rays or a CT scan is then performed, and the radiation oncologist determines the borders of the radiation fields. Upon completion of the simulation, the patient is permanently tattooed and the setup parameters are recorded and photographed in order to ensure reproducibility of daily setup.

After simulation, the radiation oncologist works with a physicist or dosimetrist to generate the treatment plan. When developing a radiation treatment plan, the radiation oncologist typically delineates certain "target volumes" on the CT scan that were performed during simulation. These include the gross target volume (GTV), clinical target volume (CTV), and planning target volume (PTV). As described by the International Committee on Radiation Units and Measurements Report 50 (ICRU-50), the GTV is the gross palpable or visible/demonstrable extent and location of the malignant growth. The CTV is an expansion of the GTV that accounts for subclinical microscopic extension of disease, or alternatively can also be a region that is at risk for microscopic disease in the absence of gross disease. This volume has to be treated adequately in order to achieve the aim of the therapy: cure or palliation. The PTV is a further expansion of the CTV, accounting for geometric uncertainties such as variation in daily patient setup and internal organ motion (e.g., respiration). The PTV is therefore the volume that is to be irradiated to account for gross and microscopic disease, with an additional and appropriate margin to account for reproducibility of patient setup from day to day. In order to treat the thyroid bed in the anterior neck and the at-risk lymph node stations laterally while limiting the dose to the spinal cord, the typical PTV in thyroid cancer is shaped like an inverted U (Figure 17.2a). With conventional two-dimensional radiation planning techniques,

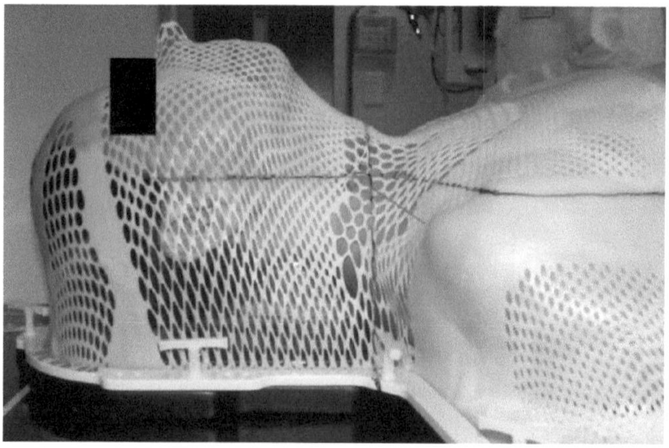

FIGURE 17.1 (See color insert.) Patient in treatment position. An aquaplast immobilization mask has been used to ensure reproduction of the head position for daily treatment. Underneath the mask, a metal wire has been placed over the scar in order for visualization on simulation CT (red arrow).

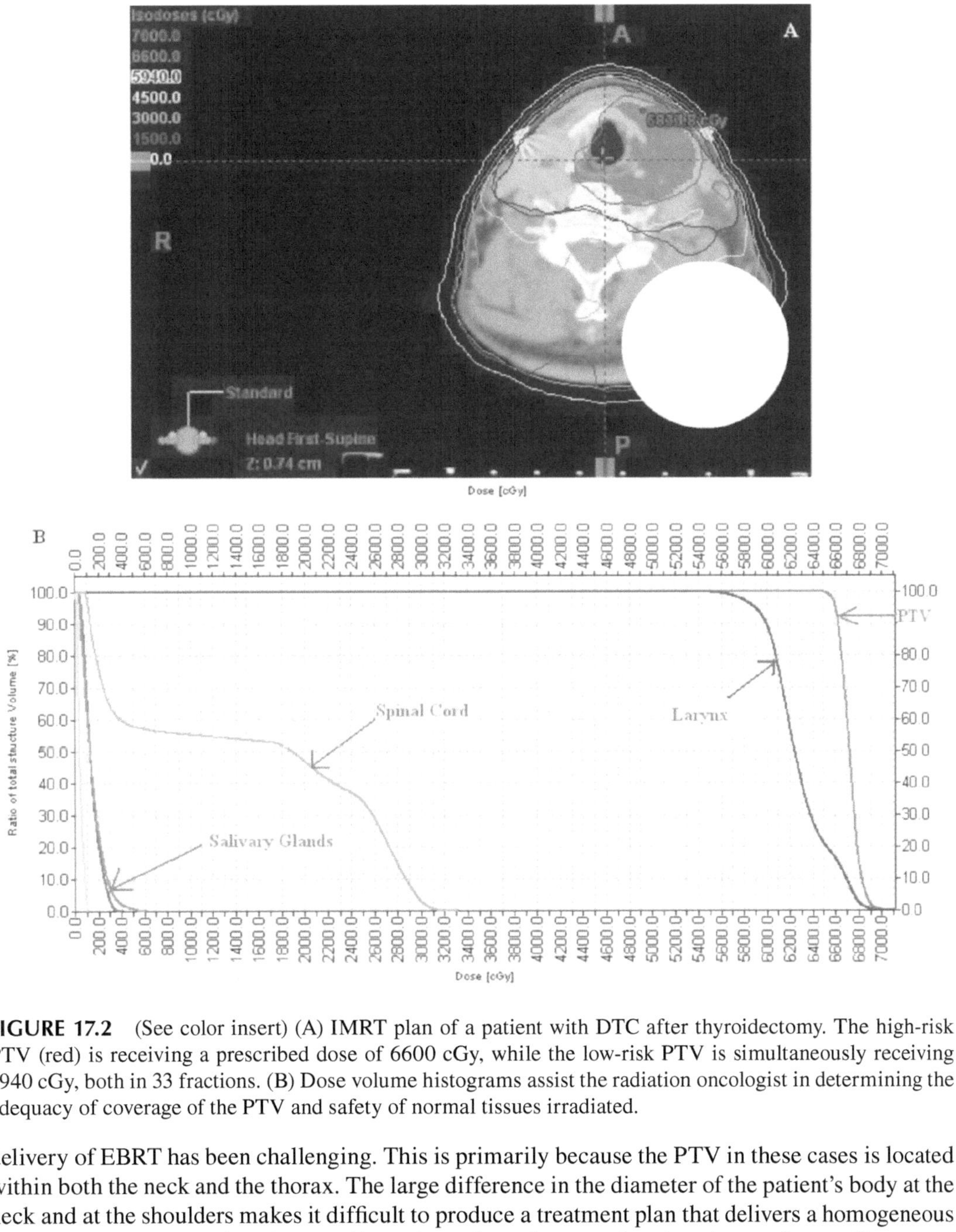

FIGURE 17.2 (See color insert) (A) IMRT plan of a patient with DTC after thyroidectomy. The high-risk PTV (red) is receiving a prescribed dose of 6600 cGy, while the low-risk PTV is simultaneously receiving 5940 cGy, both in 33 fractions. (B) Dose volume histograms assist the radiation oncologist in determining the adequacy of coverage of the PTV and safety of normal tissues irradiated.

delivery of EBRT has been challenging. This is primarily because the PTV in these cases is located within both the neck and the thorax. The large difference in the diameter of the patient's body at the neck and at the shoulders makes it difficult to produce a treatment plan that delivers a homogeneous radiation dose without high-dose gradients throughout the treated region. Areas of gross disease are typically treated with doses of 70–72 Gy; other regions at intermediate risk of recurrence are treated with 60–65 Gy, and low-risk regions receive 50–54 Gy. The maximum dose to the spinal cord must be limited to <45 Gy, and when possible, at least one salivary gland should receive a mean dose of <26 Gy.

In the modern era, the use of CT-based treatment planning and *intensity-modulated radiation therapy* (IMRT) has allowed us to greatly improve the delivery of radiation. (With IMRT, a computer-based optimization process referred to as "inverse planning" is used to generate the treatment plan. On the treatment planning CT scan, the radiation oncologist delineates on each image the various target volumes, as well as the adjacent organs at risk.) An advantage of IMRT is that in a given plan, various target volumes can simultaneously be treated with different doses, in order to account for areas at higher or lower risk of recurrence of disease. The radiation physicist/dosimetrist then instructs the computer program to generate a treatment plan that delivers the prescribed doses of radiation to the target volumes, while limiting the dose of radiation to normal tissues to be within tolerable levels. The shape and intensity of the radiation beam are modulated by the computer, depending on the various normal tissues, thickness of the patient, and target volumes in the path of the beam.

IMRT is becoming routinely adapted by the radiation oncologist in order to treat the target volume(s) with nonuniform amounts of radiation in a given beam. Multiple coplanar beam angles, typically five to nine, are used. The planning involved with IMRT is labor-intensive for the physician to draw target volumes and normal structures, and also for the physicist to generate a plan. However, IMRT creates very tight target volume coverage with rapid radiation dose fall-off, allowing for critical normal tissues such as the spinal cord and salivary glands to be spared, and has been routinely adopted by the radiation oncologists in the management of head and neck cancer. When assessing an IMRT plan, a dose volume histogram (DVH) is generated (Figure 17.2b). The DVH is a graphical display of an EBRT plan, where the volume of a PTV or organ at risk (OAR) is plotted on the y-axis and the dose delivered to that structure is plotted on the x-axis. This allows the radiation oncologist to evaluate the adequacy of treatment of the PTV, the presence of hot or cold spots within the PTV, and the dose to OARs.

After a plan is finalized, it undergoes a battery of quality assurance checks by the physicist. Once treatment starts, the patient's treatment positioning and setup is regularly verified on the linear accelerator in order to ensure that it is being reproduced reliably.

17.4 TOXICITY OF RADIATION THERAPY

The major sequelae of EBRT in thyroid cancer (Table 17.4) are similar to other malignancies of the head and neck, which have been studied extensively, and can be divided into acute and chronic side effects. These depend on total dose, fraction size, fractionation, prior or concomitant therapy (i.e., surgery or chemotherapy), and target volume. The discussion below pertains to treatment of malignancies of the head and neck as a whole, and can be applied to patients with thyroid cancer. The potential acute effects on the oral cavity and pharynx after approximately 1 week of EBRT include mucositis, sore throat, loss of taste, and xerostomia (if any of the major salivary glands are in the treatment portal). If the upper neck (and therefore the major salivary glands) is within the treatment field, the decrease in saliva changes the microflora in the mouth and can dramatically increase the number of dental caries in the patient. Fluoride gel treatments are effective in reducing the

TABLE 17.4
Potential Toxicities of External Beam Radiation Therapy

Acute	Chronic
Mucositis	Dysgeusia
Sore throat	Xerostomia
Dysgeusia	Dysphagia/aspiration
Xerostomia	Dental caries
Laryngitis	Hyperpigmentation
Esophagitis	Telangiectasia
Weight loss	Soft tissue fibrosis
	Osteonecrosis

subsequent incidence of dental caries. Approximately 5% of patients develop sialoadenitis within 24 hours of the first irradiation treatment. This resolves within 24 to 48 hours. The skin experiences erythema, peeling, and tanning. If the capacity of the basal cell layer to repopulate the epidermis is overwhelmed, the result is moist desquamation. Likewise, epilation of hair-bearing areas with accompanying loss of sweat and sebaceous gland function occurs.

The late effects of EBRT after approximately 3500–4000 cGy can include xerostomia and altered sense of taste. The importance of preserving salivary production on a patient's quality of life cannot be overstated. An average individual produces up to 1–1.5 liters of saliva per day, and loss of saliva predisposes to mucosal fissures and ulceration. Moreover, saliva aids in speech, taste, mastication, and deglutition of food, and its loss may aggravate the nutritional deficiencies that are so prevalent among patients with head and neck cancer. Saliva also plays a role in preventing caries and oral cavity infections via secretion of salivary lysozyme, IgA, and other antibacterial substances, and the production of salivary bicarbonate aids in preventing esophageal injury.

All of these issues are magnified in this patient population. Amifostine, a free-radical scavenging radioprotectant, has been shown to be effective in preventing xerostomia in clinical trials. Brizel et al.[24] randomized 315 patients to receive radiation alone or with intravenous amifostine prior to each EBRT dose. Twenty-one percent of patients discontinued amifostine before completing radiation, most often due to nausea, vomiting, or hypotension. Nonetheless, the use of amifostine was shown to reduce the incidence of grade 2 or greater acute xerostomia (51% vs. 78%, $p < 0.0001$), as well as increasing the dose associated with onset of acute xerostomia (60 Gy vs. 42 Gy, $p = 0.0001$). At 2 years, the incidence of grade 2 or greater late xerostomia was also improved (19% vs. 36%, $p = 0.038$).[25] There was no reduction in progression-free or overall survival associated with administration of amifostine. These results were not confirmed, however, by Buentzel et al.,[26] who enrolled 132 patients in a multicenter randomized placebo-controlled study in which patients received either IV amifostine or placebo prior to receiving carboplatin-based chemoradiation. In this study, there was no difference in either grade 2 or higher acute xerostomia (39% vs. 34%, $p = 0.715$) or late xerostomia (37% vs. 24%, $p = 0.235$) with or without the use of amifostine.[26] Administration of subcutaneous (sq) amifostine is generally felt to be both more convenient and better tolerated than the intravenous route. In order to compare the two methods of administration, Groupe Oncologie Radiothérapie Tête Et Cou GORTEC ran a prospective randomized study comparing IV versus sq administration.[27,28] Three hundred eleven patients receiving at least 40 Gy to the parotid glands were randomized to receive either IV amifostine (200 mg/m^2/d, 15–30 minutes prior to each RT fraction) or sq amifostine (500 mg/d, 20–60 minutes prior to each RT fraction). Preliminary results[28] suggested that the sq route was better tolerated than IV administration, with a compliance rate of 80% vs. 70%, respectively. This was felt to be particularly due to the reduced incidence of hypotension (6% vs. 0%). In an update of the results of this study,[27] there was no difference in either compliance (70% vs. 71%), and the incidence of hypotension was low in both groups (3% vs. 1%). Moreover, the sq group had an increased incidence of rash (11% vs. 22%, $p = 0.02$) and local pain at the injection site (0% vs. 9%, $p = 0.0008$).[27] Although the toxicity associated with amifostine was quite low, it should be noted that this was in the setting of a rigorously controlled clinical trial; this may not be the case in a less closely monitored setting.

Another approach to prevent xerostomia involves reduction in the dose of radiation received by the parotid glands, a major goal of head and neck IMRT. Over the past several years, many clinical studies have reported on the role of IMRT in preventing xerostomia via parotid sparing. Eisbruch et al.[29] reported that in a study of 88 patients and 152 parotid glands, limiting the mean parotid dose to <26 Gy resulted in retaining a substantial fraction of the baseline salivary output, while parotid glands receiving higher doses did not produce measurable saliva. Chao et al.[30] studied 41 patients treated with either IMRT or 3D conformal radiation treatment planning technique (3DCRT; a CT-based treatment planning technique that is less sophisticated than IMRT) and measured stimulated and unstimulated salivary flow, both prior to treatment and 6 months after completion of EBRT. They observed a 4% reduction in salivary flow per Gy of mean parotid dose, and calculated that a mean dose ≤32 Gy would achieve <25% reduction in salivary flow. Several other investigators have reported similar findings.[31–33]

Recently, there has been emerging interest in sparing the submandibular glands to further aid in the reduction of xerostomia. While the parotid glands contribute over 50% of stimulated salivary flow, the submandibular glands are primarily responsible for salivary production in the resting state,[34] which is important in patients' subjective complaints of dry mouth. Attempts to spare the submandibular gland in the IMRT plan have been described[35]; however, as this would result in undertreating level II lymph nodes in at-risk patients, extreme caution should be employed when considering this approach. Submandibular sparing should certainly be possible in patients receiving ipsilateral radiation.

An alternative approach to submandibular gland sparing involves surgical transfer of the gland into the submental space prior to radiotherapy. This was described by Seikaly et al.,[36] who reported their series of patients with head and neck squamous cell carcinoma (HNSCC) and N0–N2B (multiple involved lymph nodes less than 6 cm) disease who were undergoing surgery as part of their treatment plan. Patients with oral cavity or nasopharynx primaries were ineligible, as were patients with N2c (bilateral) or N3 (>6 cm) nodal disease, or involvement of level I nodes on either side of the neck. The reader is directed to Seikaly et al.[36] for further details regarding the technique of submandibular gland transfer. With a minimum of 2 years' follow-up,[37] 83% of the 96 patients enrolled in this study reported normal salivary flow, based on the validated University of Washington quality of life questionnaire.[38] Measurement of stimulated salivary flow rate revealed preservation of saliva; this compared favorably with a matched cohort of 38 patients who did not undergo submandibular sparing, none of whom had normal salivary function. There were no disease recurrences on the side of the transferred gland or in the submental space, and no surgical complications attributed to the transfer procedure were reported. A subsequent study by the same group showed that submandibular gland transfer was also associated with improved swallowing outcomes.[39] This method of salivary sparing has been reproduced with similar success by other groups,[40,41] and a multicenter phase II trial is currently being run by the Radiation Therapy Oncology Group (RTOG).[42]

Radiation-induced dysphagia is also an area of recent interest. Swallowing is a complex mechanism, involving the actions of 26 muscles and 5 cranial nerves. Eisbruch et al.[43] recently published a significant study examining the potential to reduce dysphagia via IMRT planning. In this study, the anatomic structures that are associated with dysphagia and aspiration were first identified in 26 patients undergoing chemoradiation. Patients were evaluated before and after treatment with videofluoroscopy. Esophagography functional abnormalities in these studies were identified. Pre- and posttreatment CT scans and endoscopic examinations were then reviewed, and anatomic abnormalities were correlated to those functional abnormalities. Based on these studies, it was found that only one patient had abnormalities in the oral phase of swallowing, whereas multiple abnormalities were found in the pharyngeal phase, affecting every patient in this study. The dysphagia- and aspiration-related structures (DARSs) were determined to be the circular pharyngeal constrictors and the glottic and supraglottic larynx. In the second phase of this study, the CT data sets of 20 consecutive patients who had been treated with one of two intensive chemoradiation regimens using standard IMRT planning techniques were obtained. In addition to a standard IMRT plan based on RTOG guidelines, two more treatment plans were generated, including a 3DCRT plan and a dysphagia/aspiration-optimized IMRT (doIMRT) plan. This plan minimized the portions of the DARS that lay *outside the PTV* receiving ≥50 Gy (referred to as the V_{50Gy}), without compromising the dose to the PTVs. This study showed that it is possible to reduce the mean V_{50Gy} using a doIMRT plan, compared with 3DCRT and standard IMRT planning, to both the pharyngeal constrictors (69% vs. 90% vs. 80%, respectively; $p < 0.001$) and the larynx (61% vs. 79% vs. 72%, respectively; $p = 0.002$ for 3DCRT, $p = 0.001$ for standard IMRT). This approach may hold great promise in reducing treatment-induced dysphagia. To date, however, this technique has not been reported clinically, and as with submandibular sparing, prudence is recommended in its utilization, especially in patients with a risk of retropharyngeal nodal metastasis.

Potential long-term complications at doses exceeding 6000 cGy include soft tissue and bone ulceration and necrosis. Radiation is believed to exert an avascular effect on tissues and epithelia that are thinner and more susceptible to injury. The process usually starts with ulceration of soft tissues, which can progress to bone exposure. If bone is then injured, bone necrosis or osteoradionecrosis can result. Treatment of this latter process is frustrating and difficult. For minor bone exposures, conservative measures are used, debridement being performed when indicated. For refractory cases, hyperbaric oxygen treatment has been advocated. Factors that can influence osteoradionecrosis include elective dental extraction after EBRT and treatment of tumors near bone.[44,45] In the modern era, osteonecrosis should be an uncommon event (<5%).[46]

17.5 EMERGING THERAPIES

In the past two decades, the field of radiation oncology has changed dramatically, largely due to emerging technologies and sophistication of treatment planning programs and computer processors. Prior to this, we were dependent on fluoroscopy and x-rays for treatment planning, using bony landmarks as surrogates for sites at risk for tumor involvement. Since then, CT scans, MRIs, and PET/CTs have vastly improved our ability to accurately delineate both target volumes and normal tissues. Simultaneously, treatment planning has progressed from simple 2D treatment plans that delivered homogenous doses of radiation to anatomic sites that are drawn out on x-rays to CT-based 3DCRT and now IMRT. We are now able to obtain a treatment planning CT, fuse MRIs and PET/CT scans so that all radiologic information is overlaid accurately, and draw out target volumes and normal organs to avoid on each slice of the scan, and deliver different doses of radiation to different targets simultaneously.

Current research in radiation oncology has attempted to continue to improve the accuracy of radiation delivery, moving forward from IMRT to *image-guided radiation therapy* (IGRT). IGRT refers to the use of various imaging modalities frequently during the course of EBRT to ensure accuracy of the radiation delivered compared to the planned treatment. Many treatment units are now capable of using either kilovoltage x-rays or CT scans while the patient is in the treatment room to ensure millimeter precision of the patient's position on a daily basis in real time. Improved accuracy of treatment delivery enables the radiation oncologist to decrease the amount of additional margin that is required in creating the PTV to account for setup errors. By being able to treat smaller PTVs, IGRT allows one to both decrease radiation doses to normal tissues and escalate radiation doses to the PTV.

IGRT may also be used to evaluate the patient for anatomic changes that might arise due to either weight loss or rapid diminishment in tumor volume. This may be important if these changes result in delivery of unplanned doses of radiation to both normal tissues (resulting in treatment toxicity) and tumor (resulting in disease recurrence). If such changes are detected, a new treatment plan can be generated so that treatment can continue safely.

Although there has been justifiably tremendous excitement about IGRT, it remains in its infancy. The next several years should bring forth improvements in quality of imaging and in treatment planning systems to further enhance our ability to deliver radiation safely and effectively.

17.6 SUMMARY

EBRT often plays an important part in the management of this complex disease. Recent advances in radiation treatment planning and delivery have improved the safety and delivery of EBRT, and current research and development continues to build on those improvements. Despite success in cases of DTC, the prognosis of ATC remains poor. Nonetheless, local EBRT can improve quality of life even in the setting of incurable disease.

REFERENCES

1. Tubiana, M., et al. 1985. External radiotherapy in thyroid cancers. *Cancer* 55(9 Suppl):2062–71.
2. Chow, S. M., et al. 2006. Local and regional control in patients with papillary thyroid carcinoma: Specific indications of external radiotherapy and radioactive iodine according to T and N categories in AJCC 6th ed. *Endocr Relat Cancer* 13(4):1159–72.
3. Farahati, J., et al. 1996. Differentiated thyroid cancer. Impact of adjuvant external radiotherapy in patients with perithyroidal tumor infiltration (stage pT4). *Cancer* 77(1):172–80.
4. Meadows, K. M., et al. 2006. External beam radiotherapy for differentiated thyroid cancer. *Am J Otolaryngol* 27(1):24–28.
5. Brierley, J., et al. 2005. Prognostic factors and the effect of treatment with radioactive iodine and external beam radiation on patients with differentiated thyroid cancer seen at a single institution over 40 years. *Clin Endocrinol* 63(4):418–27.
6. Tsang, R. W., et al. 1998. The effects of surgery, radioiodine, and external radiation therapy on the clinical outcome of patients with differentiated thyroid carcinoma. *Cancer* 82(2):375–88.
7. Biermann, M., et al. 2003. Multicenter study differentiated thyroid carcinoma (MSDS). Diminished acceptance of adjuvant external beam radiotherapy. *Nuklearmedizin* 42(6):244–50.
8. Cooper, D. S., et al. 2006. Management guidelines for patients with thyroid nodules and differentiated thyroid cancer. *Thyroid* 16(2):109–42.
9. Huang, J., et al. 2003. Does delay in starting treatment affect the outcomes of radiotherapy? A systematic review. *J Clin Oncol* 21(3):555–63.
10. Kim, J. H. and R. D. Leeper. 1983. Treatment of anaplastic giant and spindle cell carcinoma of the thyroid gland with combination adriamycin and radiation therapy. A new approach. *Cancer* 52(6):954–57.
11. Kim, J. H. and R. D. Leeper. 1987. Treatment of locally advanced thyroid carcinoma with combination doxorubicin and radiation therapy. *Cancer* 60(10):2372–75.
12. Fu, K. K., et al. 2000. A Radiation Therapy Oncology Group (RTOG) phase III randomized study to compare hyperfractionation and two variants of accelerated fractionation to standard fractionation radiotherapy for head and neck squamous cell carcinomas: First report of RTOG 9003. *Int J Radiat Oncol Biol Phys* 48(1):7–16.
13. Jeremic, B., et al. 1996. Hyperfractionated radiation therapy with or without concurrent low-dose daily carboplatin/etoposide for stage III non-small-cell lung cancer: A randomized study. *J Clin Oncol* 14(4):1065–70.
14. Saunders, M., et al. 1999. Continuous, hyperfractionated, accelerated radiotherapy (CHART) versus conventional radiotherapy in non-small-cell lung cancer: Mature data from the randomised multicentre trial. CHART Steering Committee. *Radiother Oncol* 52(2):137–48.
15. Tennvall, J., et al. 1979. Undifferentiated giant and spindle cell carcinoma of the thyroid. Report on two combined treatment modalities. *Acta Radiol Oncol Radiat Phys Biol* 18(5):408–16.
16. Tennvall, J., et al. 1994. Combined doxorubicin, hyperfractionated radiotherapy, and surgery in anaplastic thyroid carcinoma. Report on two protocols. The Swedish Anaplastic Thyroid Cancer Group. *Cancer* 74(4):1348–54.
17. Tennvall, J., et al. 2002. Anaplastic thyroid carcinoma: Three protocols combining doxorubicin, hyperfractionated radiotherapy and surgery. *Br J Cancer* 86(12):1848–53.
18. Tennvall, J., et al. 1990. Anaplastic thyroid carcinoma. Doxorubicin, hyperfractionated radiotherapy and surgery. *Acta Oncol* 29(8):1025–28.
19. De Crevoisier, R., et al. 2004. Combined treatment of anaplastic thyroid carcinoma with surgery, chemotherapy, and hyperfractionated accelerated external radiotherapy. *Int J Radiat Oncol Biol Phys* 60(4):1137–43.
20. Swaak-Kragten, A. T., et al. 2009. Multimodality treatment for anaplastic thyroid carcinoma—Treatment outcome in 75 patients. *Radiother Oncol* 92(1):100–4.
21. Haigh, P. I., et al. 2001. Completely resected anaplastic thyroid carcinoma combined with adjuvant chemotherapy and irradiation is associated with prolonged survival. *Cancer* 91(12):2335–42.
22. Chen, J., et al. 2008. Surgery and radiotherapy improves survival in patients with anaplastic thyroid carcinoma: Analysis of the surveillance, epidemiology, and end results 1983–2002. *Am J Clin Oncol* 31(5):460–64.
23. Cobin, R. H., et al. 2001. AACE/AAES medical/surgical guidelines for clinical practice: Management of thyroid carcinoma. American Association of Clinical Endocrinologists. American College of Endocrinology. *Endocr Pract* 7(3):202–20.
24. Brizel, D. M., et al. 2000. Phase III randomized trial of amifostine as a radioprotector in head and neck cancer. *J Clin Oncol* 18(19):3339–45.

25. Wasserman, T. H., et al. 2005. Influence of intravenous amifostine on xerostomia, tumor control, and survival after radiotherapy for head-and-neck cancer: 2-year follow-up of a prospective, randomized, phase III trial. *Int J Radiat Oncol Biol Phys* 63(4):985–90.
26. Buentzel, J., et al. 2006. Intravenous amifostine during chemoradiotherapy for head-and-neck cancer: A randomized placebo-controlled phase III study. *Int J Radiat Oncol Biol Phys* 64(3):684–91.
27. Bardet, E., et al. 2005. Subcutaneous (SbQ) versus intravenous (IV) administration of amifostine for head and neck (HN) cancer patients receiving radiotherapy (RT): Preliminary results of the GORTEC 2000-02 randomized trial. *Int J Radiat Oncol Biol Phys* 63(S1):S127.
28. Bardet, E., et al. 2002. Preliminary data of the GORTEC 2000–02 phase III trial comparing intravenous and subcutaneous administration of amifostine for head and neck tumors treated by external radiotherapy. *Semin Oncol* 29(6 Suppl 19):57–60.
29. Eisbruch, A., et al. 1999. Dose, volume, and function relationships in parotid salivary glands following conformal and intensity-modulated irradiation of head and neck cancer. *Int J Radiat Oncol Biol Phys* 45(3):577–87.
30. Chao, K. S., et al. 2001. A prospective study of salivary function sparing in patients with head-and-neck cancers receiving intensity-modulated or three-dimensional radiation therapy: Initial results. *Int J Radiat Oncol Biol Phys* 49(4):907–16.
31. Roesink, J. M., et al. 2001. Quantitative dose-volume response analysis of changes in parotid gland function after radiotherapy in the head-and-neck region. *Int J Radiat Oncol Biol Phys* 51(4):938–46.
32. Bussels, B., et al. 2004. Dose-response relationships within the parotid gland after radiotherapy for head and neck cancer. *Radiother Oncol* 73(3):297–306.
33. Saarilahti, K., et al. 2005. Intensity modulated radiotherapy for head and neck cancer: Evidence for preserved salivary gland function. *Radiother Oncol* 74(3):251–58.
34. Dawes, C. 1996. Factors influencing salivary flow rate and composition. In *Saliva and oral health*, ed. W. M. Edgar and D. M. O'Mullane, 27–41. 2nd ed. London: British Dental Journal.
35. Saarilahti, K., et al. 2006. Sparing of the submandibular glands by intensity modulated radiotherapy in the treatment of head and neck cancer. *Radiother Oncol* 78(3):270–75.
36. Seikaly, H., et al. 2001. Submandibular gland transfer: A new method of preventing radiation-induced xerostomia. *Laryngoscope* 111(2):347–52.
37. Seikaly, H., et al. 2004. Long-term outcomes of submandibular gland transfer for prevention of postradiation xerostomia. *Arch Otolaryngol Head Neck Surg* 130(8):956–61.
38. Hassan, S. J., and E. A. Weymuller, Jr. 1993. Assessment of quality of life in head and neck cancer patients. *Head Neck* 15(6):485–96.
39. Rieger, J., et al. 2005. Submandibular gland transfer for prevention of xerostomia after radiation therapy: Swallowing outcomes. *Arch Otolaryngol Head Neck Surg* 131(2):140–45.
40. Pathak, K. A., et al. 2004. Upfront submandibular salivary gland transfer in pharyngeal cancers. *Oral Oncol* 40(9):960–63.
41. Al-Qahtani, K., et al. 2006. The role of submandibular salivary gland transfer in preventing xerostomia in the chemoradiotherapy patient. *Oral Surg Oral Med Oral Pathol Oral Radiol Endod* 101(6):753–56.
42. RTOG. http://rtog.org/members/protocols/0244/0244.pdf (accessed February 21, 2009).
43. Eisbruch, A., et al. 2004. Dysphagia and aspiration after chemoradiotherapy for head-and-neck cancer: Which anatomic structures are affected and can they be spared by IMRT? *Int J Radiat Oncol Biol Phys* 60(5):1425–39.
44. Shukovsky, L. J., and G. H. Fletcher. 1973. Time-dose and tumor volume relationships in the irradiation of squamous cell carcinoma of the tonsillar fossa. *Radiology* 107(3):621–26.
45. Bedwinek, J. M., et al. 1976. Osteonecrosis in patients treated with definitive radiotherapy for squamous cell carcinomas of the oral cavity and naso- and oropharynx. *Radiology* 119(3):665–67.
46. Parsons, J. T., R. R. Million, and N. J. Cassisi. 1982. Carcinoma of the base of the tongue: Results of radical irradiation with surgery reserved for irradiation failure. *Laryngoscope* 92(6 Pt 1):689–96.

18 Role of Chemotherapy in Thyroid Cancer

Paolo Bossi and Laura D. Locati

CONTENTS

ABSTRACT: The role of chemotherapy in thyroid cancer has been disappointing. Doxorubicin is the only systemic agent approved for the treatment of thyroid cancer in the United States and Europe, but with limited efficacy, both in differentiated and medullary thyroid cancer. The adjunct of other antiblastic treatment has been explored in small studies, with conflicting results. Nevertheless, the integration of chemotherapy with new treatment modalities, along with a better definition of optimal timing of administration, identification of more sensitive histologies, and the development of appropriate protocols, could expand the importance of this treatment. In anaplastic thyroid carcinoma, chemotherapy may be considered for radiosensitization after complete resection and also as palliative therapy in metastatic or recurrent disease.

18.1 INTRODUCTION

The role of chemotherapy in thyroid cancer has been disappointing. Nevertheless, the integration of chemotherapy with new treatment modalities, along with a better definition of optimal timing of administration, identification of more sensitive histologies, and the development of appropriate protocols, could expand the importance of this treatment. Standard therapies for patients with advanced differentiated (nonmedullary) thyroid cancer (DTC) include surgery, radioactive iodine (RAI), external beam radiotherapy (EBRT), and thyroid-stimulating hormone (TSH) suppression. Cure rates mainly depend on histotype, tumor stage, and RAI sensitivity. For medullary thyroid cancer (MTC), surgery represents the only recognized form of curative approach. RAI and TSH suppression play no role in MTC. Also, in this case, prognosis is related to disease extension at diagnosis. Once DTC loses its ability to take up RAI (25–50%) or is intrinsically RAI resistant, antiblastic treatment is a possible strategy. The role of chemotherapy in such conditions is palliative.

18.2 DIFFERENTIATED (NONMEDULLARY) THYROID CANCER

Currently, doxorubicin is the only systemic agent approved for the treatment of thyroid cancer in the United States and Europe. The first experiences with this drug date back to the 1970s, and since then, no clinical trial has demonstrated any other chemotherapeutic agent to confer superior

TABLE 18.1
Report of the Main Studies Employing Chemotherapy in Metastatic Thyroid Cancer

	Chemotherapy	No. of Patients	Response Rate (CR + PR)	Reference
DTC	Doxorubicin	30	33%	1
	Doxorubicin	22	5%	2
	Doxorubicin	41	17%	5
	Doxorubicin + cisplatin	43	26%	5
	BAP	22	41%	4
	Thalidomide	36[a]	18%	9
	Lenalidomide	18	22%	10
MTC	Doxorubicin	9	11%	2
	Cisplatin + vindesine + doxo	10	10%	3
	CVD	7	29%	16
	Dacarbazine + streptozocin + 5FU	20	15%	17
	Capecitabine	3	2/3	19
ATC	Doxorubicin	77	22%	25
	Paclitaxel	19	53%	26

Note: BAP = bleomycin, doxorubicin (adriamycin), and cisplatin; doxo = doxorubicin; CVD = cyclophosphamide, vincristine, and dacarbazine.

[a] MTC patients were also enrolled.

[b] Literature review.

survival compared with doxorubicin alone.[1] In clinical practice, before the advent of molecular targeted therapy (MTT), chemotherapy represented the standard therapeutic option for metastatic or locally advanced incurable thyroid cancer. In Table 18.1 the main studies employing chemotherapy in the setting of metastatic thyroid cancer are reported.

The first published study of chemotherapy in RAI-resistant thyroid cancer verified the activity of doxorubicin, 45–75 mg/m^2, every 3 weeks in 30 patients: 33% showed partial response (PR) and 40% showed disease stabilization (SD).[1] In the last published study with doxorubicin monochemotherapy in 22 patients with progressive metastasizing DTC, partial response and stabilization of disease were obtained in 47% (5% PR and 42% SD).[2] Between these two studies, different criteria and methodologies in assessing tumor response have been employed, making the comparison difficult.

In order to improve the clinical response to doxorubicin, several trials have studied various agents in combination. Scherübl et al.[3] administered a combination of cisplatin (60 mg/mq), vindesine (3 mg/m^2), and doxorubicin (50 mg/mq) to 18 advanced MTC and DTC patients. Among 8 patients with DTC, there were three minor responses, lasting 9, 13, and 22 months.[3] A trial with BAP regimen (bleomycin, 30 mg a day for 3 days; adriamycin (doxorubicin), 60 mg/m^2; and cisplatin, 60 mg/m^2) obtained nine objective responses out of 22 treated patients with progressive thyroid cancer of varying histologies, but complete responses were not observed in DTC.[4] A randomized study of doxorubicin plus cisplatin versus doxorubicin alone was performed on 92 patients; no response rate advantage was obtained in the combination group, even if all complete responses were observed in this group.[5] A case report by Morris et al.[6] described the possible differentiating action (conversion of noniodine concentrating to iodine concentrating) of combination doxorubicin and cisplatin in a patient with papillary thyroid cancer and extensive pulmonary metastases.

Another recent trial in advanced nonmedullary thyroid cancer investigated the role of the immunomodulator interferon alpha-2β as an adjunct to doxorubicin.[7] Even if the rationale and preclinical data of this combination approach were compelling, the clinical responses were disappointingly low

(6% PR), with unacceptably high toxicities, namely, neutropenia and fatigue. Docetaxel has also been reported to have activity against advanced DTC, but without significant corroborating evidence.[8]

In a population of rapidly progressing thyroid cancer, mainly composed of papillary and follicular components, the activity of thalidomide has been verified, with 18% achieving a partial response and 32% achieving stable disease.[9] This agent, whose mechanism of activity is not fully understood but is based on immunomodulatory and antiangiogenic effects, was able to confer a median survival of almost 2 years in responders, compared with 11 months for nonresponders. Due to thalidomide toxicity, a subsequent phase II trial with less toxic lenalidomide was performed in a population of rapidly progressing disease (increase of more than 30% in the last year). Partial response was obtained in 7 out of 18 patients, and stabilization of disease in 9 patients. In 3 patients, pulmonary emboli were reported.[10]

In summary, DTC can demonstrate partial responses to chemotherapy, which could translate in longer survival. However, the response rates are generally not satisfactory given the attendant toxicities. Hence, it is important to choose the right time for antiblastic administration in the natural history of differentiated thyroid cancer. With the apparent outstanding results of MTT in the last few years in DTC, the role of chemotherapy will continue to be debated in the future, but eventually disappear as part of the therapeutic armamentarium.

18.3 MEDULLARY THYROID CANCER

Chemotherapy for MTC also has limited efficacy, even though it is still used in patients with rapidly progressing metastatic disease. The most used agents are doxorubicin and 5-fluorouracil (5FU), whose activity has been identified in approximately 30% of the patients. Doxorubicin alone or together with cisplatin, bleomicin, or vindesine has been administered to a relatively small number of patients with a mixed DTC and MTC population in clinical trials.[3–5] Nevertheless, there is limited experience with doxorubicin used to treat MTC alone.[11–13]

Dacarbazine has been evaluated in small series, in combination with 5FU or with vincristine and cyclophosphamide, showing a moderate activity[14–16]; when this drug has been employed in larger series with 5FU and streptozotocin, relatively low responses were noted, with 3 patients exhibiting partial responses out of 20 treated patients.[17]

Some MTC will express carcinoembryonic antigen (CEA), thus providing the possibility to use anti-CEA monoclonal antibodies as an adjunct to doxorubicin: phase I showed some evidence of antitumor responses within a well-tolerated therapeutic approach.[18]

A retrospective analysis of five cases (three MTC and two DTC) treated with capecitabine showed that two with MTC had clinical responses (one of these with a calcitonin reduction of 90% and long-lasting response of 4 years), while stable disease was obtained in the patients with DTC.[19]

18.4 ANAPLASTIC THYROID CANCER

Anaplastic thyroid carcinoma (ATC), in contrast to DTC, is a very aggressive malignancy with an extremely poor prognosis. Median overall survival ranges from 3 to 7 months from the time of diagnosis. The best prognostic factors are the amount of disease at time of presentation, complete resection, and performance status.[20,21]

At the time of diagnosis, approximately 40% of patients have distant metastases, 80% of them located in the lung. Single-modality therapeutic approaches fail to obtain a significant impact on prognosis, but due to the rarity of disease, there are no randomized data on different approaches. Even if long-term survival is rare, several studies have been published in which patients were treated with combinations of aggressive approaches, consisting mainly of surgical interventions, followed by EBRT, and chemotherapy. Current management strategies are based on single-institution phase II trials or on retrospectives series adopting such schemes. Doxorubicin is also the most used chemotherapeutic agent in ATC.

In its radiosensitizing activity, doxorubicin has been employed in one prospective study, published by Tennvall et al.[22] Patients were treated according to three different protocols of radiation, with concomitant doxorubicin, 20 mg weekly. Surgery was performed in 40 out of 55 patients after 30 or 46 Gy, according to different protocols. No severe toxicity due to combined treatment was observed. Only 9% of the patients showed a survival greater than 2 years. In particular, the benefit of adding doxorubicin was confined to local control, showing no impact on distant metastases. De Crevoisier et al.[23] conducted a prospective study of combined treatment consisting of surgery, if feasible, then two cycles of chemotherapy with doxorubicin (60 mg/m^2) and cisplatin (120 mg/m^2) every 4 weeks, followed by hyperfractionated radiotherapy (dose, 50 to 55 Gy), and then another four cycles of chemotherapy. Thirty patients were treated according to this combined strategy, with an overall survival of 46% and 27% at 1 and 3 years, respectively. Also in this study, distant metastases were the major cause of death. Concomitant chemoradiotherapy was effective in terms of local control, and therefore possibly improving the quality of life of the patients, facing a rapidly aggressive disease with potential local invasion and obstruction of airways. The role of induction chemotherapy in this setting of disease has not been investigated; instead, the use of primary concurrent chemoradiotherapy could help in selecting patients for surgery and in converting unresectable diseases into resectable ones.

As of today, the only strategy able to provide patients with ATC possible long-term survival is complete radical resection followed by chemotherapy and irradiation. No difference in outcome was noticed between patients receiving palliative resection and those treated with chemotherapy and radiotherapy.[24] In patients with ATC and metastatic or recurrent disease not amenable to surgery or EBRT, the most frequently employed drug is doxorubicin, with a response rate of about 22%[25]; other systemic approaches have been based on cisplatin, carboplatin, VP-16, bleomycin, and melphalan without greater survival benefit.

Ain et al.[26] published their experience with paclitaxel in patients with persistent or metastatic disease despite surgery or local radiation therapy. Twenty patients were treated with a 96-hour continuous infusion of paclitaxel (120 or 140 mg/m^2) every 3 weeks; the response rate in evaluable patients was 53%, although this could not translate into long-term benefit. Responses were also reported when drug administration was changed from every 3 weeks to weekly therapy. In a phase I trial of combination chemotherapy consisting of docetaxel, epirubicin, and cyclophosphamide in advanced cancer, three patients with ATC were enrolled, with two of them showing a response to treatment.[27] The combination of taxane and carboplatin with combretastatin A4, a vascular disrupting agent, is ongoing in a phase II–III study.

18.5 CONCLUSION

Chemotherapy alone has very limited activity in DTC, MTC, and ATC, with doxorubicin as the only agent approved in advanced thyroid cancer. Integration of chemotherapy with biological therapies or MTT is a possible future strategy to exploit in order to optimize antitumoral activity and overcome tumor resistance.

REFERENCES

1. Gottlieb JA, Hill CS Jr. 1974. Chemotherapy of thyroid cancer with adriamycin. Experience with 30 patients. *N Engl J Med* 290:193–97.
2. Matuszczyk A, Petersenn S, Bockisch A, Gorges R, Sheu SY, Veit P, Mann K. 2008. Chemotherapy with doxorubicin in progressive medullary and thyroid carcinoma of the follicular epithelium. *Horm Metab Res* 40:210–13.

3. Scherübl H, Raue F, Ziegler R. 1990. Combination chemotherapy of advanced medullary and differentiated thyroid cancer. Phase II study. *J Cancer Res Clin Oncol* 116:21–23.
4. De Besi P, Busnardo B, Toso S, Girelli ME, Nacamulli D, Simioni N, Casara D, Zorat P, Fiorentino MV. 1991. Combined chemotherapy with bleomycin, adriamycin, and platinum in advanced thyroid cancer. *J Endocrinol Invest* 14:475–80.
5. Shimaoka K, Schoenfeld DA, DeWys WD, Creech RH, DeConti R. 1985. A randomized trial of doxorubicin versus doxorubicin plus cisplatin in patients with advanced thyroid carcinoma. *Cancer* 56:2155–60.
6. Morris JC, Kim CK, Padilla ML, Mechanick JI. 1997. Conversion of non-iodine-concentrating differentiated thyroid carcinoma metastases into iodine-concentrating foci after anticancer chemotherapy. *Thyroid* 7:63–66.
7. Argiris A, Agarwala SS, Karamouzis MV, Burmeister LA, Carty SE. 2008. A phase II trial of doxorubicin and interferon alpha 2b in advanced, non-medullary thyroid cancer. *Invest New Drugs* 26:183–88.
8. Ikeda M, Tanaka K, Sonoo H, Miyake A, Yamamoto Y, Shiiki S, Nakashima K, Kurebayashi J. 2007. Docetaxel administration for radioiodine-resistant patients with metastatic papillary thyroid carcinoma. *Gan To Kagaku Ryoho* 34:933–36.
9. Ain KB, Lee C, Williams KD. 2007. Phase II trial of thalidomide for therapy of radioiodine-unresponsive and rapidly progressive thyroid carcinomas. *Thyroid* 17:663–70.
10. Ain KB, Lee C, Holbrook KM, Dziba JM, Williams KD. 2008. Phase II study of lenalidomide in distantly metastatic, rapidly progressive, and radioiodine-unresponsive thyroid carcinomas: Preliminary results. *J Clin Oncol* 26 (May 20 suppl; abstr 6027).
11. Porter AT, Ostrowski MJ. 1990. Medullary carcinoma of the thyroid treated by low-dose adriamycin. *Br J Clin Pract* 44:517–18.
12. Ravry MJ. 1977. Response of medullary thyroid carcinomas and carcinoid tumors to adriamycin. *Cancer Treat Rep* 61:106–7.
13. Husain M, Alsever RN, Lock JP, George WF, Katz FH. 1978. Failure of medullary carcinoma of the thyroid to respond to doxorubicin therapy. *Horm Res* 9:22–25.
14. Orlandi F, Caraci P, Berruti A, Puligheddu B, Pivano G, Dogliotti L, Angeli A. 1994. Chemotherapy with dacarbazine and 5-fluorouracil in advanced medullary thyroid cancer. *Ann Oncol* 5:763–65.
15. Petturson SR. 1988. Metastatic medullary thyroid carcinoma. Complete response to combination chemotherapy with dacarbazine and 5-fluorouracil. *Cancer* 62:1899–903.
16. Wu LT, Averbuch SD, Ball DW, de Bustros A, Baylin SB, McGuire WP 3rd. 1994. Treatment of advanced medullary thyroid carcinoma with a combination of cyclophosphamide, vincristine, and dacarbazine. *Cancer* 73:432–36.
17. Schlumberger M, Abdelmoumene N, Delisle MJ, Couette JE. 1995. Treatment of advanced medullary thyroid cancer with an alternating combination of 5 FU-streptozocin and 5 FU-dacarbazine. The Groupe d'Etude des Tumeurs a Calcitonine (GETC). *Br J Cancer* 71:363–65.
18. Sharkey RM, Hajjar G, Yeldell D, Brenner A, Burton J, Rubin A, Goldenberg DM. 2005. A phase I trial combining high-dose 90Y-labeled humanized anti-CEA monoclonal antibody with doxorubicin and peripheral blood stem cell rescue in advanced medullary thyroid cancer. *J Nucl Med* 46:620–33.
19. Gilliam LK, Kohn AD, Lalani T, Swanson PE, Vasko V, Patel A, Livingston RB, Pickett CA. 2006. Capecitabine therapy for refractory metastatic thyroid carcinoma: A case series. *Thyroid* 16(8):801–10.
20. Kihara M, Miyauchi A, Yamauchi A, Yokomise H. 2004. Prognostic factors of anaplastic thyroid carcinoma. *Surg Today* 34:394–98.
21. Besic N, Hocevar M, Zgajnar J, et al. 2005. Prognostic factors in anaplastic carcinoma of the thyroid—A multivariate survival analysis of 188 patients. *Langenbecks Arch Surg* 390:203–8.
22. Tennvall J, Lundell G, Wahlberg P, Bergenfelz A, Grimelius L, Akerman M, Hjelm Skog AL, Wallin G. 2002. Anaplastic thyroid carcinoma: Three protocols combining doxorubicin, hyperfractionated radiotherapy and surgery. *Br J Cancer* 86:1848–53.
23. De Crevoisier R, Baudin E, Bachelot A, Leboulleux S, Travagli JP, Caillou B, Schlumberger M. 2004. Combined treatment of anaplastic thyroid carcinoma with surgery, chemotherapy, and hyperfractionated accelerated external radiotherapy. *Int J Radiat Oncol Biol Phys* 60:1137–43.
24. Haigh PI, Ituarte PH, Wu HS, Treseler PA, Posner MD, Quivey JM, Duh QY, Clark OH. 2001. Completely resected anaplastic thyroid carcinoma combined with adjuvant chemotherapy and irradiation is associated with prolonged survival. *Cancer* 91:2335–42.
25. Giuffrida D, Gharib H. 2000. Anaplastic thyroid carcinoma: Current diagnosis and treatment. *Ann Oncol* 11:1083–89.

26. Ain KB, Egorin MJ, DeSimone PA. 2000. Treatment of anaplastic thyroid carcinoma with paclitaxel: Phase 2 trial using ninety-six-hour infusion. Collaborative Anaplastic Thyroid Cancer Health Intervention Trials (CATCHIT) Group. *Thyroid* 10:587–94.
27. Rischin D, Ackland SP, Smith J, Garg MB, Clarke S, Millward MJ, Toner GC, Zalcberg J. 2002. Phase I and pharmacokinetic study of docetaxel in combination with epirubicin and cyclophosphamide in advanced cancer: Dose escalation possible with granulocyte colony-stimulating factor, but not with prophylactic antibiotics. *Ann Oncol* 13:1810–18.

19 Role of NF-κB in Thyroid Cancer

Eriko Suzuki, Shunichi Yamashita, and Kazuo Umezawa

CONTENTS

Abstract: NF-κB is often constitutively activated in many carcinomas, including thyroid carcinoma. Activation of NF-κB may enhance the malignant character of cancer cells by producing cell survival proteins and inflammatory cytokines. A potent and specific NF-κB inhibitor (DHMEQ) of low molecular weight has been recently discovered. This inhibitor inhibits the constitutively activated NF-κB and induces apoptosis in thyroid carcinoma cells. It also effectively suppresses the growth of thyroid carcinoma cells in animals without showing any toxicity. The V600E mutation of *BRAF* was found to enhance NF-κB, MMP expression, and cell invasion in thyroid carcinoma dells. DHMEQ inhibits MMP expression and lowers the cell invasion activity. Thus, since NF-κB increases malignant behavior in thyroid cancer cells, inhibition of NF-κB holds promise as a novel chemotherapeutic agent.

19.1 SIGNAL TRANSDUCTION IN THYROID CARCINOMA CELLS

Papillary thyroid carcinoma (PTC) is the most common type of thyroid carcinoma. A strong correlation between thyroid carcinoma and radiation exposure has been identified, and PTC is known to frequently develop after external radiation.[1] Of the survivors of the atomic bomb explosion in Japan, 6.7% of them developed PTC, which is higher than expected in the general population. The etiology of PTC is still being delineated, but a number of molecular mechanisms have been demonstrated. Activation of receptor tyrosine kinase (RET/PTC, TRK, MET), whether by rearrangement or gene amplification, appears to be specific for the transformation of thyroid follicular cells into PTC.[2] These rearrangements produce chimeric proteins with protein-tyrosine kinase activities that contribute to the development of the malignant phenotype (Figure 19.1). Approximately 40% of adults with sporadic PTC have a *RET* gene rearrangement, and about 15% have an *NTRK1* rearrangement. This latter rearrangement is higher (60%) in children.

The *BRAF* mutation has recently emerged as a potential prognostic marker for PTC. The *BRAF* gene encodes a serine/threonine kinase acting in the RAS-RAF-MEK-MAPK signaling pathway.[3] A study of the phenotypes of thyroid follicular cell lines and transgenic mice characterized by

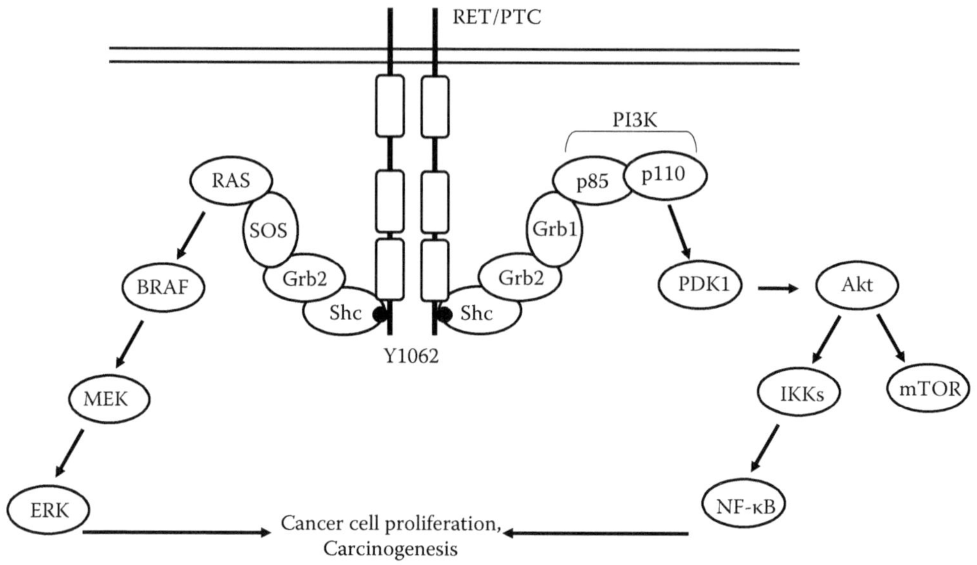

FIGURE 19.1 NF-κB activation pathway in papillary thyroid carcinoma (PTC).

targeted expression of mutated *BRAF* indicates that, at variance with *RET/PTC* rearrangement, this mutation induces genomic instability, higher invasiveness, eventually deeper tumor dedifferentiation, and more significant suppression of apoptosis. A recent study has shown that an NF-κB pathway is strictly involved in *BRAF*^{V600E}-induced cell migration and invasion, possibly via upregulated expressions of *CXCR4* and *VEGF* genes.[3] Activation of this NF-κB pathway was shown to be independent of the RAF/MAPK pathway.

Anaplastic thyroid carcinoma (ATC) is more aggressive than PTC. ATC cells often exhibit constitutive and strong activation of NF-κB, which is involved in its antiapoptotic property due to its ability to upregulate the production of antiapoptotic gene products, including Bcl-xL, XIAP, cIAPs, TRAF, and p21. The activation of NF-κB is extensively correlated with resistance to therapy and aggressiveness of thyroid cancer.

19.2 SIGNALING PATHWAY FOR THE ACTIVATION OF NF-κB

Without stimulation, NF-κB is usually inactive in the cytoplasm. It is activated by extracellular signals such as TNF-α, IL-1, lipopolysaccharide (LPS), lipopeptides, and tumor promoters such as phorbol esters.[4,5] The TNF-α signal transduction pathway (Figure 19.2) begins with activation of the TNF-α receptor (TNFR) and then TNFR-associated factor-2 (TRAF2). Then, TRAF2 activates IKKβ, which induces the phosphorylation of IκBα, an inhibitory protein that is bound to NF-κB. The phosphorylation of IκBα induces its release from its complex with NF-κB, and then IκBα is ubiquitinated and subsequently degraded by proteasomes. Liberated NF-κB molecules having a nuclear localization sequence (NLS) enter the nucleus, where they bind to the κB site of DNA.[6] Alternatively, various ligands and their cell surface receptors can activate TRAF6 instead of TRAF2 to then activate NF-κB (Figure 19.2). These receptors include the IL-1 receptor, Toll-like receptor-4 (TLR4) activated by lipopolysaccharide (LPS), TLR2 activated by lipopeptide, and receptor activator of NF-κB (RANK) activated by RANK ligand (RANKL).[7] These activation pathways are called the canonical (or classical) pathways. The canonical NF-κB pathway typically consists of the transcription factors, or subunit proteins, p65 and p50. In the noncanonical NF-κB pathway, the typical NF-κB components are RelB and p52. Before being activated, RelB is bound to p100, which is the precursor of p52 and also acts as a RelB inhibitory protein. The activation signal activates the IKKα

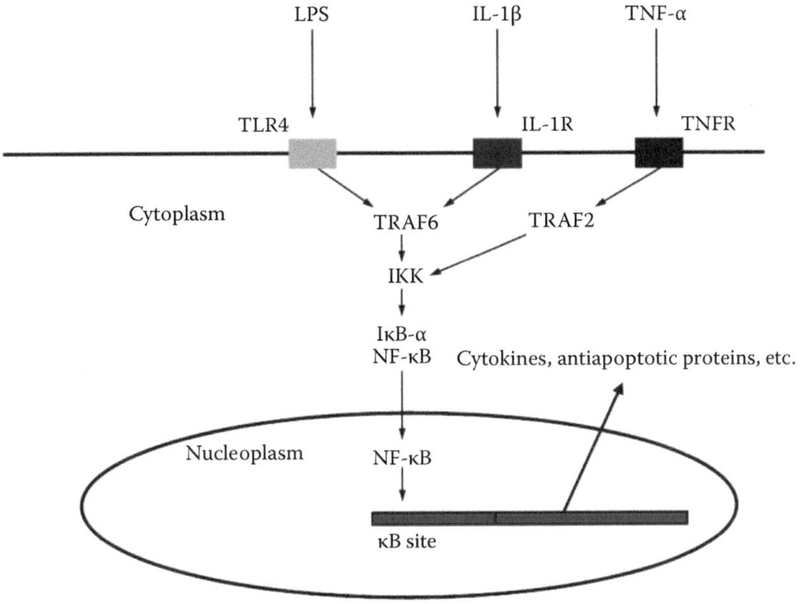

FIGURE 19.2 Activation of NF-κB through TRAF2 or TRAF6.

dimer, which then phosphorylates p100, allowing it to be degraded to p52. Then the active complex consisting of RelB and p52 is formed.[8]

19.3 NF-κB ACTIVATION IN THYROID CANCER

NF-κB is a key regulator of genes involved in the control of cellular cytokine production and apoptosis. In most cases, activation of NF-κB protects against cell death by upregulation of antiapoptotic factors:

- Inhibitor of apoptosis (IAP)
- B-cell lymphoma—extra large (Bcl-xl)
- X-linked inhibitor of apoptosis protein (XIAP)
- FLICE (a caspase 8 isoform having homology with Fas-associated protein with death domain (FADD) and interleukin-1β converting enzyme (ICE))-like inhibitory protein (FLIP)
- Bcl2-related protein a1/Bfl-1
- p21

Basal NF-κB activity is often increased in various types of human hematopoietic and solid tumors.[9–11] Persistent NF-κB activity can be a result of either chromosomal amplification, overexpression and rearrangement of genes coding for Rel family proteins, constitutive activation of upstream signaling kinases, or mutations inactivating the inhibitory proteins. It is generally believed that NF-κB-induced factors promote apoptotic resistance, transformation, cell growth, metastasis, and angiogenesis in neoplastic tissues. In addition, cell survival in some cancer types, such as hematologic malignancies, clearly depends on NF-κB. Hence, the inhibition of NF-κB activity is an intuitive approach to cancer therapy and has broad appeal, even to the extent that it has been attempted in a variety of cancer types.

In thyroid cancers, NF-κB also plays an important role, and its signaling pathways are impaired through sustained activation. This leads to the malignant potential of ATC cell lines. Inhibition of p65 protein synthesis with specific antisense oligonucleotides has been shown to greatly reduce the ability of two ATC cell lines to form colonies in soft agar, and their growth rate was also inhibited *in vitro.*[12] Additionally, the inhibition of radiation-induced activation of the NF-κB cascade potentiated the pro-apoptotic effect of radiation therapy in inoculated ATC tumors *in vivo*. Thus, the apoptosis-permissive effects of an NF-κB blockade can be considered a promising strategy for the treatment of ATC.

Activation of Akt often increases cellular malignant character and survival. Phosphatase and TENsin homolog (PTEN) is a negative regulator of Akt (protein kinase B). Akt and PTEN have been implicated in follicular thyroid cell (FTC) carcinogenesis. Germline mutations in the *PTEN* gene have been associated with Cowden's syndrome, a multiple hamartoma syndrome in which more than 50% of patients develop FTC.[13] The pro-survival ability of Akt is likely to be mediated through increased NF-κB activity.[14] Interestingly, PTEN itself is downregulated by p65 but not p50, forming a negative regulatory loop between PTEN and NF-κB. Therefore, the findings suggest that depressed PTEN or activated Akt in cancer cells might contribute to carcinogenesis by activation of the NF-κB pathway.

PPARγ may inhibit cancer progress by suppression of NF-κB. A *PAX8-PPARγ* fusion gene was first identified in FTC.[15] When fused to *PAX8*, *PPARγ* inhibits intrinsic *PPARγ* transcriptional activity. The reduced PPARγ protein leads to the activation of NF-κB, resulting in elevated cyclin D1 levels and decreased apoptosis.[16] Other signaling pathways for the progression of thyroid carcinoma are also described in Namba et al.[17]

19.4 DISCOVERY OF DHMEQ, AN INHIBITOR OF NF-κB

Low-molecular-weight inhibitors of NF-κB function could be useful as immunosuppressive, anti-inflammatory, antiviral, and anticancer agents. Many compounds, including bortezomib, a proteasome inhibitor, have been prepared as inhibitors of NF-κB activation. However, they may not be sufficiently specific for therapeutic development. While searching for clinically useful inhibitors of NF-κB function, compounds were discovered with structures similar to that of the antibiotic epoxyquinomicin.[18] Specifically, epoxyquinomicin C, panepoxydone, and cycloepoxydon possess the cyclohexylepoxydone structure. Epoxyquinomicin C does not completely inhibit TNFα-induced activation of NF-κB. A new compound was designed that had no protruding hydroxymethyl structure (Figure 19.3). The new compound, dehydroxymethylepoxyquinomicin (DHMEQ), was synthesized and shown to inhibit the activation of NF-κB in human T-cell leukemia Jurkat cells.[19] Racemic DHMEQ was synthesized from 2,5-dimethoxyaniline in five steps. Racemic DHMEQ can be separated into each of its enantiomers by one-step chiral column chromatography. After the enantiomeric separation, (−)-DHMEQ was found to be about 10 times more effective than (+)-DHMEQ.[20]

Recently, we found that the molecular targets of (−)-DHMEQ are p65 and other Rel family proteins.[21] Matrix-assisted laser desorption/ionization (MALDI)–time-of-flight (TOF) mass spectrometry (MS) analyses demonstrated that (−)-DHMEQ bound to p65 covalently with a 1:1 stoichiometry. MS analysis of the chymotrypsin-digested peptide suggested the binding of (−)-DHMEQ to a Cys residue.[21] Formation of a Cys/(−)-DHMEQ adduct in the protein was supported by chemical synthesis of the adduct. Substitution of the specific Cys in p65 and other Rel homology proteins, including p50, cRel, and RelB, resulted in the loss of (−)-DHMEQ binding. (−)-DHMEQ is the first NF-κB inhibitor proved by chemical methodology to bind to this specific Cys. Binding of

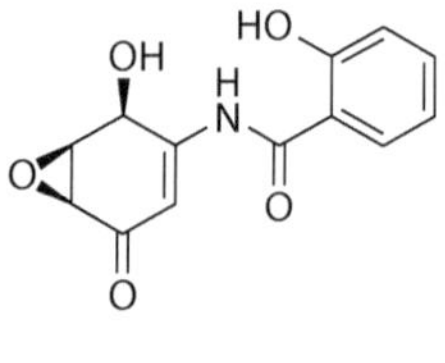

FIGURE 19.3 Structure of DHMEQ.

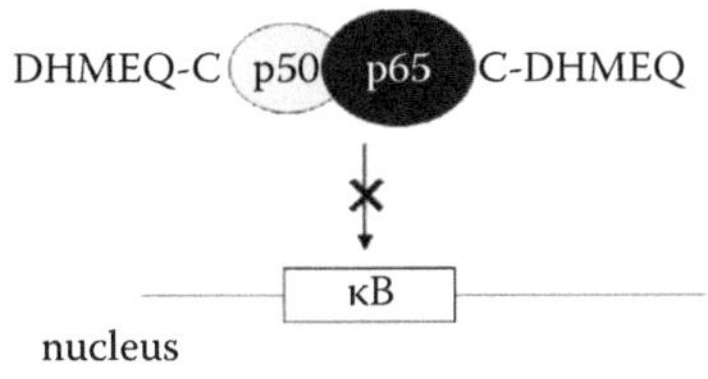

FIGURE 19.4 (–)-DHMEQ binds directly to specific cysteines of p65 and p50 (canonical NF-κB).

(–)-DHMEQ to the specific Cys diminished the DNA binding activity of p65 and p50, as shown in Figure 19.4. These findings may explain the highly selective inhibition of NF-κB and the low toxic effect of (–)-DHMEQ in cells and animals.

Since DHMEQ was shown to bind to RelB,[21] DHMEQ is likely to inhibit the noncanonical as well as the canonical activation of NF-κB. LT-$\alpha_1\beta_2$ is known to mainly induce noncanonical NF-κB. LT-$\alpha_1\beta_2$ induced the activation of NF-κB in breast carcinoma MCF-7 cells, and this activation was apparently inhibited by DHMEQ,[22] without influencing the production of p52 or processing of p100.

19.5 INVOLVEMENT OF NF-κB IN CARCINOGENESIS AND CANCER PROGRESSION

NF-κB may be involved in the regulation of cancer initiation. NF-κB enhances the expression of inducible nitrogen monoxide synthase (iNOS), the enzyme that produces NO from arginine. NO can damage DNA to induce mutations in proto-oncogenes such as K-ras and in cancer suppressor genes such as p53.[23]

It is more likely that NF-κB is involved in the promotion of tumor growth. Besides having anti-apoptotic activity, NF-κB has several activities that increase malignancy. In inflammatory sites, NF-κB enhances the production of growth factors such as TNF-α from cancer cells and activated macrophages.[24] It also enhances the expression of cyclo-oxygenase 2 (COX2), which is known to activate angiogenesis and thus promote growth of tumor. A COX2 inhibitor was found to be effective for suppressing colon cancer.[25] NF-κB also induces the expression of matrix metalloproteases (MMPs) such as MMP2 and MMP9. As these metalloproteases can degrade type 4 collagen to increase the invasiveness of cancer cells, MMP inhibitors may be useful as anticancer agents.[26] With respect to growth factor production in inflammatory sites, the deletion of the *IKKβ* gene in bone marrow cells decreased the incidence of hepatic carcinoma.[27] Also, inactivation of NF-κB inhibited colon cancer progression in animals.[28]

19.6 ANTICANCER EFFECT OF DHMEQ ON THYROID CANCER MODELS

Several years ago, researchers at Nagasaki University examined the anticancer effect of DHMEQ in animal thyroid cancer models and also studied the molecular mechanism involved in this effect.[29] They found that DHMEQ induced apoptosis in thyroid cancer cells both *in vitro* and *in vivo*. The mechanism of the DHMEQ-mediated anticancer effect was considered to be due to not only the inhibition of NF-κB but also the activation of c-Jun N-terminal kinase (JNK). This apoptosis-inducing effect of DHMEQ was shown in malignant thyroid cancer cells but not in normal thyroid cells.

The $BRAF^{V600E}$ mutation is closely linked to carcinogenesis and malignant phenotype of PTC (Figure 19.5). Signaling pathways activated by $BRAF^{V600E}$ are still unclear except for the common activation pathway, the mitogen-activated protein kinase (MAPK) cascade. To study the possible phenotypic effect of $BRAF^{V600E}$, the Nagasaki University group developed two different cell culture models:[30]

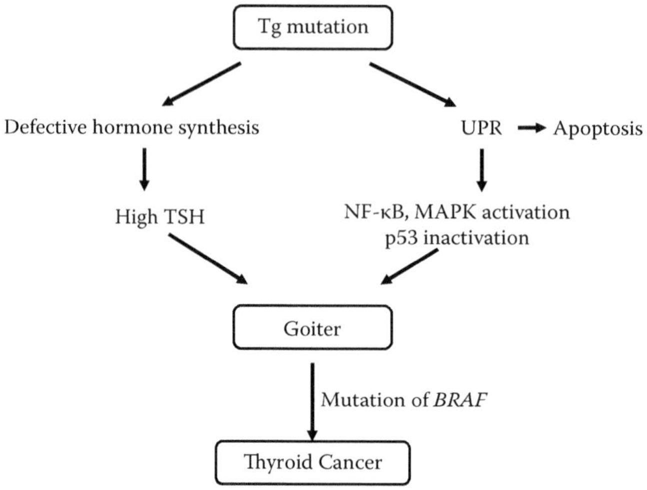

FIGURE 19.5 BRAF mutation-mediated progression of thyroid carcinoma.

1. Doxycycline-inducible *BRAF*^{V600E}-expressing clonal line derived from human thyroid cancer WRO cells originally harboring wild-type *BRAF*
2. WRO, KTC-3, and NPA cells infected with an adenovirus vector carrying *BRAF*^{V600E}

As a result, *BRAF*^{V600E} expression induced extracellular signal-regulated kinase (ERK) phosphorylation and cyclin Dl expression in these cells. The *BRAF*^{V600E}-overexpressing cells also showed an increase of NF-κB activity, resulting in upregulation of antiapoptotic c-IAP-1, c-IAP-2, and XIAP. Furthermore, *BRAF*^{V600E} expression also induced the expression of MMP and cell invasion into Matrigel- (mouse tumor cell-derived gelatinous protein mixture) through the NF-κB pathway. The increased invasiveness caused by *BRAF*^{V600E} expression was significantly inhibited by DHMEQ. These data indicate that *BRAF*^{V600E} activates not only MAPK, but also the NF-κB signaling pathway in human thyroid cancer cells, leading to an acquisition of apoptotic resistance and promotion of invasion.

NF-κB, as an antiapoptotic factor, crucially affects the outcome of cancer treatment and is one of the major causes of resistance to chemotherapy. The Nagasaki University group also studied whether DHMEQ could enhance the antitumor activities of taxanes in ATC cells.[31] Taxanes are antitumor drugs that disrupt microtubules but also induce NF-κB activation in ATC cells, which could compromise their therapeutic effect.[31] However, DHMEQ completely suppresses the DNA binding ability of NF-κB and lowers the level of nuclear NF-κB protein.[31] Compared to a single treatment with either taxane or DHMEQ, the combined treatment strongly potentiated apoptosis, which was confirmed by the results of a cell survival assay, Western blotting for PARP, caspase 3, XIAP, and survivin, and flow cytometry for annexin V.[31] Furthermore, the group also demonstrated for the first time that the combined treatment had a significantly greater inhibitory effect on tumor growth in a nude mice xenograft model.[31] These findings suggest that taxanes are able to induce NF-κB activation in ATC cells, which could attenuate antitumor activities of the drugs. In addition, inhibition of NF-κB by DHMEQ creates a chemosensitive environment and greatly enhances apoptosis in taxane-treated ATC cells *in vitro* and *in vivo*. Thus, DHMEQ may emerge as an attractive therapeutic strategy to enhance the response to taxanes by ATC cells. The NF-κB inhibitor SN50 also shows enhanced effects in combination with radiotherapy, perhaps through the same mechanism.[32]

In clinical practice, combined rather than single treatment is commonly considered to achieve a better therapeutic outcome. However, its use may cause adverse effects and systemic toxicities, which can be devastating for advanced stage cancer patients. DHMEQ, being an antibiotic derivative, has very low side effects, as shown in many *in vivo* experiments.[33–36] Thus, as the combination decreases the dosage of taxanes needed, it can circumvent the unwanted toxicity of taxanes, and above all, the therapeutic effects are increased. It is probable that the coadministration of DHMEQ and taxanes will emerge as an attractive therapeutic strategy for use in ATC patients.

Inhibition of NF-κB by inhibitors other than DHMEQ may also be useful to increase the sensitivity of thyroid cancer cells to anticancer agents. R-roscovitine, a novel cyclin-dependent kinase inhibitor, sensitized ATC cells to TNF-related apoptosis-inducing ligand (TRAIL)–induced apoptosis via regulation of the IKK/NF-κB pathway.[37] Conticello et al.[38] also identified that bortezomib, a proteasome inhibitor, sensitized ATC cells to TRAIL-induced apoptosis.[38] Additionally, Meng et al.[31] showed that TNF-α activated NF-κB in ATC cells. This activation is responsible for three-dimensional ATC cytomorphologic differentiation, which is also inhibited by SN50, an NF-κB nuclear translocation inhibitor.[31]

19.7 CONCLUSIONS AND FUTURE DIRECTIONS

NF-κB is a transcription factor that induces inflammatory cytokines and antiapoptotic proteins. NF-κB is often constitutively activated in human solid carcinomas, including thyroid carcinoma. Excess NF-κB activity in thyroid carcinoma cells should increase their viability, cellular invasiveness, metastasis, and resistance to anticancer drugs. DHMEQ is a new NF-κB inhibitor based on the structure of epoxyquinomicin C. DHMEQ inhibited the growth of thyroid carcinoma expressing NF-κB in animal experiments. It also inhibited the malignant phenotypes, such as invasiveness induced by the commonly occurring mutation of BRAF in cells. Additionally, it increased the anticancer activity of taxanes in ATC cells, both in culture and in animals. Thus, NF-κB should be considered to be functionally involved in the malignant phenotypes of thyroid carcinoma cells. Inhibition of NF-κB should be a promising new therapy for the treatment of thyroid carcinomas.

REFERENCES

1. Nikiforov YE. 2006. Radiation-induced thyroid cancer: What we have learned from chernobyl. *Endocr Pathol* 17:307–17.
2. Hamatani K, Eguchi H, Ito R, Mukai M, Takahashi K, Taga M, Imai K, Cologne J, Soda M, Arihiro K, Fujihara M, Abe K, Hayashi T, Nakashima M, Sekine I, Yasui W, Hayashi Y, Nakachi K. 2008. RET/PTC rearrangements preferentially occurred in papillary thyroid cancer among atomic bomb survivors exposed to high radiation dose. *Cancer Res* 68:7176–82.
3. Xing M. 2007. BRAF mutation in papillary thyroid cancer: Pathogenic role, molecular bases, and clinical implications. *Endocr Rev* 28:742–62.
4. Sen R, Baltimore D. 1986. Multiple nuclear factors interact with the immunoglobulin enhancer sequences. *Cell* 46:705–16.
5. Israel A, Le Bail O, Hatat D, et al. 1989. TNF stimulates expression of mouse MHC class I genes by inducing an NF-κB-like enhancer binding activity which displaces constitutive factors. *EMBO J* 8:3793–800.
6. Baldwin AS Jr. 1996. The NF-κB and IκB proteins: New discoveries and insights. *Annu Rev Immunol* 14:649–83.
7. Tanaka S, Nakamura K, Takahasi N, Suda T. 2005. Role of RANKL in physiological and pathological bone resorption and therapeutics targeting the RANKL-RANK signaling system. *Immunol Rev* 208:30–49.
8. Fagerlund R, Melén K, Cao X, Julkunen I. 2008. NF-kappaB p52, RelB and c-Rel are transported into the nucleus via a subset of importin alpha molecules. *Cell Signal* 20:1442–51.
9. Debatin KM. 2004. Apoptosis pathways in cancer and cancer therapy. *Cancer Immunol Immunother* 53:153–59.
10. Sevilla L, Zaldumbide A, Pognonec P, Boulukos KE. 2001. Transcriptional regulation of the bcl-x gene encoding the anti-apoptotic Bcl-xL protein by Ets, Rel/NFkappaB, STAT and AP1 transcription factor families. *Histol Histopathol* 16:595–601.
11. Rayet B, Gélinas C. 1999. Aberrant rel/nfkb genes and activity in human cancer. *Oncogene* 18:6938–47.

12. Zhu W, Ou Y, Li Y, Xiao R, Shu M, Zhou Y, Xie J, He S, Qiu P, Yan G. 2009. A small-molecule triptolide suppresses angiogenesis and invasion of human anaplastic thyroid carcinoma cells via down-regulation of the nuclear factor-kappaB pathway. *Mol Pharmacol* 75:812–19.
13. Liaw D, Marsh DJ, Li J, Dahia PLM, Wang SI, Zheng Z, Bose S, Call KM, Tsou HC, Peacocke M, Eng C, Parsons R. 1997. Germline mutations of the PTEN gene in Cowden disease, an inherited breast and thyroid cancer syndrome. *Nat Genet* 16:64–67.
14. Van Antwerp DJ, Martin SJ, Kafri T, Green DR, Verma IM. 1996. Suppression of TNF-a-induced apoptosis by NF-κB. *Science* 274:787–89.
15. Kroll TG, Sarraf P, Pecciarini L, Chen CJ, Mueller E, Spiegelman BM, Fletcher JA. 2000. PAX8-PPAR 1 fusion oncogene in human thyroid carcinoma. *Science* 289:1357–60.
16. Kato Y, Ying H, Zhao L, Furuya F, Araki O, Willingham MC, Cheng SY. 2006. PPARgamma insufficiency promotes follicular thyroid carcinogenesis via activation of the nuclear factor-kB signaling pathway. *Oncogene* 25:2736–47.
17. Namba H, Saenko V, Yamashita S. 2007. Nuclear factor kB in thyroid carcinogenesis and progression: A novel therapeutic target for advanced thyroid cancer. *Arq Bras Endocrinol Metab* 51:843–51.
18. Matsumoto N, Ariga A, To-e S, Nakamura H, Agata N, Hirano S, Inoue J, Umezawa K. 2000. Synthesis of NF-κB activation inhibitors derived from epoxyquinomicin C. *Bioorg Med Chem Lett* 10:865–69.
19. Ariga A, Namekawa J, Matsumoto N, Inoue J, Umezawa K. 2002. Inhibition of tumor necrosis factor-alpha-induced nuclear translocation and activation of NF-kappa B by dehydroxymethylepoxyquinomicin. *J Biol Chem* 277:24625–30.
20. Suzuki Y, Sugiyama C, Ohno O, Umezawa K. 2004. Preparation and biological activities of optically active dehydroxymethylepoxyquinomicin, a novel NF-κB inhibitor. *Tetrahedron* 60:7061–66.
21. Yamamoto M, Horie R, Takeiri M, Kozawa I, Umezawa K. 2008. Inactivation of NF-kappaB components by covalent binding of (–)-dehydroxymethylepoxyquinomicin to specific cysteine residues. *J Med Chem* 51:5780–88.
22. Matsumoto G, Namekawa J, Muta M, Nakamura T, Bando H, Tohyama K, Toi M, Umezawa K. 2005. Targeting of nuclear factor kappaB pathways by dehydroxymethylepoxyquinomicin, a novel inhibitor of breast carcinomas: Antitumor and antiangiogenic potential *in vivo*. *Clin Cancer Res* 11:1287–93.
23. Tamano S, Ward JM, Diwan BA, Keefer LK, Weghorst CM, Calvert RJ, Henneman JR, Ramljak D, Rice JM. 1996. Histogenesis and the role of p53 and K-ras mutations in hepatocarcinogenesis by glyceryl trinitrate (nitroglycerin) in male F344 rats. *Carcinogenesis* 17:2477–86.
24. Suzuki E, Umezawa K. 2006. Inhibition of macrophage activation and phagocytosis by a novel NF-kappaB inhibitor, dehydroxymethylepoxyquinomicin. *Biomed Pharmacother* 60:578–86.
25. Wang L, Chen W, Xie X, He Y, Bai X. 2008. Celecoxib inhibits tumor growth and angiogenesis in an orthotopic implantation tumor model of human colon cancer. *Exp Oncol* 30:42–51.
26. Han YP, Tuan TL, Wu H, Hughes M, Garner WL. 2001. TNF-alpha stimulates activation of pro-MMP2 in human skin through NF-(kappa)B mediated induction of MT1-MMP. *J Cell Sci* 114:131–39.
27. Sakurai T, Maeda S, Chang L, Karin M. 2006. Loss of hepatic NF-kappa B activity enhances chemical hepatocarcinogenesis through sustained c-Jun N-terminal kinase 1 activation. *Proc Natl Acad Sci USA* 103:10544–51.
28. Cusack JC Jr, Liu R, Xia L, Chao TH, Pien C, Niu W, Palombella VJ, Neuteboom ST, Palladino MA. 2006. NPI-0052 enhances tumoricidal response to conventional cancer therapy in a colon cancer model. *Clin Cancer Res* 12:6758–64.
29. Starenki DV, Namba H, Saenko VA, Ohtsuru A, Maeda S, Umezawa K, Yamashita S. 2004. Induction of thyroid cancer cell apoptosis by a novel nuclear factor kappaB inhibitor, DHMEQ. *Clin Cancer Res* 10:6821–29.
30. Palona I, Namba H, Mitsutake N, Starenki D, Podtcheko A, Sedliarou I, Ohtsuru A, Saenko V, Nagayama Y, Umezawa K, Yamashita S. 2006. BRAFV600E promotes invasiveness of thyroid cancer cells through NF-κB activation. *Endocrinology* 147:5699–707.
31. Meng Z, Mitsutake N, Nakashima M, Starenki D, Matsuse M, Takakura S, Namba H, Saenko V, Umezawa K, Ohtsuru A, Yamashita S. 2008. Dehydroxymethylepoxyquinomicin, a novel nuclear factor-kappaB inhibitor, enhances antitumor activity of taxanes in anaplastic thyroid cancer cells. *Endocrinology* 149:5357–65.
32. Starenki D, Namba H, Saenko V, Ohtsuru A, Yamashita S. 2004. Inhibition of nuclear factor-kappaB cascade potentiates the effect of a combination treatment of anaplastic thyroid cancer cells. *J Clin Endocrinol Metab* 89:410–18.

33. Watanabe M, Dewan MZ, Okamura T, Sasaki M, Itoh K, Higashihara M, Mizoguchi H, Honda M, Sata T, Watanabe T, Yamamoto N, Umezawa K, Horie R. 2005. A novel NF-kappaB inhibitor DHMEQ selectively targets constitutive NF-kappaB activity and induces apoptosis of multiple myeloma cells *in vitro* and *in vivo*. *Int J Cancer* 114:32–38.
34. Ohsugi T, Kumasaka T, Okada S, Ishida T, Yamaguchi K, Horie R, Watanabe T, Umezawa K. 2007. Dehydroxymethylepoxyquinomicin (DHMEQ) therapy reduces tumor formation in mice inoculated with tax-deficient adult T-cell leukemia-derived cell lines. *Cancer Lett* 257:206–15.
35. Kikuchi E, Horiguchi Y, Nakashima J, Kuroda K, Oya M, Ohigashi T, Takahashi N, Shima Y, Umezawa K, Murai M. 2003. Suppression of hormone-refractory prostate cancer by a novel nuclear factor kappaB inhibitor in nude mice. *Cancer Res* 63:107–10.
36. Umezawa K. 2006. Inhibition of tumor growth by NF-κB inhibitors. *Cancer Sci* 97:990–95.
37. Festa M, Petrella A, Alfano S, Parente L. 2009. R-Roscovitine sensitizes anaplastic thyroid carcinoma cells to TRAIL-induced apoptosis via regulation of IKK/NF-kappaB pathway. *Int J Cancer* 124:2728–36.
38. Conticello C, Adamo L, Giuffrida R, Vicari L, Zeuner A, Eramo A, Anastasi G, Memeo L, Giuffrida D, Iannolo G, Gulisano M, De Maria R. 2007. Proteasome inhibitors synergize with tumor necrosis factor-related apoptosis-induced ligand to induce anaplastic thyroid carcinoma cell death. *J Clin Endocrinol Metab* 92:1938–42.

20 Biologically Targeted Therapies for Thyroid Cancers*

Steven I. Sherman

CONTENTS

* Disclosure statement: The author consults for Exelixis, Bayer, Eli Lilly, Oxigene, and Veracyte; has previously consulted for AstraZeneca, Eisai, Enzon, Plexxikon, Semafore, and Celgene; has received lectureship fees from Exelixis and AstraZeneca; is on a speakers bureau for Genzyme; and receives research support from Amgen, Inc. (2005–present), AstraZeneca (2007–present), Eisai (2009–present), Genzyme (1995–present), V Foundation for Cancer Research (2009–present), and National Cancer Institute (2005–present).

ABSTRACT: Systemic chemotherapies for advanced or metastatic thyroid carcinomas have been of only limited effectiveness. For patients with differentiated thyroid epithelial or medullary thyroid carcinomas unresponsive to conventional treatments, novel therapies are needed to improve disease outcomes. Multiple novel therapies primarily targeting angiogenesis have entered clinical trials for metastatic thyroid carcinoma. Partial response rates up to 30% have been reported in single agent studies, but prolonged disease stabilization is more commonly seen. The most successful agents target the vascular endothelial growth factor receptors, with potential targets including the mutant kinases associated with papillary and medullary oncogenesis. Two drugs approved for other malignancies, sorafenib and sunitinib, have had promising preliminary results reported, and are being used selectively for patients who do not qualify for clinical trials. Additional agents targeting tumor vasculature, nuclear receptors, epigenetic abnormalities, and the immune response to neoplasia have also been investigated. Randomized trials for several agents are under way that may lead to eventual drug approval for thyroid cancer. Treatment for patients with metastatic or advanced thyroid carcinoma now emphasizes clinical trial opportunities for novel agents with considerable promise. Alternative options now exist for use of these molecular targeted therapies, especially tyrosine kinase inhibitors, which are well tolerated and may prove worthy of regulatory approval for this disease.

20.1 INTRODUCTION

Cytotoxic systemic chemotherapies for advanced, metastatic thyroid carcinomas have limited effectiveness, with response rates typically 25% or less.[1] Making matters worse, results from most clinical trials of new therapies for thyroid carcinoma that were initiated during the latter half of the twentieth century were never published.[2] A listing of the different clinical trial phases that will be referred to in this chapter is given in Table 20.1. Plaguing these early trials was the practice of lumping patients with all histologies of thyroid carcinoma (papillary (PTC), follicular (FTC), medullary (MTC), and anaplastic (ATC)), confounding interpretation of the results. The definitions of response used in these earlier studies varied as well, and none are comparable to the currently used standard RECIST methodology (for a detailed description of RECIST, see Table 20.2[3,4]). Thus, treatment with cytotoxic chemotherapy is generally limited to patients with symptomatic or rapidly progressive metastatic disease unresponsive to or unsuitable for surgery, radioiodine (RAI) therapy (for tumors derived from differentiated PTC and FTC), and external beam radiotherapy (EBRT).

TABLE 20.1
Phases of Clinical Trials

Phase I: Usually limited to a small number of patients, these trials are primarily designed to study progressively increasing doses of a new drug to determine side effects and the maximum tolerated dose. Patients with various tumor types may be recruited, with a secondary endpoint to identify evidence of possible effectiveness.

Phase II: Typically involving patients with a single tumor type, these trials have the primary objective to determine a level of effectiveness of a new drug while continuing to assess safety and toxicity. In solid tumors, effectiveness is usually assessed by serial evaluation of tumor diameter(s) on standard radiographic imaging studies, applying RECIST methodology to identify complete response, partial response, stable disease, or progressive disease.[3]

Phase III: In a large, typically multicenter study, patients are randomized to receive a new treatment or a standard treatment. Primary outcomes to be evaluated are usually survival or progression-free survival, while continuing to accrue data on safety and toxicity. Results from at least one phase III trial may be evaluated by regulatory authorities to decide whether to approve a new medication for routine clinical use.

TABLE 20.2
Standard Response Criteria (RECIST) Commonly Used in Phase II and III Trials

RECIST uses a unidimensional measure (the longest diameter) to quantify measurable tumor lesions. Overall tumor response is derived from serial radiologic studies.

Definitions

Measurable lesions: Lesions measuring 1.0 cm on spiral computed tomography scan or magnetic resonance imaging (0.5 cm section).

Target lesions (TLs): Up to 10 measurable lesions are identified as TLs for each subject, maximum 5 per organ. TLs should be representative of all organs involved in the subject's disease. TLs are selected on the basis of the size (largest) and on suitability for repetitive measurements. Bone lesions are considered to be measurable and eligible as TLs if they have identifiable soft tissue components that meet the definition of measurability as described above; bone lesions without measurable soft tissue components are not eligible. The longest diameter of each TL is recorded at baseline and then at each subsequent evaluation.

Nontarget lesions (NTLs): Lesions that do not meet criteria as measurable, or any additional measurable lesion not included as TLs (if more than 10 measurable lesions).

Response Criteria

Baseline: The longest diameter of each TL is measured, and the sum of longest diameters (LD) is calculated.

Follow-up: The longest diameter of each TL is measured using the same imaging technique as at baseline, and the sum of longest diameters for all TLs is calculated.

Best response: Classified as outlined below, taking into account both TL and NTL.

Complete response (CR): Disappearance of all TLs and NTLs, without appearance of new lesion.

Partial response (PR): Decrease of at least 30% in the sum of the longest diameters of the TLs when compared with baseline; no progression of NTL; and no appearance of new lesions.

Progressive disease (PD): Increase of at least 20% in the sum of the longest diameters of the target lesions, when compared with the nadir sum of LDs since treatment initiation (baseline or best tumor response); appearance of one or more new lesions or clear progression of existing NTLs.

Stable disease (SD): Sum of LDs did not decrease sufficiently to qualify for PR or increase sufficiently to qualify for PD, or unconfirmed objective response; no progression of NTLs.

Progression-free survival: The time from the first administration of study drug to the date of radiologic evidence of disease progression or death. Median progression-free survival is estimated using Kaplan-Meier methodology.

Objective response rate: Proportion of patients who have a best response rating of CR or PR.

Duration of response: The time between the first, subsequently confirmed, objective response and evidence of progressive disease.

Time to response: The time from the first administration of study drug to the initially documented and subsequently confirmed response.

Overall survival: Measured from the date of the first administration of study drug until the date of death. Median overall survival is estimated using Kaplan-Meier methodology.

Source: Therasse, P., et al., *J. Natl. Cancer Inst.*, 92, 205–16, 2000.

During the past decade, discoveries have sparked trials testing novel therapies for advanced thyroid carcinomas. Of prime importance has been recognition of key oncogenic mutations in PTC and MTC. *BRAF* and *RAS* genes code for kinases that activate signaling through the mitogen-activated protein kinase (MAPK) pathway, regulating growth and function in many cells, both normal and neoplastic. Evidence from various tumor models supports the contention that most PTC arise as a result of single activating somatic mutations in one of three genes: *BRAF*, *RAS*, and translocations producing *RET/PTC* oncogenes.[5] The resultant RET/PTC proteins signal upstream from RAS, thus activating the same MAPK pathway (see Figure 20.1a). For MTC, almost all familial forms of the disease arise from inheritable germline activating mutations in *RET*, whereas identical somatic mutations occurring in C-cells commonly cause sporadic disease.

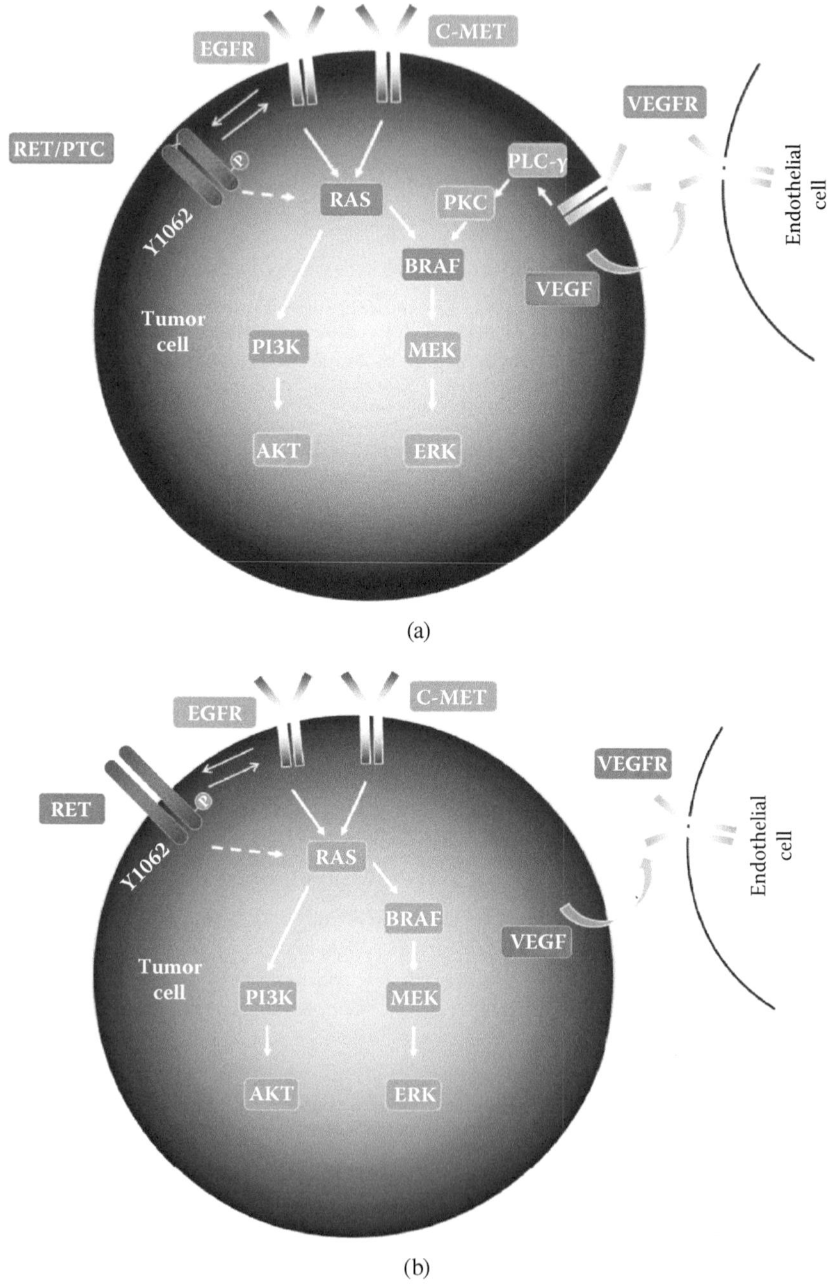

FIGURE 20.1 (See color insert.) (a) Kinase signaling pathways relevant to oncogenesis and tumor proliferation of differentiated thyroid carcinoma. (b) Kinase signaling pathways relevant to oncogenesis and tumor proliferation of medullary thyroid carcinoma.

Activated RET mutant proteins also enhance MAPK signaling (see Figure 20.1b). Consistent with the "oncogene addiction" hypothesis, inhibition of these etiologic activating mutations leads to either tumor stabilization or regression. Therefore, interest arises in the therapeutic potential of kinase inhibitors for these diseases.

A second development was recognition of processes facilitating tumor growth, reflecting either normal (such as hypoxia-inducible angiogenesis) or abnormal (such as epigenetic modifications of chromosomal DNA and histones) adaptations. Angiogenesis plays a critical role to support tumor cell growth and metastasis, supplying nutrients and oxygen, removing waste products, and facilitating distant metastasis.[6] Of the identified proangiogenic factors, vascular endothelial growth factor (VEGF) is key, binding to two receptor tyrosine kinases, VEGF receptor (VEGFR)-1 (FMS-like tyrosine kinase (Flt-1)) and VEGFR-2 (fetal liver kinase (Flk-1)/kinase insert domain receptor (KDR)), which also trigger MAPK signaling (see Figure 20.1a and b).[7] In PTC, the intensity of VEGF expression correlates with a higher risk of metastasis and recurrence, a shorter disease-free survival, and *BRAF* mutation status.[8–10]

Third, the U.S. National Cancer Institute and pharmaceutical companies have recognized that effective treatment for advanced thyroid cancers remains an unmet need. Clinical trials for thyroid cancer have evolved as hypothesis-driven protocols extending *in vitro* observations identifying a rationale for a particular drug's use. Some phase II studies were developed based on empiric observations in phase I trials that certain therapies yielded clinical benefit in patients with thyroid cancers.[11] Consensus clinical practice guidelines emerged, explicitly recommending clinical trials.[12,13] This review will focus on findings from key studies that reflect this new paradigm for research-driven treatment.[14] *

Biologically targeted therapies are defined as treatments that block the growth or metastasis of malignancies by interfering with specific molecules involved in neoplastic proliferation and progression. These are also termed molecular targeted therapies (MTTs). Small molecule inhibitors targeting signaling kinases have been of keen interest, given the oncogenic roles of mutant *BRAF*, *RET*, and *RAS*, and the contributory roles of growth factor receptors such as VEGFR.[5,15,16] These drugs partially inhibit multiple kinases at nanomolar concentrations and often affect multiple signaling pathways (Table 20.3). Orally administered, these agents have common side effects that include hypertension, diarrhea, skin lesions, and fatigue. Interest in thalidomide arose following reported responses in individual patients with ATC.[17] Abnormalities of nuclear gene regulation that affect

TABLE 20.3
Kinase Inhibitors Recently Studied in Clinical Trials for Advanced or Metastatic Thyroid Carcinomas[a]

Drug	VEGFR1 IC50 (nM)	VEGFR2 IC50 (nM)	VEGFR3 IC50 (nM)	RET IC50 (nM)	RET/PTC3 IC50 (nM)	BRAF IC50 (nM)	Other IC50 (nM)	References
Axitinib	1.2	0.25	0.29	—	—	—	C-KIT 1.7	61
Gefitinib	—	—	—	—	—	—	EGFR 33	73
Imatinib	—	>10,000	—	3,700	—	—	BCR-ABL 25 C-KIT 150	68
Motesanib	2	3	6	59	—	—	C-KIT 8	24
Pazopanib	10	30	47	2,800			PDGFR 74	126
Sorafenib	—	90	20	47	50	22	C-KIT 68	31
Sunitinib	2	9	17	41	224	—	—	45, 127
Vandetanib	1,600	40	110	130	100	—	EGFR 500	52, 53
XL184	—	0.035	—	4	—	—	C-MET 1.8	79

[a] The stronger the potency to block a specific kinase, the lower the IC50 (the concentration of drug that inhibits kinase activity by 50%).

* Online databases that can be searched to identify clinical trials currently recruiting patients can be found at www.thyroid.org and www.clinicaltrials.gov.

differentiated function stimulated interest in targeting DNA methylation, histone deacetylation, and nuclear receptors as a means to reverse these dedifferentiating steps.

20.2 TARGETING ONCOGENIC KINASES

Given the oncogenic roles of activated BRAF, RET, and RET/PTC kinases, it is theorized that specific targeting of these kinases could block tumor growth and induce senescence.[18] To date, only selective inhibitors of BRAF have entered clinical trials as a test of this hypothesis, as the agents available to target RET and RET/PTC generally also inhibit VEGFR and other kinases.[19] In contrast with the experience of treating gastrointestinal stromal tumors containing activating *c-KIT* mutations with the KIT inhibitor imatinib, use of selective BRAF inhibitors has not yet yielded impressive results in *BRAF* mutant PTC.[20]

20.2.1 PLX 4032

PLX 4032 is an orally available small molecule that specifically inhibits only the V600E mutant BRAF kinase, without significant impairment of wild type BRAF or other RAF kinases.[21] In melanoma and colon carcinoma cell lines bearing the V600E *BRAF* mutation, the IC50s for inhibiting phosphorylation of ERK were 10–30 nM, and for inhibiting cellular proliferation they were 47–126 nM.[21,22] The *RET/PTC* mutant thyroid cancer cell line TPC1, however, was poorly inhibited, with an IC50 for cellular proliferation of 10 μM. Preliminary data from a phase I study of escalating doses of PLX 4032 described the outcomes of three patients with *BRAF* mutant PTC.[22] One PTC patient experienced a partial response with shrinkage of lung metastases, whereas the other two patients had prolonged stable disease. Among the overall cohort of 55 patients with solid tumors (49 of whom had melanoma), the most common adverse events were skin rash, fatigue, pruritus, photosensitivity, and nausea. Although severe side effects were uncommon, 11% of the patients developed cutaneous squamous cell carcinomas.

20.2.2 XL 281

XL 281, an oral small molecule that inhibits both wild type and mutant BRAF kinases at low nanomolar concentrations, is currently in phase I trial.[23] Preliminary data described stable disease in five PTC patients; of the two patients whose tumors were documented to contain BRAF mutations, both remained stable after more than one year of therapy, as did a third PTC patient whose mutation status was unknown. An additional two patients with Hürthle cell carcinomas (HTCs) also were treated with prolonged stable disease, but one patient with ATC progressed despite treatment. No partial response was seen in any of the thyroid cancer patients. The most common side effect reported among all 48 solid tumor patients in the trial was fatigue in nearly half of patients, and other common toxicities included nausea, diarrhea, and vomiting, all of which were occasionally severe. Four patients were also described as having developed either cutaneous squamous cell carcinomas or premalignant keratoacanthomas.

20.3 TARGETING SIGNALING KINASES

A wide variety of multitargeted kinase inhibitors have entered clinical trials for patients with advanced or progressing metastatic thyroid cancers. Most of these agents have had the common property of inhibiting VEGFR at nanomolar concentrations, and thus have primarily targeted angiogenesis. However, given the considerable structural similarity between RET and VEGFR kinases, most of these small molecule inhibitors are capable of affecting both kinases. Because of the targeting similarities of many of these agents, common toxicities exist among them, including hypertension, diarrhea, skin rashes, and fatigue.

20.3.1 Motesanib

Motesanib (AMG706) is an oral tyrosine kinase inhibitor (TKI) targeting VEGFR-1, -2, and -3.[24] In both *in vitro* and cell-based assays, nanomolar concentrations of motesanib inhibited autophosphorylation of both wild type and mutant *RET*; growth of xenografts of TT cells bearing the C634W *RET* mutation was effectively inhibited.[25] In a phase I study, motesanib demonstrated antitumor activity in patients with advanced solid malignancies, including five patients with differentiated thyroid epithelial cancer (DTC: PTC or FTC) and one with MTC; three thyroid patients experienced >30% reductions in tumor diameters, qualifying as partial responders.[3,26] The most common toxicities included fatigue, nausea, diarrhea, and hypertension, all typical of this class of drugs.

Based on this phase I experience, a multicenter, open-label phase II trial was initiated, testing the efficacy of motesanib in separate cohorts of patients with progressive DTC[27] or progressive/symptomatic MTC,[28] starting at 125 mg daily. The eligibility criterion of progression was based upon serial radiographic imaging studies within the preceding 6 months, applying the RECIST response assessment.[3] Of 93 DTC patients who initiated therapy, one-third were still receiving drug therapy after 48 weeks. Partial response was confirmed by subsequent imaging and independent radiological review in 14% of the DTC patients, and another 35% of these previously progressive disease patients maintained stable disease for at least 24 weeks. The median progression-free survival (PFS) was 40 weeks. Although the drug does not inhibit BRAF, patients with *BRAF* mutation-bearing tumors were less likely to progress while on the drug, which may relate to a higher dependence upon VEGF-mediated angiogenesis in such tumors.[29] Of 91 patients with progressive or symptomatic MTC who initiated therapy, only 2% had confirmed partial response, but another 47% experienced stable disease for at least 24 weeks.[28] Unexpectedly, the maximum and trough plasma concentrations of the drug in MTC patients were lower than reported with other solid tumor patients, and these differing pharmacokinetics may have contributed to the lower response rate. Overall, the drug was well tolerated, with similar side effects as reported in the phase I trial. An unanticipated side effect of motesanib therapy was a 30% increase in the mean dosages of levothyroxine required to maintain TSH suppression or euthyroidism, respectively, in DTC and MTC cohorts, and 60 to 70% of patients experienced peak thyroid-stimulating hormone (TSH) concentrations above the therapeutic ranges.[30]

20.3.2 Sorafenib

Sorafenib (BAY 43-9006) is an oral small-molecule TKI targeting VEGFR-2 and -3, RET (including most mutant forms that have been examined), and BRAF.[31] In preclinical studies, sorafenib prevented the growth of the TPC-1 and TT cell lines, which contain the *RET/PTC1* and *C634W* RET mutations, respectively.[32] In four phase I trials testing varying doses and administration schedules of sorafenib, the optimal therapeutic dose was found to be 400 mg twice daily.[33] The most common or significant toxicities included hand-foot syndrome, rash, fatigue, diarrhea, and hypertension. Like other agents that inhibit BRAF, sorafenib also has been associated with development of cutaneous squamous cell carcinomas in up to 5% of treated patients, and a similar frequency of keratoacanthomas and other premalignant actinic lesions.[34]

Although no thyroid cancer patients were reported in these phase I trials, tumor shrinkage was reported in one thyroid cancer patient included in a phase II trial for advanced solid tumors.[35] Subsequently, two phase II trials were performed specifically in patients with metastatic PTC. Sponsored by the National Cancer Institute, an open-label phase II trial recruited 58 patients in a 10-month period.[36] Of 36 evaluable patients, confirmed partial response was seen in 8%, and minor response (defined as 23–29% reduction in tumor diameters) was described in another 19%. In a smaller open-label phase II study, unconfirmed partial responses were reported in 4 of 15 evaluable patients with PTC and 3 of 7 with FTC.[37] Median PFS was 84 weeks. Updated data from this latter trial were recently presented, comprising a total of 55 (25 with PTC, 19 with FTC or HTC, 4 with

MTC, and 5 with poorly differentiated or ATC).[38] Although the overall PFS remained 84 weeks, it was significantly shorter at 54 weeks in those patients without the *BRAF* activating mutation. A randomized, placebo-controlled phase III study of sorafenib as first-line therapy for progressive metastatic DTC is under development.

Anticipating synergy between sorafenib's ability to inhibit MAPK signaling and the RAS-blocking effects of the farnesyl transferase inhibitor tipifarnib, a phase I trial was performed of the combination of these drugs.[39] The maximum tolerated doses of sorafenib and tipifarnib were 200 and 100 mg twice daily, respectively. In the 21 patients with DTC treated, median progression PFS was 20 months.

The anti-RET activity of sorafenib makes MTC a potential therapeutic target for this drug as well.[16] In a small pilot study, 5 patients with metastatic MTC were treated with sorafenib. Responses were described in 2 (including one complete response) after 6 months of treatment, and symptomatic improvement was seen in all.[40] A larger, open-label phase II study has been initiated in patients with metastatic MTC. Partial responses were also reported in 4 of 9 evaluable MTC patients participating in the phase I study of the combination of sorafenib and tipifarnib.[39]

In ATC cell lines, preclinical models suggested potential efficacy of sorafenib to inhibit MAPK signaling.[41] Subsequently, a phase II trial was started, evaluating sorafenib therapy in patients who had progressed after previous cytotoxic chemotherapy.[42] Of 15 patients evaluated, 2 had experienced a partial response and 4 had stable disease as their best responses to treatment, but the overall median time to progression was only 1.5 months, and duration of survival 3.5 months.

Sorafenib is approved by the U.S. Food and Drug Administration as treatment for advanced renal cell carcinoma and unresectable hepatocellular carcinoma. Although not specifically approved for thyroid carcinomas, sorafenib is being used in selected patients with progressive metastatic PTC and MTC for whom clinical trials are not appropriate.[12] Compared with patients' rate of disease progression prior to initiation of therapy, sorafenib may prolong PFS in DTC by at least one year.[43] The drug may also be appropriate in selected pediatric cases; in one report, treatment with sorafenib yielded a marked response in a child whose lung metastases from PTC were progressing despite RAI therapy.[44] As with other antiangiogenic therapies, pediatric usage may result in bony growth plate inhibition and growth abnormalities.

20.3.3 Sunitinib

Sunitinib (SU11248) is an oral small-molecule TKI of all three VEGFR, RET, and RET/PTC subtypes 1 and 3.[45] Prolonged partial responses have been described in three patients (with PTC, FTC, and MTC) treated with sunitinib, 50 mg daily for 28 days followed by 14 days of no treatment per cycle.[46,47] Fluorodeoxyglucose (FDG) uptake by positron emission tomography (PET) imaging was markedly reduced in the DTC patients. Preliminary results from an open-label phase II trial in patients with progressive DTC or MTC report partial response in 13% of 31 DTC patients and disease stabilization in 68% of DTC and 83% of MTC patients.[48] Common or severe adverse events include fatigue, diarrhea, palmar-plantar erythrodysesthesia, neutropenia, and hypertension. Interim analysis from a second open-label phase II trial reported partial responses or stable disease for greater than 12 weeks in 2 of 12 DTC and 3 of 8 MTC patients.[49] Recently, preliminary results from a third trial, using a lower dose of 37.5 mg daily but administered continuously, were reported.[50] Of 33 patients with FDG-PET-avid metastatic thyroid cancer (23 with DTC, 7 with MTC), 29 were evaluable for response: 7% complete response (lasting at least 9 months), 25% partial response, and 48% stable disease. Like sorafenib, sunitinib is approved for treatment of renal cell carcinoma and is therefore available for use in selected thyroid cancer patients with metastatic disease warranting therapy outside of clinical trials.

20.3.4 Vandetanib

Vandetanib (ZD 6474) is an oral small-molecule TKI that targets VEGFR-2 and -3, RET, and at higher concentrations, the EGF receptor.[51,52] One of the first small molecule inhibitors to be studied in thyroid cancer cell lines, vandetanib was shown to inhibit effectively *RET/PTC3* mutations found in some PTC and *M918T* RET mutations occurring in MEN2B-associated and some sporadic MTC.[53] Growth of cell lines containing *RET/PTC1* or *RET/PTC3* was inhibited. However, the drug was not able to block RET when a hydrophobic amino acid substitution occurred at V804, as in some inherited forms of MTC.[54] In a phase I trial in 77 patients with various solid carcinomas other than thyroid, doses up to 300 mg daily were tolerated, with the most common dose-limiting side effects being diarrhea, hypertension, and skin rash.[55]

On the basis of the preclinical demonstration that vandetanib inhibited most RET kinases bearing point mutations, a multicenter, open-label phase II trial studied the efficacy of the drug in patients with metastatic familial forms of MTC.[56] Thirty patients were enrolled, starting therapy with vandetanib, 300 mg daily. Confirmed partial response was reported in 21% of these patients and unconfirmed responses in another 17%. Calcitonin (CTN) levels dropped by more than 50% in most patients, but blocking RET may lead to a direct inhibition of CTN gene expression, independent of tumor volume changes.[57] The most commonly reported side effects included rash (particularly photosensitivity), diarrhea, fatigue, and nausea, whereas the most severe toxicities included asymptomatic QT interval prolongation on electrocardiogram, rash, and diarrhea. A second phase II trial in familial MTC, starting at 100 mg daily, reported similar preliminary results.[58] Ongoing studies with vandetanib include (1) a multicenter, randomized, placebo-controlled phase III trial in patients with metastatic MTC, either sporadic or inherited; (2) an open-label phase II trial in patients under the age of 18 with familial MTC (with partial responses described in several patients, including those with aggressive tumors associated with germline M918T *RET* mutations)[59]; and (3) a randomized placebo-controlled phase II trial in patients with metastatic DTC.

Of interest have also been potential synergistic combinations of vandetanib with other agents. Given the clinical evidence of vandetanib's efficacy in MTC, and *in vitro* evidence that bortezomib triggered caspase-dependent apoptosis in MTC cells, a phase I/II trial of the combination has been initiated, with enrollment targeting patients with advanced MTC as well as other solid tumors.[60]

20.3.5 Axitinib

Axitinib (AG-013736) is an oral inhibitor that effectively blocks VEGFR at subnanomolar concentrations, but notably not the RET kinase.[61] In a phase I study of 36 patients with advanced solid malignancies, 1 of 5 thyroid cancer patients experienced tumor shrinkage, although none qualified as a partial response.[62] A multicenter, open-label phase II study examined the efficacy of axitinib in advanced or metastatic thyroid carcinoma, starting at a dose of 5 mg twice daily.[63] Of the 60 patients who started therapy, 50% had PTC, 25% had FTC (including HTC variants), and 18% had MTC. Although response assessment was not possible in 25% of the patients, confirmed partial response rate was 30% by intent-to-treat analysis (31% in DTC, 18% in MTC, 1 patient with ATC). Responses were seen in patients despite previous treatments with a variety of chemotherapeutic regimens. Median progression-free survival was 18 months. Common adverse events included fatigue, stomatitis, proteinuria, diarrhea, hypertension, and nausea. Exploratory analyses of soluble biomarkers demonstrated increases in serum VEGF levels, a recognized phenomenon of effective angiogenesis inhibition.[64] Given the absence of inhibitory activity against RET or other mutated kinases that are oncogenic in thyroid carcinoma, the efficacy of axitinib suggests that VEGFR-mediated angiogenesis is likely the primary mechanism by which the other anti-VEGFR inhibitory agents function.

Currently ongoing is a multicenter, open-label phase II study to determine the efficacy of axitinib in patients with metastatic DTC refractory to doxorubicin, or for whom doxorubicin therapy is contraindicated. Study completion is anticipated in September 2010.

20.3.6 Pazopanib

Pazopanib is a potent small-molecule inhibitor of all VEGFR subtypes as well as platelet-derived growth factor receptor (PDGFR). Like axitinib, it has no significant inhibitory activity against the oncogenic kinases RET, RET/PTC, or BRAF, and therefore its actions are expected to be primarily antiangiogenic in thyroid carcinoma. Preliminary results were recently reported for 37 patients with rapidly progressing DTC treated in a phase II trial.[65] With a starting daily dose of 800 mg, 32% of patients had confirmed partial responses, and the 6-month progression-free survival was 71%. The most common side effects of therapy included hypertension, in nearly half, as well as elevated serum transaminases, headache, and mucositis.

20.3.7 Imatinib

Imatinib (STI571), an oral small-molecule kinase inhibitor of BCR-ABL and c-KIT, inhibits RET autophosphorylation and RET-mediated cell growth.[66–68] Two small open-label phase II studies have been completed that examined a total of 24 patients with metastatic MTC treated with imatinib, starting at 600 mg daily.[69,70] No objective tumor responses were reported, and a minority of patients achieved stable disease as their best tumor response. Toxicities included diarrhea, laryngeal edema, rash, and nausea; increased thyroid hormone dosage requirements were reported in 9 of 15 patients in the larger trial.[70] No objective responses were seen in a phase I study of imatinib combined with dacarbazine and capecitabine that included 7 patients with MTC.[71] A phase II trial was recently reported on imatinib therapy, 400 mg twice daily, in patients with ATC found to overexpress either PDGFR or BCR-ABL.[72] Although 2 of 8 patients were reported with partial responses, and another 4 had stable disease, 6-month PFS and overall survival rates were still only 27 and 46%, respectively. However, the trial was stopped prematurely due to difficulty recruiting a sufficient number of patients.

20.3.8 Gefitinib

Gefitinib (ZD1839), an oral small-molecule inhibitor of the epidermal growth factor (EGF) receptor (EGFR), was initially introduced for therapy of non-small-cell lung carcinoma.[73,74] Because many PTCs and ATCs display activated EGFR signaling, and inhibitors have demonstrated efficacy in preclinical models, an open-label phase II study was initiated, examining the effectiveness of gefitinib in a mixed cohort of thyroid cancer patients.[75] The starting daily dose was 250 mg. Of 27 enrolled patients, 41% had PTC, 22% FTC, 19% ATC, and 15% MTC. There were no complete or partial responses in the 25 evaluable patients, although 8 had tumor reduction that did not qualify as partial response. One patient with ATC had stable disease beyond 12 months of therapy, similar to that reported in a phase I trial of gefitinib and docetaxel.[76] Overall, median PFS was just under 4 months, and under 3 months in the MTC cohort.

In non-small-cell lung carcinoma, the efficacy of anti-EGFR therapy is primarily seen in tumors bearing activating mutations in the kinase domains of the EGFR.[74] Generally, such mutations have not been reported in thyroid carcinomas despite a moderate frequency of ATC and poorly differentiated tumors expressing EGFR, perhaps underlying the overall lack of efficacy of gefitinib.[77] However, recently, a patient with ATC was reported whose tumor contained two distinct somatic point mutations or polymorphisms in the EGFR kinase.[78] After initial local control was achieved, she developed local recurrence and distant metastases. Treatment with the EGFR inhibitor erlotinib was initiated, titrating up to 150 mg orally every day, and marked regression of tumor was noted clinically and radiographically, both locally and distantly. Unfortunately, therapy was discontinued due to the high cost of the drug, and she died several months later.

20.3.9 XL 184

XL 184 is an oral small-molecule inhibitor of VEGFR-1 and -2, C-MET, RET, C-KIT, FLT3, and the angiopoietin receptor, Tie-2.[79] The inhibitory activity against C-MET, the cognate receptor for the hepatocyte growth factor, may provide additional synergistic benefit in thyroid carcinomas, given the enhanced expression of the receptor seen in PTC and MTC.[80–82] An ongoing phase I dose-escalation study has examined the safety and pharmacokinetics of XL 184 in patients with metastatic solid malignancies, with an expansion cohort limited to MTC.[83] Fifteen MTC patients (44%) had achieved at least 30% reduction in tumor measurements, with 10 (29%) having confirmed partial responses. No correlation was seen between *RET* mutation status (either germline or somatic) and tumor response. A phase III trial, comparing XL184 with placebo, is now under way.

20.4 OTHER APPROACHES TO TARGETING VASCULATURE AND ANGIOGENESIS

Beyond direct inhibitors of angiogenic kinases such as VEGFR, other drugs are capable of either inhibiting angiogenesis or disrupting existing tumor vasculature. Two of these agents, thalidomide and fosbretabulin (combretastatin A4 phosphate), have been of particular interest following reported responses in individual patients with ATC.

20.4.1 Thalidomide and Lenalidomide

Thalidomide was found to be an angiogenesis inhibitor decades after it achieved notoriety as a teratogenic cause of neonatal dysmelia.[84] However, the exact mechanism by which thalidomide exerts its antiangiogenic effects remains unknown. In the report that described the efficacy of paclitaxel for treatment of ATC, one patient who had progressed on the taxane was subsequently stabilized for at least 6 months while taking thalidomide.[17] Building upon this experience, an open-label phase II trial was initiated to examine the efficacy of thalidomide in patients with progressive, metastatic thyroid carcinoma of varying histologies.[85] Starting at 200 mg daily, the dose of drug was progressively increased as tolerated, with a median maximum daily dose of about 600 mg. Of 28 evaluable patients, 18% achieved a partial response and 32% had stable disease as their best response. Histology-specific partial response rates were not reported, but partial response or durable stable disease was seen in 3 PTC patients, 2 FTC patients, 3 HTC cancer patients, and 1 MTC patient, along with 4 patients with either tall-cell or insular variants of differentiated carcinoma. Toxicities were dose limiting in the majority of patients, and the most common adverse events included somnolence, peripheral neuropathy, constipation, dizziness, and infection. Given the suggested efficacy but high rate of adverse events with thalidomide, a subsequent phase II study was initiated using the presumably less toxic lenalidomide.[86] Eligibility was limited to DTC patients whose measured tumor volumes had increased by at least 30% in the past year. Of 18 evaluable patients, 7 (39%) were reported with partial responses measured by reductions in tumor volumes, with a median duration of 11 months, and another 9 (50%) were stable. However, median overall survival was less than 11 months, and 3 patients experienced pulmonary emboli.

20.4.2 Fosbretabulin

Fosbretabulin is a tubulin inhibitor whose dephosphorylated metabolite selectively inhibits proliferating endothelial cells in tumors.[87] Of four phase I studies that were performed, one patient with ATC was reported to have a complete response of more than 4 years' duration. Subsequently, in cell lines purported to derive from ATC, fosbretabulin demonstrated cytotoxicity comparable to paclitaxel, and the effects were additive in combination;[88,89] however, subsequent studies determined that these cell lines were not of thyroidal origin.[90] An open-label phase II trial of fosbretabulin in locally advanced or metastatic ATC enrolled 26 patients who received the drug intravenously, 45 mg/m^2, on days 1, 8, and 15 of every 28-day cycle.[91] Median survival was 4.7 months, and the 6-month survival was 34%; these results were probably comparable with those reported in the paclitaxel phase II trial.[17] Median duration of stable disease in 7 patients was about 1 year. Toxicities were frequent but generally not severe; they included lymphopenia, headache, tumor pain, and heart rate-corrected QT (QTc) interval prolongation. A randomized phase III trial is now under way, comparing the survival of patients treated with fosbretabulin in addition to paclitaxel and carboplatin with that of patients treated with paclitaxel and carboplatin alone.[92]

20.4.3 Celecoxib

Activation of cyclooxygenase-2 (COX-2), an enzyme overexpressed in many cancers, promotes tumor development and progression, in part through enhanced hypoxia-induced angiogenesis. Expression of COX-2 mRNA and protein levels is increased in thyroid cancer tissue compared with nonneoplastic and benign thyroid tissues, especially those expressing RET/PTC mutations, leading to the hypothesis that treatment with a COX-2 inhibitor could be therapeutically beneficial. A two-center phase II trial was performed testing this hypothesis in 32 patients with progressive differentiated thyroid carcinoma, identified radiographically or by rising serum thyroglobulin levels.[93] One patient had a partial response, and one remained stable on therapy for >12 months, but most patients progressed despite treatment. The study was terminated as a result of lack of efficacy combined with increasing concern about cardiovascular toxicity from COX-2 inhibitors.

20.5 TARGETING EPIGENETIC MECHANISMS

DNA hypermethylation and histone deacetylation are two common epigenetic mechanisms that have been implicated in the progression of thyroid carcinoma, particularly the loss of RAI avidity.[94,95] In the laboratory, treatment with DNA methylation inhibitors as well as histone deacetylase (HDAC) inhibitors has been associated with enhanced RAI uptake by nonavid cell lines, along with other markers, to suggest improved tumor cell differentiation and prompt clinical trials. Actual clinical experience, however, has been disappointing.

20.5.1 Romidepsin

The cyclic peptide romidepsin (previously known as depsipeptide) selectively inhibits four isotypes of HDAC.[96] In a variety of poorly differentiated and ATC cell lines, treatment with romidepsin led to expression of the sodium-iodide symporter, thyroglobulin, and thyroid-specific transcription factors, although tumor xenografts did not shrink.[97,98] A phase I dose-escalation trial included 9 patients with RAI-refractory thyroid cancer, of whom 6 had disease stabilization but none experienced restoration of RAI uptake on scanning.[99] Toxicities were primarily hematologic, nausea, and vomiting. Subsequently, a phase II trial was initiated in patients with RAI-unresponsive, progressive metastatic DTC.[100] Although the primary endpoint was RECIST response, restoration of RAI uptake was a secondary objective. Of 20 patients enrolled, no objective tumor responses were reported; 10 patients achieved stable disease. Two patients exhibited restoration of uptake permitting therapeutic RAI administration. Significant cardiopulmonary toxicities were seen, however, including sudden death in one patient and a grade 4 pulmonary embolus in another.

20.5.2 Vorinostat and Valproic Acid

The orally available HDAC inhibitor vorinostat, derived from hydroxamic acid, inhibits all known classes of HDAC enzymes. The drug is approved for the treatment of cutaneous T-cell lymphoma. For advanced thyroid cancer, vorinostat was studied in 16 patients.[101] No objective responses were reported, and most patients discontinued therapy due to progressive disease or adverse events, including fatigue, dehydration, ataxia, pneumonia, bruises, deep vein thrombosis, and severe thrombocytopenia. Restoration of RAI uptake was not evaluated.

Although only a weak inhibitor of several isotypes of HDAC enzymes, valproic acid has been the object of numerous preclinical studies in thyroid cancers, particularly ATC.[102] Although treatment with valproic acid alone can induce apoptosis in ATC cell lines, combinations with doxorubicin, paclitaxel, or imatinib may be significantly more potent.[103–105] An ongoing phase II trial is evaluating the effect of monotherapy with valproic acid on tumor size and RAI uptake in patients with RAI-refractory advanced DTC. Epigenetic synergy may also be expected in combination with DNA methylation inhibitors to block unregulated gene expression, as has been demonstrated in hematologic malignancies.[96] In a phase I trial combining valproic acid with 5-azacytidine, three patients with advanced thyroid cancer were among a total of 55 studied.[106] One PTC patient had prolonged stable disease beyond one year, but no objective responses were identified in any tumor type.

20.5.3 Azacytidine and Decitabine

A broad array of tumor suppressor genes are hypermethylated in PTC and FTC, leading to their decreased expression, including *PTEN*, tissue inhibitor of metalloproteinase-3, and death-associated protein kinase.[107] In various cell lines, re-expression of these genes and enhanced tumor cell differentiation has been seen following treatment with the DNA methylase inhibitor 5-azacytidine.[108,109] A phase II trial of 5-azacytidine monotherapy to restore RAI uptake was initiated, but results were never reported. Given the greater potency and tolerance of the azacytidine derivative decitabine, a phase II trial of this latter agent has been under way, evaluating the ability to restore RAI uptake in RAI non-avid metastases; results of this multicenter trial are expected shortly. One difficulty with these approaches to therapy, however, is that these agents depend upon active DNA synthesis to be capable of inhibiting the DNA methylase; in other words, they apparently do not demethylate existing hypermethylated sequences, which may limit their effectiveness in slowly replicating tumor cells like those found in most thyroid carcinomas.[110] Further research may identify approaches to combining these methylation inhibitors with other therapeutic pathways that could enhance their effectiveness.[106,111,112]

20.6 TARGETING NUCLEAR RECEPTORS

The possible role of retinoid receptors to regulate iodine uptake by thyroid follicular cells was suggested by studies demonstrating that incubation of poorly differentiated thyroid cancer cells with 13 cis-retinoic acid could partially restore RAI uptake.[113] Subsequent clinical trials yielded conflicting results.[114] Recently, a synthetic agonist of the retinoid X receptor (RXR), bexarotene, was tested in a phase II trial in patients with RAI-unresponsive metastatic disease.[115] After 6 weeks of therapy with bexarotene, 300 mg daily, RAI uptake was partially restored in 8 of 11 patients, but a clinical response with measurable tumor reduction was lacking. An ongoing phase II trial in patients with progressive metastatic PTC, FTC, or ATC is evaluating the tumor response, rather than RAI uptake, to bexarotene therapy.

The PPAR-γ agonist rosiglitazone was evaluated for the potential of restoring RAI uptake in 10 patients with unresponsive metastases.[116] In 4 patients, RAI uptake was visualized following 8 weeks of therapy with oral doses up to 8 mg daily, but clinical response was limited. The lack of major clinical effect of restoring RAI uptake may have multiple explanations, including the acquisition by tumor cells of radiation resistance.

20.7 TARGETING WITH IMMUNOTHERAPY

Following reports that interferon-α was active in the treatment of neuroendocrine malignancies, several attempts were made to define the role of interferon as an immunomodulatory therapy for thyroid carcinoma. One early study described 1 of 7 patients with marked tumor regression following monotherapy with interferon-α-2a.[117] Combining interferon-α-2b with the long-acting somatostatin analog lanreotide stabilized disease in 5 of 7 patients in a subsequent study, along with reduction in disease-related symptoms such as diarrhea and flushing, but no partial responses were reported.[118]

Given that interferon-α can induce a destructive thyroiditis and is synergistic in addition to doxorubicin in certain other solid tumors, a two-stage phase II trial evaluated the combination in patients with advanced or metastatic RAI-resistant thyroid carcinoma (other than medullary histologies).[119] In the first stage, 17 patients were treated with interferon-α-2b, 12 million units/m^2 administered subcutaneously daily for 5 days (days 1–5) of each cycle, and doxorubicin, 40 mg/m^2 intravenously administered on day 3, repeated every 28 days. Only one (6%) partial response was recorded, and 10 patients (62.5%) achieved stable disease. However, all patients eventually progressed on therapy, with median time to progression of 6 months. Nearly 3/4 of patients developed grade 3 or 4 neutropenia, and the most common other grade 3 toxicities were fatigue, nausea/vomiting, anorexia, mucositis, and neurologic symptoms. Given the low response rate and high toxicity profile, the protocol was terminated without extending to stage 2.

A novel approach to targeted immunotherapy has been the use of tumor vaccines. Dendritic cells, which are derived from bone marrow antigen-presenting cells, are capable of presenting tumor-associated antigens, thereby generating cytotoxic T-cells targeting tumor cells. This strategy has suggested efficacy in treating metastatic MTC in two recent trials. In one study, dendritic cells were obtained from each of 7 patients and stimulated in the presence of both CTN and carcinoembryonic antigen (CEA).[120] Following periodic intracutaneous injections of the stimulated dendritic cells, one patient experienced a partial response, including complete regression of hepatic metastases associated with a 70% reduction in serum tumor markers. Two other patients had mixed responses. In the second study, dendritic cells were stimulated using lysates of each individual patient's surgically resected primary tumor.[121] Three of ten patients had partial responses, including one with complete resolution of radiographic evidence of disease. Toxicities in both of these trials were minor, including low-grade fever and asymptomatic transient autoantibody development. Further small studies are under way, refining the procedures to enhance the potency of the dendritic cell vaccines.[122,123]

The expression of CEA on MTC cells has led to the exploitation of radiolabeled anti-CEA monoclonal antibodies for radioimmunotherapy. In early trials, antitumor effects have been noted using anti-CEA/anti-diethylenetriamine pentaacetic acid (DTPA)-indium recombinant bispecific antibody (BsMAb), followed 4 days later by a ^{131}I-labeled bivalent hapten. In a report of a nonrandomized trial in patients with progressive metastatic MTC (defined as a CTN doubling time of less than 2 years), median overall survival after administration of this therapy was 110 months.[124] This compared favorably with a contemporaneous untreated cohort's median survival of only 60 months. Significant toxicities included grade 4 neutropenia and thrombocytopenia, lasting up to 3 weeks, and one patient (who had received previous radiotherapies) developed myelodysplasia.

20.8 SUMMARY

The successful development of biologically targeted therapies for cancer requires several key factors: (1) identification of biologically validated targets critical to development and maintenance of the malignant phenotype; (2) development of potent inhibitors of the targets, with a broad therapeutic index separating efficacy from toxicity; (3) recognition of patient and tumor characteristics that can optimize the selection of patients for therapy; (4) identification of biomarkers that can predict patient outcome and permit optimization of drug dosing; and (5) recognition of opportunities for well-tolerated and more efficacious combinatorial treatments. As summarized in this review, such advances have been made in the past few years in the development of successful targeted therapies for thyroid cancers.

Compared with the dismal historical track record, the recent proliferation of clinical trials for thyroid cancer has been remarkable. Targeting angiogenesis (and specifically VEGFR) has produced the most impressive clinical responses to date in both DTC and MTC. Although most small molecule VEGFR antagonists also inhibit RET, the efficacy of axitinib and pazopanib to induce objective responses in the absence of any anti-RET activity suggests that RET may not be as important a target for therapy as VEGFR. Unfortunately, eventual progression despite antiangiogenic VEGFR blockade suggests emergence of alternate pathways to promote tumor growth and metastasis (including FGFR, C-MET, and angiopoietins).[125] Further studies are necessary to explore the value of effective inhibition of the MAPK pathway downstream from oncogenic mutations, as well as other pathways stimulating tumor growth and metabolism, such as PI3K-AKT-mTOR signaling. Studies of therapies targeting nuclear mechanisms of gene regulation indicate that reversal of epigenetic or nuclear receptor abnormalities can potentially reestablish the cellular capacity to take up RAI, but the clinical significance of such an effect appears limited. Immunotherapy, particularly dendritic cell vaccines, appears to be a very promising approach.

The overall goal of developing new treatments is to extend the duration of life without unduly harming the quality of that life. Presently, no novel treatment has yet demonstrated improved survival for thyroid cancer patients. Toxicities of many of these new therapies, although less life threatening than cytotoxic chemotherapies, are common and can be dose limiting, and clinicians must be familiar with recognizing and managing the side effects if they intend to use these agents. Finally, the low rate of partial responses, the absence of complete responses, and the emergence of resistance in all of the various monotherapy trials conducted warrant the need either to develop more effective single agents or to construct rational combinations of therapeutic targets (including cytotoxic chemotherapies) that have synergistic effectiveness without enhanced cross-toxicities.

REFERENCES

1. Haugen BR. 1999. Management of the patient with progressive radioiodine non-responsive disease. *Semin Surg Oncol* 16:34–41.
2. Sherman SI. 2006. Clinical trials for thyroid carcinoma: Past, present, and future. In *Practical management of thyroid cancer: A multidisciplinary approach*, ed. EL Mazzaferri, C Harmer, UK Mallick, P Kendall-Taylor, 429–34. London: Springer-Verlag.
3. Therasse P, Arbuck S, Eisenhauer E, Wanders J, Kaplan R, Rubinstein L, Verweij J, Van Glabbeke M, van Oosterom A, Christian M, Gwyther S. 2000. New guidelines to evaluate the response to treatment in solid tumors. European Organization for Research and Treatment of Cancer, National Cancer Institute of the United States, National Cancer Institute of Canada. *J Natl Cancer Inst* 92:205–16.
4. Eisenhauer EA, Therasse P, Bogaerts J, Schwartz LH, Sargent D, Ford R, Dancey J, Arbuck S, Gwyther S, Mooney M, Rubinstein L, Shankar L, Dodd L, Kaplan R, Lacombe D, Verweij J. 2009. New response evaluation criteria in solid tumours: Revised RECIST guideline (version 1.1). *Eur J Cancer* 45:228–47.
5. Fagin JA. 2004. How thyroid tumors start and why it matters: Kinase mutants as targets for solid cancer pharmacotherapy. *J Endocrinol* 183:249–56.
6. Carmeliet P. 2000. Mechanisms of angiogenesis and arteriogenesis. *Nat Med* 6:389–95.
7. Ferrara N, Kerbel RS. 2005. Angiogenesis as a therapeutic target. *Nature* 438:967–74.
8. Klein M, Vignaud JM, Hennequin V, Toussaint B, Bresler L, Plenat F, Leclere J, Duprez A, Weryha G. 2001. Increased expression of the vascular endothelial growth factor is a pejorative prognosis marker in papillary thyroid carcinoma. *J Clin Endocrinol Metab* 86:656–58.
9. Lennard CM, Patel A, Wilson J, Reinhardt B, Tuman C, Fenton C, Blair E, Francis GL, Tuttle RM. 2001. Intensity of vascular endothelial growth factor expression is associated with increased risk of recurrence and decreased disease-free survival in papillary thyroid cancer. *Surgery* 129:552–58.
10. Jo YS, Li S, Song JH, Kwon KH, Lee JC, Rha SY, Lee HJ, Sul JY, Kweon GR, Ro HK, Kim JM, Shong M. 2006. Influence of the BRAF V600E mutation on expression of vascular endothelial growth factor in papillary thyroid cancer. *J Clin Endocrinol Metab* 91:3667–70.
11. Tsimberidou AM, Vaklavas C, Wen S, Hong D, Wheler J, Ng C, Naing A, Tse S, Busaidy N, Markman M, Sherman SI, Kurzrock R. 2009. Phase I clinical trials in 56 patients with thyroid cancer: The M. D. Anderson Cancer Center experience. *J Clin Endocrinol Metab* 94:4423–32.
12. Sherman SI. 2009. NCCN practice guidelines for thyroid cancer, version 2009a. www.nccn.org (accessed June 1, 2009).

13. Cooper DS, Doherty GM, Haugen BR, Kloos RT, Lee SL, Mandel SJ, Mazzaferri EL, McIver B, Sherman SI, Tuttle RM. 2006. Management guidelines for patients with thyroid nodules and differentiated thyroid cancer. *Thyroid* 16:109–42.
14. Pfister DG, Fagin JA. 2008. Refractory thyroid cancer: A paradigm shift in treatment is not far off. *J Clin Oncol* 26:4701–4.
15. Laird AD, Cherrington JM. 2003. Small molecule tyrosine kinase inhibitors: Clinical development of anticancer agents. *Expert Opin Investig Drugs* 12:51–64.
16. Ball DW. 2007. Medullary thyroid cancer: Therapeutic targets and molecular markers. *Curr Opin Oncol* 19:18–23.
17. Ain KB, Egorin MJ, DeSimone PA. 2000. Treatment of anaplastic thyroid carcinoma with paclitaxel: Phase 2 trial using ninety-six-hour infusion. *Thyroid* 10:587–94.
18. Knauf JA, Fagin JA. 2009. Role of MAPK pathway oncoproteins in thyroid cancer pathogenesis and as drug targets. *Curr Opin Cell Biol* 21:296–303.
19. Santarpia L, Ye L, Gagel RF. 2009. Beyond RET: Potential therapeutic approaches for advanced and metastatic medullary thyroid carcinoma. *J Intern Med* 266:99–113.
20. Sleijfer S, Wiemer E, Verweij J. 2008. Drug Insight: Gastrointestinal stromal tumors (GIST)—The solid tumor model for cancer-specific treatment. *Nat Clin Pract Oncol* 5:102–11.
21. Sala E, Mologni L, Truffa S, Gaetano C, Bollag GE, Gambacorti-Passerini C. 2008. BRAF silencing by short hairpin RNA or chemical blockade by PLX4032 leads to different responses in melanoma and thyroid carcinoma cells. *Mol Cancer Res* 6:751–59.
22. Flaherty K, Puzanov I, Sosman J, Kim K, Ribas A, McArthur G, Lee R, Grippo J, Nolop K, Chapman P. 2009. Phase I study of PLX4032: Proof of concept for V600E BRAF mutation as a therapeutic target in human cancer (Meeting abstract). *J Clin Oncol* 27:9000.
23. Schwartz GK, Robertson S, Shen A, Wang E, Pace L, Dials H, Mendelson D, Shannon P, Gordon M. 2009. A phase I study of XL281, a selective oral RAF kinase inhibitor, in patients (Pts) with advanced solid tumors (Meeting abstract). *J Clin Oncol* 27:3513.
24. Polverino A, Coxon A, Starnes C, Diaz Z, DeMelfi T, Wang L, Bready J, Estrada J, Cattley R, Kaufman S, Chen D, Gan Y, Kumar G, Meyer J, Neervannan S, Alva G, Talvenheimo J, Montestruque S, Tasker A, Patel V, Radinsky R, Kendall R. 2006. AMG 706, an oral, multikinase inhibitor that selectively targets vascular endothelial growth factor, platelet-derived growth factor, and kit receptors, potently inhibits angiogenesis and induces regression in tumor xenografts. *Cancer Res* 66:8715–21.
25. Coxon A, Bready J, Fiorino M, Hughes P, Wang L, DeMelfi T, Doerr N, Radinsky R, Kendall R, Polverino T. 2006. Anti-tumor activity of AMG 706, an oral multi-kinase inhibitor, in human medullary thyroid carcinoma xenografts. *Thyroid* 16:920.
26. Rosen LS, Kurzrock R, Mulay M, Van Vugt A, Purdom M, Ng C, Silverman J, Koutsoukos A, Sun YN, Bass MB, Xu RY, Polverino A, Wiezorek JS, Chang DD, Benjamin R, Herbst RS. 2007. Safety, pharmacokinetics, and efficacy of AMG 706, an oral multikinase inhibitor, in patients with advanced solid tumors. *J Clin Oncol* 25:2369–76.
27. Sherman SI, Wirth LJ, Droz JP, Hofmann M, Bastholt L, Martins RG, Licitra L, Eschenberg MJ, Sun YN, Juan T, Stepan DE, Schlumberger MJ. 2008. Motesanib diphosphate in progressive differentiated thyroid cancer. *N Engl J Med* 359:31–42.
28. Schlumberger M, Elisei R, Bastholt L, Wirth LJ, Martins RG, Locati L, Jarzab B, Pacini F, Daumerie C, Droz J-P, Eschenberg MJ, Sun Y-N, Juan T, Stepan DE, Sherman SI. 2009. Phase II study of safety and efficacy of motesanib (AMG 706) in patients with progressive or symptomatic, advanced or metastatic medullary thyroid cancer. *J Clin Oncol* 27:3794–801.
29. Xing M. 2007. BRAF mutation in papillary thyroid cancer: Pathogenic role, molecular bases, and clinical implications. *Endocr Rev* 28:742–62.
30. Sherman SI, Schlumberger MJ, Elisei R, Wirth L, Bastholt L, Droz J-P, Martins RG, Hofmann M, Locati L, Pacini F, Eschenberg MJ, Stepan DE. 2007. Exacerbation of postsurgical hypothyroidism during treatment of thyroid carcinoma with motesanib diphosphate (AMG 706). Paper presented at 89th Annual Meeting of the Endocrine Society, Toronto, ON.
31. Wilhelm SM, Carter C, Tang L, Wilkie D, McNabola A, Rong H, Chen C, Zhang X, Vincent P, McHugh M, Cao Y, Shujath J, Gawlak S, Eveleigh D, Rowley B, Liu L, Adnane L, Lynch M, Auclair D, Taylor I, Gedrich R, Voznesensky A, Riedl B, Post LE, Bollag G, Trail PA. 2004. BAY 43-9006 exhibits broad spectrum oral antitumor activity and targets the RAF/MEK/ERK pathway and receptor tyrosine kinases involved in tumor progression and angiogenesis. *Cancer Res* 64:7099–109.
32. Carlomagno F, Anaganti S, Guida T, Salvatore G, Troncone G, Wilhelm SM, Santoro M. 2006. BAY 43-9006 inhibition of oncogenic RET mutants. *J Natl Cancer Inst* 98:326–34.

33. Strumberg D, Clark JW, Awada A, Moore MJ, Richly H, Hendlisz A, Hirte HW, Eder JP, Lenz H-J, Schwartz B. 2007. Safety, pharmacokinetics, and preliminary antitumor activity of sorafenib: A review of four phase I trials in patients with advanced refractory solid tumors. *Oncologist* 12:426–37.
34. Dubauskas Z, Kunishige J, Prieto V, Jonasch E, Hwu P, Tannir N. 2009. Cutaneous squamous cell carcinoma and inflammation of actinic keratoses associated with sorafenib. *Clin Genitourin Cancer* 7:20–23.
35. Ratain MJ, Eisen T, Stadler WM, Flaherty KT, Kaye SB, Rosner GL, Gore M, Desai AA, Patnaik A, Xiong HQ, Rowinsky E, Abbruzzese JL, Xia C, Simantov R, Schwartz B, O'Dwyer PJ. 2006. Phase II placebo-controlled randomized discontinuation trial of sorafenib in patients with metastatic renal cell carcinoma. *J Clin Oncol* 24:2505–12.
36. Kloos R, Ringel M, Knopp M, Heverhagen J, Rittenberry J, Weldy L, Arbogast D, Collamore M, King M, Young D, Shah M. 2006. Significant clinical and biologic activity of RAF/VEGF-R kinase inhibitor BAY 43-9006 in patients with metastatic papillary thyroid carcinoma (PTC): Updated results of a phase II study. *J Clin Oncol* 24:5534.
37. Gupta-Abramson V, Troxel AB, Nellore A, Puttaswamy K, Redlinger M, Ransone K, Mandel SJ, Flaherty KT, Loevner LA, O'Dwyer PJ, Brose MS. 2008. Phase II trial of sorafenib in advanced thyroid cancer. *J Clin Oncol* 26:4714–19.
38. Brose MS, Troxel AB, Redlinger M, Harlacker K, Redlinger C, Chalian AA, Flaherty KT, Loevner LA, Mandel SJ, O'Dwyer PJ. 2009. Effect of $BRAF^{V600E}$ on response to sorafenib in advanced thyroid cancer patients (Meeting abstract). *J Clin Oncol* 27:6002.
39. Cabanillas ME, Kurzrock R, Bidyasar S, Wheeler J, Fu S, Naing A, Sherman SI, Gagel R, Waguespack S, Busaidy N, Sellin R, Hu M, Kies M, Hong DS. 2009. Phase I trial of a combination of sorafenib and tipifarnib: Update on the experience in advanced thyroid malignancies (NCI Protocol 7156). In *World Congress on Thyroid Cancer 2009*, Toronto.
40. Kober F, Hermann M, Handler A, Krotla G. 2007. Effect of sorafenib in symptomatic metastatic medullary thyroid cancer. *J Clin Oncol* 25:14065.
41. Salvatore G, De Falco V, Salerno P, Nappi TC, Pepe S, Troncone G, Carlomagno F, Melillo RM, Wilhelm SM, Santoro M. 2006. BRAF is a therapeutic target in aggressive thyroid carcinoma. *Clin Cancer Res* 12:1623–29.
42. Nagaiah G, Fu P, Wasman JK, Cooney MM, Mooney C, Afshin D, Lavertu P, Bokar J, Savvides P, Remick SC. 2009. Phase II trial of sorafenib (BAY 43-9006) in patients with advanced anaplastic carcinoma of the thyroid (ATC). *J Clin Oncol* 27:6058.
43. Cabanillas ME, Waguespack SG, Bronstein Y, Williams M, Feng L, Sherman SI, Busaidy NL. 2009. Treatment (tx) with tyrosine kinase inhibitors (TKIs) for patients (pts) with differentiated thyroid cancer (DTC): The M. D. Anderson Cancer Center (MDACC) experience (Meeting abstract). *J Clin Oncol* 27:6060.
44. Waguespack SG, Williams MD, Clayman GL, Herzog CE. 2007. First successful use of sorafenib to treat pediatric papillary thyroid carcinoma (PTC). Paper presented at Pediatric Academic Societies Annual Meeting, Toronto, ON.
45. Kim DW, Jo YS, Jung HS, Chung HK, Song JH, Park KC, Park SH, Hwang JH, Rha SY, Kweon GR, Lee S-J, Jo K-W, Shong M. 2006. An orally administered multitarget tyrosine kinase inhibitor, SU11248, is a novel potent inhibitor of thyroid oncogenic RET/papillary thyroid cancer kinases. *J Clin Endocrinol Metab* 91:4070–76.
46. Kelleher FC, McDermott R. 2008. Response to sunitinib in medullary thyroid cancer. *Ann Intern Med* 148:567.
47. Dawson SJ, Conus NM, Toner GC, Raleigh JM, Hicks RJ, McArthur G, Rischin D. 2008. Sustained clinical responses to tyrosine kinase inhibitor sunitinib in thyroid carcinoma. *Anticancer Drugs* 19:547–52.
48. Cohen EEW, Needles BM, Cullen KJ, Wong S, Wade J, Ivy P, Villaflor V, Seiwert T, Nichols K, Vokes EE. 2008. Phase 2 study of sunitinib in refractory thyroid cancer (Meeting abstract). *J Clin Oncol* 26:6025.
49. Ravaud A, de la Fouchardière C, Courbon F, Asselineau J, Klein M, Nicoli-Sire P, Bournaud C, Delord J, Weryha G, Catargi B. Sunitinib in patients with refractory advanced thyroid cancer: The THYSU phase II trial. http://www.asco.org/ASCO/Abstracts+%26+Virtual+Meeting/Abstracts?&vmview=abst_detail_view&confID=55&abstractID=33144 (accessed December 15, 2008).
50. Carr L, Goulart B, Martins R, Keith E, Kell E, Wallace S, Capell P, Mankoff D. 2009. Phase II trial of continuous dosing of sunitinib in advanced, FDG-PET avid, medullary thyroid carcinoma (MTC) and well-differentiated thyroid cancer (WDTC) (Meeting abstract). *J Clin Oncol* 27:6056.
51. Herbst RS, Heymach JV, O'Reilly MS, Onn A, Ryan AJ. 2007. Vandetanib (ZD6474): An orally available receptor tyrosine kinase inhibitor that selectively targets pathways critical for tumor growth and angiogenesis. *Expert Opin Investig Drugs* 16:239–49.

52. Wedge SR, Ogilvie DJ, Dukes M, Kendrew J, Chester R, Jackson JA, Boffey SJ, Valentine PJ, Curwen JO, Musgrove HL, Graham GA, Hughes GD, Thomas AP, Stokes ES, Curry B, Richmond GH, Wadsworth PF, Bigley AL, Hennequin LF. 2002. ZD6474 inhibits vascular endothelial growth factor signaling, angiogenesis, and tumor growth following oral administration. *Cancer Res* 62:4645–55.
53. Carlomagno F, Vitagliano D, Guida T, Ciardiello F, Tortora G, Vecchio G, Ryan AJ, Fontanini G, Fusco A, Santoro M. 2002. ZD6474, an orally available inhibitor of KDR tyrosine kinase activity, efficiently blocks oncogenic RET kinases. *Cancer Res* 62:7284–90.
54. Carlomagno F, Guida T, Anaganti S, Vecchio G, Fusco A, Ryan AJ, Billaud M, Santoro M. 2004. Disease associated mutations at valine 804 in the RET receptor tyrosine kinase confer resistance to selective kinase inhibitors. *Oncogene* 23:6056–63.
55. Holden SN, Eckhardt SG, Basser R, de Boer R, Rischin D, Green M, Rosenthal MA, Wheeler C, Barge A, Hurwitz HI. 2005. Clinical evaluation of ZD6474, an orally active inhibitor of VEGF and EGF receptor signaling, in patients with solid, malignant tumors. *Ann Oncol* 16:1391–97.
56. Wells SA, Gosnell JE, Gagel RF, Moley J, Pfister D, Sosa JA, Skinner M, Krebs A, Vasselli J, Schlumberger M. 2010. Vandetanib for the treatment of patients with locally advanced or metastatic hereditary medullary thyroid cancer. *J Clin Oncol* 28:767–72.
57. Akeno-Stuart N, Croyle M, Knauf JA, Malaguarnera R, Vitagliano D, Santoro M, Stephan C, Grosios K, Wartmann M, Cozens R, Caravatti G, Fabbro D, Lane HA, Fagin JA. 2007. The RET kinase inhibitor NVP-AST487 blocks growth and calcitonin gene expression through distinct mechanisms in medullary thyroid cancer cells. *Cancer Res* 67:6956–64.
58. Haddad RI, Krebs AD, Vasselli J, Paz-Ares LG, Robinson B. 2008. A phase II open-label study of vandetanib in patients with locally advanced or metastatic hereditary medullary thyroid cancer. *J Clin Oncol* 26:6024.
59. Fox E, Widemann BC, Whitcomb PO, Aikin A, Dombi E, Lodish M, Stratakis CA, Steinberg S, Wells Jr SA, Balis FM. 2009. Phase I/II trial of vandetanib in children and adolescents with hereditary medullary thyroid carcinoma. *J Clin Oncol* 27:10014.
60. Mitsiades CS, McMillin D, Kotoula V, Poulaki V, McMullan C, Negri J, Fanourakis G, Tseleni-Balafouta S, Ain KB, Mitsiades N. 2006. Antitumor effects of the proteasome inhibitor bortezomib in medullary and anaplastic thyroid carcinoma cells *in vitro*. *J Clin Endocrinol Metab* 91:4013–21.
61. Inai T, Mancuso M, Hashizume H, Baffert F, Haskell A, Baluk P, Hu-Lowe DD, Shalinsky DR, Thurston G, Yancopoulos GD, McDonald DM. 2004. Inhibition of vascular endothelial growth factor (VEGF) signaling in cancer causes loss of endothelial fenestrations, regression of tumor vessels, and appearance of basement membrane ghosts. *Am J Pathol* 165:35–52.
62. Rugo HS, Herbst RS, Liu G, Park JW, Kies MS, Steinfeldt HM, Pithavala YK, Reich SD, Freddo JL, Wilding G. 2005. Phase I trial of the oral antiangiogenesis agent AG-013736 in patients with advanced solid tumors: Pharmacokinetic and clinical results. *J Clin Oncol* 23:5474–83.
63. Cohen EE, Rosen LS, Vokes EE, Kies MS, Forastiere AA, Worden FP, Kane MA, Sherman E, Kim S, Bycott P, Tortorici M, Shalinsky DR, Liau KF, Cohen RB. 2008. Axitinib is an active treatment for all histologic subtypes of advanced thyroid cancer: Results from a phase II study. *J Clin Oncol* 26:4708–13.
64. Bocci G, Man S, Green SK, Francia G, Ebos JM, du Manoir JM, Weinerman A, Emmenegger U, Ma L, Thorpe P, Davidoff A, Huber J, Hicklin DJ, Kerbel RS. 2004. Increased plasma vascular endothelial growth factor (VEGF) as a surrogate marker for optimal therapeutic dosing of VEGF receptor-2 monoclonal antibodies. *Cancer Res* 64:6616–25.
65. Bible KC, Smallridge RC, Maples WJ, Molina JR, Menefee ME, Suman VJ, Burton CC, Ivy SP, Erichman C. 2009. Phase II trial of pazopanib in progressive, metastatic, iodine-insensitive differentiated thyroid cancers (Meeting abstract). *J Clin Oncol* 27:3521.
66. de Groot JW, Plaza Menacho I, Schepers H, Drenth-Diephuis LJ, Osinga J, Plukker JT, Links TP, Eggen BJ, Hofstra RM. 2006. Cellular effects of imatinib on medullary thyroid cancer cells harboring multiple endocrine neoplasia type 2A and 2B associated RET mutations. *Surgery* 139:806–14.
67. Skinner MA, Safford SD, Freemerman AJ. 2003. RET tyrosine kinase and medullary thyroid cells are unaffected by clinical doses of STI571. *Anticancer Res* 23:3601–6.
68. Buchdunger E, O'Reilley T, Wood J. 2002. Pharmacology of imatinib (STI571). *Eur J Cancer* 38:S28–36.
69. Frank-Raue K, Fabel M, Delorme S, Haberkorn U, Raue F. 2007. Efficacy of imatinib mesylate in advanced medullary thyroid carcinoma. *Eur J Endocrinol* 157:215–20.
70. de Groot JW, Zonnenberg BA, Quarles van Ufford-Mannesse P, de Vries MM, Links TP, Lips CJ, Voest EE. 2007. A phase-II trial of imatinib therapy for metastatic medullary thyroid carcinoma. *J Clin Endocrinol Metab* 92:3466–69.

71. Hoff PM, Hoff AO, Phan AT, Sherman SI, Yao J, White N, Phan L, Abbruzzese JL, Gagel RF. 2006. Phase I/II trial of capecitabine (C), dacarbazine (D) and imatinib (I) (CDI) for patients (pts) metastatic medullary thyroid carcinomas (MTC). *J Clin Oncol* 24:13048.
72. Ha HT, Lee JS, Urba S, Koenig RJ, Sisson J, Giordano T, Worden FP. 2009. Phase II trial evaluating imatinib (I) in patients (pts) with anaplastic thyroid carcinoma (ATC) (Meeting abstract). *J Clin Oncol* 27:6057.
73. Wakeling AE, Guy SP, Woodburn JR, Ashton SE, Curry BJ, Barker AJ, Gibson KH. 2002. ZD1839 (Iressa): An orally active inhibitor of epidermal growth factor signaling with potential for cancer therapy. *Cancer Res* 62:5749–54.
74. Lynch TJ, Bell DW, Sordella R, Gurubhagavatula S, Okimoto RA, Brannigan BW, Harris PL, Haserlat SM, Supko JG, Haluska FG, Louis DN, Christiani DC, Settleman J, Haber DA. 2004. Activating mutations in the epidermal growth factor receptor underlying responsiveness of non-small-cell lung cancer to gefitinib. *N Engl J Med* 350:2129–39.
75. Pennell NA, Daniels GH, Haddad RI, Ross DS, Evans T, Wirth LJ, Fidias PH, Temel JS, Gurubhagavatula S, Heist RS, Clark JR, Lynch TJ. 2007. A phase II study of gefitinib in patients with advanced thyroid cancer. *Thyroid* 18:317–23.
76. Fury MG, Solit DB, Su YB, Rosen N, Sirotnak FM, Smith RP, Azzoli CG, Gomez JE, Miller VA, Kris MG, Pizzo BA, Henry R, Pfister DG, Rizvi NA. 2007. A phase I trial of intermittent high-dose gefitinib and fixed-dose docetaxel in patients with advanced solid tumors. *Cancer Chemother Pharmacol* 59:467–75.
77. Elliott DD, Sherman SI, Busaidy NL, Williams MD, Santarpia L, Clayman GL, El-Naggar AK. 2008. Growth factor receptors expression in anaplastic thyroid carcinoma: Potential markers for therapeutic stratification. *Hum Pathol* 39:15–20.
78. Hogan T, Jing Jie Y, Williams HJ, Altaha R, Xiaobing L, Qi H. 2009. Oncocytic, focally anaplastic, thyroid cancer responding to erlotinib. *J Oncol Pharm Pract* 15:111–17.
79. Cui JJ. 2007. Inhibitors targeting hepatocyte growth factor receptor and their potential therapeutic applications. *Expert Opin Ther Pat* 17:1035–45.
80. Mineo R, Costantino A, Frasca F, Sciacca L, Russo S, Vigneri R, Belfiore A. 2004. Activation of the hepatocyte growth factor (HGF)-Met system in papillary thyroid cancer: Biological effects of HGF in thyroid cancer cells depend on Met expression levels. *Endocrinology* 145:4355–65.
81. Wasenius VM, Hemmer S, Karjalainen-Lindsberg ML, Nupponen NN, Franssila K, Joensuu H. 2005. MET receptor tyrosine kinase sequence alterations in differentiated thyroid carcinoma. *Am J Surg Pathol* 29:544–49.
82. Papotti M, Olivero M, Volante M, Negro F, Prat M, Comoglio PM, DiRenzo MF. 2000. Expression of hepatocyte growth factor (HGF) and its receptor (MET) in medullary carcinoma of the thyroid. *Endocr Pathol* 11:19–30.
83. Kurzrock R, Sherman S, Pfister D, Cohen RB, Ball D, Hong D, Ng C, Frye J, Janisch L, Ratain MJ, Salgia R. 2009. Preliminary results of a phase I study of XL184, a MET, VEGFR2 and RET kinase inhibitor (TKI), administered orally to patients with medullary thyroid cancer (MTC). Paper presented at 34th Annual Meeting of the European Thyroid Association, Lisbon.
84. D'Amato RJ, Loughnan MS, Flynn E, Folkman J. 1994. Thalidomide is an inhibitor of angiogenesis. *Proc Natl Acad Sci USA* 91:4082–85.
85. Ain KB, Lee C, Williams KD. 2007. Phase II trial of thalidomide for therapy of radioiodine-unresponsive and rapidly progressive thyroid carcinomas. *Thyroid* 17:663–70.
86. Ain KB, Lee C, Holbrook KM, Dziba JM, Williams KD. Phase II study of lenalidomide in distantly metastatic, rapidly progressive, and radioiodine-unresponsive thyroid carcinomas: Preliminary results. http://www.asco.org/ASCO/Abstracts+&+Virtual+Meeting/Abstracts?&vmview=abst_detail_view&confID=55&abstractID=30832 (accessed February 2, 2009).
87. Cooney MM, Ortiz J, Bukowski RM, Remick SC. 2005. Novel vascular targeting/disrupting agents: Combretastatin A4 phosphate and related compounds. *Curr Oncol Rep* 7:90–95.
88. Yeung SC, She M, Yang H, Pan J, Sun L, Chaplin D. 2007. Combination chemotherapy including combretastatin A4 phosphate and paclitaxel is effective against anaplastic thyroid cancer in a nude mouse xenograft model. *J Clin Endocrinol Metab* 92:2902–9.
89. Dziba JM, Marcinek R, Venkataraman G, Robinson JA, Ain KB. 2002. Combretastatin a4 phosphate has primary antineoplastic activity against human anaplastic thyroid carcinoma cell lines and xenograft tumors. *Thyroid* 12:1063–70.
90. Schweppe RE, Klopper JP, Korch C, Pugazhenthi U, Benezra M, Knauf JA, Fagin JA, Marlow L, Copland JA, Smallridge RC, Haugen BR. 2008. Deoxyribonucleic acid profiling analysis of 40 human thyroid cancer cell lines reveals cross-contamination resulting in cell line redundancy and misidentification. *J Clin Endocrinol Metab* 93:4331–41.

91. Mooney CJ, Nagaiah G, Fu P, Wasman JK, Cooney MM, Savvides PS, Bokar JA, Dowlati A, Wang D, Agarwala SS, Flick SM, Hartman PH, Ortiz JD, Lavertu PN, Remick SC. 2009. A phase II trial of fosbretabulin in advanced anaplastic thyroid carcinoma and correlation of baseline serum-soluble intracellular adhesion molecule-1 with outcome. *Thyroid* 19:233–40.
92. ClinicalTrials.gov. www.clinicaltrials.gov/ct2/home (accessed January 11, 2008).
93. Mrozek E, Kloos RT, Ringel MD, Kresty L, Snider P, Arbogast D, Kies M, Munden R, Busaidy N, Klein MJ, Sherman SI, Shah MH. 2006. Phase II study of celecoxib in metastatic differentiated thyroid carcinoma. *J Clin Endocrinol Metab* 91:2201–4.
94. Kondo T, Asa SL, Ezzat S. 2008. Epigenetic dysregulation in thyroid neoplasia. *Endocrinol Metab Clin North Am* 37:389–400.
95. Haugen BR. 2004. Redifferentiation therapy in advanced thyroid cancer. *Curr Drug Targets Immune Endocr Metab Disord* 4:175–80.
96. Batty N, Malouf GG, Issa JPJ. 2009. Histone deacetylase inhibitors as anti-neoplastic agents. *Cancer Lett* 280:192–200.
97. Kitazono M, Robey R, Zhan Z, Sarlis NJ, Skarulis MC, Aikou T, Bates S, Fojo T. 2001. Low concentrations of the histone deacetylase inhibitor, depsipeptide (FR901228), increase expression of the Na(+)/I(–) symporter and iodine accumulation in poorly differentiated thyroid carcinoma cells. *J Clin Endocrinol Metab* 86:3430–35.
98. Furuya F, Shimura H, Suzuki H, Taki K, Ohta K, Haraguchi K, Onaya T, Endo T, Kobayashi T. 2004. Histone deacetylase inhibitors restore radioiodide uptake and retention in poorly differentiated and anaplastic thyroid cancer cells by expression of the sodium/iodide symporter thyroperoxidase and thyroglobulin. *Endocrinology* 145:2865–75.
99. Piekarz R, Luchenko V, Draper D, Wright JJ, Figg WD, Fojo AT, Bates SE. 2008. Phase I trial of romidepsin, a histone deacetylase inhibitor, given on days one, three and five in patients with thyroid and other advanced cancers. *J Clin Oncol* 28:3571.
100. Sherman EJ, Fury MG, Tuttle RM, Ghossein R, Stambuk H, Baum M, Lisa D, Su YB, Shaha A, Pfister DG. 2009. Phase II study of depsipeptide (DEP) in radioiodine (RAI)-refractory metastatic nonmedullary thyroid carcinoma (Meeting abstract). *J Clin Oncol* 27:6059.
101. Woyach JA, Kloos RT, Ringel MD, Arbogast D, Collamore M, Zwiebel JA, Grever M, Villalona-Calero M, Shah MH. 2009. Lack of therapeutic effect of the histone deacetylase inhibitor vorinostat in patients with metastatic radioiodine-refractory thyroid carcinoma. *J Clin Endocrinol Metab* 94:164–70.
102. Smallridge RC, Marlow LA, Copland JA. 2009. Anaplastic thyroid cancer: Molecular pathogenesis and emerging therapies. *Endocr Relat Cancer* 16:17–44.
103. Catalano MG, Poli R, Pugliese M, Fortunati N, Boccuzzi G. 2007. Valproic acid enhances tubulin acetylation and apoptotic activity of paclitaxel on anaplastic thyroid cancer cell lines. *Endocr Relat Cancer* 14:839–45.
104. Catalano MG, Pugliese M, Poli R, Bosco O, Bertieri R, Fortunati N, Boccuzzi G. 2009. Effects of the histone deacetylase inhibitor valproic acid on the sensitivity of anaplastic thyroid cancer cell lines to imatinib. *Oncol Rep* 21:515–21.
105. Kim TH, Yoo YH, Kang DY, Suh H, Park MK, Park KJ, Kim SH. 2009. Efficacy on anaplastic thyroid carcinoma of valproic acid alone or in combination with doxorubicin, a synthetic chenodeoxycholic acid derivative, or lactacystin. *Int J Oncol* 34:1353–62.
106. Braiteh F, Soriano AO, Garcia-Manero G, Hong D, Johnson MM, Silva Lde P, Yang H, Alexander S, Wolff J, Kurzrock R. 2008. Phase I study of epigenetic modulation with 5-azacytidine and valproic acid in patients with advanced cancers. *Clin Cancer Res* 14:6296–301.
107. Xing M. 2007. Gene methylation in thyroid tumorigenesis. *Endocrinology* 148:948–53.
108. Tuncel M, Aydin D, Yaman E, Tazebay UH, Guc D, Dogan AL, Tasbasan B, Ugur O. 2007. The comparative effects of gene modulators on thyroid-specific genes and radioiodine uptake. *Cancer Biother Radiopharm* 22:443–49.
109. Venkataraman GM, Yatin M, Marcinek R, Ain KB. 1999. Restoration of iodide uptake in dedifferentiated thyroid carcinoma: Relationship to human Na+I– symporter gene methylation status. *J Clin Endocrinol Metab* 84:2449–57.
110. Tallini G, Garcia-Rostan G, Herrero A, Zelterman D, Viale G, Bosari S, Carcangiu ML. 1999. Downregulation of p27KIP1 and Ki67/Mib1 labeling index support the classification of thyroid carcinoma into prognostically relevant categories. *Am J Surg Pathol* 23:678–85.
111. Provenzano MJ, Fitzgerald MP, Krager K, Domann FE. 2007. Increased iodine uptake in thyroid carcinoma after treatment with sodium butyrate and decitabine (5-Aza-dC). *Otolaryngol Head Neck Surg* 137:722–28.

112. Li W, Venkataraman GM, Ain KB. 2007. Protein synthesis inhibitors, in synergy with 5-azacytidine, restore sodium/iodide symporter gene expression in human thyroid adenoma cell line, KAK-1, suggesting trans-active transcriptional repressor. *J Clin Endocrinol Metab* 92:1080–87.
113. Van Herle AJ, Agatep ML, Padua DND, Totanes TL, Canlapan DV, Van Herle HM, Juillard GJ. 1990. Effects of 13 cis-retinoic acid on growth and differentiation of human follicular carcinoma cells (UCLA R0 82 W-1) *in vitro*. *J Clin Endocrinol Metab* 71:755–63.
114. Gruning T, Tiepolt C, Zophel K, Bredow J, Kropp J, Franke WG. 2003. Retinoic acid for redifferentiation of thyroid cancer—Does it hold its promise? *Eur J Endocrinol* 148:395–402.
115. Liu YY, Stokkel MP, Pereira AM, Corssmit EP, Morreau HA, Romijn JA, Smit JW. 2006. Bexarotene increases uptake of radioiodide in metastases of differentiated thyroid carcinoma. *Eur J Endocrinol* 154:525–31.
116. Kebebew E, Peng M, Reiff E, Treseler P, Woeber KA, Clark OH, Greenspan FS, Lindsay S, Duh QY, Morita E. 2006. A phase II trial of rosiglitazone in patients with thyroglobulin-positive and radioiodine-negative differentiated thyroid cancer. *Surgery* 140:960–66; discussion, 966–67.
117. Bajetta E, Zilembo N, Di Bartolomeo M, Di Leo A, Pilotti S, Bochicchio AM. 1993. Treatment of metastatic carcinoids and other neuroendocrine tumors with recombinant interferon-alpha-2a. *Cancer* 72:3099–105.
118. Vitale G, Tagliaferri P, Caraglia M, Rampone E, Ciccarelli A, Bianco AR, Abbruzzese A, Lupoli G. 2000. Slow release lanreotide in combination with interferon-alpha2b in the treatment of symptomatic advanced medullary thyroid carcinoma. *J Clin Endocrinol Metab* 85:983–88.
119. Argiris A, Agarwala SS, Karamouzis MV, Burmeister LA, Carty SE. 2007. A phase II trial of doxorubicin and interferon alpha 2b in advanced, non-medullary thyroid cancer. *Invest New Drugs* 26:183–88.
120. Schott M, Seissler J, Lettmann M, Fouxon V, Scherbaum WA, Feldkamp J. 2001. Immunotherapy for medullary thyroid carcinoma by dendritic cell vaccination. *J Clin Endocrinol Metab* 86:4965–69.
121. Stift A, Sachet M, Yagubian R, Bittermann C, Dubsky P, Brostjan C, Pfragner R, Niederle B, Jakesz R, Gnant M, Friedl J. 2004. Dendritic cell vaccination in medullary thyroid carcinoma. *Clin Cancer Res* 10:2944–53.
122. Papewalis C, Wuttke M, Jacobs B, Domberg J, Willenberg H, Baehring T, Cupisti K, Raffel A, Chao L, Fenk R, Seissler J, Scherbaum WA, Schott M. 2008. Dendritic cell vaccination induces tumor epitope-specific Th1 immune response in medullary thyroid carcinoma. *Horm Metab Res* 40:108–16.
123. Bachleitner-Hofmann T, Friedl J, Hassler M, Hayden H, Dubsky P, Sachet M, Rieder E, Pfragner R, Brostjan C, Riss S, Niederle B, Gnant M, Stift A. 2009. Pilot trial of autologous dendritic cells loaded with tumor lysate(s) from allogeneic tumor cell lines in patients with metastatic medullary thyroid carcinoma. *Oncol Rep* 21:1585–92.
124. Chatal JF, Campion L, Kraeber-Bodere F, Bardet S, Vuillez JP, Charbonnel B, Rohmer V, Chang CH, Sharkey RM, Goldenberg DM, Barbet J. 2006. Survival improvement in patients with medullary thyroid carcinoma who undergo pretargeted anti-carcinoembryonic-antigen radioimmunotherapy: A collaborative study with the French Endocrine Tumor Group. *J Clin Oncol* 24:1705–11.
125. Bergers G, Hanahan D. 2008. Modes of resistance to anti-angiogenic therapy. *Nat Rev Cancer* 8:592–603.
126. Kumar R, Knick VB, Rudolph SK, Johnson JH, Crosby RM, Crouthamel M-C, Hopper TM, Miller CG, Harrington LE, Onori JA, Mullin RJ, Gilmer TM, Truesdale AT, Epperly AH, Boloor A, Stafford JA, Luttrell DK, Cheung M. 2007. Pharmacokinetic-pharmacodynamic correlation from mouse to human with pazopanib, a multikinase angiogenesis inhibitor with potent antitumor and antiangiogenic activity. *Mol Cancer Ther* 6:2012–21.
127. Sun L, Liang C, Shirazian S, Zhou Y, Miller T, Cui J, Fukuda JY, Chu JY, Nematalla A, Wang X, Chen H, Sistla A, Luu TC, Tang F, Wei J, Tang C. 2003. Discovery of 5-[5-fluoro-2-oxo-1,2-dihydroindol-(3Z)-ylidenemethyl]-2,4-dimethyl-1H-pyrrole-3-carboxylic acid (2-diethylaminoethyl)amide, a novel tyrosine kinase inhibitor targeting vascular endothelial and platelet-derived growth factor receptor tyrosine kinase. *J Med Chem* 46:1116–19.

21 Molecular Targeted Therapies of Medullary Thyroid Carcinoma

Francesco Torino, Agnese Barnabei, Rosa Maria Paragliola, and Salvatore M. Corsello

CONTENTS

ABSTRACT: One of the most exciting developments in cancer research in recent years has been the clinical validation of molecular targeted therapies (MTTs) that inhibit the action of pathogenic tyrosine kinases, and in this context, medullary thyroid carcinoma (MTC) represents a promising model. In fact, it is well known that in MTC, the activated *RET* (rearranged during transfection) proto-oncogene product, the RET receptor tyrosine kinase, and its signal transduction pathways lead to subsequent neoplastic transformation. These mechanisms are highly attractive targets for selective cancer therapy. Several strategies aimed to block the activation and signaling of RET have been preclinically tested. They include the interference with formation of ligand-receptor complexes, dimerization, autophosphorylation, recruitment of adaptor proteins to various docking sites, and initiation of signal transduction cascades. Although almost all steps may be inhibited by relatively specific agents, the most advanced results have been obtained by competitive inhibition of RET-TK activity by tyrosine kinases inhibitors (TKIs). A variety of novel approaches are currently being assessed in clinical trials. However, although the inhibition of the RET pathway is actually one of the most studied for therapeutic purposes, other signal transduction pathways have been implicated as contributing to the growth and functional activity of MTC and are considered promising as therapeutic targets. In particular, inhibition of RAS-RAF-ERK, PI3K-AKT, NFκB, and the glycogen synthase kinase-3 pathways has a promising effect in cultured MTC cells. Angiogenesis, the process of new blood vessel formation from preexisting vasculature, is another important target in MTC therapy. In fact, endocrine glands are typically vascular organs, and the process of angiogenesis is also regulated by hormonal changes such as increased estrogen, insulin-like growth factor-1 (IGF-I), and thyroid-stimulating hormone (TSH) levels. Thyroid tumors are more vascular than normal thyroid tissue. There is a clear correlation between increased vascular endothelial growth factor (VEGF) expression and more aggressive thyroid tumor behavior and metastasis. Bevacizumab, a humanized monoclonal antibody against VEGF, inhibited the growth of mouse xenografts of thyroid cancer cells, but clinical trials exploring the activity of bevacizumab in thyroid cancer patients have not been conducted. In conclusion, surgery represents to date the only curative treatment of MTC but unfortunately, about two-thirds of patients experience relapse and die from the disease. A growing understanding of molecular pathogenesis of cancer has allowed the development of MTT also in MTC, but the potential relationship between toxicity and drug efficacy should also be evaluated by pharmacokinetic/genomic studies. Another essential observation is that the relevance of clinical trials depends on the power of their statistical design, which is, in turn, related to the number of patients enrolled. MTC is a rare cancer. Therefore, a major effort needs to be made by endocrinologists and oncologists to refer their patients for multi-institutional trials in order to optimize clinical trials and expedite the availability of novel beneficial therapies.

21.1 INTRODUCTION

One of the most exciting developments in cancer research in recent years has been the clinical validation of biological therapies that inhibit the action of pathogenic tyrosine kinases (TKs). Treatment of appropriately selected patients with molecular targeted therapy (MTT) can alter the natural history of their disease and improve survival. The activity of imatinib in chronic myeloid leukemia (CML) and gastrointestinal stromal tumors (GISTs), trastuzumab in breast cancer, anti-epidermal growth factor receptor (EGFR) agents in non-small-cell lung cancer (NSCLC), sunitinib in renal cancer (RCC) and GIST, and sorafenib in renal cell carcinoma (RCC) and hepatocarcinoma has validated the concept that certain tumors are oncogene dependent.[1]

Medullary thyroid carcinoma (MTC) represents a promising candidate for MTT, as oncogenic events responsible for initiating malignancy have been well characterized. In particular, activated *RET*, a receptor tyrosine kinase (RTK), and the related signal transduction pathways leading to subsequent neoplastic transformation are highly attractive targets for selective cancer therapy (Figure 21.1). Whether other proto-oncogenes, such as *RAS* or *BRAF*, identified as having a role in thyroid carcinogenesis, may be therapeutically useful targets in MTC needs to be defined. Similarly, loss of function mutations of tumor suppressor genes, such as retinoblastoma protein (pRb), tumor protein p53, and PTEN, or of cell cycle inhibitors (p27Kip1 and p18-INK4c) seem to predispose to MTC in rodents,[2] but their role remains to be confirmed in humans. A variety of novel approaches that target *RET* (rearranged during transfection) proto-oncogene directly or indirectly have recently emerged and are currently being assessed in clinical trials.

21.2 TYROSINE KINASES AS THERAPEUTIC TARGETS

21.2.1 RET Receptor Tyrosine Kinase

In 1993 and 1994, germline mutations in the *RET* proto-oncogene were identified in patients with MEN 2A, MEN 2B, or familial medullary thyroid carcinoma (fMTC).[3–5] The encoded RET protein

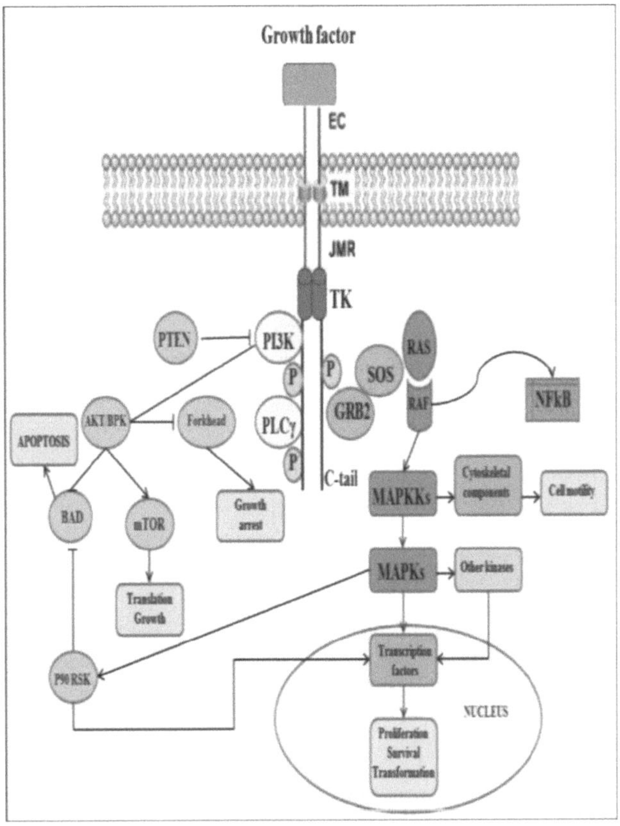

FIGURE 21.1 (See color insert.) Receptor tyrosine kinase (RTK) signaling cascade. Signaling mechanism of RTK and major biological outcomes are illustrated. Mutual transphosphorylation of tyrosine residues within active RTK dimers recruits intracellular proteins endowed with phosphotyrosine-binding domains. Proximal targets of the RTKs invoke the intracellular signaling cascades RAS-RAF-MAPK (ERK pathway) and the phosphatidylinositol 3-kinase (PI3K) AKT that ultimately lead to diverse biological responses. Abbreviations: Extracellular domain (EC), transmembrane domain (TM), juxtamembrane domain (JMR), and tyrosine kinase domain (TK). (From Castellone, M. D., et al., *Best Pract. Res. Clin. Endocrinol. Metab.*, 22, 1023–38, 2008.)

mediates downstream pathways of cell survival and mitogenesis.[6] Patients with MEN 2A have missense mutations at exon 10 (10–15%) (codons 609, 610, 611, 618, and 620) and exon 11 (80%) (codon 634), resulting in receptor dimerization and constitutive activation.[7] More than 95% of patients with MEN 2B have a mutation at exon 16 (codon 918), causing receptor autophosphorylation and activation.[7] Patients with fMTC harbor mutations in exons 10, 11, and 13 (codon 768), and exon 14 (codons 804 and 806).[7] These mutations activate the kinase function of RET and convert it into a dominantly transforming oncogene product conferring the malignant phenotype to MTC cells.[8,9] Amplification of the mutant allele or loss of the wild-type *RET* allele may act as a "second hit" in tumors of patients with MEN 2.[10] Therefore, the inhibition of the *RET* proto-oncogene pathway is considered a rational MTT. In preclinical models, blocking RET autophosphorylation led to decreased cell cycle progression and induction of cell death by apoptosis, indicating that oncogenic RET could be a useful anticancer target.[11,12]

Several strategies aimed to block the activation and signaling of RET have been preclinically tested.[9,13] They include the interference with formation of ligand-receptor complexes, dimerization, autophosphorylation, recruitment of adaptor proteins to various docking sites, and initiation of signal transduction cascades. Downregulation leading to the disappearance of RET from the

TABLE 21.1
Targeted Drugs under Clinical Evaluation in Medullary Thyroid Cancer

Drugs	Targets	Trial	Clinical Results	Main G3/4 Toxicity	References
Vandetanib (ZD6474)	VEGFR 2–3, EGFR, and RET	Phase I–II Phase IIb	20% PR; 30% SD; ↓ CTN in >50% of pts Ongoing	Rash, diarrhea, fatigue, nausea, QTc prolongation	26, 27
Imatinib mesylate (STI-571)	RET, c-KIT, PDGFR	Phase II	Minor transient activity	Fatigue, nausea, rush	30, 31
Sunitinib malate (SU 11248)	VEGFR 1–3, KIT, PDGFR α/β, RET, CSF1R, FLT3	Phase II	FDG-PET response: 44%	Leuko/neutropenia, HFS, fatigue, GI bleeding, diarrhea, mucositis, or AF	32, 33
Sorafenib tosylate (BAY 43–9006)	VEGFR 2–3, PDGFR, c-KIT, C-RAF, and BRAF	Phase II	2/5 pts: PR 3/5 pts: SD ↓ CTN: >50% of pts	HFS, hypertension musculoskeletal pain, fatigue	35
Motesanib diphosphate (AMG706)	VEGFR 1, 2, 3, PDGFR, c-KIT	Phase II	2/83 pts: PR 43/83 pts: SD	Hypertension, diarrhea, fatigue, weight loss, abdominal pain	37
Axitinib (AG-013736)	VEGFR 1, 2, 3, PDGFR-β, and c-KIT	Phase II	2/11: PR 3/11: SD	Hypertension, fatigue, proteinuria, diarrhea	88
XL184	MET, VEGFR 2, KIT, RET, FLT3, and Tie-2	Phase I Phase III	3 pts: SD and ↓ CTN Ongoing	Good safety profile	38 39
Lithium	GSK-3β, β-catenin, MAPK/ERK pathway, c-Myc, c-Jun	Phase II	Ongoing		59
17-AAG	HSP90	Phase II	Ongoing		103

Note: 17-AAG, 17-allylamino-17-demethoxygeldanamycin; AF, atrial fibrillation; FGFRs, fibroblast growth factor receptors; GI, gastrointestinal; HSP90, heat shock protein 90; IGF-1R, insulin-like growth factor receptor 1; HFS, hand-foot syndrome; NTR, neurotrophin receptor; PDGFR, platelet-derived growth factor receptor; PR, partial response; SD, stable disease; VEGFR, vascular endothelial growth factor receptor.

cell surface constitutes another important means of regulation and a potential target for therapy. Although almost all steps may be inhibited by relatively specific agents, the most advanced results have been obtained by competitive inhibition of RET activity by tyrosine kinases inhibitors (TKIs) (Table 21.1).

21.2.2 Targeting RET with TKIs: Clinical Trials

Several classes of molecules have demonstrated inhibitory properties of RET kinase activity in preclinical studies. They include the pyrazolo-pyridimidine inhibitors PP1 and PP2, the indolocarbazole derivatives CEP-701 and CEP-751, the 2-indolinone derivative RPI-1, and the anilinoquinazoline ZD6474.[6] Despite an attractive mechanism of action and except for ZD6474 (vandetanib), these molecules are not yet available for clinical trials (Table 21.2).

PP1 and PP2 inhibit the enzymatic activity and transforming ability of NIH3T3 fibroblasts transfected with almost all types of *RET/MEN 2A*, *RET/MEN 2B*, and *RET/PTC1* and *RET/PTC3*.[14–17]

TABLE 21.2
Targeted Agents under Preclinical Evaluation in Medullary Thyroid Cancer

Agents	Possible Mechanism(s) of Action	References
PP1, PP2	RET, Src kinases, proteosomal degradation of RET	14–17
CEP-701, CEP-751	Trk family kinase, RET, NTR	19, 20, 67
RPI-1	RET	14, 21, 22
NBL-1 (monoclonal antibody)	RET internalization	40
D4 aptamer	RET dimerization	43
Adenovirus-mediated gene therapy	RET gene	34, 35
RET-selective ribozyme	Mutant RET mRNA and RET-mediated cell growth and transformation	46
RET-selective phosphatases	RET activity	47
Tipifarnib (R115777)	RAS-RAF-MAPK/ERK pathway (Ras farnesylation)	48
LY294002, KP372-1	PI3K/Akt pathway	49, 57
ADW742	IGF-1R	53
PD173074	FGFR-4	79
NVP-AEE788	VEGFRs and EGFR	89–91
Vatalanib (PTK787/ZK222584)	VEGFRs	92
Cediranib (AZD2171)	VEGFRs	93
Pazopanib (GW-786034)	VEGFRs, PDGFR α/β, c-KIT	94

Note: See Table 21.1 for abbreviations.

PP1 and PP2 also inhibit the Src family of kinases, a major downstream effector of RET-mediated mitogenesis. Therefore, it remains unclear whether the inhibitory effects of PP1 and PP2 on cell cultures arise from inhibition of RET.[18] PP1 induces RET/MEN 2A and RET/MEN 2B oncoprotein degradation via proteosomal pathways, providing an additional mechanism of RET inhibition.[16]

The indolocarbazole derivatives CEP-701 and CEP-751 were originally developed to inhibit TRK family tyrosine kinase.[19] In addition, these compounds demonstrated an inhibitory effect on RET autophosphorylation and proliferation of TT cells (a MTC-derived cell line harboring a RETC634W mutation).[19] Moreover, CEP-751 inhibited TT cell growth in nude mice.[19] Interestingly, CEP-751 *in vivo* inhibitory effects increased when it was used in combination with irinotecan, a chemotherapeutic drug approved for metastatic colorectal cancer.[20] Irinotecan alone also has a strong effect on MTC xenograft growth in nude mice.[20] However, it remains to be determined whether CEP-751 alone or in combination with irinotecan is effective in patients with MTC.

RPI-1 inhibited RET in human TT cells and RET/MEN 2A- and RET/PTC1-expressing NIH3T3 cells.[21] RPI-1 decreased activation of RET downstream molecules, including PLC, ERK, and AKT, and reduced proliferation.[22] RPI-1 also showed antitumor effects in nude mice.[14,21,22] Interestingly, MTC cells showed only growth arrest rather than apoptosis induction after exposure to RPI-1.[22] Nevertheless, inhibition of MTC tumor cell proliferation seems to be sufficient to induce its antitumoral effect.

Vandetanib is the TKI with the most advanced clinical development in MTC. It is an anilinoquinazoline inhibitor of VEGF receptors (VEGFRs; VEGFR2 and VEGFR3), EGFR, and RET phosphorylation and signaling.[23,24] Vandetanib targets the enzymatic activity of both MEN 2- and PTC-related oncogenic RET and has an IC50 of 100 nmol. In addition, the compound inhibits tumor growth in RET/PTC-transformed NIH3T3 cell xenografts.[24] It has been recently demonstrated that the compound docks into the ATP-binding pocket of RET.[25] Vandetanib inhibits the wild-type enzyme and most of the activated forms of RET, with the exception of RET molecules with mutations in residue Val^{804}.[25] Mutation at this site has been proposed as a possible mechanism of acquired resistance to the drug.[17] In phase I clinical trials of patients with advanced non-small-cell lung cancer (NSCLC) and other solid tumors, including MTC, vandetanib was relatively well tolerated after oral administration.[26]

Wells et al.[27] reported the definitive results of an open-label phase II clinical trial of patients with metastatic hereditary MTC. Thirty patients with locally advanced or metastatic hereditary MTC and measurable disease (Response Evaluation Criteria in Solid Tumors (RECIST)) were enrolled. The median duration of treatment was 172 days. Twenty percent of patients experienced a partial response, and another 30% displayed stable disease, yielding a disease control rate of 50%. More than half of the patients had significant reductions in serum calcitonin (CTN) levels. The drug was overall well tolerated.

A phase II trial of vandetanib for sporadic MTC patients with locally advanced and metastatic disease is currently under way. A randomized phase II trial, comparing vandetanib to placebo in patients with hereditary or sporadic MTC, is being planned.

Imatinib mesylate (STI-571) is a 2-phenylaminopyrimidine derivative acting as an inhibitor of the c-KIT and the receptor for PDGFR TKs. It received FDA approval for clinical use as a single treatment of GIST and CML in 2004.[28] Imatinib has been shown to display inhibitory activity against RET/MEN 2A and RET/MEN 2B in MTC-derived cell lines. Imatinib induces RET degradation through nonproteasomal pathways.[29] Furthermore, it inhibits RET TK, but at high concentrations. Despite this suboptimal pharmacological property, the drug was the first TKI used in two trials with 9 and 15 patients affected by MTC, respectively. Unfortunately, imatinib resulted in limited efficacy and did not induce any tumor response.[30,31]

Sunitinib malate (SU 11248) is an indolinone derivative approved for RCC and refractory GIST. It is a multikinase inhibitor, highly active against RET, and is under investigation in phase II trials on patients affected by unresectable differentiated thyroid cancer, refractory to ^{131}I, and advanced MTC.[32] Preliminary results of 16 patients (3 with MTC) are available. Seven patients (44%) had an FDG-PET response. Response rates at 3 months, according to RECIST criteria, were not reported. No grade 4 toxicities have been registered. The main grade 3 toxicities included neutropenia (28%), leukopenia (17%), fatigue (11%), hand-foot syndrome (11%), and gastrointestinal bleeding (11%). Grade 3 diarrhea, mucositis, or atrial fibrillation was reported in 6% of the patients.[33]

Sorafenib tosylate (BAY 43-9006) was initially developed to target the Raf family of kinases, mainly B-Raf and C-Raf. It also inhibits other kinases, including VEGFR2, platelet-derived growth factor receptor (PDGFR), c-KIT, FLT3, and RET. The drug has recently been approved in the United States to treat RCC and hepatocarcinoma. Carlomagno et al.[17] showed that 20 to 50 nmol/L sorafenib inhibits 50% (IC50) of NIH 3T3 fibroblasts expressing one of three oncogenic versions of RET (RET/PTC3, RET/C634R, or RET/M918T) and showed almost complete inhibition with 100 nmol/L. Cells expressing both RET/V804L and RET/V804M, resistant to vandetanib, were sensitive to sorafenib at 110 and 147 nmol/L, respectively.[34] Very preliminary results reported that sorafenib mesilate obtained objective responses in two out of five patients with metastatic MTC.[35] Stable disease was registered in the other three patients after 6 months of treatment. Interestingly, after 2–3 months, CTN levels were decreased more than 50% from baseline in all patients.[35] Sorafenib is under investigation in two phase II trials recruiting patients that are affected by metastatic or unresectable thyroid cancers, including MTC.[36]

Motesanib diphosphate (AMG 706) is a multikinase inhibitor of VEGFRs and RET. In a phase II trial of motesanib that included 83 patients with either sporadic or hereditary MTC, two patients had a partial response and 43 had stable disease for longer than 6 months.[37]

XL184 is a potent orally available small-molecule inhibitor of MET and VEGFR2/KDR and also inhibits KIT, RET, FLT3, and Tie-2. Salgia et al.[38] reported a phase I study on 25 patients with advanced malignancies, including three patients with MTC. Using dose escalation and assessing response by RECIST criteria, as well as serologic measurements of CTN, VEGF-A, soluble VEGFR2 (sVEGFR2), and Ang2, after a median follow-up of over 12 months, the patients affected by MTC (one of whom had a documented RET mutation) had substantial reductions in plasma CTN and stable disease for over 6 months, without serious toxicities. A phase III trial is ongoing to evaluate the progression-free survival (PFS) with XL184 in comparison with placebo in subjects with unresectable, locally advanced, or metastatic MTC.[39]

21.2.3 Other Strategies to Inhibit RET

The first step in the activation of RET is the binding of the growth factor (GF)–ligand to the complex GF-receptor/coreceptor. Therefore, targeting GFs, or the binding site of the receptor/coreceptor, represents a straightforward approach for RET inhibition. Unfortunately, monoclonal antibodies against the RET ligand or receptor/coreceptors have not yet been synthesized. Yano et al.[40] generated in a neuroblastoma model an antibody capable of inducing internalization of RET, but its activity remains unknown.

The second step in RET activation is dimerization. It has been demonstrated that the introduction in the membrane of small peptides corresponding to the transmembrane domain could compete with dimerization, and thus inhibit the kinase activity of some human cancer cells (i.e., those overexpressing ErbB2 and epidermal growth factor receptor (EGFR)).[41]

Aptamers are single-stranded DNA or RNA oligonucleotides that have specific three-dimensional structures and bind to target molecules. Aptamers are usually created by selecting them from a large random sequence pool, but natural aptamers also exist in riboswitches.[42] Several unique properties of aptamers, including high binding specificity, low immunogenicity, structural stability, and ease of synthesis, have made aptamers promising agents for directed therapy against cancer targets.[42] Aptamers with antineoplastic activity against extracellular, cell membrane, and intracellular targets have been developed.[42] An aptamer that targets activated RET and blocks RET-dependent intracellular signaling by interfering with receptor dimerization has recently been described.[43] The neutralizing nuclease-resistant D4 aptamer was capable of binding and inhibiting wild-type RET and RET/MEN 2A on the cell surface inhibiting the constitutive RET dimerization.[43] However, the efficacy of the D4 aptamer against RET-associated tumors remains to be established.

Among the most fascinating treatment approach of MTC is the integration of new genetic material into the genome—"gene therapy." This approach can be used to replace defective genes or block the effects of unwanted ones by the introduction of a counteracting gene. Drosten et al.[44] reported that adenovirus-mediated transduction of dominant negative *RET* into TT cells reduced expression of oncogenic RET receptors on the cell surface. They also reported that inoculation of dominant negative RET-expressing MTC cells into nude mice led to an almost complete suppression of tumor growth.[44,45] These results suggest that inhibition of oncogenic *RET* expression by a dominant negative *RET* mutant is an effective approach for MTC treatment, although many issues remain to be resolved before viral vectors can be used *in vivo*.

Another gene therapy approach might be the introduction of a RET-selective ribozyme that specifically cuts mutant *RET* mRNA and blocks RET-mediated cell growth and transformation.[46] Also, the ectopic expression of RET-selective phosphatases has been shown to efficiently block RET activity.[47]

21.3 TARGETING OTHER KINASES

Although the inhibition of the RET pathway is actually one of the most studied for therapeutic purposes, other signal transduction pathways have been implicated as contributing to the growth and functional activity of MTC and are considered promising as therapeutic targets (Table 21.2). In particular, inhibition of RAS-RAF-ERK,[48] PI3K-AKT,[49] NFκB,[50] and the glycogen synthase kinase-3 (GSK-3)[51] pathways have a promising effect in cultured MTC cells. In addition, data from preclinical experiments have shown that the inhibition of fibroblast growth factor receptor-4 (FGFR4)[52] and the insulin-like growth factor-1 receptor (IGF-1R)[53] reduces the growth of MTC cells.

21.3.1 RAS-RAF-MAPK-ERK Pathway

Tipifarnib (R115777) is a member of a novel class of agents developed to inhibit the farnesylation of Ras and other proteins. Tipifarnib shows antiproliferative effects against many human tumor cell lines. Tipifarnib has not shown direct activity against RET kinase, but it can inhibit RET signaling through the MAPK pathway. This compound was active in myelodysplastic syndrome and in acute myelogenous leukemia.[54,55] In a patient affected by sporadic MTC, with a novel somatic deletion in exon 11 of the *RET* gene, a rapid and marked response to a regimen incorporating both sorafenib and tipifarnib was observed. It has been hypothesized that tipifarnib, which can affect downstream signaling of RET kinase, may synergistically act with sorafenib.[56]

21.3.2 PI3K-Akt Pathway

PI3K-Akt pathway activation has a central role in thyroid tumorigenesis,[57] but little is known about its role in regulating the growth of MTC tumors. The exposure of MTC cells to LY294002, a PI3K inhibitor, resulted in a dose-dependent reduction in cellular proliferation and neuroendocrine tumor markers.[49] The reduction in growth is mediated by apoptosis.[49] In addition, KP372-1, an Akt inhibitor, has been shown to inhibit cell proliferation and induce apoptosis in thyroid cancer cells.[57]

21.3.3 GSK-3 as a Potential Target for MTC Growth Regulation

GSK-3 is a serine/threonine protein kinase that was first described as playing a role in the regulation of glycogen synthesis.[58] GSK-3β, an isoform of GSK-3, is involved in many cellular processes, including metabolism, embryonic development and cell differentiation, proliferation, and survival.[59] GSK-3α has been shown to be involved in the regulation of cellular proliferation.[60,61] In contrast to other kinases, GSK-3β becomes inactivated by phosphorylation in response to signaling cascades.[58] GSK-3β regulates other molecules, such as β-catenin, MAPK kinase 1, ERK-1/2, c-Myc, c-Jun, murine double minute 2, Mcl-1, and heat shock factor by phosphorylation, and therefore modulates diverse intracellular signaling pathways that are known to play key roles in cancer biology.

The activation of the Raf-1 pathway in MTC cells leads to significant growth suppression[62] and is associated with a reduction in CTN and chromogranin A (CgA).[51] In addition, it has been shown that growth inhibition by Raf-1 activation in the MTC-TT cell line induces an autocrine-paracrine protein, leukemia inhibitory factor, and this alone could mediate differentiation and cell growth inhibition.[62] Activation of the Raf-1 pathway in these cells also leads to inactivation of GSK-3β by phosphorylation at Ser-9,[63] indicating a possible crosstalk with other pathways involved in growth regulation.

Kunnimalaiyaan et al.[59] have recently shown that inactivation of GSK-3β with lithium chloride resulted in MTC differentiation and cell growth inhibition. Based on these studies, a clinical trial using lithium for treatment of patients with metastatic MTC was initiated.

21.3.4 Nerve Growth Factor Receptor Pathway

The neurotrophin receptor (NTR) family includes three receptors with TK activity (NTR1, NTR2, and NTR3). Changes in their expression are involved in thyroid C-cell transformation.[64,65] Particularly, NTR2 activity is reduced and NTR3 is upregulated in MTC.[66] CEP-751 demonstrates inhibition of RET and NTR TK, and has a cytostatic effect in MTC cells.[67]

21.3.5 Insulin-Like Growth Factor Receptor Pathway

IGF-1R is a ubiquitous transmembrane TK, structurally similar to the insulin receptor (IR). The α-subunit of IGF-1R binds IGF-I, IGF-II, and insulin at supraphysiological doses; the β-subunits contain the TK domain.[68] IGF-II and IGF-1R are overexpressed in many cancer types. IGF-1R upregulation was found to mediate resistance to TKIs in different types of cancer cells. IGF-I and IGF-1R are overexpressed in thyroid cancer, particularly in the most aggressive variants.[69] Importantly, IGF-I and insulin are essential for the mitogenic action of TSH and EGF in thyroid follicular cells.[70] ADW742 is a specific inhibitor of IGF-1R phosphorylation and of its signaling pathway. This drug is cytotoxic for follicular and MTC-derived cancer cells.[53]

21.3.6 Fibroblast Growth Factor Receptor Pathway

Fibroblast growth factor receptors (FGFRs) comprise a subfamily of RTKs that are master regulators of a broad spectrum of cellular processes, including apoptosis, proliferation, migration, and angiogenesis. Deregulation of FGFR signaling by activating mutations or ligand/receptor overexpression could allow these receptors to become constitutively active, leading to hematopoietic and solid tumors.[71] Four FGF receptors with TK activity have been described (FGFR-1, -2, -3, and -4).[72] Mutations or genetic alterations involving FGFRs have not been identified in thyroid cancer.[72] Two ligands of FGF (FGF1 and FGF2), the basic FGF, a potent angiogenic factor, are increased in thyroid cancer.[73–75] Increased expression of FGFR-1, -3, and -4 has also been observed in malignant thyroid tumors.[76] FGFR-2 expression, instead, was downregulated in thyroid cancer.[77] FGFR-4 is mostly expressed in aggressive thyroid tumor types and MTC cells.[78]

PD173074 is an FGFR-4-TK competitive inhibitor. The drug produced the abrogation of FGF-1-mediated FGFR-4 phosphorylation in TT cells and significant inhibition of cell proliferation and tumor growth *in vivo*. Moreover, the combination of STI571 and PD173074 resulted in greater suppression of cell proliferation *in vitro* and tumor control *in vivo* than that achieved with either agent alone. These data highlight RET and FGFR-4 as therapeutic targets and suggest a potential role for the combined use of TKIs in the management of inoperable MTC.[79]

Other TKIs, such as sorafenib, sunitinib, and pazopanib, under clinical evaluation in thyroid cancer patients, exert anti-FGFR activity as well.[78]

21.4 ANGIOGENESIS AS A THERAPEUTIC TARGET

Angiogenesis is the process of forming new blood vessels from preexisting vasculature. Although vascular endothelium is usually quiescent in the adult, active angiogenesis has been shown to be an important process for new vessel development and subsequent tumoral growth, progression, and spread. The angiogenic phenotype depends on the balance of proangiogenic growth factors such as vascular endothelial growth factor (VEGF) and inhibitors, as well as interactions with the extracellular matrix, allowing for endothelial migration. Endocrine glands are typically vascular organs, and their blood supply is essential for normal function and tight control of hormone feedback loops. In addition to metabolic factors such as hypoxia, the process of angiogenesis is also regulated by hormonal changes such as increased estrogen, IGF-I, and TSH levels. Thyroid tumors are more vascular than normal thyroid tissue. There is a clear correlation between increased VEGF expression and more aggressive thyroid tumor behavior and metastasis.

Overexpression of VEGF-A was correlated with stage, tumor size, and metastasis of thyroid carcinoma. MTC cells overexpress VEGF, and RET inhibition has been correlated with VEGF downregulation.[80,81] A direct correlation between the angiogenesis pathway and RET remains to be elucidated, although some hypotheses have surfaced. In fact, loss of heterozygosity in the von Hippel–Lindau (VHL) disease tumor suppressor locus occurs at the somatic level in familial MTC.[82] RET and the protein encoded by the gene associated with VHL disease might act along the same pathway by controlling neuronal cell survival.[83] In addition, loss of the VHL protein leads to increased expression of hypoxia-inducible factor (HIF), thereby promoting expression of VEGF and tumor angiogenesis. Finally, the expression of VEGFR1 and VEGFR2[84] has been detected in thyrocytes as well as in endothelial cells.

The above findings suggest that antiangiogenic therapy of patients with MTC is rational. As reported below, all the inhibitors of the RET kinase activity that are currently under clinical evaluation have activity against the TK of VEGFRs. These compounds might, therefore, have a combined effect, inhibiting RET in tumor cells and VEGFRs in endothelial cells. Thus, VEGF targeting is a widely used anticancer therapeutic approach. VEGF-targeted therapy acts through various mechanisms: inhibition of new vessel formation, apoptosis of pre-existing vessels, blockade of endothelial cell progenitors, and vessel constriction (with reduced blood flow and ischemia).[85]

Bevacizumab, a humanized monoclonal antibody against VEGF, is the first antiangiogenic agent approved for the treatment of patients affected by metastatic colorectal cancer, breast cancer, and NSCLC, in combination with standard chemotherapy.[86] Bevacizumab inhibited the growth of mouse xenografts of thyroid cancer cells.[87] Clinical trials exploring the activity of bevacizumab in thyroid cancer patients have not been conducted.

Several multitargeting TKIs that block VEGFRs have shown promising clinical activity against several solid tumors, including MTC. Axitinib (AG-013736), a VEGFR inhibitor with no known anti-RET activity, has been tested in a phase II trial that involved 60 patients with thyroid cancer derived from either medullary or follicular cells: partial responses were observed in 18 patients (30%), and stable disease lasting more than 16 weeks was reported in another 23 patients (38%).[88] Preclinical studies with NVP-AEE788, a dual VEGFR and EGFR inhibitor, in differentiated thyroid cancer cell lines[89–91] and with vatalanib (PTK787/ZK222584), a pan-VEGFR TKI, in thyroid carcinoma mouse xenografts showed promising results.[92] Cediranib (AZD2171), another pan-VEGFR TKI, demonstrated inhibitory activity on tumor growth and prolonged animal survival in an orthotopic nude mouse model of anaplastic thyroid cancer.[93] In addition, pazopanib (GW-786034), another multitargeted VEGFR inhibitor, is undergoing clinical evaluation in thyroid cancer patients.[94]

21.5 OTHER TARGETED APPROACHES

The 26S proteasome is a large ATP-dependent multimeric complex that degrades intracellular proteins that have been targeted for proteolysis by the process of ubiquitination.[95] Several key regulators of transcription and growth/apoptosis, such as nuclear factor-κB (NFκB) inhibitor (IκB), p53, c-myc, and c-Jun N-terminal kinase (JNK), are known substrates for proteasomal degradation.[96] NFκB was implicated in the pathophysiology of both anaplastic[97] and medullary[96] carcinomas, suggesting that novel therapies targeting NFκB may be effective in these malignancies.

Proteasome inhibitors constitute a novel class of antitumor agents with preclinical evidence of activity against hematological malignancies and solid tumors.[95] Specifically bortezomib, a boronic acid dipeptide proteasome inhibitor, is approved by the U.S. FDA for use in relapsed refractory multiple myeloma,[98] and is currently being evaluated in a variety of other hematological and solid malignancies.[99]

Mitsiades et al.[100] investigated the effect of bortezomib in a panel of thyroid carcinoma cells *in vitro* and defined apoptotic pathways triggered by this novel anticancer agent. Bortezomib-induced apoptosis is mediated by caspases and may be modulated by the mitochondria and the Bcl-2 family members. The bortezomib sensitivity of TT cells was reduced in the presence of IGF-I, suggesting that the antitumor activity of bortezomib might be enhanced by inhibition of IGF-I and its downstream signaling. In addition, the combination of bortezomib with conventional chemotherapeutic agents produced a synergistic effect.

21.5.1 THE NOTCH-1–HES-1–ASCL-1 SIGNALING PATHWAY

Notch-1 signaling might have a tumor suppressor role in MTC tumors and cell lines. The activation of Notch-1 significantly reduced the growth of MTC-TT cells.[101] In addition, Notch-1 regulated CTN levels in a dose-dependent manner, and the levels of reduction in growth and hormone production depended on the amount of Notch-1 protein present in the cell.[101] A lack of active Notch-1 protein was found in tumor tissues and MTC cell lines, whereas neuroendocrine markers such as CgA and ASCL-1 were highly expressed. Activation of doxycycline-inducible Notch-1 in TT cells by varying the concentration of doxycycline led to a dose-dependent increase in Notch-1 protein and Hairy/Enhancer of Split homolog 1 (HES-1) protein. As expected, the level of Achaete-Scute Complex-Like homolog 1 (ASCL-1) was reduced with an increase in Notch-1.[101] Furthermore, the levels of reduction in growth and hormone production depended on the amount of Notch-1 protein present in the cell.[101] These observations suggest that activation of Notch-1 signaling may be a potential target to treat patients with MTC tumors.

21.5.2 Heat Shock Proteins

Many oncogenic protein kinases depend on the molecular chaperone heat shock protein 90 (HSP90) for correct maturation and activity. 17-Allylamino-17-demethoxygeldanamycin (17-AAG) is an antibiotic of the ansamycin family with specific inhibitory activity against HSP90. 17-AAG resulted in the loss of oncogenic RET activity and effective TT cell growth inhibition, suggesting a requirement for HSP90 action by activated RET.[102] However, it remains to be determined whether RET oncoproteins are real targets of HSP90.[9] A phase II trial is ongoing to determine the one-year treatment failure rate in patients with inoperable, advanced, or metastatic differentiated thyroid carcinoma or MTC, treated with 17-AAG.[103] Although preclinical results seem encouraging, the clinical development of these agents, in terms of long-term toxicity profiles, clinical benefit, and the development of drug resistance, is a subject of intense research.

21.6 CONCLUSIONS

At present, surgery represents the only curative treatment of MTC. Unfortunately, about two-thirds of patients experience relapse and die of the disease. Although the majority of patients survive for decades, there is still a need for novel therapeutic approaches to improve outcomes in terms of clinical response and quality of life.

A growing understanding of molecular pathogenesis of cancer has allowed the development of MTT in MTC. Some novel compounds, mainly TKI of RET and angiogenesis inhibitors, presently under evaluation in clinical trials, have clearly shown that they are effective with a low toxicity. However, the activity and toxicity profile of the majority of other promising MTT requires further testing in clinical trials.

We are at the beginning of the clinical evaluation of a new therapeutic approach, but researchers are aware that several hurdles remain to be overcome in the clinical evaluation of MTT. They include the recognition in preclinical studies of appropriate cell systems and animal models to define and characterize the best target(s) and the real activity of targeted compounds of interest. Another problem is the selection and validation of surrogate markers for target inhibition combination of agents against different targets (e.g., phosphorylation of receptor of interest in pre- and posttreatment tumor biopsies). In parallel, the selection of patients most suitable to receive the agent(s) of interest is a major challenge. To this end, genotyping patients enrolled in clinical trial, for example, in order to identify a mutation in the targeted kinase, such as RET, or a genome-wide profiling, is a promising approach.[104] Potential relationships between toxicity and drug efficacy should also be evaluated by pharmacokinetic/genomic studies to ensure that adequate plasma drug concentrations are achieved. Such studies are particularly relevant in patients with MTC because diarrhea can induce drug malabsorption.[105]

Among the most important methodological issues in MTC, clinical trial is the evaluation of response. In fact, the response rate according to RECIST guidelines does not necessarily predict a benefit in terms of progression-free survival or overall survival in MTC.[105] Overall survival cannot be used to assess response in patients with slowly progressing disease. Hence, the time to progression might be a better endpoint to measure the durability of treatment response, the value of stable disease, and the overall efficacy of treatment, especially if crossover from the placebo to the treatment arm is allowed. Further endpoints, such as progression-free survival, have been proposed, but need to be validated as predictive measures of efficacy.

Another essential observation, which constitutes a major problem, is that the relevance of clinical trials depends on the power of their statistical design, which is related to the number of patients enrolled. MTC is a rare cancer. Therefore, a major effort needs to be made by endocrinologists and oncologists to participate in multi-institutional trials in order to improve accrual of patients in well-designed clinical trials. This appears to be the best way to define the oncogenic mechanisms driving the progression of MTC, as well as the real efficacy of novel therapeutic strategies that can improve therapeutic options offered to patients.

REFERENCES

1. Baselga J. 2006. Targeting tyrosine kinases in cancer: The second wave. *Science* 312:1175–78.
2. Bai F, Pei XH, Pandolfi PP, Xiong Y. 2006. p18 Ink4c and Pten constrain a positive regulatory loop between cell growth and cell cycle control. *Mol Cell Biol* 26:4564–76.
3. Carlson KM, Dou S, Chi D, Scavarda N, Toshima K, Jackson CE, Wells SA Jr, Goodfellow PJ, Donis-Keller H. 1994. Single missense mutation in the tyrosine kinase catalytic domain of the RET protooncogene is associated with multiple endocrine neoplasia type 2B. *Proc Natl Acad Sci USA* 91:1579–83.
4. Donis-Keller H, Dou S, Chi D, Carlson KM, Toshima K, Lairmore TC, Howe JR, Moley JF, Goodfellow P, Wells SA Jr. 1993. Mutations in the RET proto-oncogene are associated with MEN 2A and FMTC. *Hum Mol Genet* 2:851–56.
5. Mulligan LM, Kwok JB, Healey CS, Elsdon MJ, Eng C, Gardner E, Love DR, Mole SE, Moore JK, Papi L, et al. 1993. Germline mutations of the RET proto-oncogene in multiple endocrine neoplasia type 2A. *Nature* 363:458–60.
6. You YN, Lakhani V, Wells SA. 2007. New directions in the treatment of thyroid cancer. *J Am Coll Surg* 205:S45–48.
7. Brandi ML, Gagel RF, Angeli A, Bilezikian JP, Beck-Peccoz P, Bordi C, Conte-Devolx B, Falchetti A, Gheri RG,Libroia A, Lips CJ, Lombardi G, Mannelli M, Pacini F, Ponder BA, Raue F, Skogseid B, Tamburrano G, Thakker RV, Thompson NW, Tomassetti P, Tonelli F, Wells SA Jr, Marx SJ. 2001. Guidelines for diagnosis and therapy of MEN type 1 and type 2. *J Clin Endocrinol Metab* 86:5658–71.
8. Santoro M, Carlomagno F. 2006. Drug insight: Small-molecule inhibitors of protein kinases in the treatment of thyroid cancer. *Nat Clin Pract Endocrinol Metab* 2:42–52.
9. Drosten M, Pützer BM. 2006. Mechanisms of disease: Cancer targeting and the impact of oncogenic RET for medullary thyroid carcinoma therapy. *Nat Clin Pract Oncol* 3:564–74.
10. Huang SC, Koch CA, Vortmeyer AO, Pack SD, Lichtenauer UD, Mannan P, Lubensky IA, Chrousos GP, Gagel RF, Pacak K, Zhuang Z. 2000. Duplication of the mutant RET allele in trisomy 10 or loss of the wild-type allele in multiple endocrine neoplasia type 2-associated pheochromocytomas. *Cancer Res* 60:6223–26.
11. Drosten M, Hilken G, Böckmann M, Rödicker F, Mise N, Cranston AN, Dahmen U, Ponder BA, Pützer BM. 2004. Role of MEN2A-derived RET in maintenance and proliferation of medullary thyroid carcinoma. *J Natl Cancer Inst* 96:1231–39.
12. Wang DG, Liu WH, Johnston CF, Sloan JM, Buchanan KD. 1998. Bcl-2 and c-myc, but not bax and p53, are expressed during human medullary thyroid carcinoma tumorigenesis. *Am J Pathol* 152:1407–13.
13. de Groot JW, Links TP, Plukker JT, Lips CJ, Hofstra RM. 2006. RET as a diagnostic and therapeutic target in sporadic and hereditary endocrine tumors. *Endocr Rev* 27:535–60.
14. Carlomagno F, Vitagliano D, Guida T, Napolitano M, Vecchio G, Fusco A, Gazit A, Levitzki A, Santoro M. 2002. The kinase inhibitor PP1 blocks tumorigenesis induced by RET oncogenes. *Cancer Res* 62:1077–82.
15. Carlomagno F, Vitagliano D, Guida T, Basolo F, Castellone MD, Melillo RM, Fusco A, Santoro M. 2003. Efficient inhibition of RET/papillary thyroid carcinoma oncogenic kinases by 4-amino-5-(4-chloro-phenyl)-7-(t-butyl)pyrazolo[3,4-d]pyrimidine (PP2). *J Clin Endocrinol Metab* 88:1897–902.
16. Carniti C, Perego C, Mondellini P, Pierotti MA, Bongarzone I. 2003. PP1 inhibitor induces degradation of RETMEN2A and RETMEN2B oncoproteins through proteosomal targeting. *Cancer Res* 63:2234–43.

17. Carlomagno F, Guida T, Anaganti S, Vecchio G, Fusco A, Ryan AJ, Billaud M, Santoro M. 2004. Disease associated mutations at valine 804 in the RET receptor tyrosine kinase confer resistance to selective kinase inhibitors. *Oncogene* 23:6056–63.
18. Melillo RM, Barone MV, Lupoli G, Cirafici AM, Carlomagno F, Visconti R, Matoskova B, Di Fiore PP, Vecchio G, Fusco A, Santoro M. 1999. Ret-mediated mitogenesis requires Src kinase activity. *Cancer Res* 59:1120–26.
19. Strock CJ, Park JI, Rosen M, Dionne C, Ruggeri B, Jones-Bolin S, Denmeade SR, Ball DW, Nelkin BD. 2003. CEP-701 and CEP-751 inhibit constitutively activated RET tyrosine kinase activity and block medullary thyroid carcinoma cell growth. *Cancer Res* 63:5559–63.
20. Strock CJ, Park JI, Rosen DM, Ruggeri B, Denmeade SR, Ball DW, Nelkin BD. 2006. Activity of irinotecan and the tyrosine kinase inhibitor CEP-751 in medullary thyroid cancer. *J Clin Endocrinol Metab* 91:79–84.
21. Lanzi C, Cassinelli G, Pensa T, Cassinis M, Gambetta RA, Borrello MG, Menta E, Pierotti MA, Zunino F. 2000. Inhibition of transforming activity of the RET/PTC1 oncoprotein by a 2-indolinone derivative. *Int J Cancer* 85:384–90.
22. Cuccuru G, Lanzi C, Cassinelli G, Pratesi G, Tortoreto M, Petrangolini G, Seregni E, Martinetti A, Laccabue D, Zanchi C, Zunino F. 2004. Cellular effects and antitumor activity of RET inhibitor RPI-1 on MEN2A-associated medullary thyroid carcinoma. *J Natl Cancer Inst* 96:1006–14.
23. Herbst RS, Heymach JV, O'Reilly MS, Onn A, Ryan AJ. 2007. Vandetanib (ZD6474): An orally available receptor tyrosine kinase inhibitor that selectively targets pathways critical for tumor growth and angiogenesis. *Expert Opin Investig Drugs* 16:239–49.
24. Carlomagno F, Vitagliano D, Guida T, Ciardiello F, Tortora G, Vecchio G, Ryan AJ, Fontanini G, Fusco A, Santoro M. 2002. ZD6474, an orally available inhibitor of KDR tyrosine kinase activity, efficiently blocks oncogenic RET kinases. *Cancer Res* 62:7284–90.
25. Knowles PP, Murray-Rust J, Kjaer S, Scott RP, Hanrahan S, Santoro M, Ibáñez CF, McDonald NQ. 2006. Structure and chemical inhibition of the RET tyrosine kinase domain. *J Biol Chem* 281:33577–87.
26. Holden SN, Eckhardt SG, Basser R, de Boer R, Rischin D, Green M, Rosenthal MA, Wheeler C, Barge A, Hurwitz HI. 2005. Clinical evaluation of ZD6474, an orally active inhibitor of VEGF and EGF receptor signaling, in patients with solid, malignant tumors. *Ann Oncol* 16:1391–97.
27. Wells SA, Gosnell JE, Gagel RF, Moley JF, et al. 2007. Vandetanib in metastatic hereditary medullary thyroid cancer: Follow-up results of an open-label phase II trial. In *2007 ASCO Annual Meeting*, abstract 6018.
28. Dagher R, Johnson J, Williams G, Keegan P, Pazdur R. 2004. Accelerated approval of oncology products: A decade of experience. *J Natl Cancer Inst* 96:1500–9.
29. de Groot JW, Plaza Menacho I, Schepers H, Drenth-Diephuis LJ, Osinga J, Plukker JT, Links TP, Eggen BJ, Hofstra RM. 2006. Cellular effects of imatinib on medullary thyroid cancer cells harboring multiple endocrine neoplasia type 2A and 2B associated RET mutations. *Surgery* 139:806–14.
30. Frank-Raue K, Fabel M, Delorme S, Haberkorn U, Raue F. 2007. Efficacy of imatinib mesylate in advanced medullary thyroid carcinoma. *Eur J Endocrinol* 157:215–20.
31. de Groot JW, Zonnenberg BA, van Ufford-Mannesse PQ, de Vries MM, Links TP, Lips CJ, Voest EE. 2007. A phase II trial of imatinib therapy for metastatic medullary thyroid carcinoma. *J Clin Endocrinol Metab* 92:3466–69.
32. www.clinicaltrial.gov, NCT00381641.
33. Goulart B, Carr L, Martins RG, Eaton K, Kell E, Wallace S, Capell P, Mankoff DArt was here, but was deleted.. 2008. Phase II study of sunitinib in iodine refractory, well-differentiated thyroid cancer (WDTC) and metastatic medullary thyroid carcinoma (MTC). In *2008 ASCO Annual Meeting*, abstract 6062.
34. Carlomagno F, Anaganti S, Guida T, et al. 2006. BAY 43-9006 inhibition of oncogenic RET mutants. *J Natl Cancer Inst* 98:326–34.
35. Kober F, Hermann M, Handler A, Krotla G. 2007. Effect of sorafenib in symptomatic metastatic medullary thyroid cancer. In *2007 ASCO Annual Meeting*, abstract 14065.
36. www.clinicaltrials.gov, NCT00601783, NCT00654238.
37. Schlumberger MJ, Elisei R, Sherman SI, et al. 2007. Phase 2 trial of motesanib diphosphate (AMG 706) in patients with medullary thyroid cancer (MTC). Paper presented at 89th Annual Meeting of the Endocrine Society, Toronto, ON.
38. Salgia R, Hong DS, Camacho LH, Ng CS, Janisch L, Ratain MJ, Kurzrock R. 2007. A phase I dose-escalation study of the safety and pharmacokinetics (PK) of XL184, a VEGFR and MET kinase inhibitor, administered orally to patients (pts) with advanced malignancies. *J Clin Oncol* 25(18 Suppl):14031.
39. www.clinicaltrial.gov, NCT00704730.

40. Yano L, Shimura M, Taniguchi M, Hayashi Y, Suzuki T, Hatake K, Takaku F, Ishizaka Y. 2000. Improved gene transfer to neuroblastoma cells by a monoclonal antibody targeting RET, a receptor tyrosine kinase. *Hum Gene Ther* 11:995–1004.
41. Bennasroune A, Fickova M, Gardin A, Dirrig-Grosch S, Aunis D, Cremel G, Hubert P. 2004. Transmembrane peptides as inhibitors of ErbB receptor signaling. *Mol Biol Cell* 15:3464–74.
42. Barbas AS, White RR. 2009. The development and testing of aptamers for cancer. *Curr Opin Investig Drugs* 10:572–78.
43. Cerchia L, Duconge F, Pestourie C, Boulay J, Aissouni Y, Gombert K, Tavitian B, de Franciscis V, Libri D. 2005. Neutralizing aptamers from whole-cell SELEX inhibit the RET receptor tyrosine kinase. *PLoS Biol* 3:e123.
44. Drosten M, Frilling A, Stiewe T, Putzer BM. 2002. A new therapeutic approach in medullary thyroid cancer treatment: Inhibition of oncogenic RET signaling by adenoviral vector-mediated expression of a dominant-negative RET mutant. *Surgery* 132:991–97.
45. Drosten M, Stiewe T, Putzer BM. 2003. Antitumor capacity of a dominant-negative RET proto-oncogene mutant in a medullary thyroid carcinoma model. *Hum Gene Ther* 14:971–82.
46. Parthasarathy R, Cote GJ, Gagel RF.1999. Hammerhead ribozyme mediated inactivation of mutant RET in medullary thyroid carcinoma. *Cancer Res* 59:3911–14.
47. Hennige AM, Lammers R, Höppner W, Arlt D, Strack V, Teichmann R, Machicao F, Ullrich A, Häring HU, Kellerer M. 2001. Inhibition of RET oncogene activity by the protein tyrosine phosphatase SHP1. *Endocrinology* 142:4441–47.
48. Zatelli MC, Piccin D, Tagliati F, Bottoni A, Luchin A, Degli Uberti EC. 2005. SRC homology-2-containing protein tyrosine phosphatase-1 restrains cell proliferation in human medullary thyroid carcinoma. *Endocrinology* 146:2692–98.
49. Kunnimalaiyaan M, Ndiaye M, Chen H. 2006. Apoptosis-mediated medullary thyroid cancer growth suppression by the PI3K inhibitor LY294002. *Surgery* 140:1009–15.
50. Ludwig L, Kessler H, Wagner M, Hoang-Vu C, Dralle H, Adler G, Bohm BO, Schmid RM. 2001. Nuclear factor-κB is constitutively active in C-cell carcinoma and required for RET-induced transformation. *Cancer Res* 61:4526–35.
51. Sippel RS, Carpenter JE, Kunnimalaiyaan M, Chen H. 2003. The role of human achaete-scute homolog-1 in medullary thyroid cancer cells. *Surgery* 134:866–71.
52. Ezzat S, Huang P, Dackiw A, Asa SL. 2005. Dual inhibition of RET and FGFR4 restrains medullary thyroid cancer cell growth. *Clin Cancer Res* 11:1336–41.
53. Mitsiades CS, Mitsiades NS, McMullan CJ, Poulaki V, Shringarpure R, Akiyama M, Hideshima T, Chauhan D, Joseph M, Libermann TA, García-Echeverría C, Pearson MA, Hofmann F, Anderson KC, Kung AL. 2004. Inhibition of the insulin-like growth factor receptor-1 tyrosine kinase activity as a therapeutic strategy for multiple myeloma, other hematologic malignancies, and solid tumors. *Cancer Cell* 5:221–30.
54. Fenaux P, Raza A, Mufti GJ, Aul C, Germing U, Kantarjian H, Cripe L, Kerstens R, De Porre P, Kurzrock R. 2007. A multicenter phase 2 study of the farnesyltransferase inhibitor tipifarnib in intermediate-to high-risk myeodysplastic syndrome. *Blood* 109:4158–63.
55. Kurzrock R, Kantarjian HM, Cortes JE, Singhania N, Thomas DA, Wilson EF, Wright JJ, Freireich EJ, Talpaz M, Sebti SM. 2003. Farnesyltransferase inhibitor R115777 in myelodysplastic syndrome: Clinical and biologic activities in the phase I setting. *Blood* 102:4527–34.
56. Hong D, Ye L, Gagel R, Chintala L, El Naggar AK, Wright J, Kurzrock R. 2008. Medullary thyroid cancer: Targeting the RET kinase pathway with sorafenib/tipifarnib. *Mol Cancer Ther* 7:1001–6.
57. Mandal M, Kim S, Younes MN, Jasser SA, El-Naggar AK, Mills GB, Myers JN. 2005. The Akt inhibitor KP372-1 suppresses Akt activity and cell proliferation and induces apoptosis in thyroid cancer cells. *Br J Cancer* 92:1899–905.
58. Cohen P, Frame S. 2001. The renaissance of GSK3. *Nat Rev Mol Cell Biol* 2:769–76.
59. Kunnimalaiyaan M, Vaccaro AM, Ndiaye MA, et al. 2007. Inactivation of glycogen synthase kinase-3beta, a downstream target of the Raf-1 pathway, is associated with growth suppression in medullary thyroid cancer cells. *Mol Cancer Ther* 6:1151–58.
60. Liang MH, Chuang DM. 2006. Differential roles of glycogen synthase kinase-3 isoforms in the regulation of transcriptional activation. *J Biol Chem* 281:30479–84.
61. Liang MH, Chuang DM. 2007. Regulation and function of glycogen synthase kinase-3 isoforms in neuronal survival. *J Biol Chem* 282:3904–17.

62. Park JI, Strock CJ, Ball DW, et al. 2003. The Ras/Raf/MEK/extracellular signal regulated kinase pathway induces autocrine-paracrine growth inhibition via the leukemia inhibitory factor/JAK/STAT pathway. *Mol Cell Biol* 23:543–54.
63. Harwood AJ, Plyte SE, Woodgett J, Strutt H, Kay RR, Woodgett J, et al. 1995. Glycogen synthase kinase 3 regulates cell fate in *Dictyostelium*. *Cell* 80:139–48.
64. Pierotti MA, Greco A. 2006. Oncogenic rearrangements of the NTRK1/NGF receptor. *Cancer Lett* 232:90–98.
65. Gimm O, Dziema H, Brown J, de la Puente A, Hoang-Vu C, Dralle H, Plass C, Eng C. 2001. Mutation analysis of NTRK2 and NTRK3, encoding 2 tyrosine kinase receptors, in sporadic human medullary thyroid carcinoma reveals novel sequence variants. *Int J Cancer* 92:70–74.
66. McGregor LM, McCune BK, Graff JR, McDowell PR, Romans KE, Yancopoulos GD, Ball DW, Baylin SB, Nelkin BD. 1999. Roles of trk family neurotrophin receptors in medullary thyroid carcinoma development and progression. *Proc Natl Acad Sci USA* 96:4540–45.
67. Strock CJ, Park JI, Rosen DM, Ruggeri B, Denmeade SR, Ball DW, Nelkin BD. 2006. Activity of irinotecan and the tyrosine kinase inhibitor CEP-751 in medullary thyroid cancer. *J Clin Endocrinol Metab* 91:79–84.
68. Surmacz E. 2003. Growth factor receptors as therapeutic targets: Strategies to inhibit the insulin-like growth factor I receptor. *Oncogene* 22:6589–97.
69. Wang Z, Chakravarty G, Kim S, Yazici YD, Younes MN, Jasser SA, Santillan AA, Bucana CD, El-Naggar AK, Myers JN. 2006. Growth-inhibitory effects of human anti-insulin-like growth factor-I receptor antibody (A12) in an orthotopic nude mouse model of anaplastic thyroid carcinoma. *Clin Cancer Res* 12:4755–65.
70. Coulonval K, Vandeput F, Stein RC, Kozma SC, Lamy F, Dumont JE. 2000. Phosphatidylinositol 3-kinase, protein kinase B and ribosomal S6 kinases in the stimulation of thyroid epithelial cell proliferation by cAMP and growth factors in the presence of insulin. *Biochem J* 348:351–58.
71. Acevedo VD, Ittmann M, Spencer DM. 2009. Paths of FGFR-driven tumorigenesis. *Cell Cycle* 8:580–88.
72. Kondo T, Ezzat S, Asa SL. 2006. Pathogenetic mechanisms in thyroid follicular-cell neoplasia. *Nat Rev Cancer* 6:292–306.
73. Boelaert K, McCabe CJ, Tannahill LA, Gittoes NJ, Holder RL, Watkinson JC, Bradwell AR, Sheppard MC, Franklyn JA. 2003. Pituitary tumor transforming gene and fibroblast growth factor-2 expression: Potential prognostic indicators in differentiated thyroid cancer. *J Clin Endocrinol Metab* 88:2341–47.
74. Eggo MC, Hopkins JM, Franklyn JA, Johnson GD, Sanders DS, Sheppard MC. 1995. Expression of fibroblast growth factors in thyroid cancer. *J Clin Endocrinol Metab* 80:1006–11.
75. de la Torre NJ, Buley I, Wass JA, Turner HE. 2006. Angiogenesis and lymphangiogenesis in thyroid proliferative lesions: Relationship to type and tumour behaviour. *Endocr Relat Cancer* 13:931–44.
76. St Bernard R, Zheng L, Liu W, Winer D, Asa SL, Ezzat S. 2005. Fibroblast growth factor receptors as molecular targets in thyroid carcinoma. *Endocrinology* 146:1145–53.
77. Kondo T, Zheng L, Liu W, Kurebayashi J, Asa SL, Ezzat S. 2007. Epigenetically controlled fibroblast growth factor receptor 2 signaling imposes on the RAS/BRAF/mitogen-activated protein kinase pathway to modulate thyroid cancer progression. *Cancer Res* 67:5461–70.
78. Castellone MD, Carlomagno F, Salvatore G, Santoro M. 2008. Receptor tyrosine kinase inhibitors in thyroid cancer. *Best Pract Res Clin Endocrinol Metab* 22:1023–38.
79. Ezzat S, Huang P, Dackiw A, Asa SL. 2005. Dual inhibition of RET and FGFR4 restrains medullary thyroid cancer cell growth. *Clin Cancer Res* 11:1336–41.
80. Petrangolini G, Cuccuru G, Lanzi C, Tortoreto M, Belluco S, Pratesi G, Cassinelli G, Zunino F, et al. 2006. Apoptotic cell death induction and angiogenesis inhibition in large established medullary thyroid carcinoma xenografts by Ret inhibitor RPI-1. *Biochem Pharmacol* 72:405–14.
81. Bunone G, Vigneri P, Mariani L, Butó S, Collini P, Pilotti S, Pierotti MA, Bongarzone I. 1999. Expression of angiogenesis stimulators and inhibitors in human thyroid tumors and correlation with pathological features. *Am J Pathol* 155:1967–76.
82. Koch CA, Brouwers FM, Vortmeyer AO, Tannapfel A, Libutti SK, Zhuang Z, Pacak K, Neumann HP, Paschke R. 2006. Somatic VHL gene alterations in MEN2-associated medullary thyroid carcinoma. *BMC Cancer* 6:131.
83. Lee S, Nakamura E, Yang H, Wei W, Linggi MS, Sajan MP, Farese RV, Freeman RS, Carter BD, Kaelin WG Jr, Schlisio S. 2005. Neuronal apoptosis linked to EglN3 prolyl hydroxylase and familial pheochromocytoma genes: Developmental culling and cancer. *Cancer Cell* 8:155–67.

84. Jebreel A, England J, Bedford K, Murphy J, Karsai L, Atkin S. 2007. Vascular endothelial growth factor (VEGF), VEGF receptors expression and microvascular density in benign and malignant thyroid diseases. *Int J Exp Pathol* 88:271–77.
85. Hicklin DJ, Ellis LM. 2005. Role of the vascular endothelial growth factor pathway in tumor growth and angiogenesis. *J Clin Oncol* 23:1001–27.
86. Cao Y. 2009. Tumor angiogenesis and molecular targets for therapy. *Front Biosci* 14:3962–73.
87. Soh EY, Eigelberger MS, Kim KJ, Wong MG, Young DM, Clark OH, Duh QY. 2000. Neutralizing vascular endothelial growth factor activity inhibits thyroid cancer growth *in vivo*. *Surgery* 128:1059–65.
88. Cohen EE, Rosen LS, Vokes EE, Kies MS, Forastiere AA, Worden FP, Kane MA, Sherman E, Kim S, Bycott P, Tortorici M, Shalinsky DR, Liau KF, Cohen RB. 2008. Axitinib is an active treatment for all histologic subtypes of advanced thyroid cancer: Results from a phase II study. *J Clin Oncol.* 26:4708–13.
89. Younes MN, Yigitbasi OG, Park Y dW, Kim SJ, Jasser SA, Hawthorne VS, Yazici YD, Mandal M, Bekele BN, Bucana CD, Fidler IJ, Myers JN. 2005. Antivascular therapy of human follicular thyroid cancer experimental bone metastasis by blockade of epidermal growth factor receptor and vascular growth factor receptor phosphorylation. *Cancer Res* 65:4716–27.
90. Kim S, Schiff BA, Yigitbasi OG, Doan D, Jasser SA, Bekele BN, Mandal M, Myers JN. 2005. Targeted molecular therapy of anaplastic thyroid carcinoma with AEE788. *Mol Cancer Ther* 4:632–40.
91. Hoffmann S, Burchert A, Wunderlich A, Wang Y, Lingelbach S, Hofbauer LC, Rothmund M, Zielke A. 2007. Differential effects of cetuximab and AEE 788 on epidermal growth factor receptor (EGF-R) and vascular endothelial growth factor receptor (VEGF-R) in thyroid cancer cell lines. *Endocrine* 31:105–13.
92. Schoenberger J, Grimm D, Kossmehl P, Infanger M, Kurth E, Eilles C. 2004. Effects of PTK787/ZK222584, a tyrosine kinase inhibitor, on the growth of a poorly differentiated thyroid carcinoma: An animal study. *Endocrinology* 145:1031–38.
93. Gomez-Rivera F, Santillan-Gomez AA, Younes MN, Kim S, Fooshee D, Zhao M, Jasser SA, Myers JN. 2007. The tyrosine kinase inhibitor, AZD2171, inhibits vascular endothelial growth factor receptor signaling and growth of anaplastic thyroid cancer in an orthotopic nude mouse model. *Clin Cancer Res* 13:4519–27.
94. www.clinicaltrials.gov.
95. Adams J. 2004. The proteasome: A suitable antineoplastic target. *Nat Rev Cancer* 4:349–60.
96. Gallel P, Pallares J, Dolcet X, Llobet D, Eritja N, Santacana M, Yeramian A, Palomar-Asenjo V, Lagarda H, Mauricio D, Encinas M, Matias-Guiu X. 2008. Nuclear factor-kappaB activation is associated with somatic and germ line RET mutations in medullary thyroid carcinoma. *Hum Pathos* 39:994–1001.
97. Pacifico F, Mauro C, Barone C, Crescenzi E, Mellone S, Monaco M, Chiappetta G, Terrazzano G, Liguoro D, Vito P, Consiglio E, Formisano S, Leonardi A. 2004. Oncogenic and anti-apoptotic activity of NF-kappa B in human thyroid carcinomas. *J Biol Chem* 279:54610–19.
98. Richardson PG, Barlogie B, Berenson J, Singhal S, Jagannath S, Irwin D, Rajkumar SV, Srkalovic G, Alsina M, Alexanian R, Siegel D, Orlowski RZ, Kuter D, Limentani SA, Lee S, Hideshima T, Esseltine DL, Kauffman M, Adams J, Schenkein DP, Anderson KC. 2003. A phase 2 study of bortezomib in relapsed, refractory myeloma. *N Engl J Med* 348:2609–17.
99. Orlowski RZ, Kuhn DJ. 2008. Proteasome inhibitors in cancer therapy: Lessons from the first decade. *Clin Cancer Res* 14:1649–57.
100. Mitsiades CS, McMillin D, Kotoula V, Poulaki V, McMullan C, Negri J, Fanourakis G, Tseleni-Balafouta S, Ain KB, Mitsiades N. 2006. Antitumor effects of the proteasome inhibitor bortezomib in medullary and anaplastic thyroid carcinoma cells *in vitro*. *J Clin Endocrinol Metab* 91:4013–21.
101. Kunnimalaiyaan M, Vaccaro AM, Ndiaye MA, Chen H. 2006. Overexpression of the Notch1 intracellular domain inhibits cell proliferation and alters the neuroendocrine phenotype of medullary thyroid cancer cells. *J Biol Chem* 281:39819–30.
102. Cohen MS, Hussain HB, Moley JF. 2002. Inhibition of medullary thyroid carcinoma cell proliferation and RET phosphorylation by tyrosine kinase inhibitors. *Surgery* 132:960–67.
103. www.clinicaltrials.gov, NCT00118248.
104. Castellone MD, Santoro M. 2008. Dysregulated RET signaling in thyroid cancer. *Endocrinol Metab Clin North Am* 37:363–74.
105. Schlumberger M, Carlomagno F, Baudin E, Bidart JM, Santoro M. 2008. New therapeutic approaches to treat medullary thyroid carcinoma. *Nat Clin Pract Endocrinol Metab* 4:22–32.

22 Synthesis of Emergent Biotechnologies into Future CPG

Jeffrey I. Mechanick, Angelo Carpi, Mark L. Urken, Donald A. Bergman, and Rhoda H. Cobin

CONTENTS

ABSTRACT: The focus of this book is to present emergent biotechnologies and then forecast their impact on future clinical practice guidelines (CPG). Four clinical cases have been presented at the beginning of this book in which current CPG recommendations failed to provide optimal management. Knowledge gaps, which cannot be solved by interpolation of current information, account for shortcomings in risk stratification and design of therapies that address proximate molecular pathogenic events. CPG methodology now allows for the rapid incorporation of new information, particularly molecular medicine and imaging techniques, to close these knowledge gaps. A complex molecular pathogenesis may be addressed by personalized single- or multimodality approaches involving targeted therapies, new chemotherapeutics, new radiopharmaceuticals, image-guided radiotherapy, or novel surgical procedures. Advances in bioinformatics will facilitate the compilation, analysis, and dissemination of this information. Primary and secondary prevention strategies for patients "at risk" based on genetic studies may also include targeted therapies. CPG recommendations can contain cascades in which modifications are based on resource availability in certain geographic areas or clinical settings around the world. Ultimately, being aware of potential innovations will encourage referrals for clinical investigation and earlier adoption of safe and effective biotechnologies when they become more widely available.

22.1 INTRODUCTION

The evolution of care for the thyroid cancer patient has been slow. The mainstay of therapy, surgery + radioiodine (RAI) + thyroid-stimulating hormone (TSH) suppression, has been relatively unchanged for decades. As seen in Figure 22.1, in the 1980s, typical thyroid nodule evaluations consisted of thyroid function tests, scans, and if available, ultrasound (US), but only to determine whether a lesion was a simple cyst and therefore presumably benign. The use of fine-needle aspiration (FNA) was not prevalent and, when done, frequently yielded nondiagnostic or indeterminate findings, unless performed in the hands of experts. In the ensuing years, through the 1990s and early 2000s, the evaluation shifted to using a supersensitive TSH assay (to exclude hot nodules), US, which was now more prevalent, and FNA, which was encouraged to be performed under US guidance (UG-FNA). RAI scanning was not generally performed when the supersensitive TSH was normal or elevated. Surgery was still debated, but when performed and a differentiated thyroid malignancy discovered, patients generally underwent total thyroidectomy, postoperative radioiodine (RAI) ablation, and thyroid hormone suppression therapy. Until very recently, the mainstay of treatment for differentiated thyroid epithelial cancer incorporated RAI. However, we now recognize a higher frequency of complications and risks with RAI treatment, which are limiting its use. On the other hand, if the thyroid cancer was aggressive, regardless of histological type, options were very limited and the prognosis was poor. Chemotherapy, in particular, was rather disappointing.

Almost any academic conference today still debates the same questions (Q), based on knowledge gaps (KG), regarding the patient with a thyroid nodule or cancer:

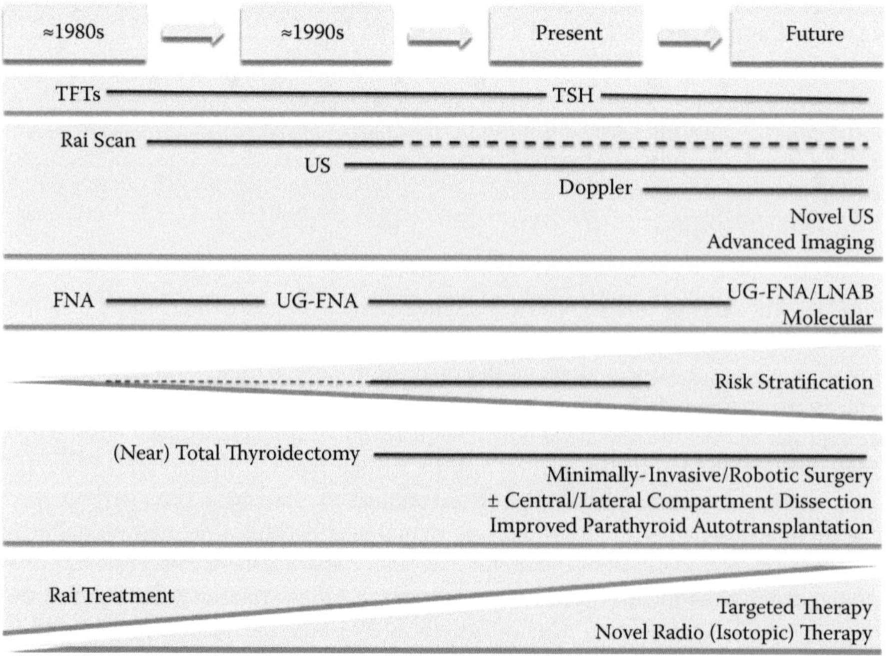

FIGURE 22.1 Evolution of clinical practice guidelines for thyroid cancer. Abbreviations: TFTs, thyroid function tests; TSH, thyroid-stimulating hormone; RAI, radioactive iodine; US, ultrasound; FNA, fine-needle aspiration; UG-FNA, US-guided FNA.

Q: How can inadequate and indeterminate FNA cytological diagnoses be reduced?
KG: UG-FNA technology, thyroid nodule and cancer natural history, and molecular biology/cytology.
Q: What is the appropriate extent of surgery?
KG: Pathophysiology determining extrathyroidal extension and extracapsular involvement in lymph nodes; determinants of aggressive biology; and mechanisms of nodal growth, degeneration, and spread.
Q: What is the significance of the microscopic or minimally invasive cancer?
KG: Epidemiology, natural history, and molecular biology of these low-risk thyroid cancers; surgical technology, especially involving cervical lymphatic system.
Q: What is the ideal role and dose of RAI?
KG: Technology of radioisotopes; risk for secondary malignancies; how to protect vital organs from concentrating effects of RAI that may lead to excess harm.
Q: What is the optimal level for thyroid hormone suppression therapy?
KG: Effect of novel management approaches on the natural history of thyroid cancer and then redefining the impact of thyroid hormone suppression therapy.
Q: What is the appropriate follow-up for patients?
KG: Natural history of low-risk, intermediate-risk, and high-risk patients receiving various clinical protocols, including watchful waiting for low-risk patients.
Q: Is there any hope for the patient with anaplastic thyroid cancer (ATC)?
KG: Molecular biology of ATC (pathophysiology) and technology (design of targeted therapies, radiation-based therapies, and surgical innovations).

Ultimately, the reason thyroidologists remain in this quagmire is that the biology of thyroid cancer has been elusive. Without truly understanding the biology of thyroid carcinogenesis, it is not possible to answer the above questions. Relying solely on epidemiological, observational, retrospective, and small prospective studies has not clarified the problem. The low lethality and slow natural history of most thyroid cancers generally precludes adequately powered randomized controlled studies. Moreover, risk stratification features need to be better understood, particularly with respect to genomic interactions with environmental factors, such as iodine and radiation exposures.

Fortunately, the current paradigm of thyroid cancer management is changing with the rapid acceleration of molecular biology information. We are migrating toward personalized medicine strategies where cancer care is individualized to intensify or de-intensify based on specific clinical and genomic/epigenomic features. Moreover, risk stratification is transitioning from risk of dying to risk of recurrence. This controversy is best exemplified in the consideration of cervical nodal involvement as clinically significant or not significant, in surgical planning.

In this chapter, we synthesize the information presented in this book and address these issues. We then show by example, referring to the case presentations in Chapter 7, how emergent biotechnologies can impact CPG and the care of the patient with thyroid cancer.

22.2 SURVEY OF EMERGENT BIOTECHNOLOGIES

22.2.1 Molecular Medicine (Chapters 1, 2, 3, 4, 12, 19, 20, 21)

The greatest technological impact on CPG is in the field of molecular medicine. Translational medicine and systems biology have identified prime strategic targets for diagnostic procedures and therapeutics. These targets include, among others, the cross-talking signaling pathways BRAF-MAPK and PTEN-AKT-PI3K as well as NFκB. In addition to molecular determinations of components of these pathways in preoperative and operative specimens, preoperative decision making can be optimized through the use of large-needle aspiration biopsies for galectin-3. Thus, the coevolution of basic molecular research and pharmaceutical companies primed for large-scale chemical synthesis has resulted in the availability of targeted therapies for investigative and clinical use in thyroid cancer.

The rapidly accelerating body of knowledge and technological advances in biomathematics and computing has popularized the paradigm of systems biology. The complex interactions of the multiple strategic molecules involved in thyroid cancer pathogenesis can be clarified with systems biology in such a way that multiple targeted therapies can be designed and administered. Thus, one can envision in a computerized clinical algorithm a decision tree node involving molecular medicine positioned after clinical and anatomical (US) risk stratification. This would optimize diagnostic accuracy, therapeutic appropriateness, and effectiveness.

22.2.2 Emergent Imaging Technologies (Chapters 8, 9, 10, 11, 13)

Historically, imaging conferred anatomical information. Current state-of-the-art imaging techniques provide correlated anatomical and functional details regarding presentation and progression of thyroid cancer. Future technologies promise greater diagnostic accuracy, facility, and anatomical resolution, which can guide nonsurgical interventions, and add dimensions of functionality and potential responsiveness to therapies. These innovations include improvements in US (speckle reduction imaging, contrast imaging with microbubbles, elastography, three-dimensional imaging, coimaging with computed tomography, US therapy, and guidance for radiofrequency ablation and percutaneous ethanol injection), scintigraphy (RAI with novel adjuvant drugs to improve retention and effectiveness [retinoids, inhibitors of methylation, thiazolidinediones, and statins], effects on the MDR gene, and novel radio-ligands and -nuclides for thyroid cancer [MIBG, c-11 hydroxyephedrine, indium-11 octreotide, seglitide, lanreatide, positron-emitting Cu-64, Ga-68, F-18, Y-86, DMSA, and CEA monoclonal antibody—pretargeted immunoscintigraphy]) and enhanced hybrid imaging with scintigraphy, PET, and CT.

These novel imaging technologies, which should be more available in the coming years, would be expected to have a key role in the initial evaluation of the higher-risk patient (such as PET-CT with RAI negativity) or after primary surgery and RAI treatment or targeted therapy to assess response.

22.2.3 Emergent Therapies

22.2.3.1 Nonsurgical Techniques for Lesion and Metastasis Ablation (Chapters 8 [Section 8.5.4], 13)

One of the unique dimensions of this book is the multinational list of contributors and diverse nature of clinical practices represented. An example of this is the use and development of nonsurgical thyroid lesion and metastasis interventions not commonly implemented in the United States. Examples of image-guided mini-invasive ablation of thyroid lesions and metastases that may become popularized, in addition to the currently practiced ethanol injection of thyroid nodules, include transarterial (chemo)embolization, percutaneous laser, radiofrequency, high-intensity US, microwave, and cryo-ablation, and electroporation. With the accumulation of more clinical evidence supporting safety and effectiveness, and a culture that nurtures learning and improved experience with these procedures, we may anticipate less need for surgical interventions and possibly a trend to even supplant surgical procedures with nonsurgical ablative and targeted therapies for low- to intermediate-risk, localized or metastatic disease. These techniques, as with all investigational modalities, must be tested scientifically through careful clinical trials and not through single clinician experience.

22.2.3.2 Emergent Surgical Techniques (Chapters 14, 15)

For nearly all patients with thyroid cancer, surgery is the primary therapy. Controversy surrounding the extent of initial thyroid surgery or management of subsequent recurrences has not compromised the central role for surgery in current CPG. With the emergence of safe and effective nonsurgical therapies, more narrowly defined indications for surgical procedures will depend on technological advances. Pre- and intraoperative sentinel node lymphoscintigraphy and improved anatomical delineation of central and lateral compartment lymphatics can characterize echelons of thyroid cancer metastases and guide selection of minimally invasive procedures and type of neck dissections. However, recent observational data have shed more confusion on the matter than clarity. For instance, even though we are now appreciating that the likelihood of cervical nodal involvement (central and lateral compartments) is greater than once suspected with low-risk PTC, it is not clear that aggressive RAI therapy, and even surgical intervention, will have a significant impact on prognosis. There is a need for carefully constructed clinical trials, as well as a mandate for surgical training programs, to ensure that future thyroid surgeons are well equipped with the skill sets to perform both prophylactic and therapeutic lymph node surgery. Having stated this, is watchful waiting (following a patient with demonstrable cancer), at any stage of thyroid cancer management, an option that most patients will feel comfortable with? Furthermore, is there a role for even more invasive initial surgery to manage nodal involvement and utilize parathyroid autotransplantation? Alternatively, the use of robotic technology, laryngeal nerve monitoring, and regional anesthetics can improve safety and allow for greater popularization of same-day/outpatient procedures. Taken together, these surgical issues may influence future CPG in such a way that there is a decreased use of surgery, but when used, surgery will be less invasive, safer, and more effective.

22.2.3.3 Emergent Radioisotopic Therapies (Chapter 16)

In addition to surgery, thyroid cancer patients have typically been counseled to receive RAI dosing. Though there has been substantial controversy regarding the timing, dose amounts, effects of stunning, and relative merits of dosimetry, recent concerns about safety, particularly with respect to sialoadenitis and secondary malignancies, have questioned whether we are overutilizing this treatment modality in lower-risk patients. The pendulum is indeed swinging away from the routine use of RAI treatment, especially in low-risk patients and those high-risk patients unlikely to trap RAI. Novel radioisotopes and imaging technologies can potentially improve the safety and effectiveness of this modality of treatment. Examples include individual lesion dosimetry with I-124 PET, bone-seeking strontium-89 or samarium-153, somatostatin analogs with gamma- (indium 111 (In-111-DTPA-D-Phe1-octreotide)) or beta-emitting isotopes (I-131, Y-90 (Y-90-DOTA-Tyr3-octreotide; Y-90-DOTA-lanreotide), Re-188), Y90 hMN-anti-CEA antibody + chemotherapy + autologous bone marrow rescue, and alpha particle emitters, such as astatine-211. Strategies to augment RAI uptake by increasing activity of the sodium-iodide symporter (NIS) are also on the horizon, such as gene therapy devices, as well as targeted therapies for *trans*-activating factors that affect NIS. Although there are no exciting strategies to prevent or treat sialoadenitis per se, this complication will hopefully be less frequent with the use of lower radiation doses to the salivary glands. Future CPG will reflect the use of these new highly selective radioisotopic protocols in intermediate-risk patients and as part of more targeted, hybridized protocols for higher-risk patients.

22.2.3.4 Emergent Radiotherapy Techniques (Chapter 17)

The use of external beam radiotherapy (EBRT) has been, currently is, and is expected to still be used primarily for those patients with advanced thyroid cancer with symptomatic disease in which palliation is an expected outcome. What can be expected with future innovations is a safer and more effective delivery of the radiation dose. Although optimization of salivary gland transfer techniques will decrease risks, most improvements will depend on technological advances. This includes dysphagia/aspiration-optimized intensity-modulated radiation therapy (doIMRT), image-guided radiation therapy (IGRT), fusion scans, and kilovoltage x-ray or CT. Future CPG will therefore continue to incorporate EBRT in the management of advanced, symptomatic, high-risk thyroid cancer.

22.2.3.5 Emergent Chemotherapeutics (Chapter 18)

Traditional chemotherapeutic agents will essentially disappear from future CPG due to a general lack of benefit with continued risks associated with adverse effects. These agents may continue to be used in investigative protocols for anaplastic thyroid cancers (ATCs) and exceptional cases of aggressive medullary thyroid cancer, or poorly differentiated thyroid epithelial cancer, but eventually targeted therapies are expected to supplant their use. In fact, chemotherapy that targets molecular pathways is more like molecular targeted therapy (MTT) than the traditional chemo, which we have become accustomed to. One example of this is the NFκB inhibitor dehydroxymethylepoxyquinomicin (DHMEQ). Thus, future CPG will probably make no mention of traditional chemotherapy, except as a footnote that risks outweigh benefits, and they may be considered in situations where no other treatment is available. The list of targeted therapies, however, will continue to expand and include novel chemicals.

22.2.3.6 Emergent Targeted Therapies (Chapters 19, 20, 21)

The burgeoning field of molecular medicine has yielded a new category of drugs: targeted therapies. Depending on how you define the terms, biologic therapies generally refer to large molecules, such as antibodies (name ends with *ab*), that are naturally found in the body, act on serum proteins or cell membrane binders/receptors, and are typically injected. On the other hand, molecular targeted therapies are small molecules that are chemically synthesized in the laboratory to gain entry into the cell, where they interact with (generally inhibit, hence name ends with *ib*) specific molecules, like kinases. MTTs only became "drugable" within the past 10 years when G-protein-coupled receptors (with seven transmembrane domains that can bind drugs) and their associated signal pathways were elucidated. In addition, there are devices that are targeted therapies, such as gene therapy with an adenovirus. A full listing of these targeted therapies, at various phases of preclinical and clinical investigation, is provided in Table 20.3 (Chapter 20) and Tables 21.1 and 21.2 (Chapter 21). Future CPG concerning intermediate- and high-risk patients will be replete with specific targeted therapy indications, contraindications, and dosing strategies as these agents become FDA approved and widely available. Furthermore, targeted therapies might even find a role in primary or secondary prevention strategies. As with other new technologies, targeted therapies will be expensive, and future CPG must include cascades of recommendations based on resource and expertise availability so that other safe and effective options are discussed.

22.3 IMPACT OF FUTURE CPG ON CLINICAL CASES (CHAPTER 7)

22.3.1 Case 1: Familial PTC

In this case, a poor outcome occurred and was traced back to an unexpected finding of ATC. The initial surgery was delayed because the risk for familial PTC was not appreciated. Earlier surgery may have discovered the ATC, perhaps at a stage where it may have been more treatable. This case illustrates the beneficial role of genetic screening in a high-risk family. With the incorporation of molecular testing to identify high-risk patients in future CPG, and then the initiation of targeted therapies to prevent or abrogate the pathogenesis of thyroid cancer, a better outcome might have occurred. Future CPG may call for the creation of a tissue or molecular bank from histological or cytological specimens to enhance the performance of a screening tool.

22.3.2 Case 2: Papillary Thyroid Microcarcinoma

In this case, papillary thyroid microcarcinoma (PTMC) was actually found to be associated with extensive nodal metastasis. However, there is insufficient evidence in the literature to reconcile divergent expert conclusions (1) that PTMC is such a low-mortality disease, treatment should be minimal, potentially ranging from simple observation to just a hemithyroidectomy without postoperative ablative RAI treatment; (2) that even though PTMC can be associated with nodal metastases and recurrences, the mortality risk is still low; and (3) that PTMC can be associated with nodal metastases and recurrences, and this does increase the mortality risk as with PTCs of larger size. The use of molecular markers can potentially clarify this dilemma, and risk stratification algorithms will be reconstructed with a priority on tissue molecular profiling. Future CPG will contain a decision tree based on a different staging system than is currently used. In addition, CPG that guide surgical decision making regarding the indications for central and lateral compartment dissections will depend on the preoperative molecular markers and not simply preoperative US (or other imaging) surveillance. Finally, the treatment modality of watchful waiting may gain in popularity.

22.3.3 Case 3: Hürthle Cell Carcinoma

In this case, a patient was found to have a minimally invasive Hürthle cell carcinoma (HCC) and a total thyroidectomy was performed based on an assessment that the patient was at higher risk. The question arises as to whether this patient was overmanaged. Future CPG based on emergent biotechnologies would be based on a more accurate and robust predictive model that will be able to identify not only patients at higher risk than expected, but also those at lower risk, in which treatments with potential adverse effects could be avoided. In the case of HCC, molecular analysis could be performed using large-needle aspiration biopsy or operative histological analysis to guide subsequent interventions. HCC can be RAI uptake negative, and this scenario highlights the future role for earlier postoperative use of PET-CT. Future CPG would incorporate these concepts.

22.3.4 Case 4: Medullary Thyroid Cancer

In this case, a small medullary thyroid cancer (MTC) is found in the setting of a family history of MTC. The question arises regarding how routine genetic screening should be for thyroid cancer. The answer, with respect to future CPG, will be broader since a trend for more widespread use of genetic screening for all heritable diseases in higher-risk individuals can be expected. Therefore, future CPG can be expected to have detailed information regarding genetic screening for all thyroid cancers in which markers are identified.

22.4 FINAL THOUGHTS

Describing a potential future can be challenging: it requires learning the past and diligently taking note of the present. In the case of thyroid cancer management, the learning curve has experienced a dramatic "elbow" with a surge of knowledge in the field of molecular medicine. This opportunity demands attention as knowledge gaps are being closed.

Clinical practice guidelines are becoming invaluable tools to assist decision making. They are not static but respond to new evidence and practice patterns. They are not rigid, but offer flexibility to physicians faced with resource constraints. In the future, the flow of information from individual patients to CPG will be facilitated through the use of advanced bioinformatics and evolving electronic medical record technology. One can easily foresee shared comprehensive information about thyroid cancer patients, easily accessible by all members of the treating team, and also interrogated by clinical researchers to update databanks on genetic, molecular, clinical, biochemical, and imaging parameters. It is hoped that a glimpse into a possible future will encourage referrals for clinical investigation and earlier adoption of safe and effective biotechnologies.

Index

A

B

D

E

F

R

S

U

V

W

X

Y